Physics of amorphous materials

Physics of amorphous materials

Second edition

S. R. ELLIOTT
Department of Physical Chemistry,
University of Cambridge

Longman Scientific & Technical
Copublished in the United States with
John Wiley & Sons, Inc., New York

Longman Scientific & Technical
Longman Group UK Limited
Longman House, Burnt Mill, Harlow,
Essex CM20 2JE, England
and Associated Companies throughout the world.

Copublished in the United States with
John Wiley & Sons Inc., 605 Third Avenue, New York, NY 10158

First published in 1984
Reprinted 1989
Second edition 1990

British Library Cataloguing in Publication data
Elliott, S. R. (Stephen Richard), 1952–
Physics of amorphous materials. – 2nd ed.
1. Amorphous materials
I. Title
530.4′1

ISBN 0-582-02160-X

Library of Congress Cataloging in Publication Data available

Produced by Longman Group (FE) Limited
Printed in Hong Kong

Contents

To Penelope

Preface to the second edition

In the six years that have elapsed since the first edition of this book appeared, our understanding of the disordered state of matter has advanced considerably. In part, this has been due to some momentous and unexpected discoveries, such as the icosahedral (quasicrystalline) phases in certain metallic alloys – a bridge between the perfect crystalline and the topologically disordered state – but mostly this advance has come about through a steady application of experimental techniques, both old and new, and the further development of theoretical ideas. It seemed, therefore, to be the appropriate time to prepare a second edition, incorporating the new developments in the field and taking the opportunity to expand somewhat the scope of the book.

The overall structure of the book has been changed slightly by distributing the material relating to amorphous metals throughout the text, where appropriate, rather than confining it to a special chapter, as before. In its place, a final chapter has been added, discussing some technological applications of amorphous materials. Other amendments are as follows. A discussion on icosahedral materials and quasicrystalline order has been included in the definitions of order and disorder given in Chapter 1. Also in this chapter, new sections have been added dealing with diffusion-related preparation methods of amorphous materials. New ideas on the glass transition based on dynamical (mode-coupling) theories of the transition are included in Chapter 2, and the section dealing with those factors conductive to glass formation has been expanded, with a more detailed discussion of constraint theory for covalent systems and strain- and structure-related criteria for glass formation in metallic systems. The third chapter dealing with structure has been very extensively rewritten. In the discussion of experimental techniques for structure determination, a new section has been added dealing with Mössbauer spectroscopy, and the sections on X-ray absorption and NMR spectroscopy, in particular, have been expanded so as to include many recent examples of the application of these techniques to amorphous solids. The section dealing with medium-range order has been completely rewritten, taking account of the latest ideas on this controversial topic. In the section dealing with macroscopic structure, fractal structures and their characteristics are now included. Chapter 4 has been retitled 'atomic dynamics'; it includes material on vibrational excitations, as before, but now also includes an extensive discussion of atomic transport (including ionic diffusion and conductivity). A discussion on analytic methods of computing vibrational characteristics is now included, and much greater prominence is given to the use of inelastic neutron scattering as a probe of vibrational excitations, in particular the low-frequency modes associated with low-temperature thermal behaviour. In addition, a discussion of 'fractons' (vibrational excitations on a fractal lattice) is given. In Chapter 5, the latest theoretical ideas relating to electron localization are now included, and a somewhat more extensive discussion of percolation theory (particularly as it relates to electronic properties) is given. In the section dealing with dc electronic transport, a discussion is now given of the effects of quantum interference (weak localization) on the conductivity of metallic systems, and a more extensive treatment of the behaviour of the conductivity pre-factor and activation energy for amorphous semiconductors. A new section dealing with frequency-dependent conduction resulting from band transport has been added, and in the section on optical properties, a discussion of the technique of fluorescence line-narrowing spectroscopy is now included. In Chapter 6, dealing with defects, a new section has been added which discusses the thermodynamic aspects of defects in amorphous materials, and an extensive discussion is now given of the various spectroscopic techniques that are used to investigate the electronic density of states in the gap region of amorphous semiconductors, particularly hydrogenated amorphous silicon.

I hope, therefore, that this extensively revised edition provides an up-to-date view of the subject matter of amorphous materials. I am most grateful to those readers and reviewers of the first edition whose suggestions have been incorporated into this revised edition. The aim of the book remains the same: it is primarily intended as an introductory text for researchers new to the field, but I hope that established workers will also find it useful for reference purposes.

Finally I would like to express my gratitude to Professors G. N. Greaves, C. N. R. Rao, K. J. Rao and J. L. Souquet, in addition to those named in the first edition, for many useful discussions of the ideas described in this book. All errors which remain, however, are my responsibility. I am also grateful to Nick and Deborah Hopkin for typing the revisions, and to Laura Cordy for preparing the new list of references.

Cambridge
May 1989

Preface to the first edition

This book has been written to be an introduction to the science of amorphous materials, a subject which receives scant mention in conventional solid-state texts. It is aimed at final year undergraduates, beginning graduate students and researchers in solid-state physics or chemistry, materials science or engineering. Some background of (crystalline) solid-state physics is assumed, say between the levels of Kittel's *Introduction to Solid-State Physics* and Ziman's *Principles of the Theory of Solids*, because it was felt that in a book of this length such background materials could not easily be incorporated; however, no knowledge of amorphous materials is assumed.

When I first considered the idea of writing such a book as this, it seemed to me that a general understanding of amorphous materials had been established, and that much of what remained to be done in the field lay in the realms of 'engineering' new materials and filling in the details of the theories proposed to account for their properties. It seemed to me then to be an appropriate time to take stock and write a general introduction to the field of amorphous solids, covering a wide range of their properties, and bridging the gap for the newcomer to the field to such specialized monographs on the subject as the classic work by Mott and Davis, *Electronic Processes in Non-Crystalline Materials*. However, as is usual in these matters, I have found myself overtaken by events. Recent work has thrown into the melting pot many of the 'established' ideas – the example of scaling theories of electron localization and their relation to mobility edges and the so-called minimum metallic conductivity, springs immediately to mind. I felt that such important developments could not be left out, but this immediately led to problems since it is difficult in an introductory textbook to deal with recent controversial ideas. I have tried to resolve this dilemma by mentioning the new concepts, certainly, but putting them into historical perspective, as it were, and pointing out their relation and relevance to previous ideas. In this way, I hope I will have given the reader a flavour of the vitality of this rapidly expanding branch of solid-state science.

I have also tried to make this book into something of a reference work by the inclusion of several tables of useful parameters, such as glass-transition, melting, superconducting and Curie temperatures, electrical and optical gaps for amorphous semiconductors, etc. These are all quantities which practising researchers use on a day-to-day basis, and in this small way, I hope that this book might be of some value to them, since I believe that such collections of data are not readily available together elsewhere. I have also given perhaps more references to the original literature than is usual in a textbook, but have kept the number down by referring, where possible, to reviews although references to original works have been given where these have been seminal to the development of their aspect of the field. I hope that in this way, the beginning researcher will find his way into the literature. A collection of problems and exercises are given at the end of each chapter which illustrate or extend points raised in the text.

The layout of this book differs somewhat from that conventionally adopted in texts dealing with the crystalline solid state. The first chapter begins with a series of definitions of the terms 'amorphous', 'glassy' etc., followed by a review of various preparative techniques which can be used to produce non-crystalline materials. Such a discussion is necessary because there is no unique structure of an amorphous solid (unlike a single crystal); non-crystalline materials produced in different ways can have very different structures and hence physical properties, even though the composition might be the same. One of the most common and important methods of producing amorphous materials is to freeze a liquid sufficiently rapidly that crystallization is precluded; in this way a 'glass' is formed. Chapter 2 is devoted to a study of the glassy state, considering the kinetic, thermodynamic and other factors which control the glass transition, and discussing those material parameters which can facilitate glass formation.

The structure of amorphous materials is one of the recurrent themes of this book. The determination of the structure of a solid is a difficult task when there is no periodicity; special experimental techniques need to be employed and these form the subject of the first part of Chapter 3. Even when such techniques are used only a limited amount of local structural information is generally obtainable, and the construction of structural models can be a most useful route to a further understanding of the structure, particularly medium-range order; modelling is also discussed in Chapter 3. Amorphous materials are seldom structurally homogeneous, and different preparative techniques can result in various growth morphologies, e.g. phase separation in melt-quenched glasses or the inclusion of voids in vapour-deposited films; this 'macroscopic' structure, and the experimental techniques which probe it, form the last part of the chapter.

The absence of a periodic lattice in amorphous materials has several effects; an immediate consequence is that there is no reciprocal lattice and so $\boldsymbol{k}$ is no longer a good quantum number for excitations in the solid, such as phonons or electrons. Thus the Bloch formalism cannot be used to simplify the mathematical description of the excitations, and the phonon (or electron) states cannot be described in terms of dispersion curves, $\omega(\boldsymbol{q})$ (or band structures, $E(\boldsymbol{k})$); instead the only quantity which is a valid description of excitations in a non-crystalline solid is the 'density of states', and the various methods which have been used to compute it are described in Chapter 4 for the case of phonons, and again in Chapter 5 for electrons.

One seemingly universal feature of amorphous solids is the existence of very low-energy excitations, describable as 'two-level systems', which affect thermal properties in a variety of ways: they can act as internal degrees of freedom, thereby giving rise to an anomalous specific heat, and they can act as scattering centres for phonons, thereby controlling the thermal conductivity and ultrasonic absorption. Two-level systems form the subject of the last part of Chapter 4.

Electronic excitations, manifested as transport or optical properties, are the other dominant theme of this book. It is convenient to divide the discussion, on structural grounds, into two: Chapter 5 deals with the electronic behaviour of 'ideal' (i.e. defect-free) amorphous semiconductors, and Chapter 6 considers the electronic (and other) properties associated with a variety of structural defects. A consequence of the presence of structural disorder is the possibility that excitations, whether vibrational or electronic, may become spatially localized. Different models for electron localization, such as scaling theories or that giving rise to a minimum metallic conductivity and a mobility edge, are introduced in Chapter 5. Electrical transport in amorphous semiconductors, discussed in terms of these models, and in terms of a rival theory involving localized, self-trapped charge carriers (small polarons), also form part of Chapter 5.

The structural, electronic, magnetic and mechanical properties of amorphous metals form the subject of Chapter 7. It was felt that amorphous metals merited a separate chapter in view of the fact that, while they exhibit certain features in common with other amorphous solids (electron localization effects for example), many of their properties are very different to those shown by amorphous semiconductors and insulators discussed earlier in the book.

Finally, I owe an enormous debt of gratitude to those in the Cavendish Laboratory in Dr A. D. Yoffe's group, principally Professors Davis and Mott, who guided my first forays in this field and whose influence has been incalculable, and more recently to those in Professor J. M. Thomas's group in the Department of Physical Chemistry, from whom I have gained many new insights. I am also extremely grateful to Professor E. A. Davis, Drs P. H. Gaskell, M. R. J. Gibbs, J. Klinowski, E. Marseglia, W. A. Phillips and A. C. Wright and Messrs T. G. Fowler and A. J. Lowe for reading parts of this book, for their valuable comments and for pointing out many errors. I am also especially grateful to Dr P. Extance for reading the galley proofs. Any mistakes that remain, however, are my sole responsibility. I am also grateful to Mrs M. Pomery, Mrs M. Crosbie and Miss J. Rowe for typing various parts of the manuscript, to Mr E. Smith and his colleagues for preparing the photographs, to Mrs K. I. Johnson for preparing the index and Mrs W. Roberts for running the literature search to generate Fig. 1.1. Last, but not least, I am grateful to Penelope for putting up with rather a unsociable author during the time that it has taken to write this book.

It is well known that in the writing of any text, the person who benefits most from the process is the author himself. I think this to be true in this case, but I hope a little rubs off on others.

The following kindly consented to the reproduction of material originally published by them:

Academic Press, Akademic-Verlag, American Institute of Physics, American Physical Society, Cambridge University Press, Centre for Industrial Consultancy and Liaison (University of Edinburgh), Elsevier Sequoia, Institute of Physics, International Union of Crystallography, Japanese Institute of Metals, John Wiley and Sons, Les Editions de Physique, Macmillan, Marcel Dekker, Metals Society, North-Holland Publishing, Oxford University Press, Pergamon Press, Plenum Publishing, Physical Society of Japan, Royal Microscopy Society, Royal Society, Scientific American, Society of Glass Technology, Springer-Verlag, Stanford University Synchrotron Radiation Laboratory, Taylor and Francis.

Acknowledgements

We are grateful to the following for permission to reproduce copyright materials:

Academic Press Inc. and the respective authors for figs. 3.10 from fig. 2, p. 248 (Suzuki), 6.24 from fig. 27, p. 49 (Cody, 1984), 6.25a, 6.25b & 6.25c from figs. 4 & 7, pp. 93 & 96 (Amer & Jackson, 1984), 6.29 from fig. 2.36, p. 57 (Madan & Shaw, 1984); Akademie-Verlag for fig. 6.37a from fig. 26, p. 234 (Engemann & Fischer, 1977); American Association for the Advancement of Science and the author, R. Orbach for fig. 4.35 from fig. 1, p. 814 (Orbach, 1986) copyright © 1986 by the AAAS; American Institute of Physics and the respective authors for figs. 1.2 from fig. 2, p. 234 (Kelton & Wu, 1985), 7.4 from fig. 3.96, p. 304 (Matsumura *et al.*, 1980); American Physical Society and the respective authors for figs. 2.10a from fig. 1, p. 572 (Metzei *et al.*, 1987), 3.27 from a figure, p. 36b (Baum *et al.*, 1986), 3.39 from a figure (Verrall & Elliott, 1988), 3.40 from fig. 4 (Gladden & Elliott, 1987), 3.56 from fig. 4, p. 2373 (Schaefer *et al.*, 1984), 4.3, 4.4a & 4.4b from figs. 1, 2 & 3, pp. 645 & 4032 (Kamitakahara *et al.*, 1984), 4.5c from fig. 10c, p. 28 (Galeener & Thorpe, 1983), 4.6a & 4.6b from figs. 1 & 2, p. 2429 (Galeener *et al.*, 1984), 4.11a from a figure (Alben *et al.*, 1975), 4.11b from fig. 12 (Maley *et al.*, 1988), 4.13a, 4.13b & 4.13c from fig. 13, p. 1063 (Galeener *et al.*, 1983), 4.16a, 4.16b & 4.18b from figs. 2, 14 & 13, pp. 10614, 10621 & 10620 (Maley *et al.*, 1988), 4.20 from fig. 2, p. 6540 (Kamitakahara *et al.*, 1987), 4.21 from fig. 17, p. 693 (Hafner, 1983), 4.28a & 4.28b from figs. 2 & 3, p. 2317 (Buchenau *et al.*, 1984), 4.29a & 4.29b from fig. 2a, p. 1319 and 4.31a from fig. 1, p. 5667 (Buchenau *et al.*, 1988), 5.15a & 5.15b from figs. 3 & 4, p. 2917 (Bergmann, 1988), 5.18 from fig. 27, p. 267 (Hess *et al.*, 1982), 5.20a from fig. 1, p. 1089 (Cochrane & Strom-Olsen, 1984), 5.20b from fig. 1, p. 4805 (Howson & Greig, 1984), 5.37b from fig. 13, p. 60 (Galeener *et al.*, 1983), 6.2a from figs. 2a & 2b, p. 1319 (Pantelides, 1986), 6.8 from fig. 2 (Thio *et al.*, 1984), 6.9b from fig. 4, p. 6022 (Tarnow *et al.*, 1988), 6.18 from fig. 2 (Smith *et al.*, 1986), 6.19a from fig. 2a, p. 3669 and 6.20b from fig. 27 (Stutzmann *et al.*, 1987), 6.22a from fig. 1, p. 1038 (Kakalious, 1987), 6.35 from fig. 1, p. 2296 (Higashi & Katner, 1981); Butterworth Scientific Ltd, for figs. 7.11a, 7.11b & 7.11c from figs. 14.1, 14.3 & 14.14, pp. 258, 260 & 273 (O'Hadley, 1983); CRC Press Inc. for figs. 7.6 & 7.7 from figs. 9 & 11, pp. 238 & 240 (Hamakawa, 1987) copyright © 1987 CRC Press Inc.; Editions de Physique for figs 3.13 from fig. 1, p. C9-31 (Wright *et al.*, 1982), 4.36 from fig. 4, p. 249 (Coutens *et al.*, 1988), 7.3 from fig. 3.78, p. 279 (Le Comber *et al.*, 1981); Elsevier Science Publishers B.V. and the respective authors for figs. 2.15 from fig. 1, p. 119 (Egami & Waseda, 1984), 2.23b from fig. 2 (Hafner *et al.*, 1984), 3.15 from fig. 5, p. 33 (Greegor *et al.*, 1983), 3.19a, 3.19b &

3.19c from figures on pp. 42 & 44 (Mobilio & Flipponi, 1987), 3.38b from fig. 6, p. 314 (Windisch & Risen, 1982a), 3.41 from fig. 5 (Greaves, 1985), 3.43 from fig. 4, p. 298 (Kamitakahara *et al.*, 1984), 3.44 from fig. 5, p. 368 (Sen & Thorpe, 1977), 4.18a & 4.19b from figs. 8 & 14b, pp. 166 & 173 (Price & Carpenter, 1987), 4.37a from fig. 2 (Angell, 1986), 4.38, 4.39 & 4.41 from figs. 8, 6 & 2, pp. 198, 193 & 187 (Martin & Angell, 1986), 4.42 from a figure (Elliott, 1988), 5.23 from fig. 6, p. 5 (Stuke, 1987), 6.2 from a figure (Hishikawa *et al.*, 1987), 6.19b from fig. 4 and 6.20a from fig. 11, p. 12 (Street, 1985); Institute of Physics Publishing Ltd. and the respective authors for figs. 2.5 from fig. 6, p. 178 (Jackle, 1986), 2.9 from fig. 25, p. 214 (Leutheusser, 1984), 5.5b from fig. 6, p. 6202 (Hickey & Morgan, 1986); International Union of Crystallography for figs. 3.25 & 3.48b from figs. 2b & 3, pp. L86 & L3 (Mozzi & Warren, 1969); Kluwer Academic Publishers for figs. 7.8a, 7.8b, 7.9a, 7.9b & 7.10 from figs. 1, 2, 5, 7 & 2, pp. 508, 510, 513, 516 & 367 (Baucke, 1985); Macmillan Magazines Ltd. and the author, R. Dupree for figs. 3.23a & 3.23b from figs. 1 & 3, pp. 402 & 404 (Dupree *et al.*, 1984) copyright © 1984 Macmillan Magazines Ltd.; Pergamon Press for figs. 2.2a, from fig. 8, p. 93 (Johnson, 1986), 2.12 from fig. 1, p. 867 (Tanaka, 1985), 3.26 from a figure on p. 30 (Reimer *et al.*, 1981), 5.30a & 5.30b from figs. 2 & 3, pp. 374–5 (Schlegel *et al.*, 1979) © 1979, 1981, 1985 & 1986 Pergamon Press PLC; Plenum Publishing Corp. and the respective authors for figs. 3.29a & 3.29b from figs. 10 & 11, pp. 239 & 240 (Boolchand, 1985), 6.12 from fig. 2, p. 93 (Robertson, 1988); Royal Swedish Academy of Sciences for figs. 2.10b, 5.36 & 5.37a from figs. 2, 9 & 12, pp. 365 & 359 (Imbusch, 1987); Scientific American Inc. for fig. 6.7 from a figure on p. 47 (Adler, 1977) copyright © 1977 Scientific American Inc.; Society of Glass Technology for figs. 3.21b from a figure (Bray & O'Keefe, 1963), 3.22a from a figure (Muller *et al.*, 1983), 4.37b from fig. 1, p. 215 (Ingram, 1987), 4.40 from fig. 9, p. 220 (Ravaine & Souquet, 1977); Springer-Verlag and the respective authors for figs. 2.2a from fig. 15, p. 187 (Kimura & Yonezawa, 1983), 2.11 from fig. 1, p. 74 (Frick *et al.*, 1988), 3.28b from fig. 4.11, p. 108 (Gonser & Preston, 1983), 5.14 from fig. 6, p. 10 (McKinnon & Kramer), 6.15b from fig. 1, p. 26 (Chaudhari *et al.*, 1983); Taylor & Francis Ltd. and the respective authors for figs. 3.51b from fig. 4, p. 273 (Gaskell & Tarrant, 1980), 5.28 from fig. 4, p. 149 (Vimas & Roilos, 1984), 6.27a, 6.27b, 6.28a & 6.28b from figs. 4, 3, 10 & 11, pp. 154 & 161 (Goodman & Fritzche, 1980), 6.33a from fig. 3 and 6.34 from fig. 7, p. 95 (Higashi & Kastner, 1983).

Whilst every effort has been made to trace the owners of copyright material, in a few cases this has proved impossible and we take this opportunity to offer our apologies to any copyright holders whose rights we may have unwittingly infringed.

Glossary of symbols

a	activity coefficient
α	optical absorption coefficient
α^{-1}	localization length
α_d	disorder parameter of CTRW theory
α_T	coefficient of thermal expansion
α_ρ	temperature coefficient of resistivity
b	neutron scattering length
C_p	heat capacity (constant pressure)
C_v	heat capacity (constant volume)
γ	temperature coefficient of the electrical gap
$\gamma(r)$	shape factor (SAXS)
γ_g	gyromagnetic ratio
Γ	Urbach edge parameter
$\tilde{d}$	fracton dimension
D	diffusion constant
$\mathbf{D}$	force constant matrix
D_H	Hausdorff (fractal) dimensionality
Δ	energy level splitting
ΔE_σ	activation energy of electrical conductivity
e	electronic charge
E_A	valence band edge
E_B	conduction band edge
E_C	conduction band mobility edge
E_F	Fermi energy (level)
E_g	band gap
E_o	optical band gap (mobility gap)
E_v	valence band mobility edge
ε_1	real part of dielectric constant
ε_2	imaginary part of dielectric constant
$f(E)$	Fermi–Dirac distribution function
$f(k)$	X-ray atomic scattering (form) factor
F	Helmholtz free energy
$F(\boldsymbol{k})$	reduced scattering intensity

g	Lande g-factor
g_s	Lande g-factor (free spin)
$g(E)$	density of states ($eV^{-1}\ cm^{-3}$)
$\mathbf{G}$	Green function
$G(r)$	reduced radial distribution function
η	viscosity
η_a	EFG asymmetry parameter
η_p	atomic packing fraction
η_q	quantum efficiency
$\boldsymbol{H}$	magnetic field
H_c	coercive magnetic field
θ	bond angle
θ_C	Curie temperature
θ_D	Debye temperature
$\boldsymbol{I}$	nuclear spin
$\mathbf{I}$	identity tensor
$I(\boldsymbol{k})$	scattering intensity
$J(r)$	radial distribution function
$\boldsymbol{k}$	scattering (or momentum) vector
k_B	Boltzmann constant
K_u	uniaxial magnetic anisotropy constant
κ	thermal conductivity
κ_T	isothermal compressibility
ℓ	mean free path
λ	electron–phonon coupling constant
λ_s	magnetostrictive constant
$\mathbf{\Lambda}$	spin–orbit interaction tensor
m_e	electron mass
$\boldsymbol{M}$	dipole moment vector
M_s	saturation magnetization
$M(\boldsymbol{k})$	modification function
$\boldsymbol{\mu}$	magnetic moment
μ_B	Bohr magneton
μ_d	drift mobility
μ_e	average band conduction mobility
μ_H	Hall mobility
$\mu(E)$	mobility at energy E
$\mu_X(E)$	X-ray absorption coefficient
$n(\omega)$	Bose occupation number
$N(E)$	integrated density of states ($eV^{-1}\ cm^{-3}$)
ν_{el}	characteristic electronic frequency
ν_0	characteristic phonon frequency

$p(\omega)$	Raman depolarization ratio
$P(\phi)$	dihedral-angle distribution
$P(\omega)$	participation ratio
Π	Peltier coefficient
q	configuration coordinate
$\boldsymbol{Q}$	scattering vector
Q_{cc}	quadrupole coupling constant
R_H	Hall coefficient
ρ^0	average density
$\rho(E)$	density of states (eV^{-1})
$\rho(r)$	atomic density function
$\rho(T)$	electrical resistivity
$\rho(\omega)$	density of vibrational states
S	thermopower
$S(\boldsymbol{Q}, \omega)$	dynamic structure factor
σ	r.m.s. bond-length fluctuation
σ_E	electrical conductivity at energy E
σ_f	fracture strength
σ_{min}	minimum metallic conductivity
$\sigma(\omega)$	a.c. conductivity
T_f	fictive temperature
T_g	glass-transition temperature
T_m	melting temperature
τ	relaxation time
U	Hubbard (correlation) energy
U_{eff}	effective Hubbard energy
W_D	disorder energy
W_H	polaron hopping energy
W_p	polaron energy
$W(T)$	Debye–Waller factor
ϕ	dihedral angle
$\chi(k)$	EXAFS amplitude
$\chi'(\omega)$	real part of dielectric susceptibility
$\chi''(\omega)$	imaginary part of dielectric susceptibility
Z	atomic number
ω	radial frequency
Ω	solid angle

Glossary of abbreviations

AM	air mass
BIS	bremsstrahlung isochromat spectroscopy
CB	conduction band
CBH	correlated barrier hopping
CFO	Cohen, Fritzsche and Ovshinsky (model)
CON	chemically ordered network
CRN	continuous random network
CTRW	continuous time random walk
CVD	chemical vapour deposition
DCR	diffusion-controlled relaxation
DLTS	deep-level transient spectroscopy
DOS	density of states
DRP	dense random packing
DRPSS	dense random packing of soft spheres
DSC	differential scanning calorimetry
DTA	differential thermal analysis
EDC	energy distribution curve
EL	electroluminescence
ESCA	electron spectroscopy for chemical analysis
ESR	electron spin resonance
EXAFS	extended X-ray absorption fine structure
EXELFS	extended energy-loss fine structure
FIC	fast-ion conducting
FID	free-induction decay
FLN	fluorescent line-narrowing
FSDP	first sharp diffraction peak
GD	glow discharge (decomposition)
HREM	high-resolution electron microscopy
ICTS	isothermal capacitance transient spectroscopy
IR	infra-red
ITO	indium tin oxide
IVAP	intimate valence alternation pair
LA	longitudinal acoustic
LCAO	linear combination of atomic orbitals
LEED	low-energy electron diffraction
LO	longitudinal optic
M	metalloid

MAS	magic-angle spinning
MD	molecular dynamics
MMI	mixed-mobile ion
MRO	medium-range order
MT	multiple trapping
NBO	non-bridging oxygen
NEXAFS	near-edge extended X-ray absorption fine structure
NMR	nuclear magnetic resonance
NQR	nuclear quadrupole resonance
ODMR	optically detected magnetic resonance
OHC	oxygen hole centre
PDS	photothermal deflection spectroscopy
PL	photoluminescence
PLE	photoluminescence excitation (spectrum)
QMT	quantum-mechanical tunnelling
RCN	random covalent network
RDF	radial distribution function
RE	rare earth metal
RPA	random-phase approximation
SAS	small-angle scattering
SANS	small-angle neutron scattering
SAXS	small-angle X-ray scattering
SRO	short-range order
TA	transverse acoustic
TCR	temperature coefficient of resistivity
TE	early transition metal
TFT	thin film transistor
TL	late transition metal
TLS	two-level system
TM	transition metal
TO	transverse optic
TOF	time of flight
UPS	ultra-violet photoemission spectroscopy
VAP	valence alternation pair
VB	valence band
VDOS	vibrational density of states
XANES	X-ray absorption near-edge structure
XAS	X-ray absorption spectroscopy
XPS	X-ray photoemission spectroscopy

List of tables

1 Preparation

1.1 Introduction

Amorphous materials *per se*, are not new; the iron-rich siliceous glassy materials recovered from the moon by the Apollo missions are some billions of years old, and man has been manufacturing glassy materials (principally from silica) for thousands of years. What is new, however, is the *scientific study* of amorphous materials, and there has been an explosion of interest recently as more new materials are produced in an amorphous form, some of which have considerable technological promise. This is obviously a fast-moving field, and any textbook on the subject runs the risk of becoming out-dated rather rapidly. Nevertheless, it is felt that sufficient is now known about the amorphous state to warrant a general, introductory text on the subject, particularly in view of the fact that amorphous materials rarely receive more than a passing mention in the conventional solid-state texts, which deal almost wholly with *crystalline* materials.

The interest in amorphous materials is perhaps twofold. Firstly, there is the materials science aspect; a wide diversity of materials can be rendered amorphous – indeed almost all materials can. This is in sharp contrast to the knowledge of the average layman, for whom the word 'glass' signifies only that transparent material (made from silica with the addition of a few alkali oxides) which is placed in windows.

The second interest in amorphous materials is in the fundamental physics of such systems: Why is window 'glass' transparent when the conventional solid-state explanation of band gaps depends crucially on the assumption of periodicity in the underlying lattice and hence on the presence of Bloch electron wavefunctions? Furthermore, amorphous materials exhibit many properties which are unique to them and are not shared by crystalline solids at all.

It is the purpose of this book, then, to guide the reader through the murky waters of the subject, beginning with a description of various ways of producing amorphous materials, giving a discussion of their structure and properties, and incorporating some technological applications. In this way, it is hoped that interest in, and appreciation of, the fascinating field of the amorphous state will be quickened.

1.2 Definitions

This book is concerned with non-crystalline (or amorphous) materials which possess randomness to some degree. But what kind of randomness, and how much? We will try to answer the first question here and leave the second for discussion in Chapter 3.

Randomness can occur in several forms, of which topological, spin, substitutional, and vibrational disorder are the most important. These types of disorder are illustrated schematically in Fig. 1.1. Disorder is not a unique property; it must be compared to some standard, and that standard is the perfect crystal. This can be defined in the following way:

A perfect crystal is that in which the atoms (or groups of atoms or 'motifs') are arranged in a pattern that repeats periodically in three dimensions to an infinite extent.

With this definition, an imperfect crystal can simply be one which is finite and ends at surfaces (non-infinite extent), or one which possesses a defect, e.g. a vacancy or interstitial foreign atom or a dislocation (non-periodic). However, the forms of disorder with which we are concerned are more drastic than these small perturbations to perfect crystallinity.

Topological (or geometric) disorder is that form of randomness in which there is no translational periodicity whatsoever (Fig. 1.1(a)), and this type of positional disorder forms the theme of this book. Nevertheless, there are 'degrees' of topological disorder: certain amorphous materials have considerable short-range (or local) order while others have little; both have no long-range order however. All amorphous or glassy solids are therefore distinguished by their lack of periodicity.

Another variety of randomness is spin (or magnetic) disorder, in which the underlying perfect crystalline lattice is preserved, but each atomic site possesses a spin or magnetic moment, and this is oriented randomly (Fig. 1.1(b)). This situation occurs in some dilute magnetic alloys such as Cu–Mn or Au–Fe, with from 0.1

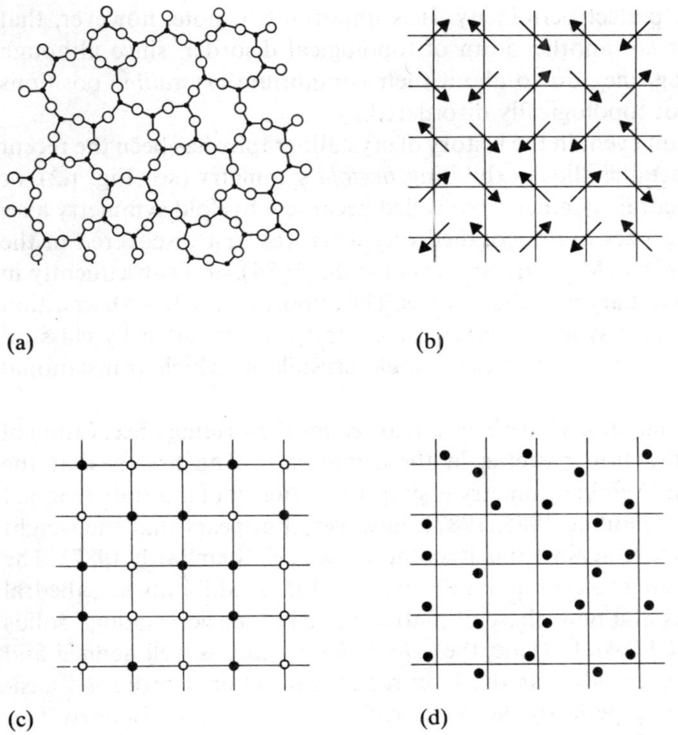

Fig. 1.1 Types of disorder:
(a) topological disorder (no long-range order);
(b) spin disorder (on regular lattice);
(c) substitutional disorder (on regular lattice);
(d) vibrational disorder (about equilibrium positions of a regular lattice).

to 10 at. % magnetic component. The local moments are frozen into particular, but random, orientations because the 'exchange' interaction between moments in a metal has an oscillatory spatial dependence, and so randomly distributed moments in a *metallic* host suffer a corresponding distribution of exchange interactions, and hence become randomly oriented. These, and those materials which are topologically disordered *and* possess a randomly oriented spin, are termed 'spin glasses', and must not be confused with true glasses which are defined below.

A further kind of randomness is substitutional disorder (Fig. 1.1(c)) in which, although the underlying crystalline lattice is preserved, the material is in fact an alloy (say Cu–Au) with one type of atom randomly substituting for the other in the lattice. These systems are of great importance in metallurgy and other branches of materials science, and for the case of binary alloys are described by the Bragg–Williams theory.

The final category of randomness considered here is vibrational disorder of a crystalline lattice (Fig. 1.1(d)). Of course the concept of a perfect crystal is only valid at the absolute zero of temperature (if zero-point motion is ignored), and at any finite temperature the random motion of atoms about their equilibrium positions destroys the perfect periodicity. It is important to note, however, that vibrational disorder is *not* another form of topological disorder, since although the atoms are vibrating, they do so about their equilibrium *crystalline* positions which of course are not topologically disordered.

A highly significant event in the history of crystallography has been the recent discovery of certain metallic alloys exhibiting *fivefold* symmetry (see Fig. 1.2), or in other words 'icosahedral' symmetry (so called because 6 fivefold symmetry axes pass through the 12 vertices of an icosahedron). This was first discovered in the rapidly quenched alloy $Al_{86}Mn_{14}$ by Shectman *et al.* (1984), and subsequently in many other binary and ternary metallic systems. The importance of this observation is that fivefold (or tenfold) symmetry axes are completely precluded by classical crystallography from occurring in perfect single crystals in which translational periodicity exists.

Three distinct explanations have been advanced for the striking observation of fivefold symmetric diffraction patterns in these materials. The first is that the appearance of apparent fivefold symmetry is simply an artefact of multiply twinned (conventional) crystals (Pauling 1985, 1987); however, it appears that the weight of experimental evidence is against this hypothesis (see e.g. Steinhardt 1987). The second is the more startling assertion that the metallic alloys exhibiting icosahedral symmetry are examples of a new phase of matter, namely 'quasicrystalline' solids (Levine and Steinhardt 1984). In these, the *orientational* order is well defined and perfect even at long range, whereas the long-range translational order is quasi-periodic rather than being perfectly periodic. Such a lattice can be obtained by a so-called Penrose tiling of *two* types of unit cell (Penrose 1974); Fig. 1.3(a) shows a 2D analogue, where the unit cells are acute and obtuse rhombi (with identical edge lengths and with interior angles of 36° and 144°, and 72° and 108°, respectively) which occur in a proportion equal to the 'golden mean', $\tau = (1 + \sqrt{5})/2$. Since all interior angles in the rhombi are multiples of 36° (i.e. one-tenth of a complete revolution), all the edges point to the vertices of a regular decagon, thereby giving

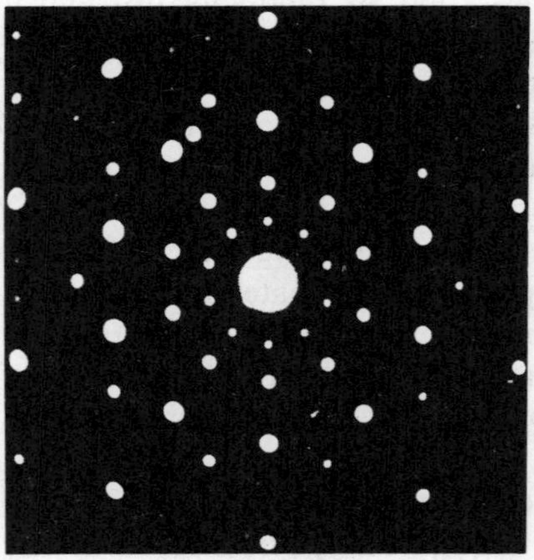

Fig. 1.2 Diffraction pattern normal to a fivefold symmetry axis in the icosahedral phase of Al–Mn. (K.F. Kelton and T.W. Wu (1985) *Appl. Phys. Lett.* **46**, 1059).

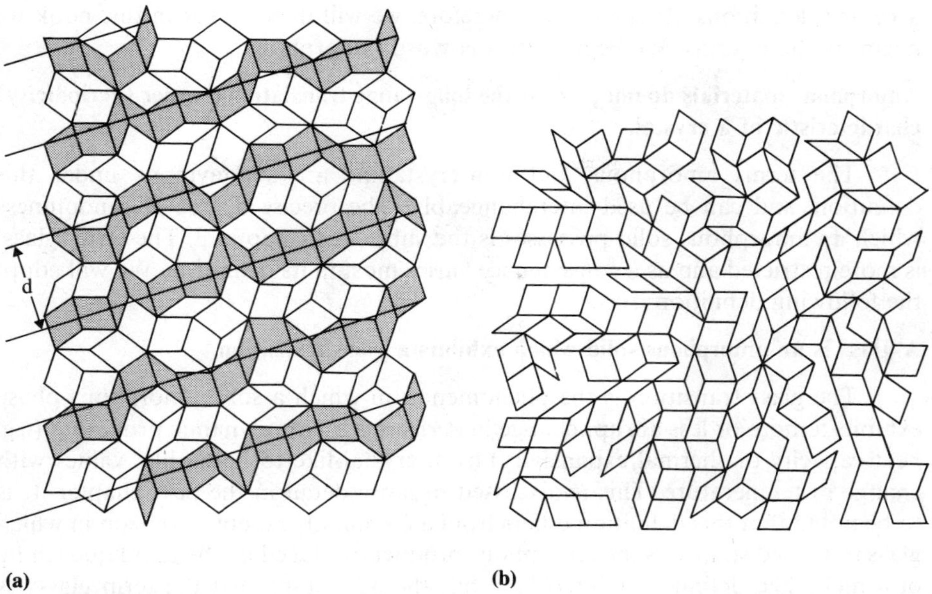

Fig. 1.3 Two-dimensional models for icosahedral structures.
(a) Ideal pentagonal quasicrystal (Penrose tiling) showing the regularly spaced lines of a Bragg lattice, each of which is associated with a shaded row of tiles.
(b) Pentagonal 'glass' model (2D analogue of 3D icosahedral glass model).

rise to the tenfold orientational order characteristic of this lattice. The quasicrystalline translational order arises from the fact that the structure involves two irrationally-related length scales, resulting from the occurrence ratio of the two types of rhombi. These considerations can readily be extended to 3D, where the two building blocks now become rhombohedra (Levine and Steinhardt 1984). Sharp Bragg peaks in a diffraction pattern are expected for such quasiperiodic structures because sets of regularly spaced grids of lines (rotated at angles of 36° and 72° with respect to each other) can nevertheless be superimposed on the lattice through lines of rhombi (see Fig. 1.3(a)); the relatively small degree of positional disorder of the rhombi with respect to the lines simply reduces the intensity of the Bragg peaks rather than broadening them. The final model proposed to account for the existence of icosahedral symmetry is at the other extreme from the quasicrystal, namely the 'icosahedral glass' model (Stephens and Goldman 1986). In this, it is assumed that structural units (e.g. Penrose tiles in 2D or icosahedra in 3D) are randomly packed (thereby obviating long-range translation periodicity) but with the constraint that local (and correspondingly long-range) orientational order is preserved (see Fig. 1.3(b)). It appears that many icosahedral metallic alloys have structural features which are in fact intermediate between those characteristic of the ideal quasicrystal and the icosahedral glass models. Thus, icosahedral alloys may be a bridge between crystalline and amorphous (metallic) structures (Nelson and Spaepen 1989).

There is a considerable amount of confusion in the literature concerning the terms 'amorphous', 'non-crystalline' and 'glassy' and there are as yet no universally accepted definitions. At the outset, therefore, we will define what in this book we mean by these terms. We begin with the word 'amorphous':

Amorphous materials do not possess the long-range translational order (periodicity) characteristic of a crystal.

The terms amorphous and non-crystalline are synonymous under this definition, and can be used interchangeably. The precise degree of randomness which an amorphous solid possesses is the subject of Chapter 3. The term 'glass' is more restricted and as a consequence varies most in its definition. We will adopt the following definition:

A glass is an amorphous solid which exhibits a glass transition.

The glass transition is the phenomenon in which a solid amorphous phase exhibits a more or less abrupt change in *derivative* thermodynamic properties (e.g. heat capacity or thermal expansivity) from crystal-like to liquid-like values with change of temperature. This is discussed in more detail in the next chapter. It is to be noted that this definition differs from a commonly accepted version in which glass is defined simply as the amorphous product produced by the rapid quenching of a melt. The definition offered here has the advantage that the term glassy is confined to those materials which can be obtained in a reproducible state (even after temperature cycling) since the material can be in a state of internal equilibrium above the glass transition. 'Glassy' materials, therefore, need not be prepared solely by quenching from the melt. Note that glassy (or equivalently, vitreous) solids are a special sub-set of amorphous materials or, to express it in a different way, all glasses are amorphous but not all amorphous solids are necessarily glasses.

The definition of the term amorphous given above does not explicitly exclude liquids, although in this book we will almost always be concerned with the solid state. Therefore, we must also define what we mean by the term 'solid', particularly in view of the fact that in the next chapter we will discuss the changes that take place when a glassy solid forms on cooling a liquid:

A solid is a material whose shear viscosity exceeds $10^{14.6}$ poise (or $10^{13.6}$ N s m^{-2}).

This rather arbitrary division of viscosity corresponds to a relaxation time of one day. This may be seen as follows. The expression for the viscosity can be written as

$$\eta = G_x/(\mathrm{d}v_x/\mathrm{d}z)$$

for the application of a shear stress G_x in the x direction causing a velocity gradient $\mathrm{d}v_x/\mathrm{d}z$, where $\mathrm{d}z$ is the thickness of an element perpendicular to the direction of the applied stress. A force of 100 N applied for one day to 1 cm^3 of material having a viscosity of $10^{13.6}$ N s m^{-2} yields a deformation of 0.02 mm, a value which is just measurable. Hence, a material is a solid if the application of a small force for one day produces no permanent deformation. For comparison, the viscosities of most common liquids at room temperature are of the order 10^{-2} poise.

The absence of long-range order, or periodicity, characteristic of an amorphous solid is most clearly evinced in a diffraction experiment, for example using X-rays, where instead of the sharp Bragg spots or rings produced by single crystal or polycrystalline samples respectively, broad diffuse haloes are observed. The origin and interpretation of such diffraction patterns form the subject of Chapter 3; suffice it to say here that the observation of diffuse haloes in a diffraction experiment is a prerequisite to characterization of an amorphous material. However, it is sometimes extremely difficult to ascertain by diffraction means alone whether a given material is truly amorphous or is simply microcrystalline or even 'nanocrystalline' (i.e. consisting of crystallites whose size is of the order of a few nanometres); in both cases broad diffraction haloes can be observed. This ambiguity is the origin of the microcrystalline model for the structure of amorphous materials (see sect. 3.3.3.).

However, very recently Chen and Spaepen (1988) have proposed a calorimetric test to distinguish between amorphous and microcrystalline samples. The crystallization of a truly amorphous material occurs by means of 'nucleation and growth' of the crystallites, whereas an already microcrystalline material will simply experience overall grain growth of the crystallites; these two processes have fundamentally different kinetics, in principle enabling the two types of materials to be distinguished.

In an isothermal nucleation-and-growth process, the fraction of material transformed after time t can be expressed as $x = 1 - \exp(-bt^n)$, where b is a constant dependent on the nucleation frequency, and $2 < n < 4$ (depending on the nucleation mechanism). The corresponding enthalpy release rate is then (Chen and Spaepen 1988)

$$-\frac{\mathrm{d}H}{\mathrm{d}t} = \Delta H(1-x)nbt^{n-1} \qquad [1.1]$$

where ΔH is the total transformation enthalpy of the sample. As seen from Fig. 1.4, the isothermal calorimetric signal for a nucleation-and-growth process consists of an exothermic *peak*, having a maximum at a non-zero time $t = [(n-1)/bn]^{1/n}$. (The solid curve in Fig. 1.4 is calculated using parameters corresponding to the crystallization of glassy $Fe_{80}B_{20}$ (Chen and Spaepen 1988).

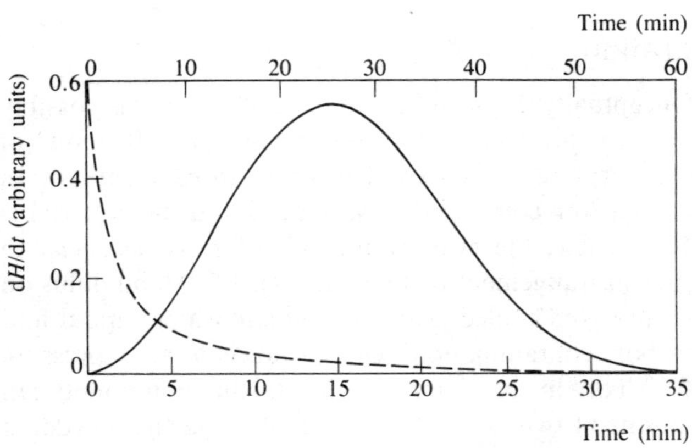

Fig. 1.4 Isothermal enthalpy release rates for crystallite nucleation and growth (solid line) and crystallite grain-coarsening mechanisms (dashed line) (Chen and Spaepen 1988).

In a simple grain-coarsening process, where the average grain size increases at a rate given by $\mathrm{d}r/\mathrm{d}t = M\gamma/r^m$, where M is the atomic mobility, γ is the interfacial surface tension and $0.5 < m < 3$, the corresponding isothermal enthalpy release rate becomes (Chen and Spaepen 1988)

$$-\frac{\mathrm{d}H}{\mathrm{d}t} = H(0)r(0)M\gamma/r^{m+2} \qquad [1.2]$$

where $H(0)$ is the zero-time interfacial enthalpy of a grain of size $r(0)$. As seen from Fig. 1.4, the isothermal calorimetric characteristic of grain coarsening is very different to that for the nucleation-and-growth case, consisting of a monotonically decreasing curve. (The dashed curve shown in Fig. 1.4 is calculated using parameters for the growth of a micro-(quasi) crystalline $Al_{83}Mn_{17}$ film (Chen and Spaepen 1988).)

1.3 Preparation of amorphous materials

There are at least a dozen different techniques that can be used to prepare materials in an amorphous state. Of these, five are commonly used in one form or another to produce most non-crystalline materials of commercial or academic interest. This section details these various methods, in approximate order of importance, together with their advantages and disadvantages and the types of material that are commonly produced using them. The different techniques use variously all three phases of matter, vapour, liquid and solid, as the starting materials in producing amorphous solids, but only deposition from vapour or liquid phases is really important.

Since the amorphous phase is less thermodynamically stable than the corresponding crystalline form (i.e. it possesses a greater free energy), the preparation of amorphous materials can be regarded as the addition of excess free energy in some manner to the crystalline polymorph. How this is done can vary widely, but it is a rule of thumb that the faster the rate of deposition (or cooling), the further the amorphous solid lies from 'equilibrium'.

1.3.1 Thermal evaporation

This technique is perhaps conceptually the most easy to understand, and is possibly the most widely used method for producing amorphous thin films. It is one of several ways of producing amorphous solids by deposition from a vapour. In essence, it is very simple; the starting compound is vaporized and the material is collected on a substrate. In practice, the evaporation is performed *in vacuo* to reduce contamination; a typical arrangement is shown in Fig. 1.5. An oil diffusion pump, together with a liquid nitrogen cooled trap (to eliminate water vapour and to a lesser extent hydrocarbon contamination) can produce a base pressure typically of the order of 10^{-6} Torr in the chamber. The starting compound can be heated to vaporization in one of two ways. A 'boat' containing the powdered material acts as a resistive element between two electrodes for the passage of a large d.c. current; high melting-point metals, e.g. molybdenum or tungsten, are

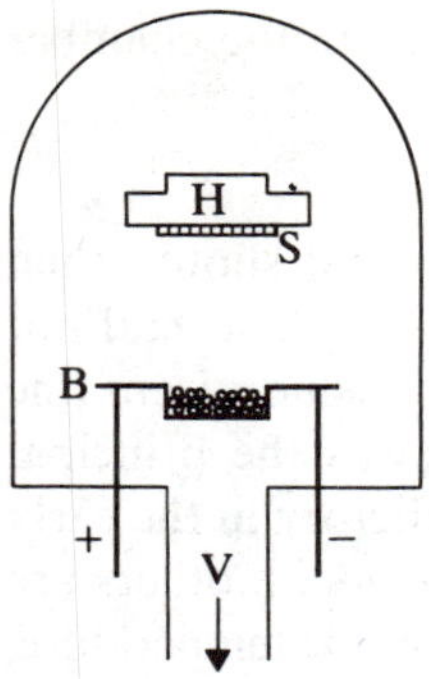

Fig. 1.5 Schematic illustration of a thermal evaporation chamber. The material to be evaporated lies in a high-melting point metallic boat (B) through which a heating current is passed, and the evaporated film is collected on a substrate (S) attached to a heater (H). The entire apparatus is placed in a vacuum chamber pumped to a high vacuum (V).

used as boat materials. This method is suitable for relatively low melting-point compounds; those materials with higher melting points are best vaporized by bombardment with high-energy electrons from an electron gun placed in the chamber. The purpose of the electron-beam bombardment is simply to heat the sample to its melting point; compare this with the mechanism that takes place in 'sputtering' which is the topic of the next section. One major problem with the evaporation technique is differential evaporation, which is discussed more fully later; a remedy is to use 'flash evaporation', in which powdered material is dropped steadily on to a heated ribbon, thereby almost instantaneously vaporizing it.

Many materials can be rendered amorphous using thermal evaporation. The essential feature is that atomic surface mobility is greatly diminished because of the cold substrate, causing the adatoms to be 'frozen' in the random positions at which they arrive. Most amorphous semiconductors of interest to be discussed later can be produced in this way on substrates held at, or near, room temperature. Examples include Si and Ge and the tetrahedrally bonded III–V alloys (e.g. GaAs), 'chalcogenide' materials (those containing wholly or mainly chalcogen group VI elements), and SiO_x (where x can be varied between 0 and 2). These are all examples of solids containing principally covalent bonds. Certain metals can also be produced in an amorphous form by thermal evaporation without the need of stabilizing 'metalloid' atoms (see Ch. 2), but this must be done using substrates held at very low temperatures (4 K), otherwise polycrystalline films are produced. Amorphous Ga and Bi can be made in this way, although other elements invariably crystallize even at these low temperatures, and other methods must be employed.

The principal disadvantage of thermal evaporation as a preparative technique lies in the variability in purity and composition of the resulting films (in the case of alloys). Many factors combine to produce this variability, and some are listed below:

(a) Substrate temperature.
(b) Substrate–source separation and orientation.
(c) Base gas pressure in chamber.

(d) Impurities from the evaporation boat, or desorbed gases from the chamber surfaces.
(c) Boat or filament temperature.

Some of these parameters the experimenter can control (and should then specify in a description of the material preparation), others, including several not listed, cannot be varied at will, and so can change in a random fashion from one preparation to another of ostensibly the *same* material. This was the principal reason why so many experimental data were apparently contradictory in the early days of the subject. Some of the variables listed above are self-evident, others are more subtle. The substrate temperature is important since the higher the temperature, the higher is the surface mobility which can lead to a material with considerably fewer structural defects. However, if the substrate temperature is too high, the material will crystallize. The relative orientation of substrate with respect to source is of importance since evaporation at oblique angles can produce materials which are macroscopically inhomogeneous (Leamy *et al.* 1980). A 'shadowing' effect causes a columnar growth morphology to be formed with dense regions approximately parallel to the source–substrate direction (see Fig. 1.6); this has been observed by scanning electron microscopy in e.g. thin films of a-Ge and a-Ge–Se alloys. In general, the column direction and the evaporant beam direction do not lie parallel to each other; rather, the columns lie more nearly perpendicular to the film surface. The respective angles of incidence, β and α, are found to obey the empirical equation (Leamy *et al.* 1980):

$$\tan \alpha = 2 \tan \beta \qquad [1.3]$$

A major problem with conventional thermal evaporation is differential evaporation and the consequent lack of control over the resulting composition of the deposited film. In conventional vacuum evaporation, the vapour is removed immediately from the vicinity of the molten source, thereby preventing the establishment of equilibrium. In this case the element in the source alloy with the highest vapour pressure evaporates preferentially, depleting the source and leading to composition inhomogeneities in the deposited film. Flash evaporation can be used to circumvent this problem to a certain extent, as mentioned earlier. A further complication which can arise in both flash and conventional evaporation is that the dominant (equilibrium) vapour species may not have the same composition

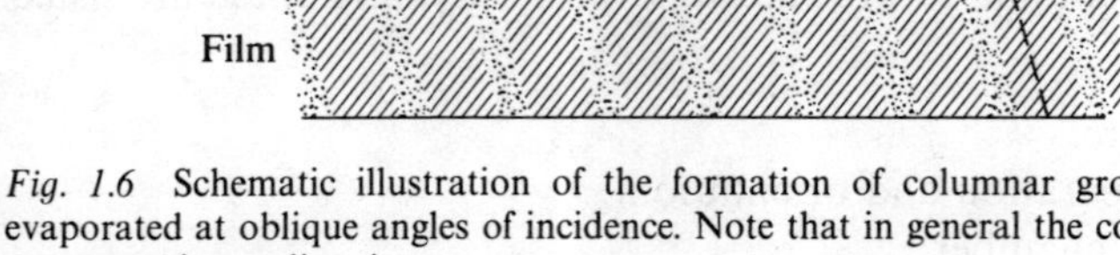

Fig. 1.6 Schematic illustration of the formation of columnar growth morphology of thin films evaporated at oblique angles of incidence. Note that in general the columns do not lie parallel to the evaporant beam direction.

as the source material. A case in point is that of arsenic chalcogenide alloys (e.g. As–S), in which the dominant vapour species are molecules of As_4S_4 even though the starting material may be stoichiometric As_2S_3. If the substrate is at a low enough temperature (and room temperature might be sufficient for this purpose) the deposited film will tend to consist of discrete As_4S_4 molecules bonded together by Van der Waal's forces. Annealing the film at a higher temperature will polymerize and cross-link these units yielding a fully coordinated random structure similar to that obtained by quenching from the melt (see sect. 1.3.5). Alternatively, the substrate can be held at an elevated temperature and the annealing then occurs simultaneously with the deposition process. Certain vapour species (e.g. As_4) have very low 'sticking coefficients' because of their shape, and consequently produce poor quality films.

It must be stressed that thermal evaporation, like other vacuum techniques is a *thin-film* preparative technique; typically film thicknesses of the order of tens of microns are produced for deposition rates of the order of 0.1–1.0 μm per second. Nevertheless, this is sufficient for certain applications, and indeed the amorphous selenium at the heart of the Xerox process is produced by evaporation.

1.3.2 Sputtering

The sputtering process is rather more complicated than thermal evaporation but enjoys the advantage of being far more flexible. In essence, it consists of the bombardment of a target by energetic ions from a low-pressure plasma, causing erosion of material, either atom by atom or as clusters of atoms, and subsequent deposition of a film on a substrate. The simplest way to induce sputtering is to apply a high negative voltage to the target surface, thereby attracting positive ions from the plasma. However, this d.c. sputtering process is only feasible for targets composed of metals, or at least consisting of material sufficiently electrically conducting that the target can act as an electrode; certain of the *crystalline* arsenic chalcogenides are sufficiently conducting to act in this way to produce amorphous films.

A more common approach is to apply an r.f. field (at typically $\simeq$13 MHz), to the target (Fig. 1.7). In this case, both metallic and insulating materials may be

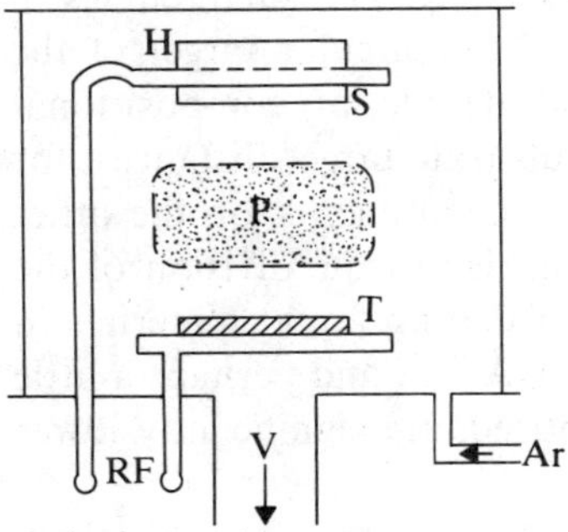

Fig. 1.7 Schematic illustration of an r.f. sputtering apparatus. The r.f. field is applied between the target (T) electrode and the substrate (S) electrode. A sputtering gas (e.g. Ar) is introduced into the chamber and a plasma (P) is struck. The gas is pumped away by a vacuum system (V); the substrate may also be heated by an electrically insulated heater (H).

used as targets; all that is required is that for the r.f. voltage to be capacitatively coupled to the target surface. For metallic targets, this is achieved by connecting a capacitor in series with the target; insulating targets are normally bonded to a metal backing electrode which itself acts as a capacitive component. As in evaporation a residual gas pressure in the chamber of less than 10^{-6} Torr is desirable and achieved by similar means. The sputtering gas, commonly an inert gas such as argon, is introduced at a higher pressure, say 1–20 mTorr. Application of the r.f. voltage causes the striking of a plasma and positive ions are attracted to the target during each negative half-cycle. However, since the mobility of electrons is higher than that of ions in the plasma, more electrons than ions are attracted to the front surface of the target in their respective half-cycles, resulting in the build-up of a negative bias. After some time, therefore, the ions are attracted from the plasma essentially by the bias potential, rather than the r.f. potential. The ejected material from the target is then carried to the substrate by the following half-cycle of the r.f. field.

The material deposited on the substrate can form an amorphous film for the same reason as in evaporation; namely, adatom mobility is sufficiently low, because of the relatively low temperature of the substrate, that crystallization is prevented. However, the substrate must be cooled and the r.f. power limited so that heating due to electron bombardment is minimized. This method generally produces reasonably homogeneous and uniformly thick samples; for $\pm 5\%$ uniformity of film thickness the diameter of the target needs to be about 5 cm larger than the maximum linear dimension of the substrate. The process can be operated in either sputter-up or sputter-down modes, the former having the advantage of being able to use targets in powder form.

Sputtering, particularly using r.f. fields, can produce amorphous samples of most materials of interest, Si, Ge, SiO_2, chalcogenide glasses and amorphous metals being examples. Sputtering is significantly superior to evaporation for the production of multicomponent systems. For targets already in compound form this is because the sputtering rates for different elements do not vary widely, unlike melting points, vapour pressures and, therefore, evaporation rates. As a consequence, sputtered films tend to preserve the stoichiometry of the starting material. In addition, amorphous alloys can be sputtered simply by using a target composed of the individual components of the alloy. For targets comprising, say, small circles of one component placed uniformly over the surface of the circular target of the predominant component, films possessing a good degree of compositional homogeneity can be achieved for sufficiently large substrate-target distances; for 'split' targets consisting of two semicircles of different components, on the other hand, films having a composition gradient (perpendicular to the division of the target) can be obtained. Usual sputtering rates for insulating target materials in an inert gas atmosphere (e.g. Ar) are in the range 1–10 Å s^{-1}, and perhaps a little higher for metallic targets. These rates, it will be noticed, are significantly lower than those encountered in evaporation.

A further option offered by sputtering is the use of gases other than Ar which chemically *react* with the target, resulting in 'reactive' sputtering. This can significantly increase the sputtering rate, as well as incorporate chosen additives into the films. This has been used in producing hydrogenated amorphous silicon (by sputtering in an $Ar:H_2$ mixture), doped amorphous silicon films ($Ar:H_2:N_2/PH_3$), and amorphous transition-metal oxide materials ($Ar:O_2$).

Most of the factors which cause variations in the films produced by thermal evaporation are also significant in the sputtering process. Additional factors which need to be considered are the following:

(a) Sputtering gas pressure.
(b) Ratio of partial pressures of reactive gas to inert gas (in reactive sputtering).
(c) R.F. power applied to target.
(d) Bias voltage of target or substrate.

Close control over all these parameters is necessary to ensure that films with reproducible characteristics are produced.

1.3.3 Glow-discharge decomposition

Another technique which can produce thin films of amorphous materials is glow-discharge (GD) decomposition in the vapour phase. This has sprung into prominence recently with the discovery by Spear and Le Comber (1975) that prepared in this manner, certain amorphous semiconductors can be substitutionally doped to control their electronic properties.

This technique, like sputtering, relies on the production of a plasma in a low-pressure gas, but instead of ions from the plasma ejecting (sputtering) material from a target, chemical decomposition of the gas itself takes place, leading to deposition of a solid film on a substrate placed in the plasma (Fig. 1.8(a)). The plasma is produced by the application of an r.f. field (as in sputtering) in one of

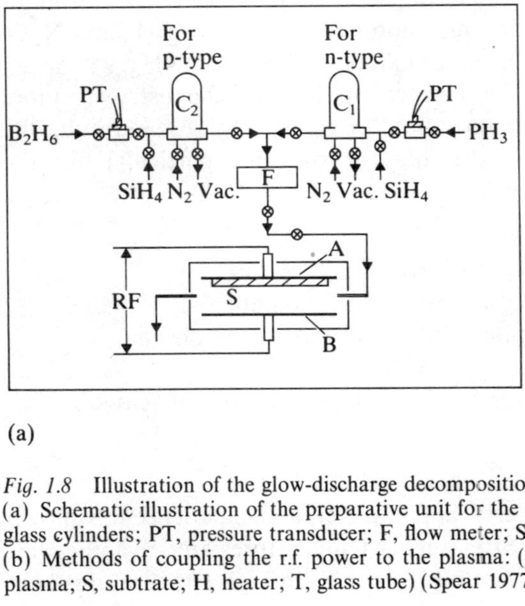

(a)

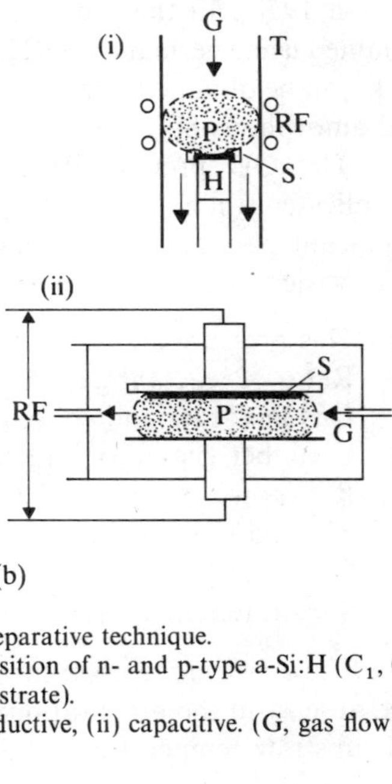

(b)

Fig. 1.8 Illustration of the glow-discharge decomposition preparative technique.
(a) Schematic illustration of the preparative unit for the deposition of n- and p-type a-Si:H (C_1, C_2, glass cylinders; PT, pressure transducer; F, flow meter; S, substrate).
(b) Methods of coupling the r.f. power to the plasma: (i) inductive, (ii) capacitive. (G, gas flow; P, plasma; S, subtrate; H, heater; T, glass tube) (Spear 1977).

two ways, either using inductive coupling or capacitive coupling; these configurations are shown schematically in Fig. 1.8(b). The capacitive configuration is the more important since it can readily be scaled up to produce large surface area material for device applications. The reactant gas may be used either by itself, or mixed with a carrier gas (e.g. Ar), or another reactant gas to form alloys, and is admitted to the reaction chamber through a mixing tank or flowmeter. Typically, reactant gas pressures between 0.05 and 1.0 Torr are used with about 1–10 W of, say, 14 MHz r.f. power being supplied to the plasma, resulting in deposition rates of 1–10 Å s^{-1}.

The reactant gases that are used are generally hydrides and produce polymers under certain conditions; this is the 'glow discharge polymerization' of monomers utilized in chemistry. Solid-state physicists refer to the process as 'glow-discharge decomposition', particularly when non-polymeric material is produced. The material most often produced by the GD technique is amorphous silicon (a-Si); this is important because of its device prospects in view of the fact that it can be readily doped (see sect. 5.4.1). Silane (SiH_4) has long been the reactant gas used to produce a-Si, with or without Ar as diluent. This produces *hydrogenated* amorphous silicon (a-Si:H) films under optimum conditions (with of the order of 5–10 at. % H contained within the material), but polymeric polysilane or amorphous silicon hydride (a-SiH_x) films with much higher hydrogen contents (and significantly worse electronic characteristics) are formed under other conditions. The a-Si:H materials are doped either n-type or p-type by admitting phosphine (PH_3) or diborane (B_2H_6) respectively, into the silane gas stream; the dopant atoms are incorporated into the amorphous structure substitutionally, thereby liberating excess electrons (for P) or holes (for B). Recently, fluorinated material produced by the GD decomposition of SiF_4 and H_2 (a-Si:F:H) has been shown to possess superior properties to simply hydrogenated silicon (Ovshinsky and Madan 1978). In the same manner, hydrogenated germanium (a-Ge:H) can be obtained using germane (GeH_4) and can be doped in a similar fashion. Amorphous SiO_2 can be obtained from the GD decomposition of a mixture of SiH_4 and N_2O, and amorphous Si_3N_4 is produced from a mixture of SiH_4 and NH_3.

The GD process, depending in essence on plasma chemistry, is more complicated than either thermal evaporation or sputtering. As a result, the significant parameters that can affect the characteristics of a particular film are more varied. Some of these are given below:

(a) Gas pressure and temperature.
(b) Ratio of reactant gas (e.g. SiH_4) to carrier gas (e.g. Ar).
(c) Gas flow rate and consequent dwell time in the chamber.
(d) Chamber geometry and substrate position relative to the plasma.
(e) R.F. power applied to plasma.
(f) Method of coupling of r.f. power to plasma (inductive or capacitive).
(g) Substrate bias.
(h) Substrate temperature.

The substrate temperature is a critical parameter, and experience has shown that, at least in the preparation of a-Si:H, higher temperatures are advantageous, and substrate temperatures of 520 K are often used. Such high temperatures help

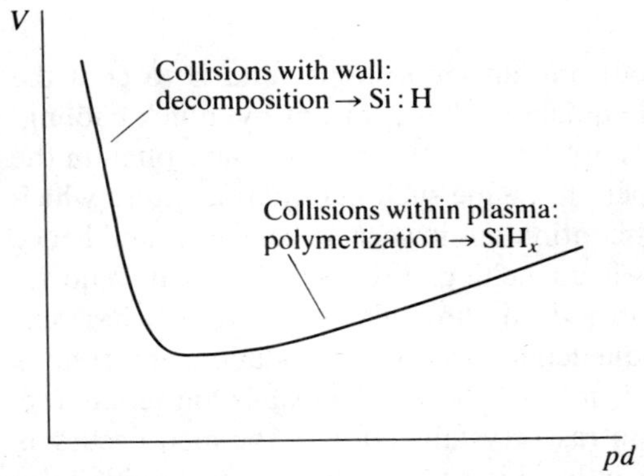

Fig. 1.9 Schematic illustration of the characteristics of a glow-discharge plasma (Paschen's law): V is the applied r.f. voltage, p is the pressure and d is a characteristic dimension of the chamber (Brodsky 1978).

to ensure that volatile side-products are not incorporated into the films. Another crucial parameter is the pressure of the reactant gas. In an informative review of the GD process, Brodsky (1978) points out that there are two distinct pressure regimes that govern the behaviour of the plasma discharge; these are shown schematically in Fig. 1.9. This behaviour is analogous to Paschen's law, which applies strictly only for the threshold characteristics of the GD, but can be applied to the operating region if the r.f. power levels are sufficiently low. At high pressures, the electron energies within the plasma are collision-limited within the plasma itself, and this is found to promote polymerization within the vapour phase leading to (amorphous) silicon hydrides (a-SiH_x). In contrast, at low pressures the electron energies are limited by collisions near or at surfaces, and this is found to enhance decomposition resulting in the desired hydrogenated material (a-Si:H).

1.3.4 Chemical vapour deposition

The technique of chemical vapour deposition (CVD) is analogous to the GD method in that both depend on the decomposition of a vapour species. The difference is that the CVD process relies on thermal energy for the decomposition (i.e. it is pyrolytic), and the applied r.f. field (if used) simply serves to heat up the substrate upon which the vapour decomposes. Temperatures of the order of 1000 K are commonly used. The CVD process is often used to produce polycrystalline material, particularly silicon from SiH_4. However, under the appropriate conditions true amorphous hydrogenated silicon (a-Si:H) can also be deposited by the CVD technique from SiH_4, and doped samples using mixtures with PH_3 or B_2H_6.

Another important application which can be regarded as a variant of CVD is the production of amorphous silica (SiO_2) by the reaction:

$$SiCl_4 + 2H_2O \rightarrow SiO_2 + 4HCl$$

This results in so-called 'vapour phase hydrolysis', which is carried out in an oxyhydrogen flame producing 'synthetic grade' high purity amorphous silica.

1.3.5 Melt quenching

The oldest established method of producing an amorphous solid is to cool the molten form of the material sufficiently quickly; what is meant by 'quick cooling' will emerge later. Amorphous materials produced in this manner have often in the past been termed 'glasses'; we prefer here to define such materials as those which exhibit glass-transition phenomena (discontinuities in specific heat, etc.) and hence are not necessarily formed only by melt quenching, although the great majority of melt-quenched amorphous solids do indeed show glass-transition behaviour. The distinguishing feature of the melt-quenching process of producing amorphous materials is that the amorphous solid is formed by the *continuous* hardening (i.e. increase in viscosity) of the melt. In contrast, crystallization of the melt occurs as a *discontinuous* solidification, solid growth taking place only at the liquid–solid interface, with the result that crystallites grow in the body of the melt.

An essential prerequisite for 'glass' formation from the melt, therefore, is that the cooling be sufficiently fast to preclude crystal nucleation and growth; the crystalline phase is thermodynamically more stable and crystal growth will always dominate over the formation of the amorphous phase if allowed to take place. A good discussion of crystal nucleation and growth kinetics can be found in Turnbull (1969).

The crystallization rate of an undercooled liquid depends on the rate of crystal nucleation and on the speed, u, with which the crystal–liquid interface moves. Both in turn are strongly dependent on the reduced temperature $T_r = T/T_m$, and the undercooling $\Delta T_r = (T_m - T)/T_m$. It is expected that u should be proportional to some function of ΔT_r (linear for small ΔT_r), and inversely proportional to an average jump time, τ, of atoms in the interfacial region; τ in turn might be expected to scale as the viscosity. Thus, high values of viscosity reduce the crystallization-front velocity and hence the crystallization rate itself; this is in accordance with common experience in which it is often found that the liquids of 'easy' glass-formers have high viscosities for temperatures at and below T_m (e.g. the viscosity of silica $= 10^7$ poise at the melting point of crystobalite). Nucleation of crystallites in a liquid may be either 'heterogeneous' or 'homogeneous'; the former arises when 'seeds', such as crystals of the material suspended in the liquid or the container walls are introduced into the system, and the latter takes place when all heterogeneous centres have been removed.

What then is a condition for glass formation? It is that the nucleation rate I_n should be less than a certain value, say, $10^{-6}\,\mathrm{cm^{-3}\,s^{-1}}$, not to be observable in practicable time-scales and liquid volumes. Simple nucleation theory dictates that crystal nucleation is *promoted* by the change in free energy on crystallization (dependent on the *volume* of the crystallite), and at the same time *retarded* by the work needed to form the crystal–liquid interface (proportional to the *area* times the surface tension γ); the interplay between these two competing factors ensures that only those crystals which contain more than some critical number of atoms can exist as crystal nuclei; all others remelt. Turnbull (1969) has shown that the homogeneous nucleation rate can be expressed in the form

$$I_n = \frac{k}{\eta}\exp[-b\alpha_n^3\beta_n/T_r(\Delta T_r)^2] \qquad [1.4]$$

where k is a constant, b is a constant dependent on the shape of the nucleus ($16\pi/3$ for a sphere) and α_n and β_n are dimensionless parameters defined as:

$$\alpha_n = \frac{(N\bar{V}^2)^{1/3}\gamma}{\Delta H_m} \quad [1.5]$$

(where N is Avogadro's number, $\bar{V}$ is the molar volume of the crystal and ΔH_m is the molar heat of fusion), and as:

$$\beta_n = \frac{\Delta H_m}{RT_m} = \frac{\Delta S_m}{R} \quad [1.6]$$

The quantity α_n represents the number of monolayers per unit area of crystal which would be melted at T_m by an enthalpy equal in magnitude to γ. A plot of [1.4] is shown in Fig. 1.10, showing the nucleation rate I_n as a function of reduced temperature T_r for different values of the parameter $\alpha_n\beta_n^{1/3}$. The value chosen for the pre-exponential factor in [1.4] was 10^{32} with a temperature independent value of $\eta = 10^{-2}$ poise; this factor can be written equivalently as the product of the atomic density (10^{21} cm^{-3}) and a typical atomic jump frequency (10^{11} s^{-1}). It can be seen from Fig. 1.10 that for values of $\alpha_n\beta_n^{1/3} \gtrsim 0.9$ the crystal nucleation rate becomes less than that which is readily observable, and these liquids would not crystallize (unless seeded), forming glasses if sufficiently undercooled. Using a more realistic model for the temperature dependence of the viscosity (see [2.6]), it is found that the peak in the $I - T_r$ relation moves to higher values of T_r and

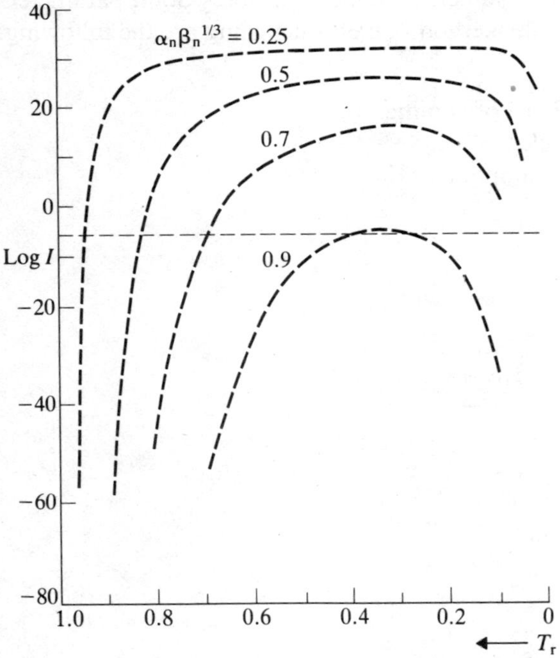

Fig. 1.10 Homogeneous nucleation rate as a function of reduced temperature $T_r = T/T_m$ calculated for various values of $\alpha_n\beta_n^{1/3}$ (Turnbull 1969).

becomes lower and narrower for increasing values of the reduced glass-transition temperature T_g/T_m. The consequence of this is that glass-forming tendency should increase with T_g/T_m.

In this respect, then, we may regard 'glass' formation as a kinetic phenomenon; a glass is simply a supercooled liquid in which no time for crystal growth has been allowed. In fact, 'glass' is considerably more complicated than this simple picture implies, and many factors contribute to whether a particular glass will form; this topic forms the subject of Chapter 2.

We have seen that the cooling rate is often a critical factor in determining glass formation. Certain 'easy' glass-formers such as B_2O_3, will form an amorphous solid even under conditions of very slow cooling (say $1\ K\ s^{-1}$), whereas other materials, notably metallic glasses, require very high rates of cooling indeed before they form an amorphous phase; otherwise polycrystalline material is produced. The most usual way of producing samples of 'easy' glass-formers is to seal a charge (1–10 g) in a fused silica ampoule under a good vacuum (10^{-6} Torr), and keep the ampoule in a rocking or rotating oven at a sufficiently elevated temperature that the constituents become molten and can react; the rocking motion ensures that a thorough mixing of the mixture takes place. The melt can then be quenched, either slowly by simply switching off the oven, or more rapidly by bringing the ampoule into the air, or yet more rapidly by plunging the ampoule immediately from the oven into a liquid (preferably one with a high thermal conductivity and high latent heat of vaporization so that heat is conducted away from the sample as fast as possible without the formation of a thermally insulating vapour layer around the ampoule). In this manner cooling rates of the order 10^{-2} to $10^3\ K\ s^{-1}$ are achievable (see Table 1.1), and are sufficient for most purposes. Some parameters which are important in the ampoule method of melt-quenching are the following:

(a) Temperature of the oven.
(b) Rate (or equivalently, method) of cooling.
(c) Volume of charge in ampoule.
(d) Thickness of the wall of the ampoule.

Table 1.1 Quenching techniques and their characteristic rates of cooling

Technique	*Cooling rate* ($K\ s^{-1}$)
Annealing	
large telescope mirror	10^{-5}
optical 'glass'	3×10^{-4}
ordinary 'glass'	10^{-3}–10^{-2}
Air quenching	1–10
Liquid quenching	10^2–10^3
Chill-block	
splat-cooling	10^5
melt-spinning, extraction	10^6–10^8
Evaporation, sputtering	10^9 ?

Material produced in this manner is often in the form of a 'plug' or rod, but can be made into thinner sections by either sawing and polishing or by heating the glass up to its softening point (glass-transition temperature) and then compressing it between two parallel plates or by blowing a thin film.

For certain materials however, notably metals, even the fastest cooling rates achieved by the above methods are insufficient to prevent crystallization, and still faster cooling methods are required. Such methods have been pioneered by Duwez who was the first to demonstrate, in 1959, that cooling rates in excess of 10^6 K s^{-1} can be achieved. The technique relies on the rapid cooling afforded by a 'chill-block' in intimate contact with a thin film of the liquid, in which the heat is swiftly and continually transported away from the interface; chill-blocks of copper are commonly used. The liquid melt may be brought into contact with the chill-block in one of several ways. The first that was used projected small droplets of liquid at a Cu sheet ('splat cooling'); in this manner the first amorphous metal $Au_{75}Si_{25}$ was produced by melt quenching (Klement *et al.* 1960). This and a related technique, the aptly named hammer-and-anvil 'drop smasher', can give cooling rates of the order of 10^5 to 10^6 K s^{-1}, but produce thin foils of material of irregular area. These amorphous metallic materials have many interesting properties (e.g. see Cahn 1980) which lend themselves to several technological applications (see sect. 7.5), but material produced by the hammer-and-anvil or projection techniques is not generally in a usable form. However, there are certain chill-block cooling techniques more recently developed which do produce amorphous metals in a useful geometry, and with a faster cooling rate as an added bonus; these are the 'melt-spinning' and 'melt-extraction' techniques (Figs 1.11(a) and (b)). Both use a rapidly spinning copper disc as the chill-block. The former uses a jet of liquid impinging on the disc producing thin ribbons, whereas the latter removes liquid from a reservoir producing fine wires; in both cases, cooling rates of the order 10^6–10^8 K s^{-1} can be achieved. These deceptively simple techniques yield cooling rates which depend in a complicated way on a variety of parameters, some of which for melt spinning as an example, are given below:

(a) Rotation speed of Cu disc.
(b) Coating and surface finish of disc.
(c) Melt jet diameter, velocity and angle of attack.
(d) Ribbon geometry and uniformity, particularly surface smoothness.
(e) Nature and pressure of ambient gas.

Despite this apparent complexity, ribbons up to 15 cm wide and 30 μm thick can be produced by careful attention to conditions of preparation, although 1 cm by 15 μm is more usual. Notice that in contrast to other melt-quenching methods, chill-block techniques cannot produce material in bulk form, but only in the form of thin ribbons (or wires). If attempts are made to produce ribbons of thickness greater than a few tens of microns, only polycrystalline material is produced, since beyond a critical thickness conduction of heat *through* the ribbon becomes the critical factor, and this rate cannot be made large enough to produce amorphous metals.

A summary of various melt-quenching techniques and their cooling rates are given in Table 1.1 together for comparison with those processes normally regarded

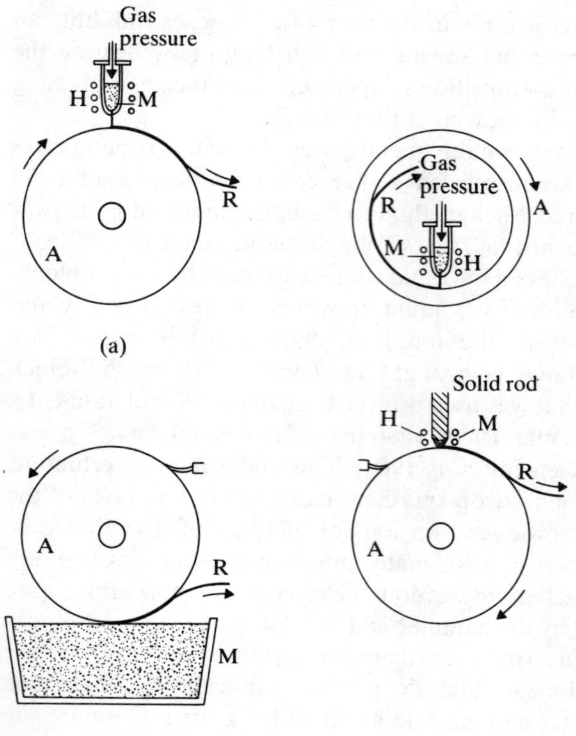

Fig. 1.11 Schematic illustration of the different techniques to form continuous ribbons of glassy metals (Cahn 1980).
(a) Melt spinning, both external and internal.
(b) Melt extraction, both crucible and pendant drop.
(A, rotating Cu disc or cylinder; M, melt; H, induction heater; R, ribbon of metallic glass.)

as annealing. It will be seen that thermal evaporation and sputtering (i.e. *vapour* quenching) are also included; the precise rates of 'cooling' are impossible to quantify in these cases, but must certainly be well in excess of those achievable using chill-block methods. This makes understandable the fact that if a material cannot be rendered amorphous any other way, then some form of vapour deposition will almost certainly be successful.

1.3.6 Gel desiccation

The technique for producing amorphous materials via the sol–gel process has considerable technological promise (e.g. see Mukherjee 1980). The method has its greatest usefulness for those systems which give rise to very viscous melts near the melting point (and consequently have considerable difficulty in achieving homogeneous mixing), or alternatively which have extremely high melting points and hence pose considerable technical problems in actually being able to make a glass by melt quenching.

There are two essentially different routes to produce an amorphous material via a gel, using either aqueous or organic materials as the starting components:

(a) Destabilization of a sol (commonly of silica), with other components being added in the form of appropriate aqueous solutions.
(b) Hydrolysis and polymerization of mixtures of organometallic compounds.

In both methods, however, a multicomponent 'gel' (the elastic solid product produced abruptly from a viscous liquid by a process of continuing polymerization), being non-crystalline and homogeneous, is heated to remove volatile components and cause an initial densification, which is completed by a final process of sintering or fusion to produce the amorphous solid. The two methods are represented schematically in Figs 1.12(a) and (b).

The method employing aqueous components starts with a silica 'sol' of silicic acid, $Si(OH)_4$ (a dispersion at the molecular level in aqueous solution) to which are added other components (e.g. metal salts in aqueous solution). The formation of the gel then involves polymerization of the monomeric $Si(OH)_4$ to form particles, growth and interconnection of particles to form branched chains, finally forming a 3-D network. The gelling process is influenced by several factors, e.g. pH of the medium, particle size and temperature. The gel can then be converted to an amorphous solid in one of three ways: heating at temperatures *below* the glass-transition temperature, T_g, resulting in chemical polymerization; sintering (with or without the application of pressure) at temperatures above T_g but much below the melting point; fusing the gel particulates to form a glass. In this manner, amorphous materials in, for example, the SiO_2–P_2O_5 or SiO_2–B_2O_3 series may readily be produced by using phosphoric acid or ammonium tetraborate additions respectively.

The second approach uses alkoxysilanes (e.g. tetraethoxysilane $Si(OC_2H_5)_4$) mixed with other metal alkoxides (or salts) as the starting materials which, in the presence of a small amount of water, partially hydrolyse replacing alkoxide groups by hydroxyl progressively, while polymerizing and incorporating the various components homogeneously into the network. As an example, consider the reaction between silicon, boron and sodium alkoxides, $Si(OR)_4$, $B(OR)_3$ and NaOR (where R is an alkyl group). On addition of water these hydrolyse partially and polymerize to give, say,

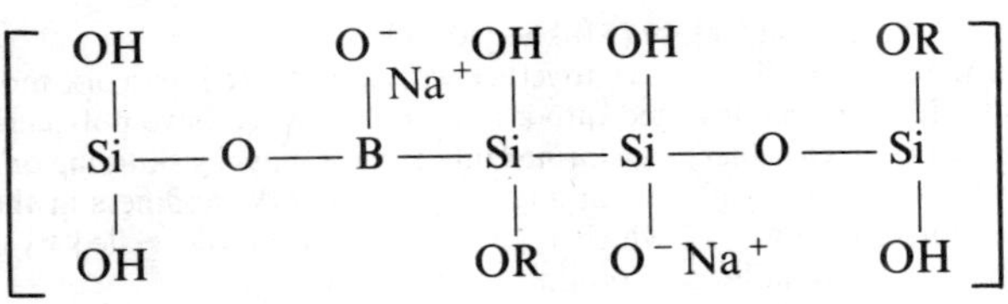

which then gels. Heating then causes dehydration and loss of alkyl groups leaving a mixed oxide gel, which can be transformed to an amorphous material by one of the processes mentioned earlier. Compounds such as $Si(OC_2H_5)_4$, $B(OCH_3)_3$, $Al(OC_4H_9)_3$ and $Na(OCH_3)$ can be used in this way.

The principal advantage of the gel desiccation technique is that very refractory materials, such as those in the TiO_2–SiO_2 system, can be prepared homogeneously

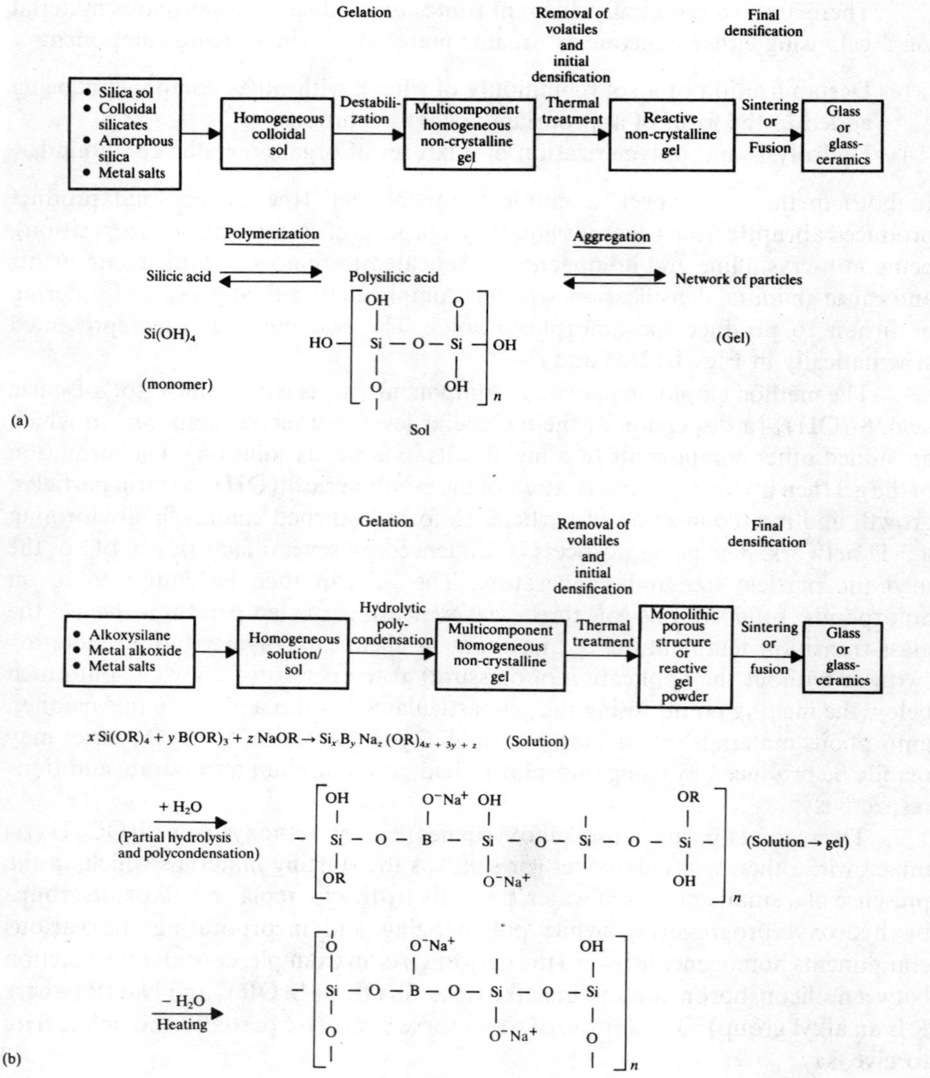

Fig. 1.12 The sol–gel process for glass manufacture (Mukherjee 1980).
(a) Aqueous process. Silica (in the form of silicic acid), together with any desired network modifiers, are formed into an aqueous sol. This is then converted into a rigid gel by progressive polymerization and formation of a branched network, which is then transformed into a dense glass by sintering or fusion.
(b) Organo-metallic process. Alkoxysilane, together with any desired network modifiers in the form of metal alkoxides or salts, are formed into a sol which is then polymerized and gelled by partial hydrolysis. Heat treatment removes any volatiles and produces a dense glass.

at relative *low* temperatures, in some cases much below the melting point. Very pure products can also be made, since the starting components can readily be purified. The final amorphous products, although not prepared by melt quenching, behave in a very similar manner to glasses, exhibiting the same kinds of glass-transition phenomena.

1.3.7 Electrolytic deposition

Another technique which uses solutions of the desired material is electrolytic deposition. In this manner, for example, amorphous Ge films can be deposited on a Cu cathode by the electrolysis of $GeCl_4$ in glycol. Although comparatively thick films ($\simeq 30\ \mu m$) can be prepared in this way, contamination by the solution and reaction products is a problem.

1.3.8 Reaction amorphization

In certain cases the precipitate thrown out of solution as a result of a chemical reaction proves to be amorphous. For example, amorphous As_2S_3 can be formed by passing H_2S gas through a solution of As_2O_3 in dilute hydrochloric acid. The product shows very similar glass-transition behaviour to that observed in melt-quenched glasses. Equally, gel formation (sect. 1.3.6) could be regarded as a form of reaction amorphization.

Thin films of amorphous SiO_2 can result from the anodic oxidation of Si; the use of non-aqueous electrolytes prevents the formation of the porous films that result when aqueous electrolytes are used. Mixed silicate systems, such as SiO_2–P_2O_5 and SiO_2–B_2O_3, can be formed using appropriate electrolyte solutions. Thermal oxidation of Si with dry oxygen or steam at high temperatures also leads to the formation of a-SiO_2 thin films.

1.3.9 Irradiation

The interaction between high-energy ionizing particles and crystalline solids can often produce enough structural damage to amorphize the material. As an example, crystalline quartz is transformed to a-SiO_2 by irradiation with $\simeq 10^{20}$ neutrons cm^{-2}. This form, interestingly, has a density which is *higher* ($2.26\ g\,cm^{-3}$) than that produced by conventional melt quenching ($2.205\ g\,cm^{-3}$); the density decreases to the lower value on annealing.

Ion bombardment of crystalline solids produces amorphous surface layers, some hundreds of ångströms in thickness. This is of great significance in the current semiconductor industry (based on *crystalline* silicon), where doping to control the electronic properties is often accomplished by implanting the dopant atoms (e.g. P, B) by ion bombardment. The resulting amorphous layers are very deleterious to the performance of the device, but can be removed by 'laser annealing'. Illumination of the amorphous surface by pulsed, high-power laser light causes the surface to *recrystallize* and heal the bombardment damage; the laser-annealing process is complicated, and believed to be not simply a surface melting phenomenon, but possibly due to bonds reforming as a result of the creation of an intense electron-hole plasma caused by the absorption of the photons.

Bombardment by high-energy ions, neutrons or electrons has been found to cause amorphization of both crystalline intermetallic compounds (e.g. $CuTi_2$) and so-called 'diffusion couples' consisting of alternating thin layers of two pure crystalline metals, such as Ru or Cu, together with a variety of transition metals (see Johnson 1986 for a review). The ion-beam-induced amorphization process for the diffusion couples can be regarded as being simply a very-high-temperature diffusion process ($T \simeq 10^4$ K) taking place in the very short time ($t \simeq 10^{-11}$ s) for which the displacement cascade exists after the impact of an ion; the mechanism for solid-state diffusion-induced amorphization will be discussed in a little more detail in sect. 1.3.11. The process for the intermetallic compounds is a little different. Here, two opposing radiation-induced processes can be operative: radiative-induced chemical disordering of atoms on lattice sites will tend to promote amorphization, whereas radiation-induced defect (e.g. vacancy) migration will tend to restore ordering. Metallic glass formation by irradiation of crystalline intermetallic compounds would therefore appear to be favoured if the irradiation is carried out at sufficiently low temperatures to suppress the defect migration (Johnson 1986).

Electron-beam irradiation can also result in amorphization in certain cases. Some organic crystals became amorphous on exposure to electrons in an electron microscope. The same effect is also seen in silicate-based materials; quartz itself is rendered amorphous as are various complicated alumino-silicate minerals (zeolites) under the action of electron irradiation.

The crystalline-to-amorphous transition as a result of irradiation can also occur naturally. Certain minerals, although possessing the external form of crystals (facets, etc.) are, in fact, *amorphous* as determined by X-ray diffraction, etc. These materials, termed 'metamict' (Pabst 1952), are always found to possess radioactive elements such as uranium or thorium, sometimes only in trace amounts, and over geological time have been rendered amorphous by the radioactive decay products of the U or Th.

1.3.10 Pressure-induced amorphization

The application of pressure to certain crystalline materials can cause them to become non-crystalline under certain circumstances. Thus, shock-wave loading resulting from explosive impact can cause amorphization. For example, crystalline silica (quartz) is rendered amorphous by the impact of a shock-wave front producing local pressures of the order of 600 kbar and temperatures of the order of 1200 K (Di Carli and Jamieson 1959). Local melting and subsequent rapid quenching are thought to be responsible for the amorphization process.

However, much more remarkable is the *solid-phase* crystal-to-amorphous phase transition which can occur at high pressures and yet simultaneously at *low* temperatures. The first such example found was for H_2O in which the phase ice I transforms to an amorphous solid at 77 K at pressures above 10 kbar (Mishima *et al.* 1984). Ice I is one example of a solid which melts with a decrease in volume; in other words, the melting temperature falls with increasing pressure. At sufficiently high pressures, therefore, the material can 'melt' (at temperatures low enough to preclude transformation to another crystalline phase) to an amorphous solid if that temperature is below the glass-transition temperature of the amorphous phase. Another apparent example of this solid-state pressure-induced amorphization process is crystalline silica, in which both quartz and coesite are observed to transform irreversibly to an amorphous phase when subjected to (quasihydrostatic) pressures above 350 kbar at 300 K (Hemley *et al.* 1988), even though the crystalline polymorph stishovite is the thermodynamically stable phase at these temperatures and pressures. Although the thermodynamic situation is not as clear, it appears that the crystalline polymorphs of silica might also melt with a decrease in volume, and so the same mechanism for amorphization as for H_2O could also hold for SiO_2.

Pressure-induced transformations occur much more readily in the presence of shear. Correspondingly, many crystalline materials can be rendered amorphous by the simple act of grinding. Thus, quartz, crystobalite and chalcogenide materials (compounds of the chalcogens S, Se and Te with As, Ge, etc.) are susceptible to this type of crystalline-to-amorphous transformation. It is interesting to note that cold-rolling at room temperature of lamellar composites of two metals, e.g. Er and Cu (Atzmon *et al.* 1985), also gives rise to an amorphous composite product. Interfacial melting is unlikely to be the cause of the amorphization; instead, it is likely that mechanical-deformation-induced chemical mixing at the interfaces is responsible, with the transformation being driven by the large negative heat of mixing of the constituent metals (Johnson 1986), as in the case of solid-state diffusional amorphization (sect. 1.3.11).

A related method is 'ball-milling'. In this, a mixture of different crystalline powders is thoroughly agitated in a steel enclosure with a toughened steel ball. Remarkably, at the end of this ball-milling process, the resulting product can be not only a chemically homogeneous alloy, but amorphous as well. Thus, metallic alloys such as Ni–Nb (Koch *et al.* 1983) and Ni–Ti (Schwarz *et al.* 1985) can be produced in an amorphous form in this way starting from the crystalline elemental powders. One advantage of this technique appears to be that the composition range for which amorphous metallic alloys may be made can be significantly larger than that for conventional melt quenching, where the glass-forming concentrations are centred around the eutectic composition (see sect. 2.5.2). Although the high pressures encountered between particles during ball-milling could obviously play a significant role in the amorphization process, as could local melting and subsequent quenching at grain interfaces, there is evidence that another important factor is solid-state diffusion (particularly along grain boundaries), as discussed in sect. 1.3.11. One significant potential technological attraction of the ball-milling process is that the resulting amorphous powders may subsequently be consolidated by hot pressing at temperatures below the crystallization temperature of the amorphous alloy, thereby offering the prospect of the manufacture of bulk objects made from amorphous metallic alloys, without loss of their unique physical properties, a circumstance which is impossible using melt-quenching techniques (sect. 1.3.5).

1.3.11 Solid-state diffusional amorphization

A remarkable new technique for the preparation of amorphous (metallic) systems was discovered by Schwarz and Johnson in 1983. In this, alternate layers of crystalline metallic films ('diffusion couples') interdiffuse under isothermal conditions, with the eventual amorphization of the *entire* multilayer. Couples such as Au with La, Zr or Y, Cu with Zr or Er, and Ni with Er, Ti or Hf have all been found to exhibit this behaviour (Johnson 1986), amorphization taking place after diffusion has been allowed to occur for several hours whilst the temperature is maintained at a few hundred degrees Celsius.

Two features, one thermodynamic and one kinetic, appear to be necessary for solid-state diffusional amorphization to take place. The thermodynamic aspect is that, in those systems which exhibit this behaviour, interdiffusion occurs with a *negative* heat of mixing, indicating that the amorphous phase, if formed, can be thermodynamically stable with respect to the starting metals. Two separate cases can be distinguished (Johnson 1986): in the first, direct nucleation of the amorphous phase occurs at the original interface(s) of the couples (e.g. Au–La); in the second, appreciable mutual dissolution of the constituent metals can take place, leading to unstable solid solutions from which the amorphous phase can grow (e.g. Ni–Zr). These thermodynamic considerations are not the whole explanation, however, since it is invariably found that crystalline intermetallic phases have even larger negative heats of formation than the glassy phase, and therefore they should be thermodynamically the most stable product. Obviously, therefore, kinetic factors must also be important in inhibiting the nucleation and growth of these intermetallic compounds. This inhibition can arise if the atomic mobilities of the two metals concerned are very different in magnitude; the formation of chemically homogeneous intermetallic phases can thereby be precluded. Indeed, it does appear that for, say, couples consisting of an early transition metal (e.g. Zr, Hf) and a late transition metal (e.g. Ni, Cu, Au), it is only the smaller, late transition metal atoms that are mobile; the larger, early transition metals are essentially immobile (Johnson 1986).

A related phenomenon is the hydrogen-diffusion-induced amorphization of crystalline transition metal alloys, e.g. Zr–Rh and Zr–Pd (Yeh *et al.* 1983, Johnson 1986). Exposure to hydrogen gas at moderate temperatures ($\simeq 200°C$) causes the formation of a homogeneous amorphous hydrided product, e.g. a-$Zr_{0.75}Rh_{0.25}H_{1.14}$, indistinguishable from that formed by the hydriding of an already glassy alloy prepared by, say, melt-spinning. At higher temperatures, the equilibrium products c-ZrH_2 and c-Rh are formed. It seems likely that the very low atomic mobility of Zr does not allow the formation of the equilibrium crystalline phases at low temperatures, exactly as for the metal–metal diffusion couples discussed above, and the high mobility of H facilitates the formation of a homogeneous (amorphous) phase.

1.4 Summary

Despite the welter of techniques that can produce the amorphous state, only the first five, or possibly six, of these described here are at all important commercially or otherwise. It is very important at the outset of a study of the amorphous state to notice that although many materials can be prepared in an amorphous form by more than one technique, *the different forms need not, and indeed often do not, have the same properties.* As we shall see in Chapter 3, there is no *unique* structure for an amorphous solid (an infinite number of possible structures being in principle possible), and hence there is no reason *a priori* to suppose that different preparative techniques produce materials with the same or similar structure. In addition, the various techniques differ in the degree to which their products depart from equilibrium, mainly as a result of the rate of (effective) cooling; vapour-deposited material is therefore expected to be less near an equilibrium state than melt-quenched material. This difference can be evinced at two levels: *macroscopic* structural inhomogeneities (e.g. voids, density fluctuations) are more prevalent in vapour-deposited material as are *microscopic* structural defects, such as broken bonds and vacancies. These can have dramatic effects on various properties, particularly electronic, as we shall see later.

A further important point to note about the preparation of amorphous solids is that, *even for the same preparative technique*, there may be differences in the nature of the material produced unless great care and attention is paid to control of the various experimental parameters, some of which we have seen. This non-reproducibility of samples bedevilled the field in its early days, especially as far as vapour-deposited materials were concerned. It is therefore of paramount importance to give as many details as possible of the controllable parameters in the preparation process in order that future studies may attempt to reproduce the material in question.

2 Glasses

2.1 Introduction

We have seen in the last chapter that many different techniques can be used to prepare amorphous materials. Perhaps the most widely used, and certainly the most important historically, is melt quenching. Materials prepared in this way are often referred to as 'glasses' because they exhibit glass-transition phenomena (of which more later), although it should be noted that the widely used term 'metallic glass' is somewhat of a misnomer, since these materials often crystallize *before* exhibiting a glass transition on heating.

Materials which can be quenched from a melt to form an amorphous solid are represented by all the major types of bonding interactions found in solids, viz.

Table 2.1 Examples of glass-formers for the major types of bonding

Type of bond	*Material*
Covalent	As_2Se_3, Se
Ionic	KNO_3–$Ca(NO_3)_2$
Metallic	Zr–Cu, Ni–Nb
Van der Waals'	*ortho* terphenyl
Hydrogen bond	$KHSO_4$, ice, aqueous solution of ionic salts (e.g. LiCl)

covalent, ionic, metallic, Van der Waals' and hydrogen bonds; some representative examples of glass-formers from each category are listed in Table 2.1. Inevitably, there are some examples which do not fit into any one category. For example, the most widely known glass of all, silica, has some ionic contribution to the predominant covalent bonding. Since many materials can be rendered glassy (some authors would maintain that practically all liquids, with the exception of the quantum liquid of helium, can be made glassy given high enough cooling rates), it is obviously of interest to discuss these particular materials in a little more detail.

The first part of this chapter, then, attempts to answer the question '*how* do glasses form?' In so doing, the fascinating and complex field of glass-transition phenomena is introduced. The second part of the chapter tackles the knotty problem of '*why* do glasses form?' In particular, why are some materials so much better glass-formers than others?

2.2 The glass transition

When a liquid is cooled, one of two events may occur. Either *crystallization* may take place at the melting point T_m, or else the liquid will become 'supercooled' for temperatures below T_m, becoming more viscous with decreasing temperature, and may ultimately form a *glass*. These changes can be observed readily by monitoring the volume as a function of temperature by means of a dilatometer, and a typical result is shown schematically in Fig. 2.1. The crystallization process is manifested by an *abrupt* change in volume at T_m, whereas glass formation is characterized by a *gradual* break in slope. The *region* over which the change of slope occurs is termed the 'glass-transition temperature' T_g. (Similar behaviour would also be exhibited by other extensive thermodynamic variables, entropy S, and enthalpy H.) Since the transition to the glassy state is continuous, and the glass-transition temperature is not well defined (albeit often used), it is often convenient also to use the 'fictive' temperature T_f (Jones 1956), which is defined as that specific temperature obtained as the intersection of the extrapolated liquid and glass curves (Fig. 2.2(a)). It is the temperature at which the glass would be in metastable equilibrium if it could be brought to T_f instantaneously. Although it appears that T_f defined in this way is a precise temperature, in fact this is not so; it depends on the *rate* of cooling of the supercooled liquid. It is found that the *slower* the rate of cooling, the larger is the region for which the liquid may be

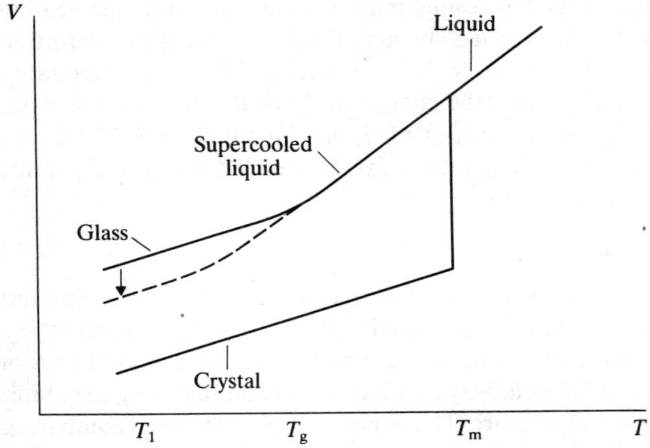

Fig. 2.1 Schematic illustration of the change in volume with temperature as a supercooled liquid is cooled through the glass-transition temperature T_g. The first-order phase transition accompanying crystallization from the melt is also shown. The vertical arrow illustrates the volume change accompanying the structural relaxation or stabilization of the glass if held at temperature T_1.

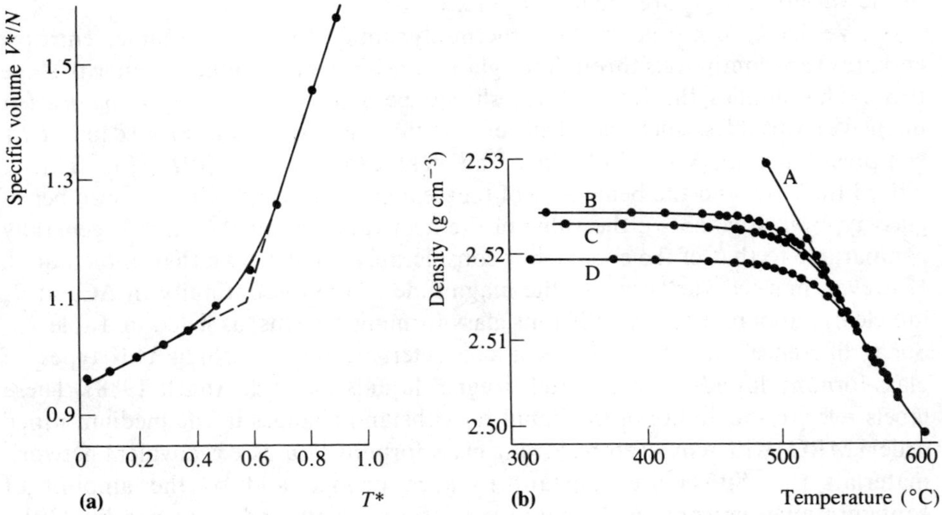

Fig. 2.2 (a) Computer simulation of a glass transition in a Lennard-Jones liquid manifested in the temperature dependence of the specific volume V^*/N (M. Kimura and F. Yonezawa (1983), in *Topological Disorder in Condensed Matter* (*Springer Series in Solid-State Sciences*, vol. 46) (Springer-Verlag) p. 80. The intersection of the dashed extrapolated lines marks the glass-transition temperature (or fictive temperature).
(b) The glass transition for a borosilicate glass manifested in temperature-dependent density curves (H.N. Ritland (1954), *J. Am. Ceram. Soc.* **37**, 370). The dependence on cooling rate, q, is clearly apparent (for B, C and D, $q = 1$, 2 and 10 K h^{-1}, respectively). Curve A corresponds to the equilibrium liquid.

supercooled, and hence the *lower* is the fictive or glass-transition temperature; this is shown for the case of a borosilicate glass in Fig. 2.2(b). Thus, the glass-transition temperature of a particular material partly depends on its thermal history; if the T_g of a glass is determined by heating the material through the transition while observing a given parameter, spurious results may be obtained unless the heating rate is equal to the original cooling rate. The actual value of the glass-transition temperature may vary by as much as 10 to 20% for widely differing cooling rates. As an example, for silica-based glasses, the change in T_g for different cooling rates may be as much as 100–200 K for values of T_g in the range 600–900 K. An expression relating T_g and the cooling rate q can be derived using free-volume theory (see problem 2.2) and is:

$$q = q_0 \exp[-1/C(1/T_g - 1/T_m)] \qquad [2.1]$$

where C is a constant ($\simeq 3 \times 10^{-5}$), and for chalcogenide glasses (Se, As_2Se_3 and As_2S_3), q_0 varies between $10^{23}\ K\ s^{-1}$ (Se) and $10^4\ K\ s^{-1}$ (As_2S_3) (Owen 1973). The outcome of this discussion, therefore, is that one should not expect glasses of the same material prepared in different ways, with different cooling rates, to exhibit a glass transition at the same temperature. This is particularly true when comparing the same material prepared either by melt quenching or by vacuum deposition. Nevertheless, for a given class of materials which all have the same type of bonding and consequently are approximately isostructural, systematic trends in the variation of T_g can be discerned if one neglects the relatively small variations due to cooling-rate differences discussed above. The factors determining such gross trends in the variation of T_g are the subject of sect. 2.4.

We have seen that certain thermodynamic variables (volume, entropy, enthalpy) are continuous through the glass transition, but exhibit a change of slope there. This implies that at T_g there should be a discontinuity in *derivative* (or intensive) variables, such as coefficient of thermal expansion $\alpha_T = (\partial \ln V/\partial T)$, compressibility $\kappa_T = -(\partial \ln V/\partial p)_T$ and heat capacity $C_p = (\partial H/\partial T)_p$. This is indeed the case, and the behaviour of the heat capacity is shown for a number of glass types in Fig. 2.3(a); the value of the heat capacity for the glass is generally comparable to that of the crystal, but considerably smaller than that of the liquid. However, marked variations in the magnitude of the discontinuity in ΔC_p at T_g are clearly apparent for the different glass-forming systems, as listed in Table 2.1. Such differences can be used as a characteristic distinguishing two types of glass-forming liquids: 'strong' and 'fragile' liquids (see e.g. Angell 1988). These labels refer to the ability of the liquid to withstand changes in the medium-range order (MRO) with temperature. Strong glass-forming liquids, e.g. covalent network materials (i.e. SiO_2) are resistant to such change and so the amount of configurational entropy in the liquid is relatively small, and correspondingly the change in heat capacity at T_g, ΔC_p, is also small. Fragile liquids, being characterized by predominantly non-directional (ionic or Van der Waals') bonding have a high configurational entropy in the liquid state, and correspondingly ΔC_p at T_g is also large (see Fig. 2.3(a)). Another distinction between strong and fragile glass-forming liquids is afforded by the temperature dependence of the viscosity (see sect. 2.3.3 and Fig. 2.6). It can be seen that in general calorimetric experiments offer a much better marker of the glass-transition temperature than, for example, simple dilatometric measurements. A convenient way of monitoring glass-transition phenomena is by means of differential scanning calorimetry (DSC) or differential thermal analysis (DTA), in which the sample is heated at a constant rate and the changes in heat (DSC) or temperature (DTA) with respect to an empty reference

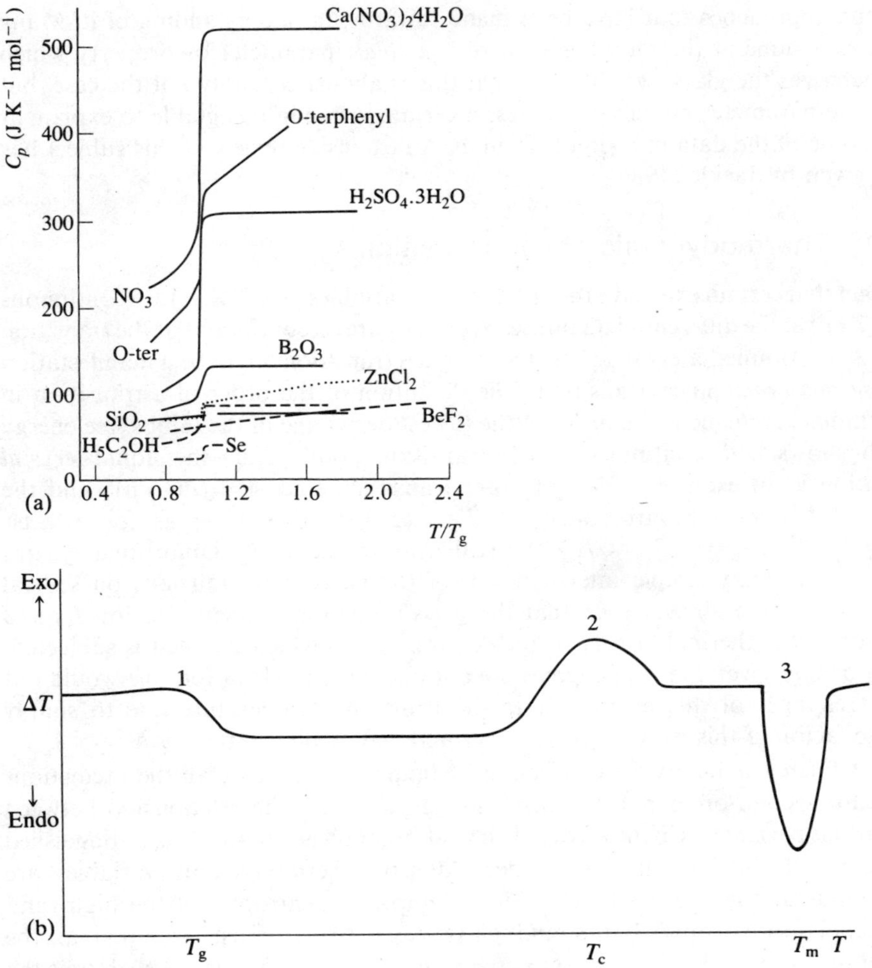

Fig. 2.3 Manifestations of the glass transition in thermodynamic measurements.
(a) Specific heat C_p vs. reduced temperature T/T_g for various glass-forming systems (S.A. Brauer (1983), *Relaxation in Viscous Liquids and Glasses* (Am. Ceram. Soc.)). 'Fragile' glass-forming liquids are characterized by a large jump in C_p at T_g, and 'strong' glass formers by a small jump.
(b) Schematic DTA trace showing the glass transition (1), crystallization (2) and melting (3).

pan are measured; a schematic diagram of a typical DTA trace is shown in Fig. 2.3(b), exhibiting both a glass transition and subsequent crystallization and melting. Controlled cooling is also possible with such instruments, and hence detailed studies of the dependence of T_g on the heating rate and on the previous thermal history are possible.

2.3 Theories for the glass transition

The nature of the glass transition is very complex and even now is poorly understood. Nevertheless, we shall review the various aspects of this phenomenon, and the approaches that have been made towards an understanding of it. Many theories assume at the outset that there is a *single* parameter or property which characterizes the glass; we shall see that this is almost certainly not the case, but such one-parameter models do possess a certain virtue in being able to explain at least some of the data in a simple fashion. An extensive review of this subject has been given by Jäckle (1986).

2.3.1 Thermodynamic phase transition

The fact that certain extensive thermodynamic variables (e.g. V, S, H) are continuous (Fig. 2.1), yet the differential quantities α_T, C_p, κ_T are discontinuous at the transition (Fig. 2.3(a)) immediately suggests that the glass transition might be a manifestation of a *second-order* phase transition. The definition of the order of a transition in the Ehrenfest scheme is the order of the lowest derivative of the Gibbs free energy which shows a discontinuity at the transition point. Thus the liquid–*crystal* transition is an example of a first-order transition, since $V=(\partial G/\partial p)_T$ and the volume changes discontinuously at T_m (see Fig. 2.1), whereas for a glass $C_p(=T(\partial S/\partial T)_p=-T(\partial^2 G/\partial T^2)_p)$ is discontinuous at T_g. Unfortunately, this rather appealingly simple interpretation of the glass transition fails on several counts. We have already seen that the glass-transition temperature (or T_f) is a function of the thermal history or the cooling rate to which the melt is subjected; variations by several tens of degrees are not uncommon. However, one would not expect changes of this magnitude in the transition temperature due to simply kinetic factors if this were a genuine thermodynamic phase transition.

A further difficulty arises when one examines in more detail the predictions made for second-order phase transitions, in particular the relationship between the various parameters in the theory. A second-order phase transition is distinguished by virtue of the fact that first-order extensive thermodynamic variables are continuous at the transition. Thus, for example the entropies of the high- and low-temperature forms (liquid and glass, respectively) must be equal at the transition, i.e. $S_1 = S_2$. Changes in temperature or pressure must be such that the two forms remain in equilibrium so that $dS_1 = dS_2$, or in terms of partial derivatives:

$$\left(\frac{\partial S_1}{\partial T}\right)_p dT + \left(\frac{\partial S_1}{\partial p}\right)_T dp = \left(\frac{\partial S_2}{\partial T}\right)_p dT + \left(\frac{\partial S_2}{\partial p}\right)_T dp \qquad [2.2]$$

Since $C_p = T(\partial S/\partial T)_p$, and using one of Maxwell's thermodynamic relations $(\partial S/\partial p)_T = -(\partial V/\partial T)_p$, where $\alpha_T = 1/V(\partial V/\partial T)_p$, one obtains finally an expression for the shift in transition temperature with pressure:

$$\frac{dT_g}{dp} = \frac{TV(\alpha_{T_2} - \alpha_{T_1})}{(C_{p_2} - C_{p_1})} = TV\frac{\Delta\alpha_T}{\Delta C_p} \qquad [2.3]$$

Similarly, by considering continuity of volume at the transition, a relation involving the bulk compressibility $\kappa_T = -(1/V)(\partial V/\partial p)_T$ is obtained:

$$\frac{dT_g}{dp} = \frac{\Delta\kappa_T}{\Delta\alpha_T} \qquad [2.4]$$

It is found by measuring the discontinuities $\Delta\alpha_T$, ΔC_p, $\Delta\kappa_T$ at the glass transition that [2.3] is almost always obeyed within experimental error, but that values for $\Delta\kappa_T/\Delta\alpha_T$ are generally appreciably higher than those of $\mathrm{d}T_g/\mathrm{d}p$ ([2.4]). It therefore appears on this evidence that the glass transition is not a simple second-order phase transition. However, without invoking a second-order thermodynamic transition, Prigogine and Defay (1954) showed generally that the ratio

$$R = \frac{\Delta\kappa_T \Delta C_p}{TV(\Delta\alpha_T)^2} \qquad [2.5]$$

is equal to unity if a single ordering parameter determines the position of equilibrium in a relaxing system, but if more than one ordering parameter is responsible, then $R > 1$; the latter case seems to describe most glasses. Goldstein (1973) has suggested that the reason for the breakdown of the Ehrenfest relation [2.4] could be that the specific volume V_g of the glass depends not only on the temperature, being continuous through the transition, but also on the *pressure* of formation p_f. Jäckle (1989) has shown that, taking this extra dependence into account, [2.4] becomes:

$$\frac{\mathrm{d}T_g}{\mathrm{d}p} = \frac{\Delta\kappa_T + \partial(\ln V_g)/\partial p_f}{\Delta\alpha_T} \qquad [2.6]$$

The extra term in [2.6] is the reason why the Prigogine–Defay ratio is greater than unity. It should be noted in passing that an additional consequence of the experimental verification of [2.3] but not [2.4], is that one expects glasses prepared under high pressures to have higher than normal densities but normal entropies or enthalpies, since continuity of entropy leads to [2.3] whereas continuity of volume leads to [2.4].

2.3.2 Entropy

One would expect intuitively that entropy might play a role in determining the glass transition since we are considering disordered systems. We have already seen one instance of this in the previous section in which glass formation may be assumed to occur at a given value of *excess* entropy, and by further supposing that the entropy is continuous at the transition, [2.3] results.

A further influence of entropy was first pointed out by Kauzmann (1948), and concerns the precipitous decrease in heat capacity at the glass-transition temperature (Fig. 2.3(a)). The heat capacities for both the crystalline and glassy ($T < T_g$) states of most materials are essentially the same, and arise from vibrational contributions. The excess heat capacity measured for the glass above T_g is due to configurational degrees of freedom which the material possesses in the supercooled liquid state. We have already seen that the rate of cooling influences the glass-transition temperature; the slower the rate, the lower T_g. The point at issue is whether there is a lower limit to this decrease in T_g – i.e. is there an 'ideal' glass-transition temperature? The argument advanced by Kauzmann suggests that such a limit does exist. The area under the curve of C_p plotted versus the *logarithm* of temperature yields the entropy, and such a plot for the case of an ionic system,

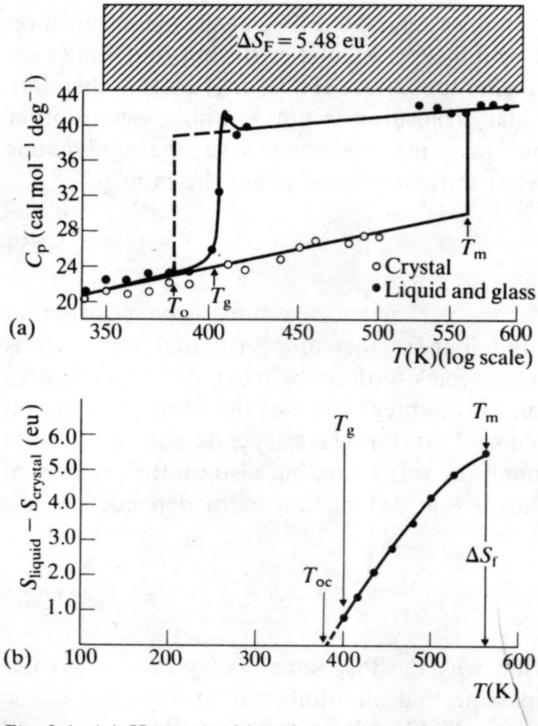

Fig. 2.4 (a) Heat capacities of glassy, liquid and crystalline phases of lithium acetate. The data are plotted against ln T so that integrated areas under the curves yield entropies directly, and the entropy of fusion is shown shaded in the upper part of the figure.
(b) The difference in entropy between liquid and crystalline phases as a function of temperature. The vanishing excess entropy temperature is termed the 'ideal' glass-transition temperature T_{oc} (Wong and Angell 1976).

lithium acetate, is shown in Fig. 2.4 (Wong and Angell 1976). The change in entropy upon fusion of the crystal which occurs at T_m is ΔS_f (5.48 entropy units (e.u.) in this case). The *lowest* temperature to which the liquid could possibly be supercooled by progressively increasing the experimental time-scale, yet maintaining the large liquid-like value of C_p, is governed by the requirement that the area under the $C_p - \ln T$ curve for the supercooled liquid be equal to or greater than ΔS_f. The temperature, defined by the equality of the areas, is called the calorimetric ideal glass-transition temperature, T_{oc}. The experimental glass-transition temperature cannot decrease below T_{oc}, since if it did, the total entropy of the liquid would fall *below* that of the crystal, thereby violating the Third Law of Thermodynamics; this situation is termed 'Kauzmann's paradox'. It can be seen that for the example shown in Fig. 2.4, the experimentally determined value of T_g occurs just 20 K above T_{oc}; in many cases, the difference is considerably larger, particularly for silicate or similarly based glasses. We conclude from this discussion, therefore, that even if no other factor contributed to the glass transition, the heat capacity should decrease sharply at a certain temperature as a result solely of *equilibrium* thermodynamic considerations.

The existence of an ideal glassy state possessing zero residual entropy, having a glass-transition temperature T_{oc}, is probably hypothetical. This is because near T_g the probability of crystal growth and nucleation increases very rapidly, and thus for very long experimental time-scales crystallization is more probable than relaxation to a lower energy, fully amorphous state. Nevertheless, there are certain materials, 'atactic' polymers, which have a random arrangement of side-groups, which *cannot* crystallize and which, therefore, offer a practical prospect of investigating whether an 'ideal' amorphous solid may exist; unfortunately, the relevant data needed to determine the excess entropy in the liquid are not available.

The ideas above have been employed in a discussion limited to linear polymer chains (Gibbs and Di Marzio 1958, Gibbs 1960), and the transition temperature was defined there as that point at which the configurational entropy should vanish. The number of configurations available for a polymer chain decreases with decreasing temperature because of the hindrance to intramolecular rotation caused by the presence of side-groups which produce rotational potential barriers; a chain becomes effectively stiffer as the temperature decreases. The configurational entropy was calculated using the partition function for a quasi-lattice model and a relationship between the transition temperature and the potential barriers was established, leading to the qualitative result that T_g should increase with increasing barrier height. This entropy theory concentrates solely on the stiffness factor of a polymer chain, and does not take into account intermolecular interactions which are almost certainly important. A more complete discussion of glass-transition phenomena in polymers can be found in Gee (1970).

2.3.3 Relaxation processes

Another important facet of the glass transition – some would argue the most important aspect – concerns the relaxation processes that occur as a supercooled liquid cools. We have already seen that the experimentally measured value of T_g is not unique; it depends on the time-scale of the experiment used to observe it. The configurational changes that cause the relaxation of the supercooled liquid become increasingly slow with decreasing temperature until, at a given temperature (the glass-transition temperature), the material behaves as a solid; for times of observation, t_o, long compared with the 'structural' relaxation time τ_r ($t_o > \tau_r$) the material appears 'liquid-like' whereas for $t_o < \tau_r$ the material behaves as if it were 'solid-like'. A 'transition' will appear to have taken place if values of liquid-like parameters differ significantly from solid-like ones, as is the case for the heat capacity. T_g occurs therefore when $t_o \simeq \tau_r$, viewed in this way. As shown in Fig. 2.5, a 'decoupling' of vibrational (phonon) modes from relaxational modes (involving diffuse motion) occurs progressively as the temperature of a fluid is decreased; this is because the viscosity, say, increases markedly with decreasing temperature (see [2.8]).

It should be remarked here that the slow processes of relaxation involving shear (but also dielectric polarizability and other relaxing quantities – see Ngai 1985) very often exhibit a markedly non-exponential time dependence in which the response function $\phi(t)$ obeys a so-called 'stretched-exponential' or Kohlrausch–Williams–Watts law (Kohlrausch 1854, Williams and Watts 1970):

$$\phi(t) = \exp[-(t/\tau_o)^\beta], \quad 0 < \beta < 1 \qquad [2.7]$$

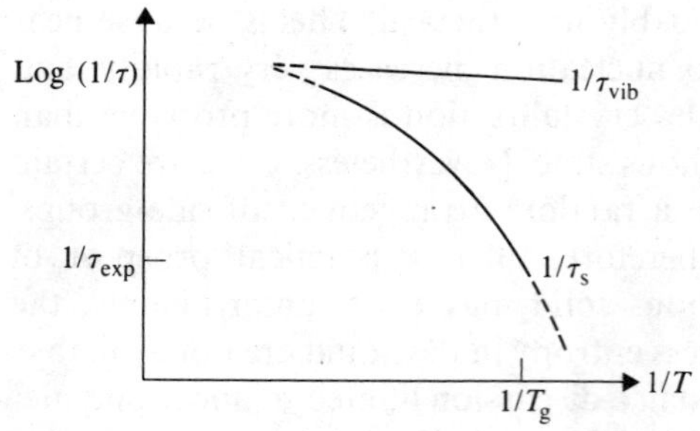

Fig. 2.5 Schematic illustration of the decoupling of the vibrational and structural relaxation times for temperatures near T_g (Jäckle 1986).

This formula will be discussed in more detail in connection with dielectric relaxation associated with ionic transport in glasses in sect. 4.5.3.

Another (but rather unsatisfactory) definition of T_g, again in terms of the experimental time-scale, is that temperature at which the liquid attains a certain viscosity (say, 10^{13} poise, a value characteristic of solids – sect. 1.2). This brings us to the subject of viscosity of fluids, which is important since many glass-forming liquids have high viscosities at elevated temperatures. The viscosity of these liquids changes very rapidly in the region of the glass-transition temperature, as can be seen from Fig. 2.6. The temperature dependence of the viscosity for the two types of behaviour commonly encountered differs quite markedly. 'Strong' network glass-forming liquids such as SiO_2, GeO_2 or BeF_2 obey accurately an Arrhenius behaviour over a large temperature range, whereas many so-called 'fragile' glass-forming liquids, including ionic systems such as $Ca(NO_3)_2$, and organic materials such as *o*-terphenyl or i-butyl bromide, show strong deviations from a simple Arrhenius behaviour. Over much of the range, the empirical Vogel–Tammann–Fulcher equation for the 'fluidity' (inverse viscosity)

$$\eta^{-1} = A \exp[-a/(T - T_{o\eta})] \qquad [2.8]$$

is obeyed, although for very high viscosities near T_g, the dependence is often observed to become Arrhenius-like again. It is a matter of considerable interest to note that in those glasses for which Fulcher's law is obeyed, $T_{o\eta} \simeq T_{oc}$, seeming to indicate that the ideal glass-transition temperatures observed kinetically and thermodynamically might have a common origin. It is important to note that the viscosity is continuous through T_g, exhibiting none of the discontinuities observed in heat capacity, etc., although there is a singularity at the (inaccessible) temperature $T_{o\eta}$. The particular temperature dependence embodied in Fulcher's law is rather unusual, and as Anderson (1979) has pointed out, it cannot be obtained by an appropriate averaging of the energy in an Arrhenius expression nor by a suitable temperature dependence of the energy, and its physical origin remains relatively obscure. The form of [2.8] is in fact common to many transport phenomena of the liquid phase, e.g. diffusion rate, nuclear spin relaxation rate and ionic conduction, and it is found that the values for the different T_o (and a) are very similar in all cases. An important source of information on this area is 'relaxation'

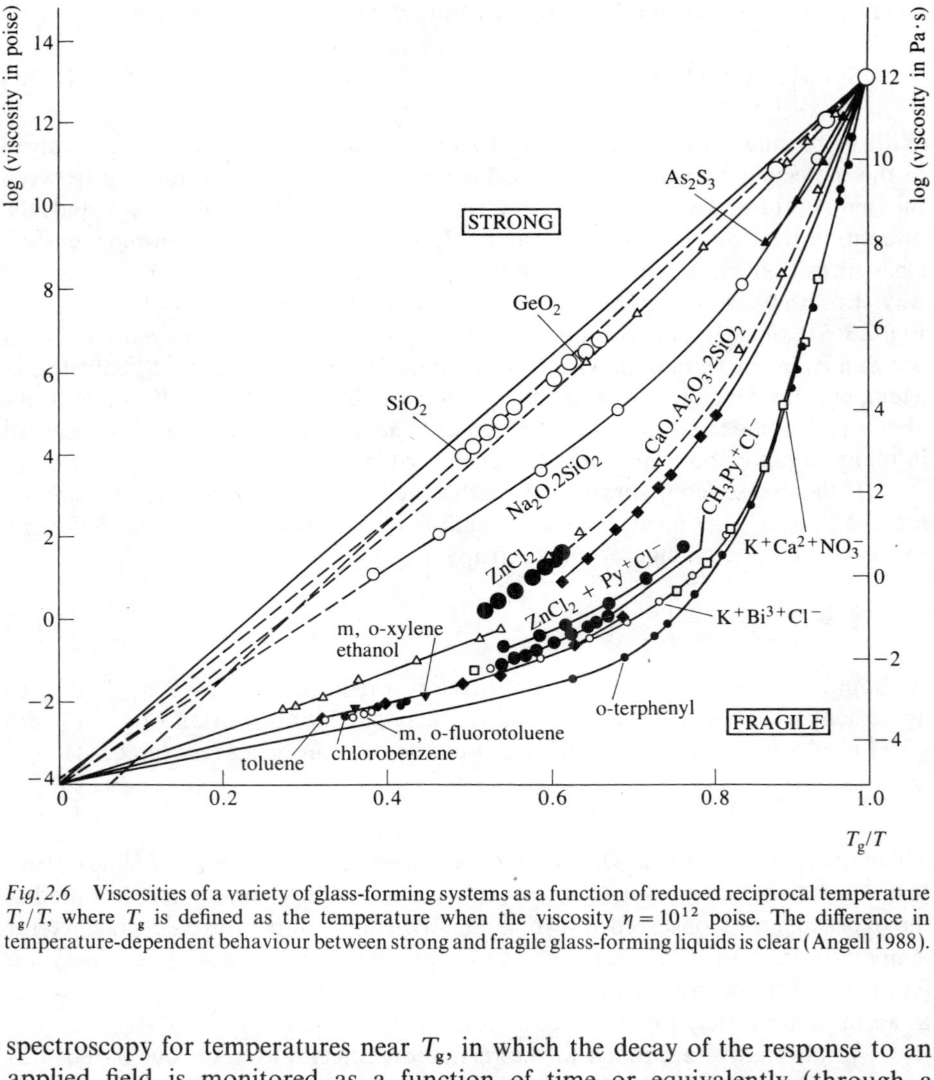

Fig. 2.6 Viscosities of a variety of glass-forming systems as a function of reduced reciprocal temperature T_g/T, where T_g is defined as the temperature when the viscosity $\eta = 10^{12}$ poise. The difference in temperature-dependent behaviour between strong and fragile glass-forming liquids is clear (Angell 1988).

spectroscopy for temperatures near T_g, in which the decay of the response to an applied field is monitored as a function of time or equivalently (through a Kramers–Kronig or Hilbert transform) the continuous dissipation caused by an externally applied sinusoidal field. Examples of the use of this technique are for stress relaxation (or mechanical loss) and dielectric loss; a fuller discussion with many examples is given by Wong and Angell (1976).

A theory which attempts to link relaxation aspects with entropy considerations has been proposed by Adam and Gibbs (1965). They accounted for mass transport (diffusion, viscosity, etc.) by considering cooperative atomic rearrangements and, by finding the smallest size of a molecular group capable of rearrangement at a given temperature, showed that the group size and therefore the probability of rearrangement and hence, say, the viscosity could be expressed as a function of the total configurational entropy S_c:

$$\eta = \eta_o \exp(B/TS_c) \qquad [2.9]$$

where η_o and B are constants. The configurational entropy is given by the equation:

$$S_c = \int_{T_o}^{T} \Delta C_p \mathrm{d} \ln T \qquad [2.10]$$

With increasing viscosity, an increasing number of molecular units are involved in the molecular rearrangements. In this manner, a natural relationship between the temperature dependence of the viscosity ($\eta \simeq 10^{14}$ poise at T_g) and the discontinuities in C_p observed also at T_g is obtained. Furthermore, the ideal glass-transition temperature enters into expressions for mass transport in a natural way. For the case when ΔC_p is very small, as it is for SiO_2, GeO_2, etc. (see Fig. 2.3(a)), then S_c is almost independent of temperature, and the viscosity should obey an Arrhenius equation. On the other hand, when ΔC_p is large, S_c contributes a temperature-dependent term to the equation for viscosity [2.9] at all temperatures above T_o, leading to strongly non-Arrhenius behaviour of the viscosity, as observed in many organic and ionic ('fragile') glass-forming liquids.

If the excess (configurational) heat capacity is taken to have the hyperbolic form $\Delta C_p = D/T$ (Angell 1988), evaluation of [2.10] results in the following expression for the configurational entropy:

$$S_c = \mathrm{D}\left(\frac{1}{T_o} - \frac{1}{T}\right) \qquad [2.11]$$

It should be noted that this theory therefore predicts that $S_c \to 0$ as $T \to T_o$, in agreement with the Kauzmann paradox (sect. 2.3.2). Substitution of [2.11] into [2.9] results in the Vogel–Tammann–Fulcher relation being recovered:

$$\eta = \eta_o \exp[\alpha T_o/(T - T_o)] \qquad [2.12]$$

where the coefficient $\alpha = B/D$. A similar treatment has been given by Jäckle (1986). It should be noted, however, that this picture is incomplete since it does not predict the experimentally observed return to an Arrhenius temperature dependence for temperatures close to T_g. Scherer (1984) has used the Adam–Gibbs expression [2.9], together with another relation for the excess heat capacity $\Delta C_p = C_o + C_1 T$, to account accurately for the viscosity of silicate glass.

An alternative approach to the relaxation behaviour of viscous liquids has been given by Goldstein (1969). In this, a fluid system composed of N particles is characterized by a $(3N + 1)$-dimensional potential energy surface containing many minima of different depths. This space is then explored as the system evolves in time, but at low temperatures the system is restricted to the deeper minima which are fewer in number and hence contribute to a lower value of configurational entropy. The activation energy needed to move between minima is therefore temperature dependent. The glass transition, in this picture, then corresponds to the system becoming trapped in one of the deep minima. An important feature of the theory is the prediction of 'islands of mobility' in which local rearrangement is possible (between neighbouring similarly deep minima) but long-range flow is impossible.

This circumstance can explain the weak secondary (or sub-glassy) relaxations which are observed experimentally in many glasses at temperatures well below the major glass-transition temperature T_g. These are most often observed in polymeric systems, where they are termed β-processes to differentiate secondary relaxations from the primary (α-) relaxations occurring at T_g. The β-relaxations are perhaps most clearly seen in plots versus temperature of dielectric loss, ε_2 or loss tangent $\tan\delta = \varepsilon_2/\varepsilon_1$ (see sect. 6.3.4) in which T increases through T_g; an example is shown schematically in Fig. 2.7. The β-processes can be observed also by other relaxation spectroscopic techniques (McCrum *et al.* 1967). For the case of glassy polymers, β-relaxations can be understood physically as the (rotational or librational) rearrangement of small side-groups which can occur for $T < T_g$, in contrast to the conformational changes of the polymer backbone chain involving many tens of atoms, which occur at temperatures above T_g.

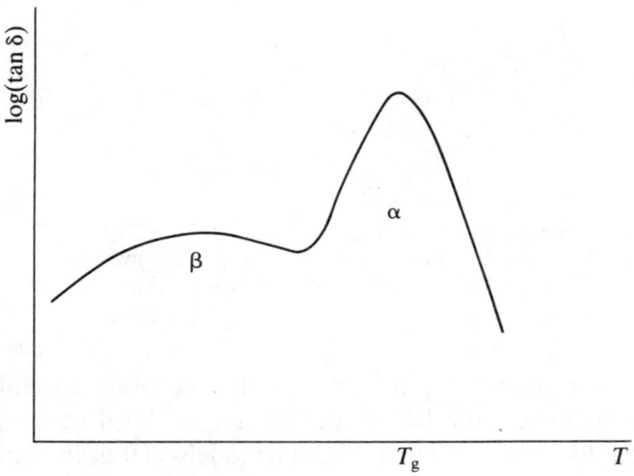

Fig. 2.7 Schematic illustration of the primary (α-) and secondary (β-) relaxations which can occur in glasses at and below T_g, manifested by measurement of the dielectric loss, $\tan\delta = \varepsilon_2/\varepsilon_1$, as a function of temperature at constant frequency.

An intermediate type of motion has been advanced to account for the so-called 'γ-transition' in polyethylene (and other polymers containing more than four consecutive —CH_2 groups in the chain) in which four polymer groups rotate simultaneously around an axis parallel to the chain direction, termed crankshaft motion. This is shown schematically in Fig. 2.8, together with the rotational or librational modes of side-groups believed to be responsible for β-relaxations. However, the fact that sub-glassy relaxations are not restricted to polymeric materials alone, but have also been observed experimentally in other systems, such as ionic or simple molecular liquids in which there are no freely rotating side-groups or sub-units (Johari and Goldstein 1970, 1971), lends support to the general notions of Goldstein's potential minima picture.

2.3.4 Dynamical theories

A resurgence in theoretical interest in glass-transition phenomena has occurred recently through interest in *dynamical* aspects of the transition (see Fredrickson

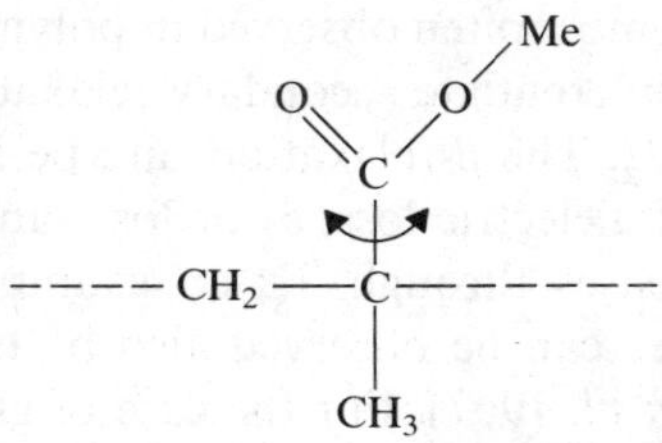

(a)

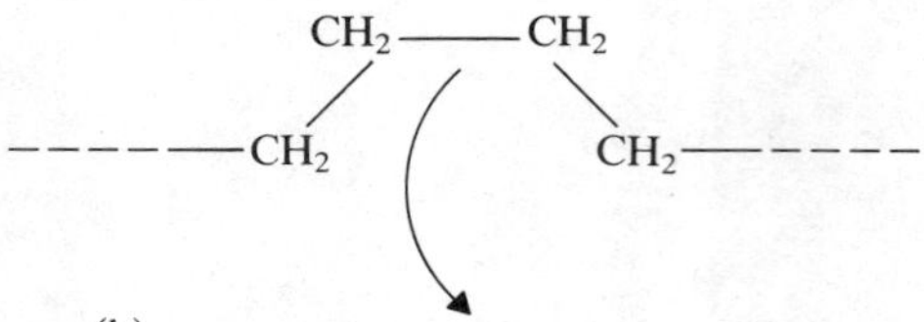

(b)

Fig. 2.8 Molecular motions giving rise to sub-glassy relaxations.
(a) β-Relaxation in polymethyl methacrylate caused by librational motion of side-groups.
(b) γ-Relaxation in polyethylene caused by 'crank-shaft' motion of the polymer backbone.

(1988) for a recent review); this progress has arisen from the generally sound theoretical understanding now achieved for the physics of simple (hard-sphere) liquids in terms of generalized hydrodynamic or kinetic models (Hansen and McDonald 1976, 1986).

At the heart of this approach is the idea that there is another independent relation between the shear viscosity η and the self-diffusion coefficient D, in addition to the Stokes–Einstein equation

$$\eta = \frac{k_B T}{4\pi a D} \qquad [2.13]$$

(where a is the interatomic spacing), and this extra relationship acts like a feedback mechanism in controlling the structural relaxation time, τ_s (Geszti 1983). The viscosity is related to the shear-stress relaxation time by

$$\eta = G_\infty \tau \qquad [2.14]$$

where G_∞ is the instantaneous shear modulus. In a 'soft-sphere' system, where G_∞ must remain finite on vitrification (i.e. the glass remains elastic like any crystalline solid), it must therefore be the relaxation time τ that diverges, in turn causing the viscosity to increase rapidly, marking the onset of the glass transition (from the liquid side). For liquids in the pre-vitrification region, the vibrational relaxation time τ_{vib} and the structural relaxation time τ_s become decoupled (see Fig. 2.5), so that, approximately,

$$\tau \approx \tau_{vib} + \tau_s$$

Since $\eta_o \propto \tau_{vib}$ and $\eta \propto \tau_s$, and furthermore $\tau_s \propto D^{-1}$ with $\eta = b(T)/D$, then $\eta = \eta_o + B(T)\eta$ or (Geszti 1983):

$$\eta = \eta_o/(1 - B(t)) \qquad [2.15]$$

where [2.13] has been used, and $B(T) = 4\pi ab(T)/k_B T$, a monotonically decreasing function of temperature. Thus at some temperature T_o, $\eta \to \infty$, i.e. a glass transition occurs.

This simple phenomenological model has been superseded by more sophisticated treatments based on mode-coupling theories (Leutheusser 1982, 1984, Bengtzelius *et al.* 1984). In *simple* liquids (i.e. those with spherically symmetric interaction parameters), several correlation functions are sufficient to describe the dynamics of such systems; these involve fluctuations in density, and transverse and longitudinal currents (Hansen and McDonald 1976, 1986). The idea of mode coupling is that damping of such modes occurs by the decay into two (or more) modes of the same (or similar) type. With the glass transition, it is the density–density correlation function that is important since, in the solid state, energy fluctuations can still be dissipated, but density fluctuations are frozen-in if diffusion becomes negligible. The mode-coupling mode of decay then produces the extra coupling between η and D mentioned above. A detailed discussion of these mode-coupling theories is in general beyond the level of this book, but the treatment by Leutheusser (1984) is the most understandable. In this, the time-dependent decay of the density–density correlation function $\Phi(t)$ can be written in the form of a non-linear, damped second-order differential equation:

$$\left(\frac{d^2}{dt^2} + \gamma\frac{d}{dt} + \omega_o^2\right)\Phi(t) + 4\lambda\omega_o^2 \int_0^t M(t-t')\Phi(t')\,dt' = 0 \qquad [2.16]$$

where ω_o is the harmonic (longitudinal sound-wave) frequency, γ a damping term and λ a parameter indicating the strength of the non-linear coupling. This coupling is provided by the 'memory function' $M(t)$, which acts as another damping function for longitudinal sound waves. In the simplest version of the theory, the memory function providing the feedback is taken to be

$$M(t) = \Phi^2(t) \qquad [2.17]$$

by analogy with simple mode-coupling theories, and the wavevector dependence is neglected.

Equation [2.16], together with [2.17], has several interesting properties. The first is that it predicts stretched-exponential time decays for the response function $\Phi(t)$ as the coupling constant λ approaches the critical value $\lambda_C = 1$ from below, i.e. the fluid side (Bengtzelius 1986) – see Fig. 2.9. The more startling property is that, at a critical value $\lambda_C = 1$, the density–density fluctuations do *not* decay to zero in a finite time, but are instead frozen-in (see Fig. 2.9). This simple model predicts, therefore, the existence of a purely *dynamic* glass transition, a type of ideal glass transition, controlled entirely by kinetic factors. This phenomenon can be regarded as a *localization* phase transition, akin to that which occurs for electrons (see Chapter 5), but here the localization involves the atomic self-diffusion

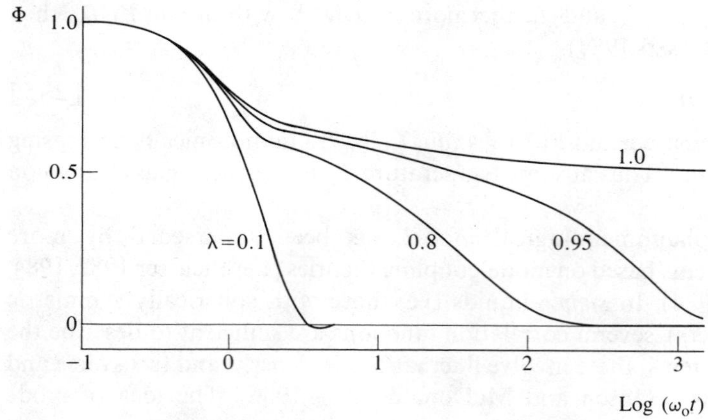

Fig. 2.9 Slowing down of decay of density response function Φ for increasing values of the coupling constant λ approaching from below the critical value $\lambda_c = 1$, for which the glassy state occurs (Leutheusser 1984).

coefficient D. At the transition, the long-time limit of $\Phi(t)$ changes discontinuously from being zero above the transition to a non-zero value below (see Fig. 2.9) – a signature of a *first-order* transition. Finally, on approaching the transition from the fluid side (i.e. for $\lambda < \lambda_C = 1$), the theory predicts that a power-law singularity of the relaxation time, and correspondingly also of the viscosity, should occur:

$$\tau \sim |1 - \lambda/\lambda_C|^{-\mu} \qquad [2.18]$$

where $\mu = 1.765$. (This should be compared with the Fulcher expression [2.8], where $\tau \sim \exp[1/(\lambda_C - \lambda)]$.) Similar power-law singularities are expected as the density is increased or the temperature is decreased.

A number of modifications have been made to the simple theory outlined above. Jacobs (1986) has used a *linear* dependence for the memory function, i.e. $M(t) \propto \Phi(t)$ (cf. [2.17]), based on the assumption that relaxation at a site is controlled by the stress present elsewhere, which in turn decays with the same time dependence. (Note that this does *not* employ mode-coupling arguments to infer a form for $M(t)$.) A (glass) transition is predicted by this model too, but with the difference that the transition is *second order* (Jacobs 1986). Extensions of mode-coupling to other modes (e.g. currents, Götze and Sjögren (1987)) have produced similar results to that of the simple model of Leutheusser (1984), except that the ideal glass-transition singularity is removed, leaving a rounded transition, such as is normally observed experimentally. Finally, proper treatment of the wavevector dependence involved in the mode-coupling theory, neglected by Leutheusser (1984), indicates that only those density fluctuations with wavevectors comparable to that corresponding to the first peak in the static structure factor ([3.23]) are important in contributing to the feedback mechanism leading to vitrification (Bengtzelius *et al.* 1984).

Although the above dynamical theory of the glass transition is appealing in many respects, there are some shortcomings. One major limitation is that so far only spherically symmetric interatomic interactions have been considered, as for instance in the computer calculations by Bengtzelius (1986) on a 'Lennard-Jones' system. Thus, the complexities in real glass-forming systems arising from non-central (covalent) interactions are neglected. Furthermore, it appears that the mode-coupling approximation describes the dynamics of fluids up to viscosities of $\simeq 10$ poise reasonably well, but predicts a glass transition at temperatures where $\eta \simeq 100$ poise, instead of the experimental values of $\eta \simeq 10^{13}$ poise (Frederickson 1988). We have already referred to the fact that scaling predicted by the dynamical theories near the transition on the liquid side [2.18] is *not* of the Vogel–Tammann–Fulcher type [2.8] observed experimentally.

Recent experimental tests of mode-coupling predictions have employed neutron scattering, since the scattering function $S(Q, t)$ corresponds to the density correlation function $\Phi(t)$ (weighted by the neutron scattering amplitudes of the nuclei). The observed time-dependent structure factor for glassy $Ca_{0.4}K_{0.6}(NO_3)_{1.4}$ is shown in Fig. 2.10(a) (Mezei *et al.* 1987). These data can be plotted against a reduced time variable t/τ, where $\tau \propto \eta/T$, and the resulting master plot can be

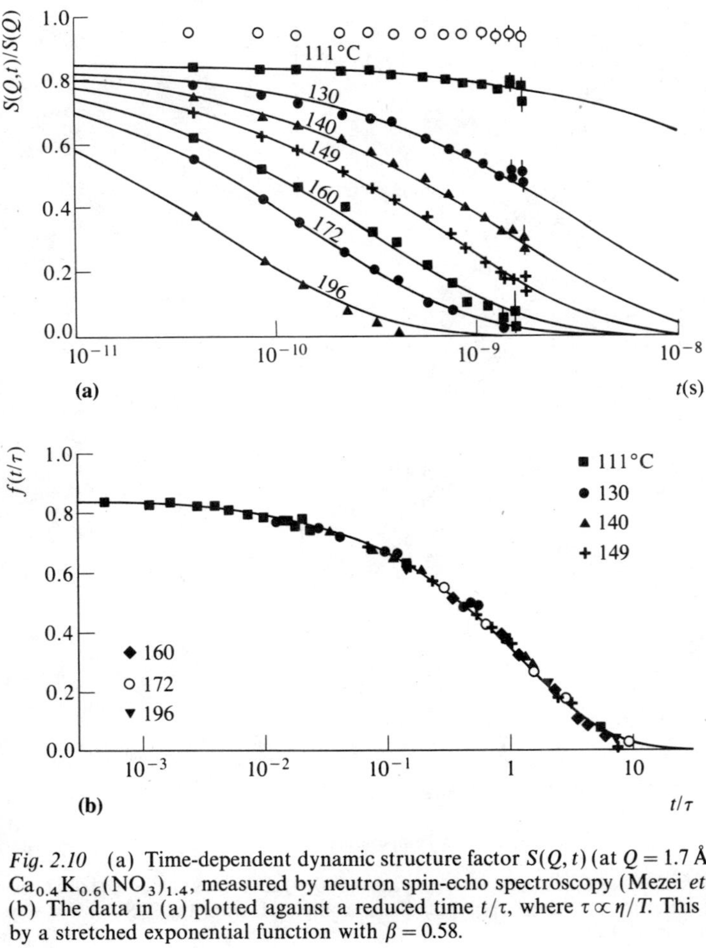

Fig. 2.10 (a) Time-dependent dynamic structure factor $S(Q, t)$ (at $Q = 1.7$ Å^{-1}) for the glassy system $Ca_{0.4}K_{0.6}(NO_3)_{1.4}$, measured by neutron spin-echo spectroscopy (Mezei *et al.* 1987).
(b) The data in (a) plotted against a reduced time t/τ, where $\tau \propto \eta/T$. This master plot can be fitted by a stretched exponential function with $\beta = 0.58$.

fitted by a stretched exponential function [2.7], with $\beta = 0.58$ (Fig. 2.10(b)). Rather larger values of β ($\simeq 0.7$–0.9) were predicted from mode-coupling numerical simulations for a single-component Lennard-Jones system (Bengtzelius 1986), although a mode-coupling analysis for a two-component model simulating an ionic salt (Bosse and Krieger 1986) results in a functional form which is practically indistinguishable from a simple stretched exponential with $\beta = 0.56$ (Krieger and Bosse 1987), i.e. very close to that obtained by fitting the experimental data (Fig. 2.10(b)).

Although the time dependence of the density correlation functions near T_g observed experimentally seems to be in accord with theoretical predictions from mode-coupling theories, this appears *not* to be the case for the temperature dependence. The neutron scattering function as a function of temperature for glassy polybutadiene (Frick *et al.* 1988) is shown in Fig. 2.11, together with a DSC trace showing the glass transition at 185 K. The near-vertical slope corresponding to

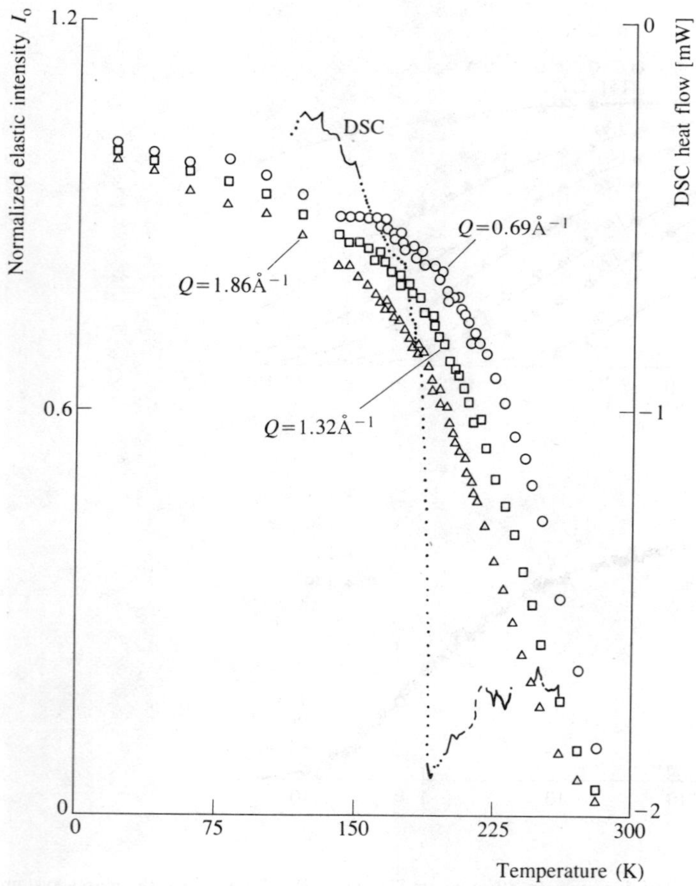

Fig. 2.11 Temperature dependence of the elastic structure factor $S(Q, 0)$ at various values of Q obtained from inelastic neutron scattering for glassy polybutadiene, compared with the DSC trace, for temperatures near T_g (Frick *et al.* 1988).

the square-root singularity $S(Q, \omega) \propto (T_C - T)^{1/2}$ predicted from the mode-coupling model (Götze 1986) is clearly not observed experimentally.

2.3.5 Free volume

A popular theory describing liquids and certain aspects of the glass transition is that concerned with 'free volume', which has many similarities with aspects of the relaxation theories discussed in the previous section. Nevertheless, we accord it the honour of a separate section in view of its historical importance in the development of the subject. The model was originally developed for fluids assumed to be composed of hard spheres (Cohen and Turnbull 1959, Turnbull and Cohen 1961, 1970). The spheres oscillate thermally in their own cages corresponding to their Voronoi polyhedra (sect. 3.3). The total volume of a liquid is supposed to be divided into the part 'occupied' by the spheres or molecules V_{occ}, and that part in which the molecules are free to move. The latter volume permitting diffusive motion is termed the 'free volume' V_f. It is further assumed that molecular transport can occur only when the voids have a volume greater than a critical volume. The free volume is shared communally; in other words it is divided into holes or voids of varying size and position by the random motion of the molecules, and no local free energy is required for redistribution of the free volume. As the temperature of the liquid is lowered, both the occupied volume and the free volume are expected to contract. The essential difference between a glass and a liquid on this model is, therefore, that (a) the free volume is independent of temperature, and (b) redistribution of free volume no longer occurs, i.e. the free volume is 'frozen-in' in the locations occupied when the glass was formed. The idea that the glass transition occurred when the free volume decreased below some critical value was first proposed by Fox and Flory (1950, 1951, 1954).

It is interesting to note that simulations of the glass transition using the mode-coupling approach applied to a hard-sphere system predict that the phase transition occurs at a critical packing fraction $\eta_p = 0.516$ (Bengtzelius *et al.* 1984), which should be compared with the value obtained from molecular-dynamics simulations, $\eta_p = 0.56$ (Woodcock and Angell 1981); for comparison, the packing fraction of a dense random packing of hard spheres is $\eta_p = 0.636$ (sect. 3.5.2).

Some idea of the magnitude of the free volume can be gauged by consideration of the coefficients of cubic expansion α_{Tl} and α_{Tg} of the liquid and glass, respectively (Gee 1970). The expansion of the free volume is $\Delta\alpha_T = \alpha_{Tl} - \alpha_{Tg}$, since α_{Tg} represents the expansion of the occupied volume. The occupied volume (at $T = 0$) can be obtained by extrapolating the liquid volume, giving $V_{occ}(T=0) = V_g(1 - \alpha_{Tl} T_g)$, where V_g is the total volume of the glass at T_g. Similarly, the total volume of the glass at $T = 0$ is $V = V_g(1 - \alpha_{Tg} T_g)$ and hence the free volume of the glass is $V_{fg} = V - V_{occ}$, giving an expression for the fractional free volume $\mathscr{R}$:

$$\mathscr{R} = \frac{V_{fg}}{V_g} = T_g \Delta\alpha_T \qquad [2.19]$$

Application of this formula to a variety of glasses, including organic high polymers and chalcogenide glasses, indicates that at T_g some 10% of the total volume is 'free'; the value for B_2O_3 (a very good glass-former which has never been observed to devitrify), interestingly, is much higher at 34% (Suzuki and Abe 1981).

Free-volume theory has also been applied to the problem of viscosity of liquids. This stems from the discovery of an empirical relation between viscosity and free volume, the Doolittle (1951) equation:

$$\eta = A \exp(BV_{occ}/V_f) \qquad [2.20]$$

where A and B are empirical constants. Assuming a linear form for the thermal expansion, such that the free volume of the *liquid* is $V_f = V_{fg} + V_g \Delta\alpha_T (T - T_g)$ at $T > T_g$ and inserting this into [2.20], one recovers the Vogel–Tammann–Fulcher expression [2.8]. Alternatively, by additional algebraic manipulation, substitution of the equation for the temperature dependence of the free volume of the liquid into the Doolittle expression yields an equation of the same form as that of the empirical Williams–Landel–Ferry (1955) equation (see Problem 2.1) which has been very successful in accounting for relaxational behaviour of many glasses:

$$\ln\left(\frac{\eta_g}{\eta}\right) = \frac{C(T - T_g)}{D + (T - T_g)} \qquad [2.21]$$

where C and D are empirical constants.

More recently, Cohen and Grest (1979) have enlarged on earlier work on the free-volume model, and used 'percolation' theory (see sect. 5.3.6) to account for the exchange of free volume between nearest-neighbour 'liquid-like' cells which have a sufficient number of neighbouring liquid-like cells that the volumes of any neighbouring *solid*-like cells do not change volume simultaneously. In this manner, they were able to calculate the communal entropy, associated with the accessibility of all the configurational volume inside the finite liquid clusters, and hence calculate the heat capacity. They conclude that the glass transition is predicted to be more likely a *first*-order phase transition rather than second order. It is perhaps significant that the simple mode-coupling theory of the glass transition proposed by Leutheusser (1984) also predicts that the transition is first order (sect. 2.3.4); the reason for this similarity may be an underlying common percolative behaviour.

Free-volume theory is a conceptually appealing and relatively simple theory for the liquid state and the glass transition. Nevertheless, it does have its drawbacks and inconsistencies, as indeed do all the theories mentioned so far. A serious objection is the assumption that the thermal expansion remains linear even when V_f becomes very small, and that although diffusive motion ceases when $V_f \to 0$ the thermal contraction is continuous up to that point and then ceases (Anderson 1979). A further problem lies in the fact that free-volume theory is derived on the basis of the hard-sphere model of liquids in which non-directional bonds are predominant, but is then applied to certain glass-formers which are strongly covalent and are almost certain to have directed bonds for $T > T_g$.

2.3.6 Summary

From the discussion above, it is apparent that the phenomenon of the glass transition is very complex and that no single theory that has been advanced so far is capable of accounting for all aspects of it. The glass transition, as observed experimentally, is certainly at least partly a relaxation phenomenon, in that the experimental time-scale profoundly affects the transition observed as the supercooled liquid falls out of equilibrium; very different results are obtained if the glass transition is probed by calorimetric (DSC or DTA) experiments in which the time-scale is of the order of seconds, than if, say, ultrasonic relaxation spectroscopy is employed. Nevertheless, there is also very definitely a thermodynamic aspect in addition to the relaxation character of the transition; this is clearly evinced by Kauzmann's paradox. Although the glass transition appears superficially to be a second-order thermodynamic phase transition, kinetic effects play an important role. Any complete theory describing the formation of a glass from the liquid state must therefore combine both relaxation and thermodynamic aspects in a natural way; the recent developments of mode-coupling theory perhaps point the way to this goal.

2.4 Factors that determine the glass-transition temperature

We have seen in the last section that the value of the glass-transition temperature, T_g, observed experimentally is, to a considerable degree, influenced by the prevailing experimental conditions (e.g. cooling rate of the melt). Nevertheless, if measured under standard conditions (e.g. calorimetrically at a fixed heating rate), is T_g dependent on any material parameter or the structure? This is a very important question technologically (for example, the glass–rubber transition for polymers marks the *lower* limit for the use of an amorphous rubber but the *upper* limit for a thermoplastic, such as polystyrene or PVC), and although the question has long been posed there is as yet no consensus as to which factors are mainly responsible for determining T_g. Various proposals made in the past have suggested that T_g scales with the melting temperature T_m, the cohesive energy (scaling approximately as the normal boiling temperature T_b), the Debye temperature of the phonon spectrum, or the band gap (for amorphous semiconductors). For example, many glass-formers (particularly organic materials) have been observed to have $T_g \simeq 2/3T_m$ (Kauzmann 1948). Many of the suggestions stem from an attempt to correlate the large-scale molecular or atomic motions taking place above T_g with characteristic energies needed to give rise to such motions, either through bond breaking or the creation of free volume. Other characteristic energies that have been suggested are less directly related. It is assumed that the optical band gap, for instance, scales with the cohesive energy; the stronger the chemical covalent bond, the larger the band gap (see Ch. 5). Such a correlation between T_g and optical gap has been found by de Neufville and Rockstad (1974) for a variety of chalcogenide-based amorphous semiconductors, but a simple comparison has to be treated with caution since, for example, amorphous Se, As_2Se_3 and GeAsSe all have approximately the *same* value for the optical gap, yet the glass-transition temperatures vary between 318 K and 677 K! However, within chemically similar families, e.g. the arsenic chalcogenide series As_2X_3, where X = S, Se or Te, a reasonably good correlation is found. De Neufville and Rockstad in fact grouped together materials with the same average number of outer valence electrons per atom, $\bar{N}$, or equivalently average coordination number (equal to $8 - \bar{N}$). Such attempts at correlation depend crucially therefore on subtle differences in the composition of the valence or conduction bands. It obviously makes a great difference whether the top of the valence band, say, is composed of bonding or non-bonding electron states, since the optical band gap is determined by the energy difference between the highest occupied (valence) and lowest unoccupied (conduction) states (see sect. 5.6), whereas the cohesive energy is related to the energy difference between bonding and antibonding states.

However, this simple picture becomes more complicated if the variation in T_g with composition in a given alloy system is examined. For example in the simple binary chalcogenide alloy As_xSe_{1-x} $(0 < x < 1)$, Myers and Felty (1967) found that T_g did not vary monotonically with composition; instead, a maximum in T_g was found for $x \simeq 0.4$, corresponding to the stoichiometric composition As_2Se_3. Obviously in this case T_g is sensitive to the chemistry involved, and a maximum in T_g at the stoichiometric composition may result from the formation of a 'chemically ordered network' (see sect. 3.3) in which only the stronger As—Se bonds (rather than the weaker As—As and Se—Se bonds) are present. A further note of caution must be sounded in correlating T_g with the cohesive energy (or a monitor of it) since it is likely that the weaker secondary 'back bonds' binding structural units together, particularly important in chalcogenide glasses, may be ruptured preferentially at T_g, rather than the primary covalent bonds, thereby allowing large-scale motion of clusters or groups of atoms to take place. Nevertheless, Tanaka (1985) has discovered an empirical relationship between T_g and Z, the average coordination number per atom (calculated in terms of the covalent bonding):

$$\ln T_g \simeq 1.6Z + 2.3 \qquad [2.22]$$

Figure 2.12 shows this correlation for a number of glass-forming systems including chalcogenide materials and organic polymeric materials.

Alternatively, adopting the free-volume picture of the glass transition, it might be supposed that the important factor in determining T_g is the energy required to

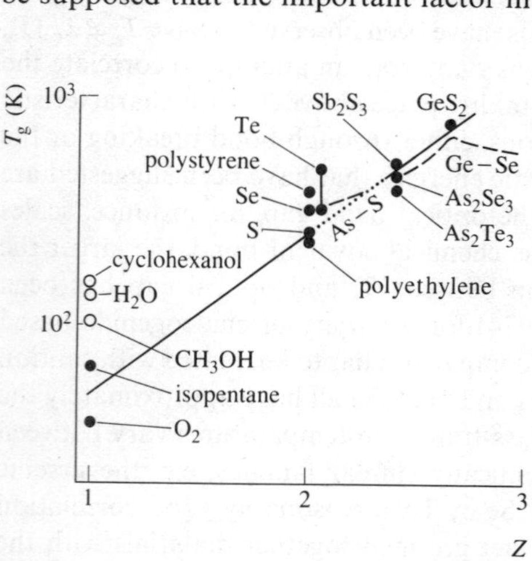

Fig. 2.12 Correlation between the logarithm of the glass-transition temperature T_g and the average coordination number Z for a variety of glass-forming materials (Tanaka 1985).

produce atomic mobility in a volume v (Gee 1970). This can be expressed as v times the cohesive energy density, where the cohesive energy density, U, of the liquid can be related to the molar latent heat of evaporation L_m and volume V_m by the equation $U=(L_m-RT)/V_m$. However, the flexibility of a polymer chain, for example, will be important; the stiffer the chain, the larger will be v and hence the higher the glass-transition temperature. Such a dependence is observed in glassy polymers, and many empirical rules relating chain parameters to T_g have been devised. (Note that the entropy theory for the glass transition of Gibbs *et al.* concentrates solely on the flexibility factor, and ignores rather unphysically the intermolecular interactions.)

In conclusion, there appear as yet to be no firm rules for predicting the glass transition, no ready prescriptions with which to calculate T_g for a given glassy material. With this in mind, we give in Table 2.2 for the benefit of the reader a table of values for T_g for a variety of materials in common use or which frequently form the subject of research. The values given in Table 2.2 must be treated with a certain degree of caution, since in practice appreciable departures from them may be observed for substantially differing heating rates, etc., but the values given are chosen to be as representative as possible.

2.5 Glass-forming systems and ease of glass formation

The question '*why* do certain materials readily form glasses on cooling a melt?' is one of considerable practical and technological importance. In many cases, this question may be reformulated as 'why do certain chemical compositions of materials have a greater glass-forming tendency than others?' This remains one of the great unsolved mysteries of glass science, and although empirical prescriptions have been developed which are reasonably successful in accounting for the glass-forming tendencies in certain specific cases, there is no general rule which may be used universally to predict glass-forming ability in any given system. Of course, the question of the ease of glass formation on cooling a melt (i.e. the cooling rate required) is intimately related to, and inseparable from, the problem of *how* do glasses form. As we have seen in the previous sections, the subject of the glass transition is poorly understood, and then often only in a phenomenological rather than a quantitative sense. A comprehensive theory of the glass transition is needed before categorical answers to the questions surrounding glass-forming ability can be given. Nevertheless, there is a virtue in rather arbitrarily divorcing a discussion on the ease of glass formation from that of the glass transition itself in view of the fact that this is a very real practical problem, encountered at the laboratory bench. Profound consideration of the nature of the glass transition rather fades into insignificance when faced with the question of, say, which chemical composition to use in order that glass formation is facilitated and crystal growth is inhibited! This aspect of the subject, as the reader may well anticipate, is still very much at the 'alchemy' stage. Frequently, and particularly in technological applications, small amounts of additives are found to be necessary so that glass formation is improved in some way, but it is often not known precisely why they should act in the way they do. For example, trace amounts of As and Cl stabilize the Se films

Table 2.2 Glass-transition and melting temperatures for some glassy materials

1. *Inorganic materials*

Material	T_g(K)	*Reference*	T_m(K)	*Reference*
S	246	1	392	1
Se	318	2	490	1
As_2S_3	478	2	573	5
As_2Se_3	468	2	633	5
As_2Te_3	379	3	633	6
$GeSe_2$	695	2	980	5
GeS_2	765	4	1073	6
SiO_2	1453	1	2003	1
GeO_2	853	4	1388	1
BeF_2	598	4	821	1
$ZnCl_2$	380	1	590	1
B_2O_3	530	1	793	1

References: (1) L. G. Van Uitert (1979), *J. Appl. Phys.* **50**, 8052; (2) J. P. de Neufville and H. K. Rockstad (1974), *Proc. 5th Int. Conf. on Amorphous and Liquid Semiconductors* J. Stuke and W. Brenig (eds) (Taylor and Francis) p. 419; (3) C. H. Seeger and R. K. Quinn (1975), *J. Non-Cryst. Sol.* **17**, 386; (4) A. C. Wright, G. Etherington, J. A. Erwin Desa, R. N. Sinclair, G. A. N. Connell and J. C. Mikkelson (1982), *J. Non-Cryst. Sol.* **49**, 63; (5) *Handbook of Chemistry and Physics* (61st edn), R. C. Weast and M. J. Astle (eds) (CRC:1980); (6) *Comprehensive Inorganic Chemistry* Vol. 2, A. F. Trotman-Dickensen (ed.) (Pergamon: 1973).

2. *Organic polymers*

Material	T_g(K)	T_m(K)
Polystyrene (isotactic)	373	503
Polymethyl methacrylate (isotactic)	318	433
Poly-*N*-vinyl carbazole	423	593
Nylon-6	323	488
Poly(ethylene oxide)	218	339
Poly(propylene oxide)	211	338
Cellulose triacetate	380	573
Polyethylene	253	393
Polypropylene	278	423
Polytetrafluoroethylene (PTFE)	388	600
Poly(vinyl chloride) (PVC)	353	—

Reference: J. A. Brydson in *Polymer Science*, A. D. Jenkins (ed.) (North Holland: 1972).

used in xerography against crystallization, as does the presence of As and Ge or Si in chalcogenides used as electronic 'switching' devices. Alternatively, the presence of a third (alkali) oxide is found to be necessary to prevent phase separation and immiscibility in binary oxide melts and hence in the resulting glasses (e.g. alkaline earth oxide–B_2O_3 or –SiO_2 systems – sect. 3.7.2).

Despite the apparent arbitrariness and uncertainty alluded to in the foregoing, none the less there are several factors which experience has shown do play a significant role in determining the ease of glass formation. These generally relate to chemical or structural properties of the system; thermodynamic or free-volume aspects have been discussed previously. They include the occurrence of eutectic compositions, the existence of several crystalline polymorphs of the material, the (average) atomic coordination and the topology of the amorphous structure. These are discussed in turn in what follows, but it must be stressed that the 'rules' enunciated often refer only to particular restricted systems of glassy materials, and are not generally universally applicable.

2.5.1 Structure and topology

The first really successful attempt to categorize materials into glass-formers and non-glass-formers was by Zachariasen (1932). This is one of the truly seminal papers in the field of glass science, and has had an enormous impact on the subject in areas other than just glass formation. Some 50 years ago, when Zachariasen wrote his paper, practically the only known glass-forming materials were oxides, of which only five formed glasses by themselves: SiO_2, GeO_2, B_2O_3, As_2O_3 and P_2O_5. In addition, these five oxides can also form glasses when mixed (up to a limiting percentage) with another oxide or oxides, which are not by themselves glass-formers. The rules that Zachariasen formulated were then capable of explaining why, for example, SiO_2 is a glass-former and Na_2O is not, but why $xNa_2O{:}SiO_2$ should also be a glass-former.

Zachariasen argued that those materials most likely to form glasses readily would have an internal energy only slightly larger in the amorphous state than when crystalline. (A hypothetical amorphous phase possessing an internal energy considerably higher than the crystal would be prone to rapid devitrification.) The internal energy of a solid is related to its structure (e.g. the Madelung energy of ionic materials). Since it is reasonable to assume that the interatomic interactions in both amorphous and crystalline phases would be very similar, this implies that the atomic structure of the two phases must be similar in certain regards. The structure of amorphous solids is, however, aperiodic. Zachariasen assumed therefore that the oxygen polyhedra found in oxide crystals (triangular, tetrahedral or octahedral) would also occur in glasses, the only difference being that the relative orientation of polyhedra should be variable in glasses, giving rise to a non-periodic structure. For example, crystalline forms of SiO_2 contain SiO_4 tetrahedra joined at the corners, and it was assumed that this should also prevail in vitreous SiO_2. The most important aspect of Zachariasen's approach which transcended the importance of his 'rules' for glass formation was that a non-periodic arrangement of atoms could be attained solely as a result of the incorporation of variations in bond angles (for two dimensions) or bond angles and 'dihedral' angles (the relative angle of twist between neighbouring units – see Fig. 3.36) for three dimensions. In this manner, a *continuous random network* (CRN) can be constructed, in which the approach of atoms closer than a normal bond length can be avoided, thereby minimizing repulsive energies and hence the total internal energy. The difference in structures for the glassy and crystalline forms of a hypothetical, two-dimensional oxide, A_2O_3, is shown schematically in Fig. 2.13 in which each oxygen is corner-shared by *two* AO_3 triangles. (The important concept of a CRN is discussed

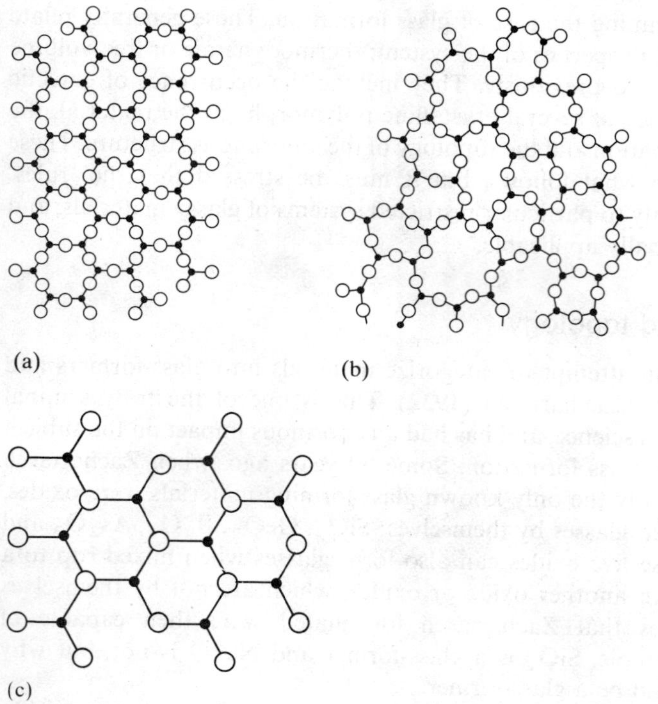

Fig. 2.13 Schematic two-dimensional representation of the structure of
(a) a hypothetical crystalline compound A_2O_3
(b) the Zachariasen model for the glassy form of the same compound and
(c) the hypothetical crystalline compound AO.

in considerably more detail in sect. 3.5.3.) Contrast this with the case shown schematically in Fig. 2.13(c) which is for the hypothetical two-dimensional crystalline oxide AO, where each oxygen is corner-shared by *three* AO_3 triangles; structural disorder cannot be introduced as in Fig. 2.13(a) simply by bond-angle distortions without considerably increasing the internal energy. The amorphous state, as represented by a CRN, is metastable since for crystallization to occur a substantial topological rearrangement of the structural units (having a large activation energy) must occur. Based on these and similar arguments, Zachariasen proposed four rules which an oxide must obey if it is to form a glass:

1. An oxygen atom may be linked to not more than two A atoms.
2. The number of oxygen atoms surrounding an A atom must be small (three or four).
3. The oxygen polyhedra share corners with each other, not edges or faces.
4. At least three corners in each oxygen polyhedron must be shared if the network is to be three-dimensional.

Thus, oxides of the type AO and A_2O should not, and indeed do not, form glasses. The rules are satisfied for oxides of the form A_2O_3 provided the oxygens form triangles around each A atom, and for oxides of the form AO_2 and A_2O_5 if the oxygens form tetrahedra. Cooper (1978) has pointed out that Zachariasen's rules are based more on topological than on energetic grounds, since it is found that corner-sharing regular octahedra or cubes can only form crystal-like, periodic arrays, leaving triangular and tetrahedral coordination of oxygens about A atoms the only contenders for glass formation. He has further shown that rule 3 is either redundant or inoperative (depending on the example used), and so should not be included. Consider firstly triangles sharing more than one edge, or tetrahedra sharing more than one face or two edges; these cases violate rule 1, thereby making rule 3 unnecessary. Triangles sharing an edge and a corner, and tetrahedra sharing either two opposite edges or a face and a corner, both form infinite chains (or loops) which can easily be formed into a non-periodic array. Alternatively, tetrahedra sharing one edge and the two opposite corners are equivalent topologically to a single corner-sharing tetrahedron which is therefore able to form a random structure. These two examples show that random structures can result even if rule 3 is violated, and that it should not therefore be considered. Furthermore, in the materials SiS_2 and $SiSe_2$, which are chemically analogous to SiO_2, edge-sharing of SiX_4 ($X = S, Se$) tetrahedra is the dominant structural motif, with the result that chain-like amorphous structures are produced (see Fig. 3.40).

The discussion thus far has centred on the glass-forming ability of simple oxide substances, known as 'network-formers'. Zachariasen also modified the original rules to take account of complex glasses, which contained non-glass-forming oxides in addition. The modified rules are:

1M. A high proportion of (network-forming) cations are surrounded by oxygen tetrahedra or triangles.
2M. The oxygen polyhedra only share corners with each other.
3M. Some oxygen atoms are linked to only two cations, and do not form additional bonds with any other cations.

These rules imply that oxide glasses must contain substantial proportions of glass-forming cations, or other cations which can substitute isomorphously (e.g. Al^{3+} for Si^{4+}). Any other cations (e.g. Na^+) that are present are termed 'network modifiers', since they depolymerize the otherwise perfectly connected CRN. Although properly the subject of the next chapter, the question of the structure of modified oxide glasses is mentioned here for the sake of continuity. Consider the case of the addition of Na_2O to SiO_2. The action of the modifier is to break up the continuous silica network by introducing 'dangling' or 'non-bridging' oxygens; this process is shown schematically in Fig. 2.14 and is not limited to silicate glasses only, but can occur in other oxide (or sulphide or selenide) systems. The non-bridging oxygens carry a single negative charge and are compensated by a Na^+ ion nearby. The chemical formula of an oxide glass may be written generally as A_mB_nO, where m and n may not be integers; A represents the network-forming cation as before, and B is the network-modifying cation. An interesting intermediate class of oxides, including TeO_2, WO_3, MoO_3, Bi_2O_3, Al_2O_3, Ga_2O_3 and V_2O_5, do not by themselves readily form glasses, but will do so when mixed with other (modifier) oxides. They are obviously incipient network-formers.

Structure-related glass-forming criteria have also been invoked for metallic systems. One of the first was put forward by Polk (1970) for transition-metal–metalloid (TM–M) alloy glasses. He proposed that glass formation in such alloys is

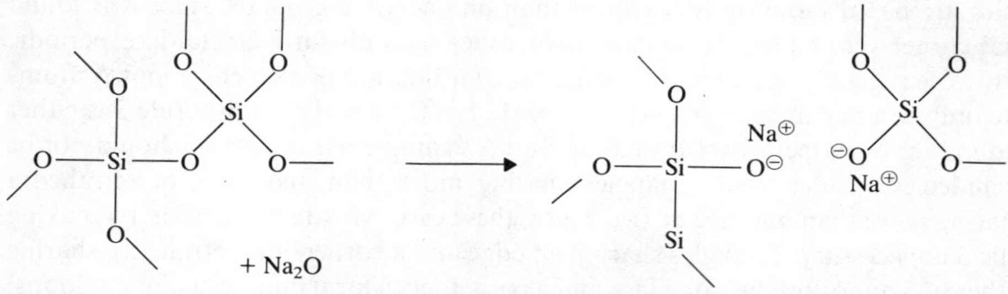

Fig. 2.14 Schematic illustration of the effect of the addition of alkali oxide (e.g. Na_2O) to silica. Each molecule of Na_2O added converts a bridging oxygen to two non-bridging oxygens (negatively charged). Diffusion in the melt would ensure that separation of the non-bridging oxygen atoms would occur.

promoted by the M atoms filling the larger interstitial holes in a dense-random packed arrangement of the TM atoms (see sect. 3.5.2 for a discussion of the DRP model), thereby stabilizing the random structure against crystallization. Although this particular structural arrangement does predict the correct composition range in which glass formation is found (namely for $\simeq 20$ at. % metalloid), glass-forming systems have been discovered subsequently which do not satisfy these conditions, and recent structural studies have ruled out the possibility that the M atoms occupy interstitial sites in such a random fashion (e.g. see sect. 3.3).

Another structural criterion concerns the relative atomic sizes of the alloy constituents. An empirical rule is that the atomic radii of two elements must differ by more than 10% for their alloys to be glass-forming. A rather more precise criterion has been given by Egami and Waseda (1984) in terms of the minimum concentration C_B^* of solute B in a binary glass-forming alloy system A_xB_{1-x}. From an examination of more than 60 glass-forming systems, it was found empirically that C_B^* appears to be inversely correlated with the atomic-volume mismatch:

$$C_B^* \approx \lambda[(V_B - V_A)/V_A]^{-1} \qquad [2.23]$$

where V_A and V_B are the atomic volumes of matrix and solute species, respectively. Figure 2.15 shows the experimental correlation, from which it can be deduced that

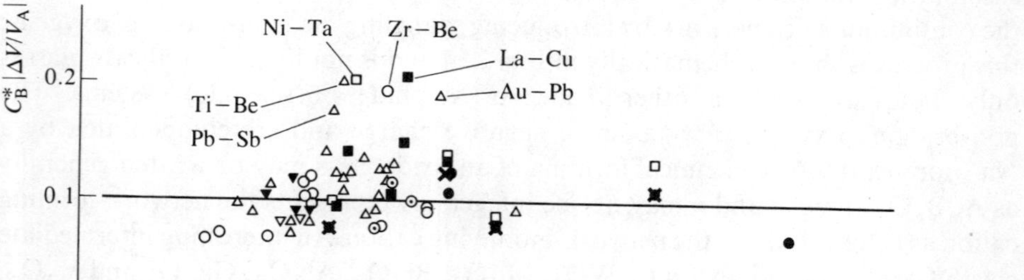

Fig. 2.15 Correlation of the minimum concentration C_B^* of solute B in a glass-forming alloy system A–B with the atomic volume mismatch $\Delta V = |V_B - V_A|$ (Egami and Waseda 1984).

empirically $\lambda \simeq 0.1$ (Egami and Waseda 1984). By performing an analysis of the local strain induced by the incorporation of different-sized solute atoms into a matrix, it was found that a crystalline solid solution becomes topologically unstable above a certain solute concentration. In other words, for solute concentrations greater than C_B^* the amorphous phase becomes stable with respect to a crystalline solid solution. This theoretical analysis yielded an expression of the same form as [2.23], with $\lambda \simeq 0.13$.

2.5.2 Eutectic compositions

For two compounds, A and B, miscible in the liquid state but completely immiscible as solids, the phase diagram will be as shown schematically in Fig. 2.16, provided that no intermediate compound formation occurs. The melting (or liquidus) temperature is depressed and is a minimum at the so-called 'eutectic' composition. It is well known that glass formation is generally favoured for compositions at which there is a deep eutectic. This is as true for metallic glass systems as it is for inorganic materials; an example is given in Fig. 2.17 of the glass-forming region for the system V_2O_5–PbO (an example of an 'incipient' network-former), which is seen to straddle the eutectic point. Compositions near the eutectic favour glass formation since the melting point is depressed and so the liquid is less supercooled at T_g, thereby reducing the possibility of crystallization (see sect. 1.3.5). Nevertheless, kinetic factors still play an important role in the formation of the glassy state, since if the cooling rate is not fast enough the liquid eutectic alloy crystallizes out into an intimate mixture of, in the simplest case, A and B.

In a study of the glass-forming system $Au_{1-x}Si_x$, which is attractive since it exhibits no intermediate compound formation, Allen *et al.* (1980) have demonstrated that the eutectic temperature is lowered considerably from the value

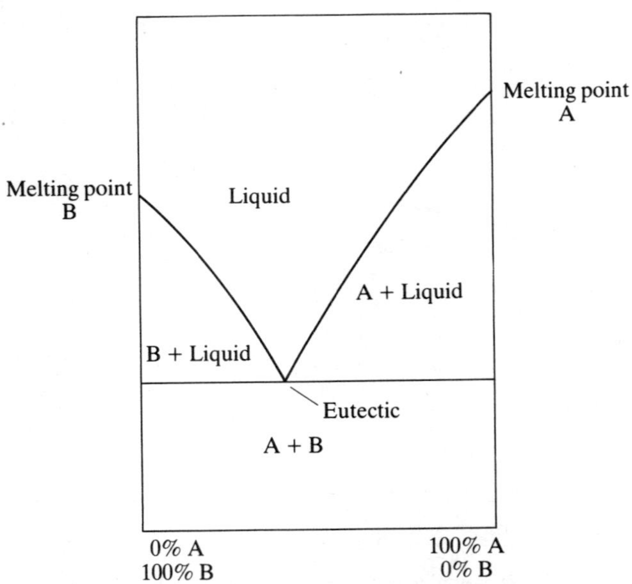

Fig. 2.16 Phase diagram of a simple binary eutectic system.

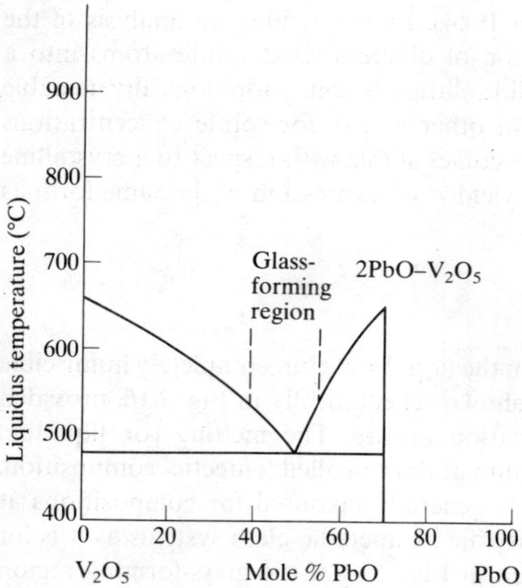

Fig. 2.17 Glass-forming region of the system V_2O_5–PbO in relation to the phase diagram (Rawson 1967).

expected for an ideal solution (Fig. 2.18) although the calculated eutectic composition, $x = 0.198$ (and indeed the observed eutectic point $x = 0.186$), is little altered from the ideal value ($x = 0.175$). The large decrease in eutectic temperature is shown to result from a large negative excess free energy of mixing, for which the dominant contribution is the enthalpy, as can be seen from Fig. 2.19(a) where the enthalpy departs much more from ideality than does the entropy. The eutectic point is determined by the condition that the free energies of the liquid and solid are equal, $G^L = G^S$, where

$$G^L = (1 - x)G^L_{Au} + xG^L_{Si} + G^L_{AuSi} \qquad [2.24]$$

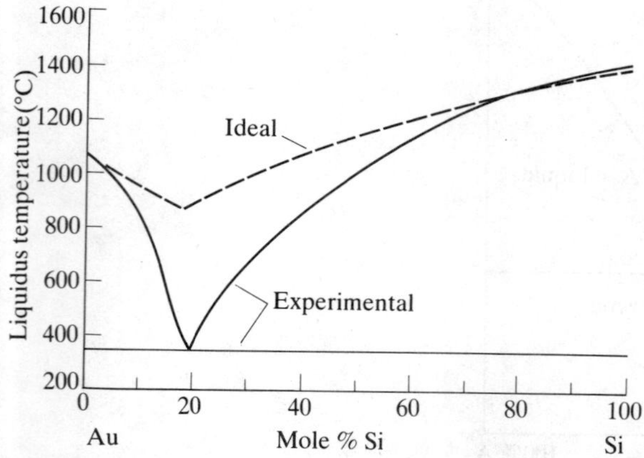

Fig. 2.18 Comparison of Au–Si eutectic phase boundaries determined experimentally and calculated for an ideal liquid (Allen *et al.* 1980).

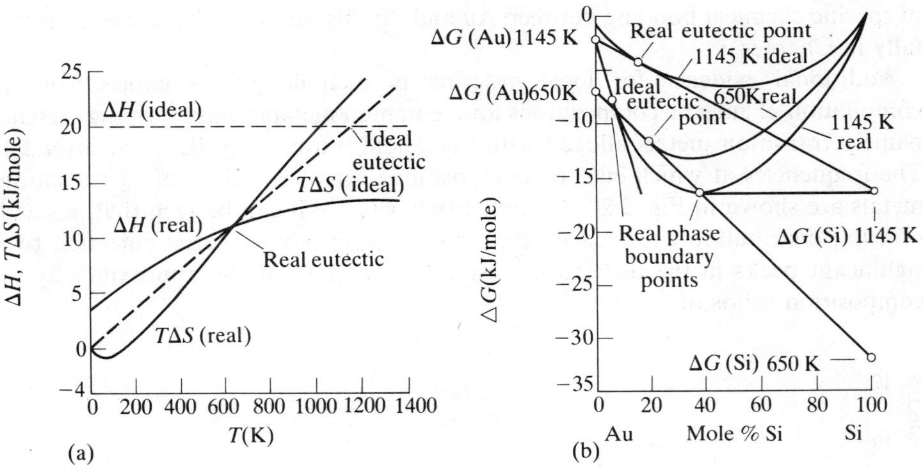

Fig. 2.19 (a) Temperature dependence of the entropy and enthalpy contributions to the free energy balance at the eutectic point of real and ideal liquids.
(b) The common tangent construction leading to the phase boundary shown in Fig. 2.18 (Allen *et al.* 1980).

and

$$G^{L}_{AuSi} = G^{I}_{AuSi} + G^{E}_{AuSi} \qquad [2.25]$$

with

$$G^{I}_{AuSi}(x, T) = -TS^{I}(x) = RT[x \ln x + (1-x)\ln(1-x)] \qquad [2.26]$$

where G^{I} and G^{E} are the ideal and excess mixing terms, respectively. Allen *et al.* further point out that the eutectic *composition* is determined primarily by the large difference in free energies of solid formation (i.e. $G^{S} - G^{L}$) of Si and Au; this is also true for the Ag–Si and Cu–Si systems. This can be seen with reference to Fig. 2.19(b) in which is shown the composition dependence of the free energy of mixing of the liquid for given T. A tangent to the free energy curve at an arbitrary composition intersects the left and right ordinates at the values for the solid–liquid free energy differences (which are themselves temperature dependent) for Au and Si, respectively. For a general case, at some temperature T, the straight line cuts the mixing curve at two compositions; the eutectic point is given by the mixing curve characterized by the eutectic temperature T_e, for which there is only one intersection, i.e. a tangent, at the eutectic composition, x_e. It can be seen from Fig. 2.19(b) that by this construction, there is little difference in x_e between the ideal and real cases if the solid–liquid free energy differences for the two components differ substantially; if the free energies were more nearly equal, a considerable disparity in values for x_e would occur. The fact that x_e is determined predominantly by the properties of the *pure* components, Au and Si, implies that it is not necessary to invoke solely structural arguments to account for the value of x_e, as had been attempted previously (Polk 1972). Allen *et al.* also show that in the vicinity of the eutectic point T_e, the excess entropy S^{E} is negative (see Fig. 2.19(a)), implying a considerable degree of chemical or structural ordering on a local scale, indicative of specific chemical bonding between Au and Si; this point will be discussed more fully in Chapter 3.

Additional evidence for local ordering in metallic glasses comes from a compilation of eutectic compositions for a common metallic glass-forming system, namely transition metals alloyed with 'metalloid' atoms (e.g. B, C, N, Si or P). The frequencies at which eutectic compositions occur for alloys of 23 transition metals are shown in Fig. 2.20 (Gilman 1980), whence it can be seen that, instead of a flat distribution expected if arbitrary compositions produced eutectics, two significant peaks in the distribution are found, corresponding approximately to composition ratios of 6:1 and 5:1.

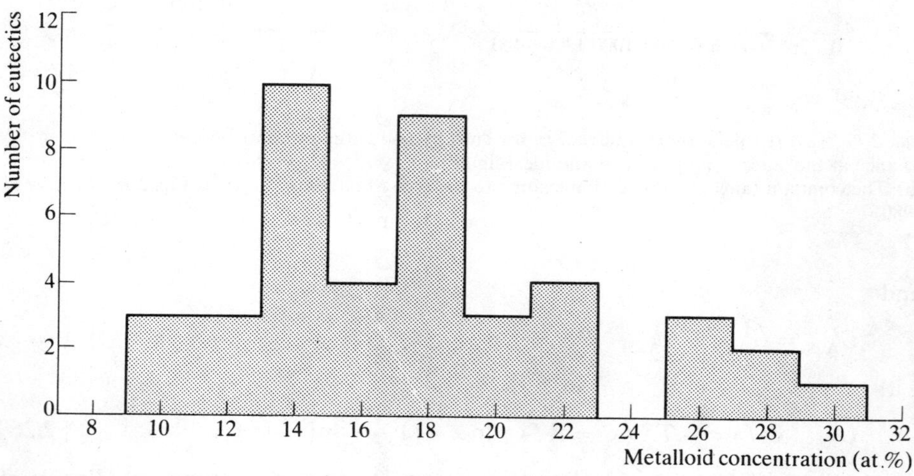

Fig. 2.20 Frequency distribution of eutectic composition for 23 transition-metal–metalloid metallic glasses (Gilman 1980).

A different influence of eutectic formation on glass-forming ability has been proposed by Rawson (1967). Following an earlier suggestion by Sun (1947) that since crystallization involves substantial atomic rearrangement and consequent rupture of bonds, those (oxide) materials having high bond strengths should be less prone to crystallization (and hence better glass-formers), Rawson suggested that the amount of thermal energy available at the melting (or liquidus) temperature might also be a significant factor. The lower the thermal energy (proportional to T) available for breaking bonds, the more difficult should be crystallization, and hence the more easy for a supercooled melt to transform into a glass. More specifically, the parameter relating to ease of glass formation is the ratio of the single bond strength to the melting temperature, B/T_m. 'Easy' glass-formers, such as B_2O_3 or SiO_2, have higher values of B/T_m than incipient network-formers (e.g. TiO_2, TeO_2) and considerably higher values than those of network modifiers (e.g. MgO and CaO), seeming to support the suggestion. The influence of eutectic formation in complex alloy glasses is immediately apparent since the melting or liquidus temperature is significantly lowered, thereby increasing the ratio B/T_m even more and hence favouring glass formation. Thus, the reason why the V_2O_5–PbO system should be glass forming in the vicinity of the eutectic composition, although V_2O_5 by itself is not a glass-former, is made understandable. Both Rawson's and Sun's criteria suffer from the limitation that only primary bonds are considered, even though atomic mobility at and above T_g is more likely to be due to rupture and reformation of secondary weaker bonds, such as 'back bonds' or Van der Waals' bonds between molecular units (e.g. between Se chains). The criteria are probably applicable, however, if there is only one type of bonding in the glass, as is probably the case, for example, in non-ionic silicate glasses.

2.5.3 Crystalline polymorphs

It has been pointed out (Goodman 1975, Wang and Merz 1976) that glass formation is often prevalent for those materials which in the crystalline state are found in a variety of polymorphic modifications. An obvious example is silica, the archetypal glass-former, which is found as quartz, cristobalite and tridymite, as well as the high-pressure forms coesite and stishovite. Goodman proposed that glass formation takes place by the generation of clusters of structurally non-related polymorphs which associate on cooling, but cannot act as nucleation centres for the growth of macroscopic regions of one polymorph or another. While being a useful empirical marker for likely glass-forming systems, the structure predicted for the amorphous solid is based on the 'microcrystallite' model, albeit composed of different types of microcrystals, and as such is at variance with the commonly held belief that amorphous *covalent* solids are better described in terms of random networks (sect. 3.5.3). Nevertheless, there is almost certainly some deep connection between the facility of glass formation and the existence of several crystalline polymorphs of the same material. The fact that several structurally unrelated polymorphs of comparable energy exist (e.g. quartz, cristobalite and tridymite) implies that the structural unit involved (in this case the SiO_4 tetrahedron) can occur in several conformations with little difference in strain energy, a circumstance needed if an amorphous solid, construed as a CRN of the structural units, is to exist having a sufficiently low internal (strain) energy that it can be metastable.

2.5.4 Constraint theory

Most inorganic covalently bonded glasses have relatively low values of atomic coordination number (or average coordination number in the case of alloys). An atom which has all covalent bonds satisfied, obeys the so-called '$8-N$' rule (where N is the number of valence electrons, and the coordination number N_c is given by the number $(8-N)$). Thus Se (a glass-former), has $N_c=2$, since it is in group VI of the periodic table. For a binary alloy A_xB_{1-x}, the average coordination, m, is given by:

$$m = xN_c(A) + (1-x)N_c(B) \qquad [2.27]$$

In general, of course, m is non-integral; for example, for SiO_2 $m=2.67$ and for As_2S_3 $m=2.4$. Thus, m can be regarded as the coordination of a hypothetical 'pseudoatom' forming a structure whose topology is identical to that of the real system.

Phillips (1979) has proposed an interesting connection between glass-forming ability and the average coordination number. He supposed that the glass-forming tendency is maximized when the number of (mechanical) constraints, N_c, experienced by each atom (due to the interatomic forces acting on it) is equal to the number of degrees of freedom, N_d, available to it; the number of constraints per atom is related to the coordination number, and the number of degrees of freedom depends on the dimensionality. A system which has a number of constraints greater than the available degrees of freedom is 'overconstrained' and cannot (easily) form a glass, although an amorphous structure may still be obtained by employing rapid cooling techniques, such as evaporation or sputtering (e.g. as in the case of Ge or Si). A homogeneous, isotropic structure is unlikely to result in this case, however; instead, 'domains' may preferentially form, with the excess strain relieved to a certain extent within each domain, but being concentrated at the domain boundaries (sect. 3.7.1).

Phillips considered the mechanical constraints experienced by an atom to result from the interatomic forces acting on an atom in the 'valence-force-field' model. In this, the (strain) potential energy can be expressed as a sum of contributions from bond-stretching and from bond-bending forces:

$$U_s = \tfrac{1}{2}\alpha \Delta r^2 + \tfrac{1}{2}\beta r_o^2 \Delta\theta^2 \qquad [2.28]$$

where α and β are the bond-stretching and bond-bending force constants, respectively, and Δr and $\Delta\theta$ represent small deviations in bond length and bond angle from the equilibrium values for the bond length, r_o, and the bond angle, θ_o. For a binary alloy, A_xB_{1-x}, there is only one bond-stretching interaction (α), but there are two bond-bending force constants, β(BAB) and β(ABA), for bending motions centred on atoms A and B respectively. It is supposed for simplicity that β(BAB) = β(ABA), and that α,β(BAB) and β(ABA) all act as rigid mechanical constraints. This is somewhat idealistic (implying, for instance, that the nearest-neighbour bond length r_{AB} and the next-nearest-neighbour bond lengths r_{AA} and r_{BB} are perfectly defined and have zero deviations from equilibrium values), but if this assumption is adopted, then the number of constraints per 'pseudoatom' is given by:

$$N_c = m/2 + m(m-1)/2 = \tfrac{1}{2}m^2, \quad m \leqslant N_d - 1 \qquad [2.29]$$

where the first term is associated with α interactions (the factor of $\frac{1}{2}$ arises because a bond-stretching mode involves two atoms) and the second term is for β interactions. Note that [2.29] is valid only for values of effective coordination number $m \leqslant N_d - 1$ (Döhler *et al.* 1980). Otherwise, the fact that $N_d + 1$ bond angles (in N_d dimensions) are *not* linearly independent must be taken into account (Döhler *et al.* 1980); for a twofold-coordinated atom, there is one angular constraint associated with the second term in [2.28], whereas adding each additional bond gives two more constraints because the angles with two existing bonds must be specified (Thorpe 1983). This yields

$$N_c = m/2 + (N_d - 1)(2m - N_d)/2, \quad m \geqslant N_d - 1 \qquad [2.30]$$

The number of degrees of freedom per atom is simply $N_d = 3$ (for a system in 3D space), and we have already seen that in constraint theory the condition for 'ideal' glass formation is given by $N_c = N_d$. For most inorganic covalent materials of interest, the effective coordination number $m \geqslant 2$, and so [2.30] must be used instead of [2.29], giving for the optimum average coordination number for glass formation

$$m_c = N_d(N_d + 1)/(2N_d - 1) \qquad [2.31]$$

which, for $N_d = 3$, results in $m_c = 2.4$. In the case of Ge_xSe_{1-x}, for example, use of [2.27] together with $m_c = 2.4$ yields for the optimum glass-forming composition $x_c = 0.2$, i.e. $GeSe_4$.

The tendency for glass formation in the Ge–Se system has been explored experimentally by finding the minimum quenching rate needed for each composition to avoid crystallization; this was done in a qualitative manner by using three quenching techniques, i.e. quenching in water, air and slow cooling, respectively in order of decreasing quench rate. The data are shown in Fig. 2.21 (Phillips 1979), from which it can be seen that indeed the minimum in glass-forming difficulty (or maximum in glass-forming tendency) occurs near a composition corresponding to $x = 1/5$, seemingly in accord with the predictions of the simple valence-force-field model. The peak at $x = 1/3$ results from compound formation ($GeSe_2$) and represents the position of *chemical* stability, in contrast to the position of *mechanical stability* at the minimum of the curve. A similar analysis for the As–Se alloy system yields $x = 0.4$ for the composition at which glass formation is most favoured; this corresponds of course to the compound As_2Se_3, for which chemical and mechanical stabilities coincide. This appealingly simple approach should be used with some caution, however. For example, it predicts that the system Si_xO_{1-x} should have

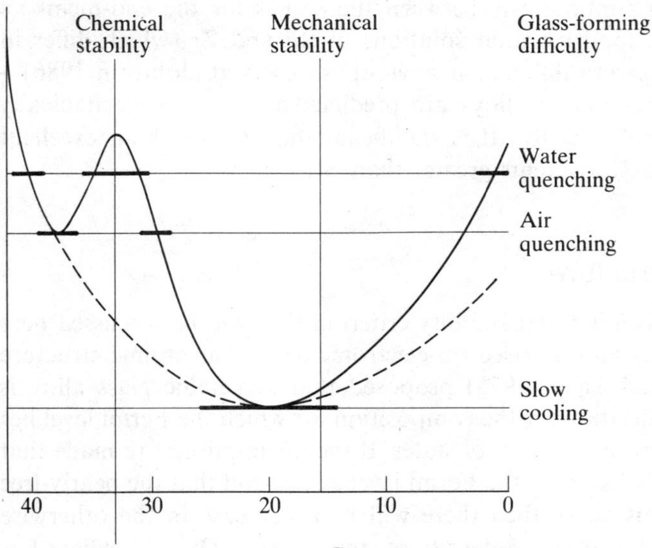

Fig. 2.21 Plot of glass-forming ability (tendency) of the Ge_xSe_{1-x} system. Experimental data are indicated by solid horizontal bars for various rates of cooling (water, air and slow quenching). The solid line is drawn through the data to guide the eye; the prediction of Phillips's mechanical constraint theory is shown by the dashed line.

the greatest glass-forming tendency at $x = 0.2$, as for Ge_xSe_{1-x} discussed above, whereas it is common knowledge that SiO_2 is a very good glass-forming composition. This difficulty can be circumvented by noting that the Si–O–Si bond-angle distribution is rather wide (see Fig. 3.25) and so the constraints associated with oxygen bond angles should be regarded as rather weak; if these are neglected, [2.30] suitably modified then yields $x_c = 0.33$ (Thorpe 1983).

These ideas based on mechanical constraint theory, have led Thorpe (1983) to propose that network glasses can undergo 'rigidity percolation' as the degree of cross-linking increases; structures for which $m < m_c = 2.4$ are termed 'floppy', and those for which $m > 2.4$ are termed 'rigid', since the vibrational characteristics of such networks change in character at $m = m_c$.

A final criterion for glass-forming ability based on strain is applicable to metallic systems. (It should be noted that the Egami–Waseda criterion in terms of atomic-volume mismatch (sect. 2.5.1) could also be regarded as a strain-related glass-forming condition.) For an elastic continuum in three dimensions, an effective elastic modulus M can be defined in terms of the shear modulus μ and bulk modulus B as $M \propto (2\mu^{3/2} + B^{3/2})^{2/3}$, and (Johnson 1986)

$$M = \left(\frac{k_B \theta_D}{h q_D}\right)^2 \rho \qquad [2.32]$$

where θ_D and q_D are the Debye temperature and wavenumber, respectively, and ρ is the mass density of the material. A condition for glass-forming ability can then be given in terms of a *mechanical instability* of crystalline solid solutions, i.e. when $M \to 0$. For alloys (solid solutions) whose constituents have very similar atomic sizes (e.g. Nb–Mo), the elastic modulus behaves in a normal fashion, the value of M varying continuously between the values for the end-members (Fig. 2.22(a)). However, for b.c.c solid solutions of Rh and Zr (which differ in atomic radii by $\simeq 20\%$), very different behaviour is observed (Johnson 1986) – see Fig. 2.22(b). Such crystalline alloys are predicted to become mechanically unstable ($M \to 0$) for $\simeq 12$ at. % Rh. It is significant that Zr–Rh is an excellent glass-forming system for Rh contents greater than $\simeq 15$ at. %.

2.5.5 Electronic structure

The final type of glass-forming and stability criterion that will be discussed here concerns metallic systems and is based on electronic as well as atomic structure considerations. Nagel and Tauc (1975) proposed that a metallic glass alloy is stabilized against crystallization for the composition for which the Fermi level lies in a *minimum* in the electronic density of states. If the assumptions are made that the electron character is s-like near the Fermi level at E_F, and that the nearly-free electron approximation is valid, then there will be a decrease in the otherwise parabolically increasing free electron density of states at $|\boldsymbol{k} + \boldsymbol{Q}| = |\boldsymbol{k}|$, where $\boldsymbol{k}$ is the electron wavevector and $\boldsymbol{Q}$, for the case of a *crystalline* metal, is the wavevector spanning the zone boundaries. For a crystal, of course, a gap opens up at the zone boundary and the density of states decreases to zero there. Nagel and Tauc (1975) assumed that for an amorphous or liquid metal a 'pseudo-gap', rather than a true

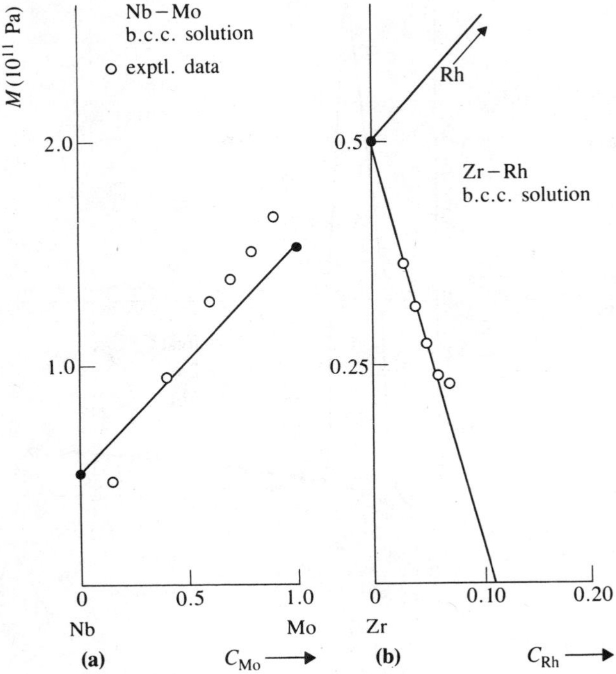

Fig. 2.22 Compositional dependence of the effective elastic modulus M for b.c.c. solid solutions of 4d transition-metal alloys (Johnson 1986) for
(a) Nb–Mo, and
(b) Zr–Rh.
The mechanical instability of the crystalline solid solution, leading to glass formation, is clearly seen for Zr–Rh.

gap, opens up at E_F (Fig. 2.23(a)). Furthermore, they associated the 'zone-boundary' wavevector in the glass with the wavevector corresponding to the position of the first peak in the structure factor $S(Q)$, i.e. to Q_p. Thus, the pseudo-gap occurs for all states at $|k_F| = \frac{1}{2}Q_p$ because of the spherical symmetry characteristic of non-crystalline systems (which is correspondingly manifested also in $S(Q)$ and the density of states). The first peak of $S(Q)$ in amorphous metals is often much narrower than for covalent materials because the effective random atomic packing in metals means that the hard-sphere diameter, and multiples of it, are the dominant features in the real-space radial distribution function. Hence, to a first approximation, a single Fourier component, with wavevector Q_p, dominates $S(Q)$ (see sect. 3.5.1), and this takes on the significance of the reciprocal lattice vector in crystalline materials.

It is well established (see e.g. Mott and Davis 1979) that for monovalent metals $2k_F$ lies slightly below, and for divalent metals slightly beyond, Q_p. Nagel and Tauc (1975) argued that on alloying of elements of higher valence with a monatomic metal (and in this respect transition metals are regarded as being monatomic), the position corresponding to $2k_F$ would move through the first peak in $S(Q)$, and, at the composition corresponding to $2k_F = Q_p$, E_F would lie at the centre of the pseudo-gap.

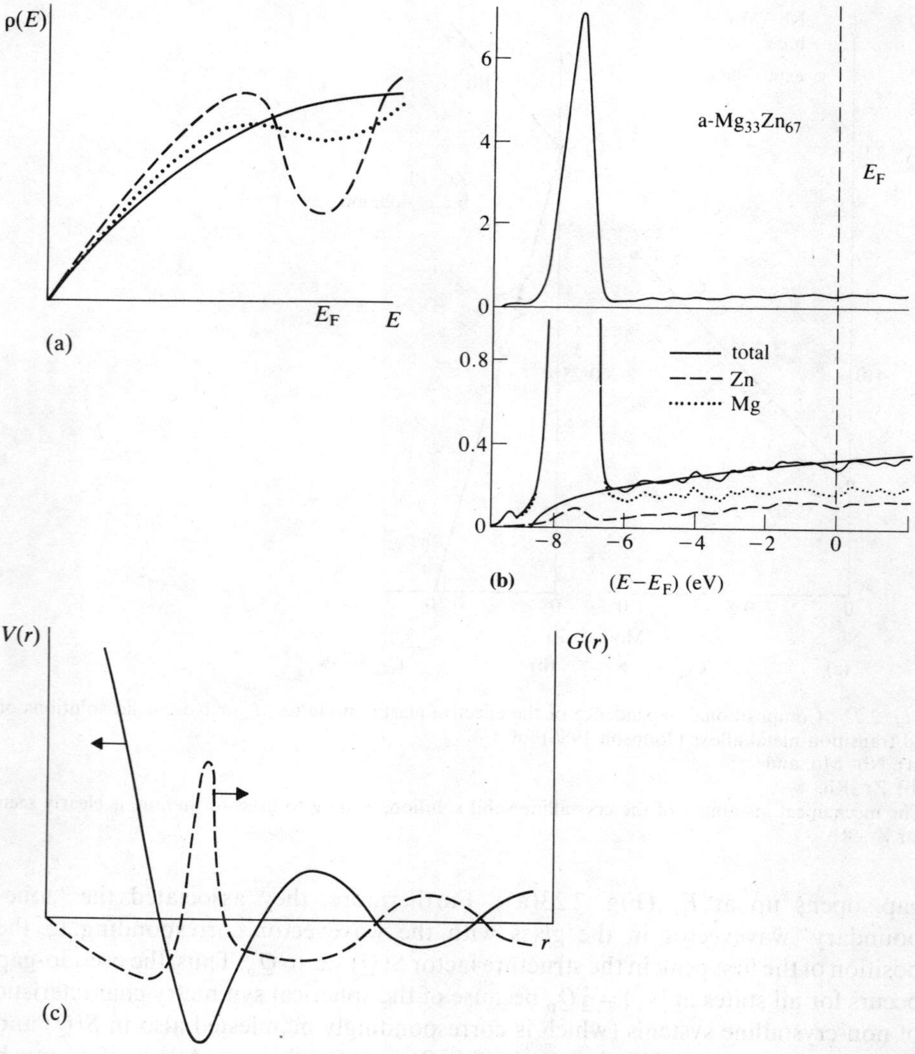

Fig. 2.23 Electronic structure criteria for metallic glass formation.
(a) Schematic illustration of the Nagel–Tauc (1975) model, in which E_F is postulated to lie at or near the minimum in the DOS, $\rho(E)$. The solid curve is the free-electron parabolic dependence of $\rho(E)$, the dashed curve represents the DOS for an isotropic (i.e. glassy) system, and the dotted curve illustrates the changes in $\rho(E)$ expected if the system becomes anisotropic (i.e. starts to crystallize).
(b) Calculated electronic DOS for glassy $Mg_{33}Zn_{67}$ (Hafner *et al.* 1988). It can be seen that there is only a very small dip in the DOS at E_F, in contrast to (a).
(c) The Hafner (1981) model, in which maxima in the atomic potential $V(r)$ of a metallic system coincide with minima in the (reduced) RDF, and vice versa, for optimum glass-forming ability.

Stability against crystallization in this model is supposed to arise through an electronic effect: if a fluctuation towards crystalline order should occur, this would tend to destroy the isotropy of $S(Q)$, thereby causing the pseudo-gap to become filled; thus, the electronic energy would be raised and the ordering fluctuation destabilized (see Fig. 2.23(a)).

Appealing though this theory is, experimental verification has been lacking. The s-like character of states near E_F in, say, $Pd_{0.8}Si_{0.2}$ has not been verified by photoemission experiments, nor have pseudo-gaps in the density of states been found experimentally (see e.g. Oelhofen 1983). *Ab initio* calculations of the electronic (and atomic) structure of simple metallic glass-forming systems, e.g. Mg–Zn, show the existence of only a very small dip in the density of states at E_F (Fig. 2.23(b)), a vestige of the Nagel–Tauc effect (Hafner *et al.* 1988). Nevertheless, the importance of electronic effects in determining glass formation and stability has been retained in those theories (see e.g. Hafner 1981) which ascribe relative stability to those materials having minima in the real-space dependence of the pair potential coinciding with peaks in the RDF (Fig. 2.23(c)). The pair potentials can be calculated accurately from pseudopotential theory, at least for simple metallic systems, using realistic values of atomic radii and taking account of the variation in size on alloying (e.g. Ca effectively contracts when alloyed with Al, but expands when alloyed with Li); in this way *quantitative* assessments of glass stability can be made.

2.5.6 Summary

In discussing those factors which favour glass formation, an alternative approach is to consider those factors which *inhibit crystallization* on cooling a melt. Thus, the presence of a deep eutectic (i.e. lowering of the melting temperature) at a given composition for a liquid alloy will reduce the rate of crystallite nucleation and growth at that composition, since the liquid will be then less supercooled at the glass-transition temperature. In addition, many covalent glass-forming materials are found to have a large number of stable crystalline polymorphs (which have comparable free energies); crystallization into a single crystalline polymorph on cooling the melt becomes therefore less likely than glass formation because the structural unit involved can be frozen randomly into a variety of conformations with little extra cost in strain energy. Such energetic considerations also underlie the Zachariasen rules for glass formation in oxide systems. However, another aspect of the Zachariasen rules (for oxides) is the flexibility inherent in a random network, which is a consequence of the particular coordination of the oxygen atoms. In general, it is found that glass formation is easiest for those covalent materials which have a low average coordination number, or in other words which have a principal constituent which itself has a low coordination number. A reason for this may have to do with the fact that the excess free volume characteristic of a liquid, and which is also present in a glass (but not in a crystal), can be distributed in a continuous fashion throughout the bulk in a random network by virtue of the flexibility afforded by the low-coordination atoms; otherwise, an amorphous or polycrystalline aggregate can form, with all the strain and free volume concentrated at the grain boundaries.

Problems

2.1 Show that the Doolittle relation ([2.20]) is equivalent to the Vogel–Tammann–Fulcher law ([2.8]), if the assumption is made that the thermal expansion is linear in temperature. Show further that insertion of the relation for the free-volume temperature dependence into the Doolittle equation gives an equation of the Williams–Landel–Ferry type ([2.21]).

2.2 Deduce the dependence of the glass-transition temperature on the cooling rate (see A. J. Kovacs (1963) *Adv. Polym. Sci.* **3**, 394 for details).

(a) If V is the volume of the glass at time t corresponding to the fictive temperature T_f, and V_∞ is the equilibrium volume at temperature T, then the time dependence of the volume change can be written as:

$$\left(\frac{\mathrm{d}V}{\mathrm{d}t}\right)_p + \frac{(V - V_\infty)}{\tau} = -\alpha_{Tg} q V_\infty \qquad [2.1P]$$

where τ is the relaxation time, and the term on the right-hand side represents the time-dependent volume change upon cooling at a constant rate $q = -\mathrm{d}T/\mathrm{d}t$, where $\alpha_{Tg} = (1/V)(\mathrm{d}V/\mathrm{d}T)$. Show that [2.1P] can be transformed into

$$\frac{\mathrm{d}\delta}{\mathrm{d}t} + \frac{\delta}{\tau} = \Delta\alpha_T q \qquad [2.2P]$$

where $\delta = (V - V_\infty)/V_\infty$, and $\Delta\alpha_T = (\alpha_T - \alpha_{Tg})$ with $\alpha_T = (1/V_\infty)(\mathrm{d}V_\infty/\mathrm{d}T)$. The relaxation time can be written as:

$$\tau = A \exp(-K_1 T - K_2 T_f) \qquad [2.3P]$$

where K_1 and K_2 are constants, and the apparent activation energy is (over a small temperature range)

$$E_A = RT^2(K_1 + K_2) \qquad [2.4P]$$

(b) By using the expression for τ in the form $(\mathrm{d}\ln\tau)/\mathrm{d}T = -\theta$ where $\theta = (K_1 + K_2)$ to give $\tau = (1/\theta q)\exp(\theta q t)$, hence solve [2.2P] to give

$$\delta = -\frac{\Delta\alpha_T}{\theta} \mathrm{e}^u E_i(-u) \qquad [2.5P]$$

where $u = \exp(-\theta q t)$, and $E_i(-u) = \int_\infty^u \mathrm{e}^x \mathrm{d}x/x$ which can be approximated as $\lim_{u\to 0}\{E_i(-u)\} \simeq \ln(\gamma u)$ where γ is the Euler constant ($=1.781$). By considering the asymptotic linear form for δ, and calling T_g the temperature at which $t = (1/\theta q)\ln q$, hence show that

$$\frac{\mathrm{d}T_g}{\mathrm{d}\ln q} = \frac{1}{\theta} \qquad [2.6P]$$

Hence derive [2.1] and show that the constant $C(\simeq 3 \times 10^{-5})$ can be obtained using reasonable physical parameters.

Bibliography

C. A. Angell, J. H. R. Clarke and L. V. Woodcock (1981), in *Advances in Chemical Physics*, vol. 48. I. Prigogine and S. A. Rice (eds) (Wiley), p. 397.

S. A. Brawer (1983), *Relaxation in viscous liquids and glasses* (Am. Ceram. Soc.)
G. S. Grest and M. H. Cohen (1981), in *Advances in Chemical Physics*, vol. 48, I. Prigogine and S. A. Rice (eds) (Wiley), p. 455.
J. Jäckle (1986), *Rep. Progr. Phys.* **49**, 171.
J. D. Mackenzie (ed.) (1960), *Modern Aspects of the Vitreous State* (Butterworth).
H. Rawson (1967), *Inorganic Glass-Forming Systems* (Academic).
J. Wong and C. A. Angell (1976), *Glass: Structure by Spectroscopy* (Dekker).
R. Zallen (1983), *The Physics of Amorphous Solids* (Wiley).

3 Structure

3.1 Introduction

Knowledge of the structural arrangement of atoms of a solid substance is an essential prerequisite to a detailed understanding of other physical or chemical properties; this is as true for amorphous solids as for crystalline materials. Determination of the structure of a crystalline solid is made straightforward by the need only to solve the structure within the 'unit cell', containing relatively few atoms in most cases and which is the fundamental 'building unit' of the structure; the structure of the crystal as a whole is then generated by repeating in a periodic fashion the position of the unit cell in space. Such a procedure is impossible for a non-periodic amorphous solid, for which the unit cell may be regarded as being infinite in extent.

The innocent reader at this stage may well then ask: 'why attempt to study the structure of amorphous materials at all (and devote a chapter of this book to it) when, almost by definition, the structure is random and chaotic, and hence structural information, if any, can only be obtained in the form of statistical averages?' The operational definition of an amorphous solid, however, is merely that the structure is *non-periodic* (i.e. non-crystalline), not that it is necessarily truly random in a statistical sense. We shall find in many cases that the structure of many amorphous solids is in fact *non-random*, at least on certain length scales, e.g. there may be a considerable degree of local ordering despite the overall lack of periodicity. We have seen already in passing an example of this in section 2.5.1 where, for example, the structure of silica and silicates in the glassy state was anticipated as being composed of SiO_4 tetrahedra connected together in some random manner. The presence of discrete SiO_4 tetrahedra, identifiable by a number of techniques as we shall see later, certainly signifies the lack of complete randomness in the structure; a similar occurrence of local order is to be found whenever chemical bonding takes place between atoms in the solid, whether formally as directed covalent bonds or as a weaker component in addition to non-directional forces, such as those involved in metallic or ionic systems. It is readily realized, therefore, that this leads to a very large number of amorphous solids for which the question 'what is the structure?' is very pertinent and important.

This question can only be addressed when the length scale under consideration is stipulated. Quite obviously the determination of atomic positions (relative to an arbitrary origin atom) is generally altogether a different proposition from a discussion of possible phase separation in a glass, and the two problems necessitate the use of different experimental techniques. For this reason, we somewhat arbitrarily divide this chapter into two main sections, the first dealing with 'microscopic' structure (or the disposition of individual atoms or groups of atoms at a local level), and the second concerning itself with the 'macroscopic' structure of amorphous systems (such as phase separation). To a certain degree the distinction, particularly concerning phase separation, is semantic.

We therefore arbitrarily define a length scale which separates *microscopic* structure from *macroscopic* structure, and we take this to be in the region of 100 Å. This division has an added advantage in that completely different and mutually exclusive experimental techniques are used as structural probes in the two domains, and can therefore be discussed separately; e.g. conventional X-ray diffraction is only sensitive to structural variations of the order of a few ångströms, i.e. in the microscopic region, whereas optical microscopy, say, is only useful for detection of structural inhomogeneities on the scale of thousands of ångströms or more, which are in the macroscopic domain. We will also find it convenient to subdivide the section on microscopic structure into two parts, one dealing with 'short-range order' (of the order of a few ångströms) and the other dealing with 'medium-range order' (on length scales of the order of 5–20 Å); there is no 'long-range order' of an amorphous solid by definition. This is very much an operational definition, in so far as again different techniques are in general used to probe the structure at these two length scales.

We begin with a discussion of the structure of amorphous materials at a microscopic level because this is the more fundamental, and a knowledge of how atoms or groups of atoms are arranged in an amorphous system can often be used to infer the nature of medium-range structure.

I MICROSCOPIC STRUCTURE

We begin this section with a discussion of the techniques that can be employed to obtain structural information for solids when there is no periodicity in atomic positions to aid in solving the structure. In so doing, we introduce a variety of concepts and mathematical functions which will help us to describe the structure of a non-crystalline solid. In addition, we will consider first those methods which give information on short-range structure, that which extends over a few ångströms or so, and follow it with a discussion of techniques which probe the structure out to further distances, so-called medium-range order. These have been reviewed by Wright (1974), Wright and Leadbetter (1976), Wagner (1978), Gaskell (1979a) and Waseda (1980).

3.2 Experimental techniques

3.2.1 Definitions

It is perhaps helpful at the outset to introduce some functions which will be used to describe the microscopic structure of non-crystalline materials, particularly at short-range distances (2–5 Å). Perhaps the most widely used function is the radial distribution function (RDF), which is important since it can be obtained directly from diffraction experiments (see sect. 3.2.2). The RDF, $J(r)$, is defined as the number of atoms lying at distances between r and $r + dr$ from the centre of an arbitrary origin atom. This can be written as:

$$J(r) = 4\pi r^2 \rho(r) \qquad [3.1]$$

where the density function $\rho(r)$ is an atomic pair correlation function, being equal to *zero* at values of r less than the average nearest-neighbour interatomic separation r_1 (where there can be no atom–atom correlations, by definition), and equal to the average value of density, ρ°, at very large values of r where the material becomes homogeneous. In between these two limits, $\rho(r)$ for an amorphous solid will exhibit an oscillatory behaviour, with the peaks in the probability function representing average interatomic separations (see Fig. 3.1). Correspondingly, the RDF will also exhibit oscillations about the average density parabola given by the curve $4\pi r^2 \rho^\circ$, and $J(r)$ is likewise zero below the first peak lying at r_1.

The importance of the RDF lies principally in the fact that, by its definition ([3.1]), the *area* under a given peak gives the effective coordination number for that particular shell of atoms (see Fig. 3.1). The first peak is generally sharp; its mean square width, σ_1^2, ideally has only a thermal disorder component σ_T^2 for crystals (resulting from the thermal fluctuations in the bonds), but amorphous materials may also have an additional broadening contribution resulting from static (topological) disorder in the bond length, σ_D^2, such that

$$\sigma_{am}^2 = \sigma_T^2 + \sigma_D^2 \qquad [3.2]$$

The position of the first peak in the RDF gives a value for the average nearest-neighbour bond length r_1, and similarly the position of the second peak gives the next-nearest-neighbour distance r_2 (Fig. 3.2); a knowledge of both immediately yields a value for the bond angle θ, given by

$$\theta = 2 \sin^{-1}\left(\frac{r_2}{2r_1}\right) \qquad [3.3]$$

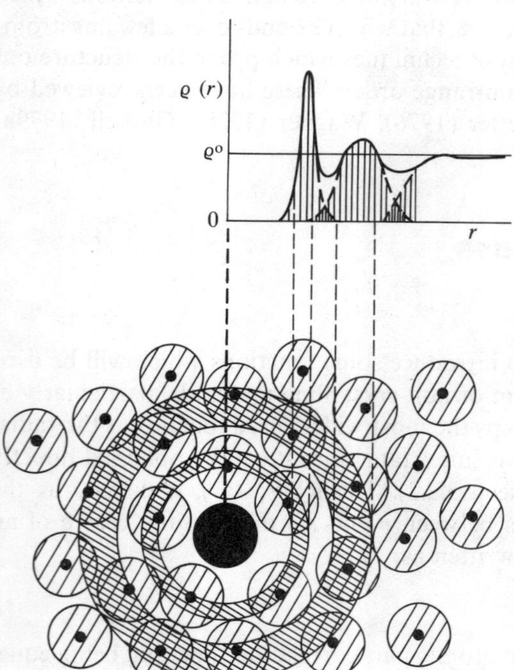

Fig. 3.1 Schematic illustration of the structural origin of certain features in the density function $\rho(r)$ for an amorphous solid.

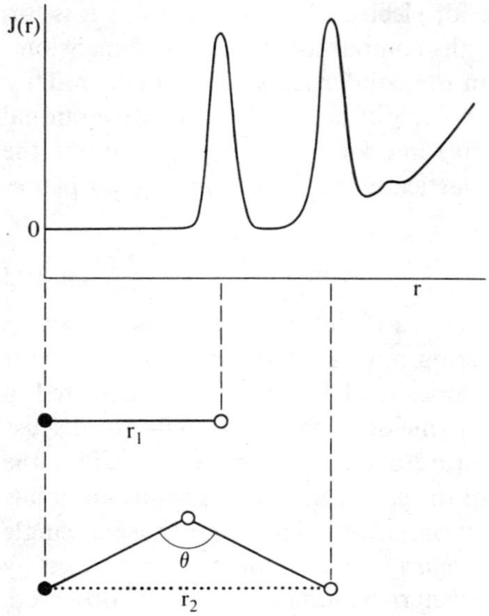

Fig. 3.2 Schematic illustration showing the relationship between short-range structural parameters: first and second nearest-neighbour bond lengths, r_1 and r_2, and bond angle θ for first and second shells, for the first two peaks in the RDF.

Note that the second peak is generally wider than the first for covalent amorphous solids; this reflects a contribution due to the presence of a static variation in the bond angles (typically $\pm 10\%$) in addition to that for thermal disorder (see [3.2]). In crystals, on the other hand, *no* static bond-angle distortion exists in general, and so the widths of the first two peaks in the RDF should be equal, particularly since there the second peak arises solely from next nearest-neighbour atomic pair correlations. However, this is not generally true for amorphous materials where higher-order correlations also contribute to the area under the second peak, thereby also contributing to the overall width; the influence of higher-order contributions can be seen in the fact that $J(r)$ does not decrease to zero on the high-r side of the second peak. The proportion of higher-order pair correlations contributing to a given peak in the RDF of amorphous materials increases significantly with increasing r, making it impossible to interpret simply even the third and fourth peaks. This ambiguity therefore makes structural modelling studies essential (see sect. 3.5).

Finally, note that the RDF is a *one-dimensional representation of a three-dimensional structure*, and as such is spatially averaged (which, strictly, is valid only if the material is isotropic). It can therefore carry only a limited amount of structural information, and this is another reason why modelling studies are so valuable.

3.2.2 Diffraction

It is a fundamental property of all waves that they diffract on meeting an obstacle, and this effect is most pronounced when the size of the obstacle is comparable to the wavelength of the wave. This is as true for electrons and neutrons as it is for, say, X-rays, and since all can have wavelengths comparable to atomic dimensions, it is expected that diffraction from atoms in the condensed state will occur. This, of course, is the basis of the technique of X-ray diffraction used in conventional single crystal studies for which, if constructive interference is to occur among the outgoing beams of X-rays of wavelength λ, elastically scattered from regular planes of atoms (Fig. 3.3), the Bragg condition

$$2d \sin \theta = n\lambda \qquad [3.4]$$

is obeyed. Here d is the spacing of planes and n is an integer. Thus diffracted beams of, say, X-rays are observed only for scattering angles 2θ and wavelengths λ for which [3.4] is satisfied; this gives rise to a series of sharp spots being detected in a plane perpendicular to the monochromatic incident beam. We will not discuss further the subtleties of single crystal structure determination, which utilize the intensity variations as well as the position of the spots, since amorphous materials do *not* produce an X-ray or electron diffraction pattern similar to those of single crystals (nor even the pattern of concentric *sharp* circles of intensity produced by *randomly* oriented crystals) but instead characteristic *diffuse* haloes are observed. In this case, where is no periodic structure, Bragg's law ([3.4]) is inappropriate, and we must derive afresh the conditions for diffraction to occur from a random structure.

We consider first, for simplicity, the case of X-ray diffraction by an amorphous sample; there are several good reviews of the application of scattering techniques (particularly by X-rays) in the literature (Warren 1969, Wright 1974, Wright and Leadbetter 1976, Wagner 1978), and we will primarily follow the treatment of Warren here. X-rays are scattered by the electrons present in the atoms of the sample (an important point to which we will return later), and although a full quantum-mechanical calculation should be used, a simpler classical treatment gives equivalent results.

The scattering from an electron, regarded as a point charge, is described by the (classical) Thomson equation, for which the differential scattering cross-section

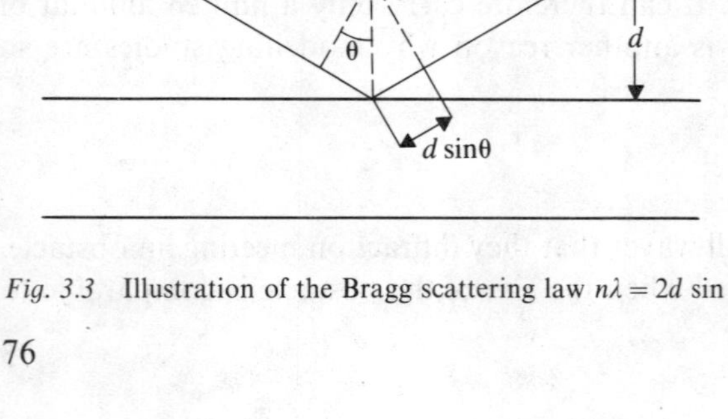

Fig. 3.3 Illustration of the Bragg scattering law $n\lambda = 2d \sin \theta$.

in SI units is given by:

$$\frac{d\sigma}{d\Omega}=\left(\frac{e^2}{4\pi\varepsilon_0 m_e c^2}\right)^2\frac{(1+\cos^2\theta)}{2} \qquad [3.5]$$

This describes the 'unmodified' or 'coherent' scattering where the photon momentum (and hence the wavelength) is conserved (in contrast to Compton or 'modified' scattering where momentum is exchanged with the electron); the factor $\frac{1}{2}(1+\cos^2\theta)$ arises if the incident beam is unpolarized. Thus, incident radiation of intensity I_o falling on a point electron at O and scattered through an angle 2θ (Fig. 3.4(a)) has an intensity at a point P, distance R from O, given by

$$I=\frac{I_o}{R^2}\left(\frac{e^2}{4\pi\varepsilon_o m_e c^2}\right)^2\frac{(1+\cos^2\theta)}{2} \qquad [3.6]$$

Scattered intensities in units of the right-hand side of [3.6] are termed 'electron units' (e.u.) of intensity, and these units will be adopted in the following discussion.

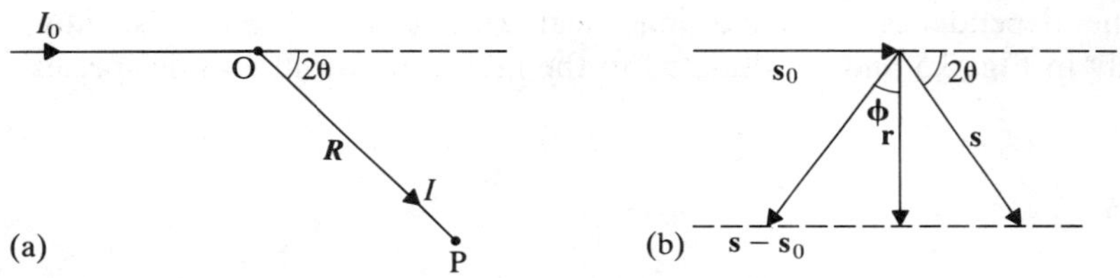

Fig. 3.4 (a) Scattering geometry for a point charge; (b) relation between scattering vectors.

We next consider the scattering from an electron whose charge is distributed in a certain volume dV as $\rho\, dV$, rather than being confined to a point; the scattering is then $\rho\, dV$ times the amplitude of the classical scattering from a single electron. The path differences between the various scattered rays from different parts of the charge distribution at **r** from an arbitrary origin must be considered, giving the following expression for the scattering intensity in e.u. as:

$$I_{eu}=f_e f_e^* \qquad [3.7]$$

where the 'scattering factor' f_e is given by:

$$f_e=\int \exp\left[\frac{2\pi i}{\lambda}(\boldsymbol{s}-\boldsymbol{s}_o)\cdot\boldsymbol{r}\right]\rho(\boldsymbol{r})\,dV \qquad [3.8]$$

The relation of the scattering vectors $\boldsymbol{s}$ and $\boldsymbol{s}_o$ to the vector $\boldsymbol{r}$ is shown in Fig. 3.4(b), from which it can readily be deduced that $(\boldsymbol{s}-\boldsymbol{s}_o)\cdot\boldsymbol{r}=2r\sin\theta\cos\phi$. We also introduce the widely used abbreviation (termed the scattering vector) given by:

$$k=\frac{4\pi\sin\theta}{\lambda} \qquad [3.9]$$

(Note that k is conventionally used to represent the scattering vector in X-ray diffraction, whereas Q is normally reserved for neutron diffraction.) If we assume spherical symmetry (true for an eventual summation over closed shells of electrons) then $\rho(\boldsymbol{r}) = \rho(r)$, and the scattering factor per electron can be written as

$$f_e = \int_{r=0}^{\infty} \int_{\phi=0}^{\pi} \exp[ikr \cos \phi] \rho(r) 2\pi r^2 \sin \phi \, d\phi \, dr$$

Integration over ϕ gives:

$$f_e = \int_0^{\infty} 4\pi r^2 \rho(r) \frac{\sin kr}{kr} \, dr \qquad [3.10]$$

For an atom containing several electrons, the 'atomic scattering factor' or 'form factor' is simply the sum of the individual amplitudes:

$$f = \sum_n f_{en} = \sum_n \int_0^{\infty} 4\pi r^2 \rho_n(r) \frac{\sin kr}{kr} \, dr \qquad [3.11]$$

and the total scattered intensity is $I_{eu} = ff^*$. Clearly, the quantity $\Sigma_n \int_0^{\infty} 4\pi r^2 \rho_n(r) \, dr$ is simply the total number of electrons in the atom (i.e. the atomic number Z); and since $\sin kr/kr \to 1$ for small kr, f tends to the value Z as k (or $\sin \theta/\lambda$) tends to zero. The dependence of the atomic scattering factor f on k is shown schematically in Fig. 3.5 and is tabulated in the literature for all atomic species.

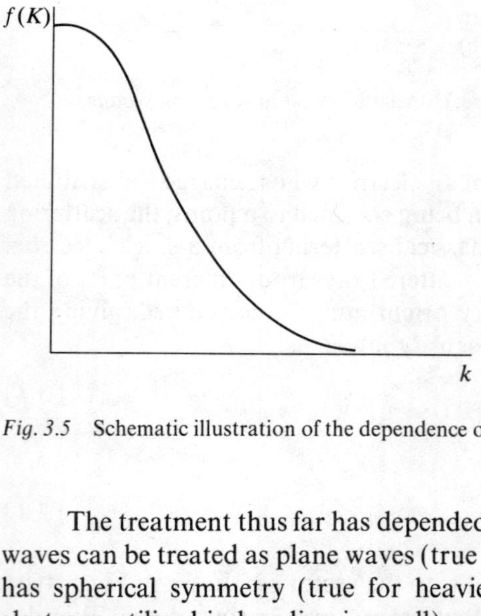

Fig. 3.5 Schematic illustration of the dependence of the X-ray atomic form factor on scattering vector.

The treatment thus far has depended on three assumptions: that the scattered waves can be treated as plane waves (true if $|\boldsymbol{R}| \gg |\boldsymbol{r}|$), that the electron distribution has spherical symmetry (true for heavier atoms where the fraction of valence electrons utilized in bonding is small), and finally that the X-ray wavelength used is far removed from the wavelengths corresponding to absorption edges of the atom. The latter case can be allowed for by the introduction of 'dispersion corrections' to the atomic scattering factor for wavelengths near an absorption edge, $f = f_o + \Delta f' + i\Delta f''$. These corrections act as the basis of the technique utilizing 'anomalous dispersion' which will be discussed later in this section.

For an assembly of atoms, the scattered intensity (in electron units) is given by the summation of the scattered amplitude from each atom m, having position

$\boldsymbol{r}_{\rm m}$, multiplied by its complex conjugate:

$$I_{\rm eu} = \sum_{\rm m} f_{\rm m} \exp\left[\frac{2\pi i}{\lambda}(\boldsymbol{s}-\boldsymbol{s}_{\rm o})\cdot\boldsymbol{r}_{\rm m}\right] \sum_{\rm m} f_{\rm n} \exp\left[-\frac{2\pi i}{\lambda}(\boldsymbol{s}-\boldsymbol{s}_{\rm o})\cdot\boldsymbol{r}_{\rm n}\right] \qquad [3.12]$$

where we have assumed the atomic scattering factors to be real. This expression is true for *all* materials, whether crystalline or amorphous. In the case of scattering from crystals, where the atoms are arranged in a periodic manner such that the position vectors of the atoms are simply related through a lattice vector $\boldsymbol{r}_{\rm o}(=\boldsymbol{r}_{\rm m}-\boldsymbol{r}_{\rm n})$, the summations in [3.12] form a geometric progression which can be evaluated. For amorphous materials, however, no such simple relationship exists between the position vectors, and the summations must be left in the analysis. Equation [3.12] can be written more succinctly by introducing the difference vector $\boldsymbol{r}_{\rm mn}=\boldsymbol{r}_{\rm m}-\boldsymbol{r}_{\rm n}$, such that

$$I_{\rm eu} = \sum_{\rm m}\sum_{\rm n} f_{\rm m} f_{\rm n} \exp\left[\frac{2\pi i}{\lambda}(\boldsymbol{s}-\boldsymbol{s}_{\rm o})\cdot\boldsymbol{r}_{\rm mn}\right] \qquad [3.13]$$

A simplifying assumption that is usually valid for amorphous solids is that the material is isotropic, so that the vector $\boldsymbol{r}_{\rm mn}$ may adopt all orientations with equal probability, or equivalently can take all positions on the surface of the sphere shown in Fig. 3.6. In this case the average of the exponential term in [3.13] may be written as:

$$\left\langle \exp\left[\frac{2\pi i}{\lambda}(\boldsymbol{s}-\boldsymbol{s}_{\rm o})\cdot\boldsymbol{r}_{\rm mn}\right]\right\rangle = \frac{1}{4\pi r_{\rm mn}^2}\int_{\phi=\rm o}^{\pi} \exp[ikr_{\rm mn}\cos\phi]2\pi r_{\rm mn}^2 \sin\phi \, {\rm d}\phi$$

$$= \frac{\sin kr_{\rm mn}}{kr_{\rm mn}} \qquad [3.14]$$

Substituting this result in [3.13] yields the 'Debye equation' for a random array of scattering atoms

$$I_{\rm eu} = \sum_{\rm m}\sum_{\rm n} f_{\rm m} f_{\rm n} \frac{\sin kr_{\rm mn}}{kr_{\rm mn}} \qquad [3.15]$$

To proceed further, one must take into account the type of atom within the sample; the analysis is considerably simplified if it is assumed, as we do at first,

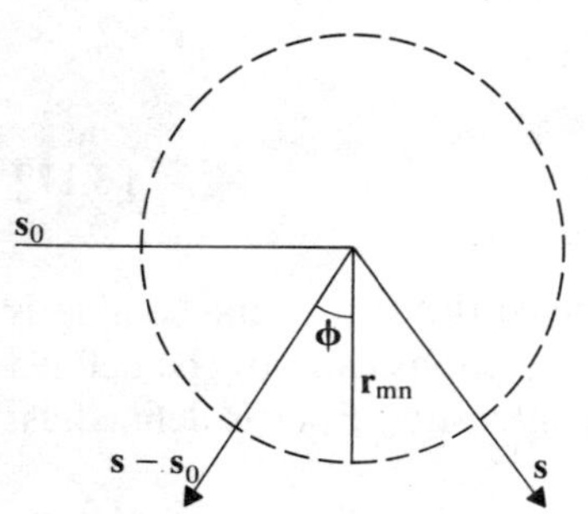

Fig. 3.6 Illustration of the positions that $r_{\rm mn}$ may take for an isotropic material.

that the material is monatomic. Rewriting [3.15] as

$$I_{eu} = \sum_{m} f^2 + \sum_{m} \sum_{n \neq m} f^2 \frac{\sin kr_{mn}}{kr_{mn}}$$

and introducing the density function $\rho_m(r_{mn})$ (for the origin atom m) gives

$$I_{eu} = \sum_{m} f^2 + \sum_{m} f^2 \int \rho_m(r_{mn}) \frac{\sin kr_{mn}}{kr_{mn}} dV_n$$

where the summation over n has been replaced by the integral over the sample volume and spherical symmetry has been assumed as before. The (microscopic) density function averaged over all atoms, m, in the sample can be written as $\rho(r) = \langle \rho_m(r_{mn}) \rangle$, and if the (macroscopic) average density is written as ρ^o, then adding and subtracting a term in ρ^o yields:

$$I_{eu} = \sum_{m} f^2 + \sum_{m} f^2 \int 4\pi r^2 [\rho(r) - \rho^o] \frac{\sin kr}{kr} dr + \sum_{m} f^2 \int 4\pi r^2 \rho^o \frac{\sin kr}{kr} dr \quad [3.16]$$

The summation over m yields simply N, the total number of atoms in the sample.

Amorphous materials possess no long-range order, and an equivalent way of expressing this is to note that the density function $\rho(r)$ tends to ρ^o at large r (see Fig. 3.1). Thus, the quantity $[\rho(r) - \rho^o]$ tends to zero for distances greater than a few primary atom separations, and hence the integral in the second term of [3.16] is dominated by scattering predominantly from close scattering centres, and therefore the Fraunhofer approximation assumed so far holds. This is not so for the third term in [3.16] which involves interactions between distant points in the sample, and Fresnel diffraction theory must be used. It can be shown (see e.g. Warren 1969) that the third term represents small-angle scattering due to the finite sample size, which is generally limited to scattering vectors $k \lesssim 2\pi/R$ where R is the specimen size. For R of the order of millimetres and λ a few ångströms, this corresponds to scattering at such extremely small angles that the primary beam completely obscures it. For sample geometries such that this scattering is observed at finite, but small angles, it can be calculated and subtracted from the total observed intensity to leave only that scattering intensity resulting from atomic correlations.

These considerations transform [3.16] into:

$$I_{eu} = Nf^2 + Nf^2 \int_0^\infty 4\pi r^2 [\rho(r) - \rho^o] \frac{\sin kr}{kr} dr \quad [3.17]$$

where the upper limit to the integration is valid since the size of the sample is much greater than atomic dimensions. Two further functions can now be defined in order to simplify [3.17]. The 'reduced scattering intensity', $F(k)$, is defined as:

$$F(k) = k \left[\frac{I_{eu}/N - f^2}{f^2} \right] \quad [3.18]$$

and the 'reduced radial distribution function', $G(r)$, is defined as:

$$G(r) = 4\pi r[\rho(r) - \rho^{\circ}] = J(r)/r - 4\pi r\rho^{\circ} \qquad [3.19]$$

Both functions are termed 'reduced' because they oscillate about zero instead of being an increasing or decreasing function of k or r. (Note that in the region below the first peak in the RDF, where $J(r) = 0$, $G(r)$ decreases linearly as $-4\pi r\rho^{\circ}$.) Thus [3.17] can be written in the form:

$$F(k) = \int_0^{\infty} G(r) \sin kr \, \mathrm{d}r \qquad [3.20]$$

This equation is important since it links a quantity which is directly obtainable from experiment, namely $F(k)$, with a function which describes the real-space structure of the amorphous solid, i.e. $G(r)$. The form of [3.20] is immediately suggestive of a Fourier transform, and thus the equation can be inverted:

$$G(r) = \frac{2}{\pi} \int_0^{\infty} F(k) \sin kr \, \mathrm{d}k \qquad [3.21]$$

Note that the sine function which arises in [3.20] results solely from the analysis for the scattering intensity (cf. the Debye equation, [3.15]), whereas that which appears in [3.21] arises from the Fourier transform procedure. Note too that the equation by which the real-space correlation function is obtained from experimental data ([3.21]) requires the data for a very large range of values of k (ideally infinite in extent); the problems that arise in practice when this condition is not met will be discussed later.

For illustration, various real-space distribution functions and their reciprocal-space analogues are shown for a-Ge in Fig. 3.7, together with their interrelationships and definitions. It should be noted that these curves strictly refer to X-ray scattering; this is particularly true for the total intensity curve ($I(k)$), where the background curve ($f^2(k)$) is k-dependent, in contrast to (non-magnetic) neutron diffraction as will be seen later.

The reduced scattering intensity, $F(k)$, is important as the precursor to obtaining the RDF (Fig. 3.7), but is often informative in its own right since subtle structural differences between samples (say, prepared under different conditions) may often be more readily discerned in a comparison of the data in reciprocal space than by comparing the RDFs in real space. This is particularly true since $F(k)$ is weighted for high k-values, as can be seen from the expression

$$F(k) = k[S(k) - 1] \qquad [3.22]$$

where the 'structure factor', $S(k)$, is defined as

$$S(k) = I_{\mathrm{eu}}(k)/Nf^2 \qquad [3.23]$$

i.e. it is independent of the type of scattering radiation used (because of the normalization by the factor f^2) and it oscillates about an average value of unity. Further, the curves are obtained directly from the measured data and hence do not suffer from spurious features that can be introduced by the Fourier transform process (see later). A major disadvantage in the examination of correlation curves

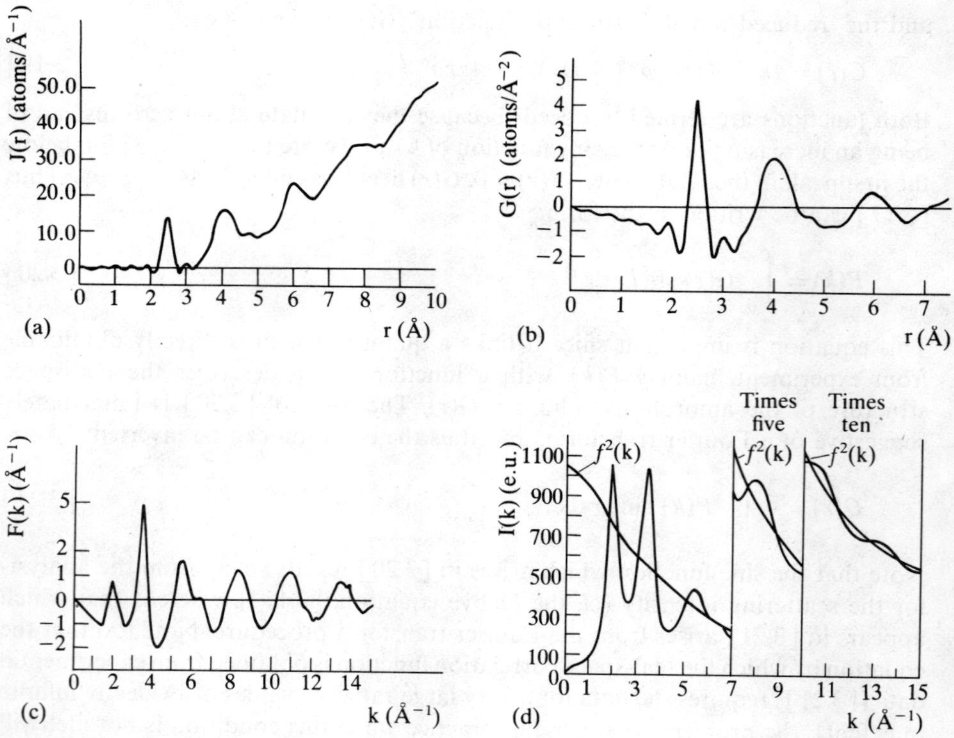

Fig. 3.7 Various structural correlation functions in real and reciprocal space for sputtered a-Ge films deposited on to substrates at 150°C (Temkin *et al.* 1973).
(a) RDF, $J(r) = 4\pi r^2 \rho(r)$;
(b) reduced RDF, $G(r) = (J(r)/r - 4\pi r \rho^0)$;
(c) reduced scattering intensity, $F(k) = [(I/N - f^2)/f^2]k$;
(d) scattering intensity, $I(k)$.

in reciprocal space, however, is that it is difficult to identify real-space structural correlations simply by inspection of the curves. This can be seen more clearly by a consideration of how the $F(k)$ curve is built up.

Consider first the case of a random system in which only nearest-neighbour correlations are well defined; beyond the first shell there are no correlations. This would correspond to a single peak in $G(r)$ at r_1, having a width σ_1^2 resulting in general from both thermal and static disorder contributions. Assuming this peak can be approximated by a Gaussian, the Fourier transform of $G(r)$ gives a damped, simple sinusoidal form for $F(k)$ or $S(k)$. The degree of damping is determined by σ_1^2, and the periodicity k_1 of the sinusoid by the nearest-neighbour separation r_1, ($k_1 = 2\pi/r_1$). If it is now assumed that a second well-defined shell of atoms of radius r_2 exists, then this would by itself give rise to a damped sinusoid, but having a *different* damping parameter and periodicity. Thus, at low values of k, there will be considerable mixing of these two contributions giving rise to a complex form for $F(k)$. However, as we have already seen, higher order shells usually have widths considerably larger than that of the first shell and so the attenuation of the contribution to $F(k)$ of the second and higher shells will be considerably greater than that of the first (since $\sigma_2^2 > \sigma_1^2$, etc.). Hence the behaviour of $F(k)$ at large values of k should solely be determined by the first peak in $G(r)$ and should be a simple sinusoid of period $2\pi/r_1$. Thus, if data can be taken to sufficiently large values of k *and* $F(k)$ is sinusoidal in this region, then a limited amount of structural information (namely the value for the bond length r_1) can be obtained by inspection from the period of the oscillations. It should be obvious to the reader that the first major peak in $F(k)$ therefore does *not* correspond in a simple manner in general to any real-space structural correlation, and it is incorrect to assume (as is sometimes done) that its position is uniquely determined by the nearest-neighbour separation. An exception to this case appears to be for transition-metal–metalloid glasses, where the total structure factor (or reduced scattering intensity) is found to exhibit a very sharp peak at $k \simeq 2$–3 Å^{-1} (Fig. 3.8(a)), whilst at large distances (greater than $\simeq 2r_1$) the RDF appears to exhibit a near-periodic oscillatory behaviour (Fig. 3.8(b)), reflecting the characteristic repeat distance of the hard-sphere diameter (and multiples of it) that occurs in a (quasi-)random packing of spheres and is a first approximation to the structure. In fact, if the first peak in $F(k)$ of, say, a-$Ni_{76}P_{24}$ is approximated by a gaussian (the dotted curve in Fig. 3.8(a)), on Fourier transformation the resulting real-space function, which is just a damped sinusoid, matches the experimental reduced RDF remarkably well at distances beyond $\simeq 6$ Å (Fig. 3.8(b)).

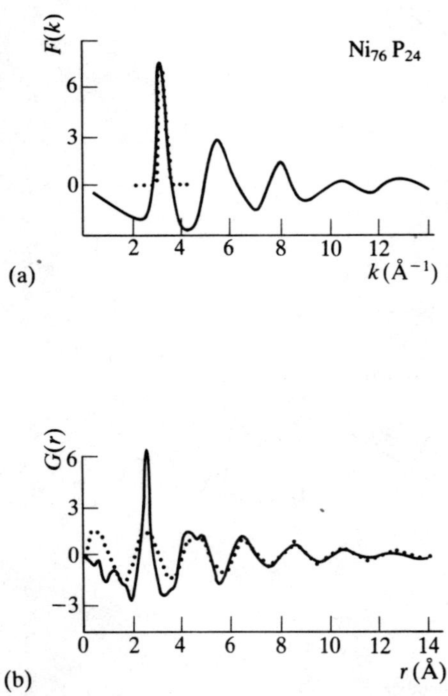

Fig. 3.8 (a) $F(k)$ for a-$Ni_{76}P_{24}$ (solid line) and gaussian fitted to first maximum (dotted line); (b) $G(r)$ (solid line) and Fourier transform of the gaussian peak shown in (a) (dotted curve) (Cargill 1975).

The total scattering intensity, $I(k)$, is the curve directly obtained from experiment after appropriate subtraction of incoherent contributions to the scattering such as Compton scattering. For the case of X-ray diffraction, $I(k)$ oscillates about the curve of the square of the atomic scattering amplitude f^2, as can be seen by inspection of [3.17] and Fig. 3.7(d), where f^2 varies strongly with scattering vector. This is in marked contrast to the case of neutron diffraction, where the scattering amplitude is *independent* of k (or Q) for closed electron shells, i.e. in the absence of *magnetic* scattering by unpaired electrons (see later).

The experimental arrangement employed in an X-ray diffraction study is shown schematically in Fig. 3.9. X-radiation from a source (e.g. MoK_α, $\lambda = 0.71$ Å or CuK_α, $\lambda = 1.54$ Å) is collimated by the first slit and may then be monochromatized by the first single crystal (or filter) monochromator and diffracted so as to strike the sample in either reflection or transmission mode. The diffracted radiation from the sample, scattered through 2θ, is then further collimated and possibly monochromatized before entering a detector (e.g. a scintillation counter). The scattered intensity is measured as a function of 2θ which can be converted via [3.9] into wavevector k.

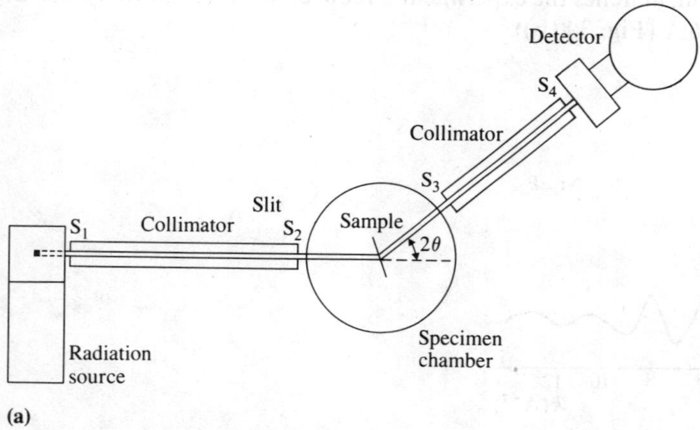

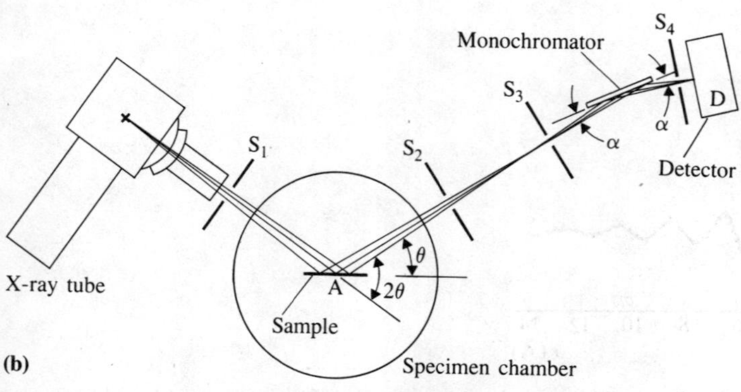

Fig. 3.9 Two experimental arrangements for performing X-ray diffraction measurements in (a) transmission and (b) reflection geometries (Sadoc and Wagner 1983).

The scattered intensity as measured contains contributions from the substrate of the sample (if present), noise and includes a polarization factor. These can be computed and eliminated, leaving the total unnormalized scattered intensity which is a sum of coherently and incoherently (Compton) scattered intensities, $I'(k) = I'_{coh}(k) + I'_{incoh}(k)$, of which we are only concerned with $I'_{coh}(k)$. The Compton contribution is difficult to calculate and hence to correct for; a better procedure is to attempt to eliminate it directly. An early method developed by Warren and Mavel (1965) is based on fluorescence; the diffracted beam strikes a thin foil which has an X-ray absorption edge at an energy only very slightly smaller than the photon energy of the incident beam, in which case coherently scattered radiation excites fluorescence which can be detected, while the modified, Compton down-scattered radiation cannot do so. Nowadays, energy dispersive detectors are used, which are capable of measuring only the elastically scattered intensity. The resulting coherent scattered intensity can be converted from arbitrary units to electron units by arranging that $I'(k)$ oscillates equally above and below the curve of $f^2(k)$ at large values of k, whereupon it is normalized automatically, or by using the more complicated Krogh-Moe–Norman method (see Warren 1969).

A considerable problem exists in the conversion of reciprocal-space data to real-space correlation functions via Fourier transforms, since the range of values of k accessible to experiment is strictly limited. This may be seen by reference to the defining equation ([3.9]) for $k = (4\pi \sin\theta)/\lambda$; the maximum value possible for k is when $\sin\theta = 1$. For MoK_α radiation ($\lambda = 0.71$ Å) this gives $k_{max} = 17.7$ Å^{-1}, but for CuK_α ($\lambda = 1.54$ Å), the maximum value of k is only $k_{max} = 8.2$ Å^{-1}. (Synchrotron radiation on the other hand, can give X-ray wavelengths down to $\simeq 0.1$ Å.) Transformation between reciprocal and real space is only perfect if $F(k)$ is known for an *infinite* range of k values (cf. [3.21]); in practice, this is obviously not the case and the data are terminated at some finite k_{max}. This is equivalent to $F(k)$ being multiplied by a 'modification function' $M(k)$, where $M(k) = 1$ for $k \leqslant k_{max}$ and 0 for $k > k_{max}$; i.e. the effective result is

$$G'(r) = \frac{2}{\pi}\int_0^{k_{max}} F(k) \sin kr \, dk = \frac{2}{\pi}\int_0^{\infty} F(k)M(k) \sin kr \, dk.$$

This leads to the introduction of 'termination errors' in the Fourier transform (i.e. in $G(r)$), which for the step-function form of $M(k)$ given above is equivalent to peaks in $G(r)$ being convoluted with the cosine transform of $M(k)$, i.e. the sinc function $(\sin rk_{max})/\pi r$. The effect of this is twofold: a loss in resolution is incurred since wavelengths in real space less than $2\pi/k_{max}$ are lost, resulting in a peak width $\simeq \pi/k_{max}$, and spurious 'termination' ripples at values $rk_{max} \simeq \pm((4n+1)\pi)/2$ are introduced (where n is an integer), which decay with increasing r.

The problem can be alleviated to a certain extent experimentally by taking data to high values of k_{max} (i.e. using smaller wavelengths), a situation more easily achievable in neutron scattering than in X-ray diffraction; this approach is particularly successful if, due to sufficiently broad peaks in the real-space correlation functions, the oscillations in $F(k)$ have died away at $k = k_{max}$, there being therefore no discontinuity in the experimental data at k_{max}. If this cannot be achieved satisfactorily, a variety of procedures may be used to minimize termination error.

The usual method employs a damping (or convergence) factor which multiplies the finitely ranged $F(k)$ in the Fourier transform, replacing a sharp discontinuity at k_{max} by a smoothly varying function. Two kinds of function are commonly used; the artificial temperature factor due to Bragg and West (1930):

$$M(k) = \exp\left(-\frac{2.3k^2}{k_{max}^2}\right) \qquad [3.24]$$

and that due to Lorch (1969), which is more favourable since it reduces the ripples more rapidly:

$$M(k) = \frac{k_{max}}{\pi k}\sin\left(\frac{\pi k}{k_{max}}\right) \qquad [3.25]$$

Both functions reduce the termination ripples but at the expense of worsening the real-space resolution.

Thus far, we have considered only scattering by X-rays. While this technique has many advantages including the ready availability of sources and ancillary equipment in the laboratory, there are a number of distinct disadvantages and drawbacks, among which number the low values of k_{max} achievable using common X-ray sources, the fact that intensity variations (in $I(k)$) are superimposed on a rapidly decreasing background of $f^2(k)$ (so data at high values of k are subject to considerable error) and that *electron* correlations are determined which are not the same as atomic correlations for low Z atoms.

An alternative technique, which resolves some of the drawbacks of X-ray diffraction but at the cost of necessitating a large facility as a source and needing large volumes ($\simeq 1\ cm^3$) of sample, is neutron scattering. Neutrons emerging from a nuclear reactor pile have wavelengths typically in the range $\lambda \simeq 0.1$ to 1 Å or larger, corresponding to energies ranging from 8 eV to 80 meV, the latter being 'thermal' neutrons produced by the use of a moderator. In practice, the minimum wavelengths for which useful fluxes occur are $\simeq 0.7$ Å (for an ambient source) and $\simeq 0.5$ Å (for a 'hot' source). In any scattering event, there is generally an energy transfer $\hbar\omega = E_o - E$ and a momentum transfer $\hbar \boldsymbol{Q} = \hbar \boldsymbol{k}_o - \hbar \boldsymbol{k}$ between the scattered wave and the scattering object. For the case of thermal neutrons, their characteristic energy is of the same order of magnitude as that of lattice vibrations in solids and so considerable energy transfers (inelastic events) are possible. X-rays have much higher characteristic energies and so the fractional energy changes due to phonon processes are very small (and hence $k \simeq k_o$), although different inelastic (Compton) processes are possible. Thus, neutron scattering may be used in two completely different ways, since *elastic* scattering ($\hbar\omega = 0$; $\boldsymbol{Q} = \boldsymbol{Q}_o$) is related to the spatial distribution of centres (atoms) and *inelastic* scattering ($\hbar\omega \neq 0$) is related to the dynamics of the centres (see sect. 4.4.4).

The scattering law $S(\boldsymbol{Q}, \omega)$ is a function of both momentum and energy transfer, and is related to a generalized pair correlation function $G(\boldsymbol{r}, t)$ by a double Fourier transform

$$S(\boldsymbol{Q}, \omega) = \frac{1}{2\pi}\int_0^\infty \int_{-\infty}^\infty G(\boldsymbol{r}, t)e^{i(\boldsymbol{Q}\cdot\boldsymbol{r} - \omega t)}\, d\boldsymbol{r}\, dt \qquad [3.26]$$

and

$$G(\boldsymbol{r}, t)=\frac{1}{(2\pi)^3}\int_0^{\infty}\int_{-\infty}^{\infty} S(\boldsymbol{Q}, \omega)\mathrm{e}^{i(\omega t-\boldsymbol{Q}\cdot\boldsymbol{r})}\,\mathrm{d}\boldsymbol{Q}\,\mathrm{d}\omega \qquad [3.27]$$

In the experimental arrangement conventionally used in neutron diffraction, namely a 'twin-axis spectrometer' (identical to the configuration shown in Fig. 3.9 with the omission of the second monochromator M_2), *all* neutrons are detected after scattering at a given angle irrespective of their energy; this is termed 'total scattering' and is the same for X-ray diffraction. Thus, this involves an integration over all values of ω, which by reciprocity is equivalent to taking $t=0$. Thus in this case, the so-called 'static approximation', [3.27] reduces to

$$G(\boldsymbol{r}, 0)=\frac{1}{(2\pi)^3}\int_0^{\infty} S(\boldsymbol{Q})\mathrm{e}^{-i\boldsymbol{Q}\cdot\boldsymbol{r}}\,\mathrm{d}\boldsymbol{Q} \qquad [3.28]$$

where $S(\boldsymbol{Q})=\int_{-\infty}^{\infty} S(\boldsymbol{Q}, \omega)\,\mathrm{d}\omega$ and $G(\boldsymbol{r}, 0)$ is the instantaneous pair correlation function. A full quantum-mechanical calculation of the scattering (see, e.g., Wright 1974) leads to the following expression for the total neutron scattering intensity from a monatomic isotropic system in the *static* approximation:

$$I_{\mathrm{N}}^{\mathrm{T}}(Q)=N\overline{b^2}+N\bar{b}^2\int_0^{\infty} 4\pi r^2\rho(r)\frac{\sin Qr}{Qr}\,\mathrm{d}r \qquad [3.29]$$

where $\bar{b}$ is the neutron 'scattering length' averaged over all isotopes of the particular element concerned and N is the number of atoms. Defining

$$i_{\mathrm{N}}(Q)=I_{\mathrm{N}}^{\mathrm{T}}(Q)-N\overline{b^2}-I_{\mathrm{N}}^{\circ}(Q) \qquad [3.30]$$

where $I_{\mathrm{N}}^{\circ}(Q)$ is the experimentally inaccessible contribution to the scattering at $Q=0$ from the average density (cf. the last term of [3.16]), then:

$$Qi_{\mathrm{N}}(Q)=N\bar{b}^2\int_0^{\infty} G(r)\sin Qr\,\mathrm{d}r \qquad [3.31]$$

This equation is entirely equivalent to that derived for coherent X-ray scattering ([3.17]) if the following equivalents in terminology between neutron and X-ray scattering are made, $Q\equiv k$, $Qi(Q)\equiv F(k)$; the scattering length b is analogous to the atomic form factor $f(k)$. In practice deviations from the static approximation lead to [3.29] being no longer obeyed, and the so-called Placzek corrections have to be applied to the first term in [3.29] (see Wright 1974).

The static approximation is avoided by detecting the *elastically* scattered neutrons by incorporating a second monochromator, i.e. using a *triple* axis spectrometer which passes only those neutrons having the same energy as the incident beam. In this case the scattering function which is relevant is $S(\boldsymbol{Q}, 0)$ which is related to $G(\boldsymbol{r}, \infty)$, or the time-averaged correlation function when the averaging time is much longer than the period of an atomic vibration. The elastically scattered intensity can be written (Leadbetter and Wright 1972) as:

$$I_{\mathrm{N}}^{\mathrm{el}}(Q)=N\overline{b^2}\mathrm{e}^{-2W}+N\bar{b}^2\int_0^{\infty} 4\pi r^2\rho^{\mathrm{eq}}(r)\mathrm{e}^{-2W}\frac{\sin Qr}{Qr}\,\mathrm{d}r \qquad [3.32]$$

where e^{-2W} is the so-called 'Debye–Waller' factor, and W is a function of the mean square atomic displacement $\overline{u^2}$ due solely to thermal vibrations given in the harmonic approximation by:

$$W = \tfrac{1}{6}Q^2\overline{u^2} \qquad [3.33]$$

By comparison, the expression for the total diffracted intensity can also be rewritten in terms of a separation of atomic positions into equilibrium and time-dependent thermal displacement terms:

$$I_N^T(Q) = N\overline{b^2} + N\bar{b}^2\int_0^\infty 4\pi r^2 \rho^{eq}(r)e^{-2W}\frac{\sin Qr}{Qr}\,dr \qquad [3.34]$$

A comparison of the two sets of data for a given material therefore yields an estimate for atomic displacements. Note, however, that only atomic displacements of purely thermal origin are obtained in this way; any static distortions in atomic positions will not be detected, since for this case both elastic and total diffracted intensities take the form of [3.34].

Both experimental arrangements discussed so far, namely double and triple axis spectrometers, use a fixed wavelength incident beam and vary 2θ in order to trace out the Q (or k) dependence of the scattering. An alternative approach is to use the so-called 'time-of-flight' method in which white, unmonochromatized radiation containing all wavelengths is incident upon the sample, and the scattered intensity is monitored at fixed scattering angles (for reviews see Windsor (1981) and Carpenter (1985)). In practice, the incident neutron beam is pulsed, either by employing a mechanical chopper with a steady-state reactor or by using a pulsed neutron source such as that produced by electrons from an electron linear accelerator impinging on to a uranium target ('spallation' source). The experimental configuration is shown schematically in Fig. 3.10; the two flight paths, source–sample (l_0) and sample–detector (l_1), are of the order of metres in length with $l_1 \ll l_0$. The elastic momentum transfer is given by the expression:

$$Q_{el} = \left[\frac{2m_n(l_0 + l_1)}{\hbar}\right]\frac{\sin\theta}{\tau} \qquad [3.35]$$

where τ is the flight time and m_n is the neutron mass. The technique offers the advantages over conventional fixed λ, variable 2θ experiments of much higher values of Q_{max} (values of $\simeq 60$ Å^{-1} are attainable) resulting in an increase in resolution, higher count rates (at least for pulsed sources) and since all values of Q are measured simultaneously, the possibility of performing time-dependent experiments, e.g. monitoring kinetic changes. Similar experiments are possible for X-rays using the broad-band radiation emitted by synchrotrons. However, the problem of the corrections needed to eliminate the Compton X-ray scattering at high k-values make the technique potentially less attractive than for neutrons.

A very important feature of neutron scattering is the behaviour of the scattering length, $\bar{b}$. This is independent of Q (or k) for closed shells, in marked contrast to the equivalent entity in X-ray scattering, the atomic form factor $f(k)$, which is strongly dependent on k (see Fig. 3.5). In addition, $\bar{b}$ varies in a random fashion between the different elements, whereas $f(k)$ varies in a systematic way

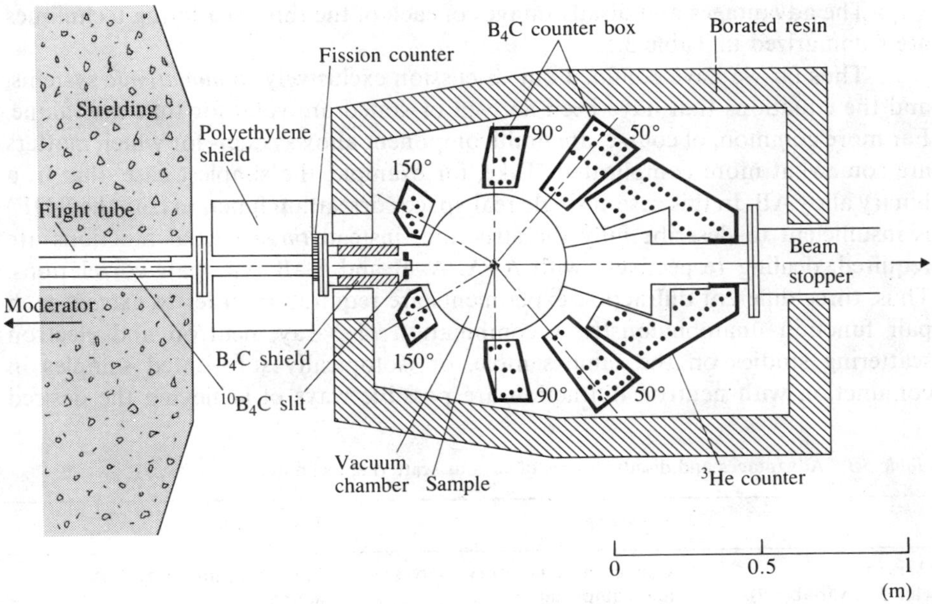

Fig. 3.10 Illustration of a time-of-flight (TOF) neutron scattering spectrometer for use with a pulsed (spallation) neutron source (K. Suzuki in *Methods of Experimental Physics, vol. 23B*: *Neutron Scattering*, D.L. Price and K. Sköld (eds.) (Academic: 1987), p. 243).

from element to element since $f(0)$ is equal to Z, the atomic number. Furthermore, b is different for different isotopes of the *same* element, again varying in a random fashion, although $f(k)$ of course is the same for all isotopes, being just dependent on the electronic structure of the atom. These features will prove to be of importance in the discussion of multicomponent systems later.

The final type of scattering probe that has been used in the study of the structure of amorphous materials is the electron; a review of the subject has been given by Dove (1973). The experiment is performed in the electron microscope with typically 50–100 keV electrons, which correspond to wavelengths of the order of 0.05 Å; again λ is kept constant and the scattering angle varied by deflecting the electrons with a scanning coil. The electrons are scattered both by the electrons and the nuclei of the atoms in a sample, and as such the technique formally lies between X-ray and neutron scattering. The form factor for electron scattering is given by

$$f^{e}(k) = \frac{2m_e e^2}{\hbar^2}[(Z - f^{x}(k))/k^2]. \qquad [3.36]$$

Because of the coulombic interaction the scattering is much stronger than for X-rays, and multiple scattering becomes important for film thicknesses greater than a few hundred ångströms. In addition, there is a large background due to inelastic loss processes such as plasmon formation, etc. The necessity of using very thin films raises the doubt, furthermore, of whether the structure of such thin films is truly representative of the structure of bulk specimens.

The advantages and disadvantages of each of the three scattering techniques are summarized in Table 3.1.

Thus far we have confined our discussion exclusively to *monatomic* systems, and the equations that have been derived or stated are valid for that case alone. Far more common, of course, are multicomponent alloy systems for which matters are somewhat more complicated. Take, for example, the simplest case, that of a binary alloy AB. In this case, a single real-space correlation function (say the RDF) is insufficient to describe fully the structure; instead *three* partial functions are required, dealing respectively with A–A, A–B and B–B type pair correlations. Thus, three different diffraction experiments are required in order to extract each pair function unambiguously. A combination of X-ray, neutron and electron scattering studies on the same sample, or isotopically substituted samples in conjunction with neutron diffraction are possible ways of achieving the desired

Table 3.1 Advantages and disadvantages of various scattering techniques

	Advantages	*Disadvantages*
X-rays (fixed λ, variable θ)	Convenient laboratory sources and equipment. Powdered or thick film samples can be used. Atomic form factor can be calculated.	Small k_{max} for many X-ray sources. Significant Compton contribution to scattering at large k. Partial correlation functions difficult to extract for multicomponent systems. Atomic form factor strongly decreasing function of k.
Neutrons (fixed λ, variable θ, double or triple axis)	Scattering length independent of k. Partial correlation functions easier to extract for multicomponent systems. Isotopic substitution possible to determine partial correlation functions.	Reactor source required. Placzek corrections required for double-axis experiments (static approximation). Large volume of material required. b must be determined experimentally for each isotope. Low count rates.
(fixed θ, variable λ, time of flight)	(*In addition to the above*) Large values of k_{max} attainable. Fixed geometry simplifies experimental set-up, e.g. for pressure studies. Simultaneous collection of data for all k-values facilitates time-resolved experiments. Higher count rates (if LINAC used).	Pulsed source required (chopper + steady-state reactor or LINAC source).
Electrons (fixed λ, variable θ)	Uses scanning electron microscope. *In situ* grown thin films can be examined. High count rates.	Scattering very strong and multiple scattering important for thicknesses $\geqslant 100$ Å. Thin films may not be representative of the bulk. Large inelastic scattering background (due to plasmons).

goal of three scattering experiments each with different scattering factors ($f(k)$ or b) in order that the partial correlation functions may be extracted. While this may be possible for a binary alloy it is not feasible in general, since for an n-component system $n(n+1)/2$ partial functions are required. For these complicated systems a chemically specific structural probe (e.g. EXAFS – see sect. 3.2.3) is required. A simple diffraction experiment obviously yields a gross average, both spatially and chemically, and although the nearest-neighbour bond length may be well defined, it is not always possible to obtain unambiguously other structural parameters (even including the nearest-neighbour coordination number) from a Fourier transform of the intensity.

For a polyatomic system consisting of n components, the equation previously derived for X-ray scattering from a monatomic material can be generalized to give:

$$\frac{I_{eu}}{N} = \sum_{i=1}^{n} x_i f_i^2 + \sum_{i=1}^{n} \sum_{j=1}^{n} x_i x_j f_i f_j \int_0^\infty 4\pi r^2 \frac{\rho_{ij}(r)}{x_j} \frac{\sin kr}{kr} \mathrm{d}r - \left[\sum_{i=1}^{n} x_i f_i\right]^2 \int_0^\infty 4\pi r^2 \rho^0 \frac{\sin kr}{kr} \mathrm{d}r \qquad [3.37]$$

where x_i is the atomic fraction of element i, f_i is the scattering factor of element i and $\rho_{ij}(r)$ is the average number of j atoms per unit volume at distance r from any i atom. Defining the following functions $\langle f \rangle = \Sigma_{i=1}^{n} x_i f_i$, $\langle f^2 \rangle = \Sigma_{i=1}^{n} x_i f^2$, and

$$\rho(r) = \sum_{ij} W_{ij} \rho_{ij}(r)/x_j \qquad [3.38]$$

where the weighting factors are given by

$$W_{ij} = x_i x_j f_i f_j / \langle f \rangle^2 \qquad [3.39]$$

leads to a reformulation of [3.37]

$$\frac{I_{eu}}{N} = \langle f^2 \rangle + \langle f \rangle^2 \int_0^\infty 4\pi r^2 [\rho(r) - \rho^0] \frac{\sin kr}{kr} \mathrm{d}r \qquad [3.40]$$

The reduced scattering intensity is now:

$$F(k) = k\left[\frac{I_{eu}/N - \langle f^2 \rangle}{\langle f \rangle^2}\right] = \int_0^\infty 4\pi r [\rho(r) - \rho^0] \sin kr \, \mathrm{d}r \qquad [3.41]$$

This expression is superficially similar to that for monatomic systems ([3.18]) except that weighted values for the scattering factors are used instead, and that $\rho(r)$ is now a function of the scattering factors through the term in W_{ij}. Equation [3.41] can be rewritten in another form in terms of partial interference functions $I_{ij}(k)$ (see e.g. Wagner 1980, 1985) as:

$$I/N - \langle f^2 \rangle = x_1^2 f_1^2 [I_{11}(k) - 1] + x_2^2 f_2^2 [I_{22} - 1] + 2x_1 x_2 f_1 f_2 [I_{12} - 1] \qquad [3.42]$$

where

$$I_{ij}(k)-1=\int_0^\infty 4\pi r^2\{[\rho_{ij}(r)/x_j]-\rho^\circ\}\frac{\sin kr}{kr}\,dr \qquad [3.43]$$

and

$$I_{ij}=I_{ji}, \quad \text{since } \rho_{ij}/x_j=\rho_{ji}/x_i$$

A useful reformulation of the expression for the scattering intensity from a *binary* alloy has been given by Bhatia and Thornton (1970). In this, the individual atom–atom partial structure factors (in the so-called Faber–Ziman formalism) are replaced by *number–number*, *number–concentration* and *concentration–concentration* correlation functions, viz. $S_{NN}(k)$, $S_{NC}(k)$ and $S_{CC}(k)$ respectively, where

$$S_{NN}(k)=x_1^2I_{11}(k)+x_2^2I_{22}(k)+2x_1x_2I_{12}(k) \qquad [3.44]$$

$$S_{NC}(k)=x_1x_2\{[x_1I_{11}(k)+x_2I_{12}(k)]-[x_1I_{21}(k)+x_2I_{22}(k)]\} \qquad [3.45]$$

$$S_{CC}(k)=x_1x_2\{1+x_1x_2[I_{11}(k)+I_{22}(k)-2I_{12}(k)]\} \qquad [3.46]$$

where the I_{ij} are given by [3.43]. The expression for the scattering intensity for a binary system ([3.42]) can then be rewritten in terms of the Bhatia–Thornton partial functions:

$$I/N=\langle f\rangle^2S_{NN}(k)+2\langle f\rangle\Delta f\,S_{NC}(k)+\frac{\langle f^2\rangle-\langle f\rangle^2}{x_1x_2}S_{CC}(k) \qquad [3.47]$$

where

$$\Delta f=f_1(k)-f_2(k)$$

The Bhatia–Thornton correlation functions have the following physical significance. The *number–number* function, $S_{NN}(k)$, represents the *topological* short-range order (analogous to the crystalline Bragg peaks due to the average lattice occupied by A and B atoms in a binary alloy). $S_{NN}(k)$ oscillates around unity and tends to that value as $k\to\infty$. The Fourier transform of $S_{NN}(k)$,

$$4\pi r[\rho_{NN}(r)-\rho^\circ]=\frac{2}{\pi}\int_0^\infty k[S_{NN}(k)-1]\sin kr\,dk \qquad [3.48]$$

gives the real-space number–number distribution function, which is related to the corresponding partial quantities in the Faber–Ziman representation by

$$\rho_{NN}(r)=x_1\rho_1(r)+x_2\rho_2(r) \qquad [3.49]$$

where

$$\rho_i(r)=\sum_{j=1}^{2}\rho_{ij}(r), \quad i=1,2 \qquad [3.50]$$

The concentration–concentration term, $S_{CC}(k)$, describes the *chemical* short-range order and, for a random distribution of A and B atoms, $S_{CC}(k)=x_1x_2$; otherwise, it oscillates about this value and tends to it as $k\to\infty$. *Positive* values of

$S_{CC}(0)-x_1x_2$ indicate a preference for *like* nearest neighbours (or clustering), whereas *negative* values indicate a preference for *unlike* neighbours (or chemical short-range order). The Fourier transform,

$$4\pi r\rho_{CC}(r)=\frac{2}{\pi}\int_0^\infty k[[S_{CC}(k)-x_1x_2]/x_1x_2]\sin kr\,dk \qquad [3.51]$$

yields $\rho_{CC}(r)$, which is related to the atom–atom correlation functions by

$$\rho_{CC}(r)=x_2\rho_1(r)+x_1\rho_2(r)-\rho_{12}(r)/x_2 \qquad [3.52]$$

where ρ_1 and ρ_2 are defined as in [3.50].

The final Bhatia–Thornton correlation function, $S_{NC}(k)$, is zero for random mixtures; otherwise, it oscillates about this value and vanishes as $k\to\infty$. The Fourier transform of this quantity,

$$4\pi r\rho_{NC}(r)=\frac{2}{\pi}\int_0^\infty kS_{NC}(k)\sin kr\,dk \qquad [3.53]$$

gives the real-space function related to the Faber–Ziman quantities by

$$\rho_{NC}(r)=x_1x_2[\rho_1(r)-\rho_2(r)] \qquad [3.54]$$

The equation for the reduced scattering intensity ([3.41]) can only be inverted simply and straightforwardly without approximation to give real-space information if W_{ij}, and hence f_i and f_j, are *independent of k* or have the same k-dependence. This is the case for neutrons ($f_i=b_i$) but not in general for X-rays ($f_i=f_i(k)$); thus for neutrons, the reduced RDF is given by the same equation as derived previously ([3.21]), but with $\rho(r)=\Sigma_{ij}W_{ij}\rho_{ij}(r)$.

The first technique to be discussed which is capable of yielding partial structural correlation information is the so-called 'isotopic substitution' method employed in neutron diffraction. This technique was pioneered by Enderby and co-workers (Neilson and Enderby 1983) in studies of ionic solutions, and has subsequently been applied to a number of binary solid-phase amorphous systems. In this, the isotopic components of the elements involved are varied, whilst keeping the overall chemical composition constant. In binary system A–B, the three partial correlation functions involving A–A, A–B and B–B correlations (either $S_{ij}(k)$ or $G_{ij}(r)$) are obtained with the best accuracy from three sets of experimental data when the weighting factors W_{ij} ([3.39]), or equivalently the neutron scattering lengths b_i etc., differ from each other by the largest possible amount. This will obviously be the case if the *sign* of b could change for different isotopes of a given element. Table 3.2 gives some examples of the very few elements with isotopes having negative values of b (where scattering of a neutron from the nucleus involves a phase change of π); other examples are ^{152}Sm, ^{162}Dy and ^{186}W. It can be seen, then, that the isotopic substitution method is likely to be most effective for metallic amorphous systems.

The existence of negative scattering lengths for some isotopes means that certain simplifying conditions are possible. Thus, for 'null elements' (e.g. where $\bar{b}_1=0$), $I_{22}(k)$ (or $G_{22}(r)$) is obtained *directly* from experiment (see [3.42]). An example of this approach is shown in Fig. 3.11 for glassy $Ni_{81}B_{19}$ (Lamparter

Table 3.2 Neutron scattering lengths for potential 'zero-alloy' components

	b (10^{-12} cm)		b (10^{-12} cm)		b (10^{-12} cm)
Ni (natural)	1.03	Ti (natural)	−0.335	^{1}H	−0.374
^{58}Ni	1.44	^{46}Ti	0.48	^{2}H	0.667
^{60}Ni	0.282	^{47}Ti	0.33		
^{61}Ni	0.76	^{48}Ti	−0.58	^{6}Li	0.18*
^{62}Ni	−0.87	^{49}Ti	0.08	^{7}Li	−0.233
^{64}Ni	−0.037	^{50}Ti	0.55		

* Complex value of b.

et al. 1982). The experimental structure factor for the null-element isotopic mixture $^{\varnothing}$Ni gives $S_{BB}(Q)$ directly.

Other simplifications occur for 'zero alloys', where $\langle b \rangle = x_1 b_1 + x_2 b_2 = 0$; (here the experimental structure factor gives directly $S_{CC}(Q)$ in the Bhatia–Thornton formulation ([3.47]), and the choice $b_1 = b_2$ (i.e. where $\Delta f = 0$), the experimental structure factor leads directly to $S_{NN}(Q)$ (see problem 3.1).

A second method for obtaining partial structural information also employs neutron scattering for amorphous metallic systems. Since the neutron is a spin $\frac{1}{2}$ particle, it may be polarized (by means of a reflecting crystal) such that its spin is either up or down. If the amorphous metal is ferromagnetic, its magnetic moment may also be oriented by an external magnetic field, with the result that the neutron scattering length is different (b^+ or b^-) depending on whether the neutron spin polarization is parallel or antiparallel, respectively, to the metal magnetic moment direction. The polarized scattering length is related to the normal scattering length b_o by:

$$b^{\pm} = b_o \pm P \qquad [3.55]$$

where the magnetic scattering amplitude is given by

$$P = \frac{r_o \gamma_n M_n f(Q) \sin \delta}{2} \qquad [3.56]$$

where r_o is the classical electron radius ($= e^2/m_e c^2$), γ_n and M_n are neutron and atomic magnetic moments, respectively, δ is the angle between the atomic magnetic moment and the scattering vector Q, and $f(Q)$ is the 'magnetic form factor', which is a rapidly decreasing function of Q (like the normal X-ray form factor) because the scattering is now from the unpaired electrons which are spatially non-isotropic. The resultant drop-off in scattering intensity at large Q-values for this magnetic scattering consequently restricts the spatial resolution achievable.

For a binary amorphous metallic alloy, the three independent experimental diffraction measurements necessary to determine the separate partial pair correlation functions can therefore involve scattering intensities from polarized magnetic neutron diffraction (in practice $I^+ - I^-$), conventional non-polarized neutron diffraction (I_o) and X-ray diffraction (I_X). Sadoc and Dixmier (1976) employed this technique in a study of glassy $Co_{80}P_{20}$, the results for which are illustrated in Fig. 3.12, where the measured scattering intensities, together with the partial Faber–Ziman correlation functions obtained from them, are shown.

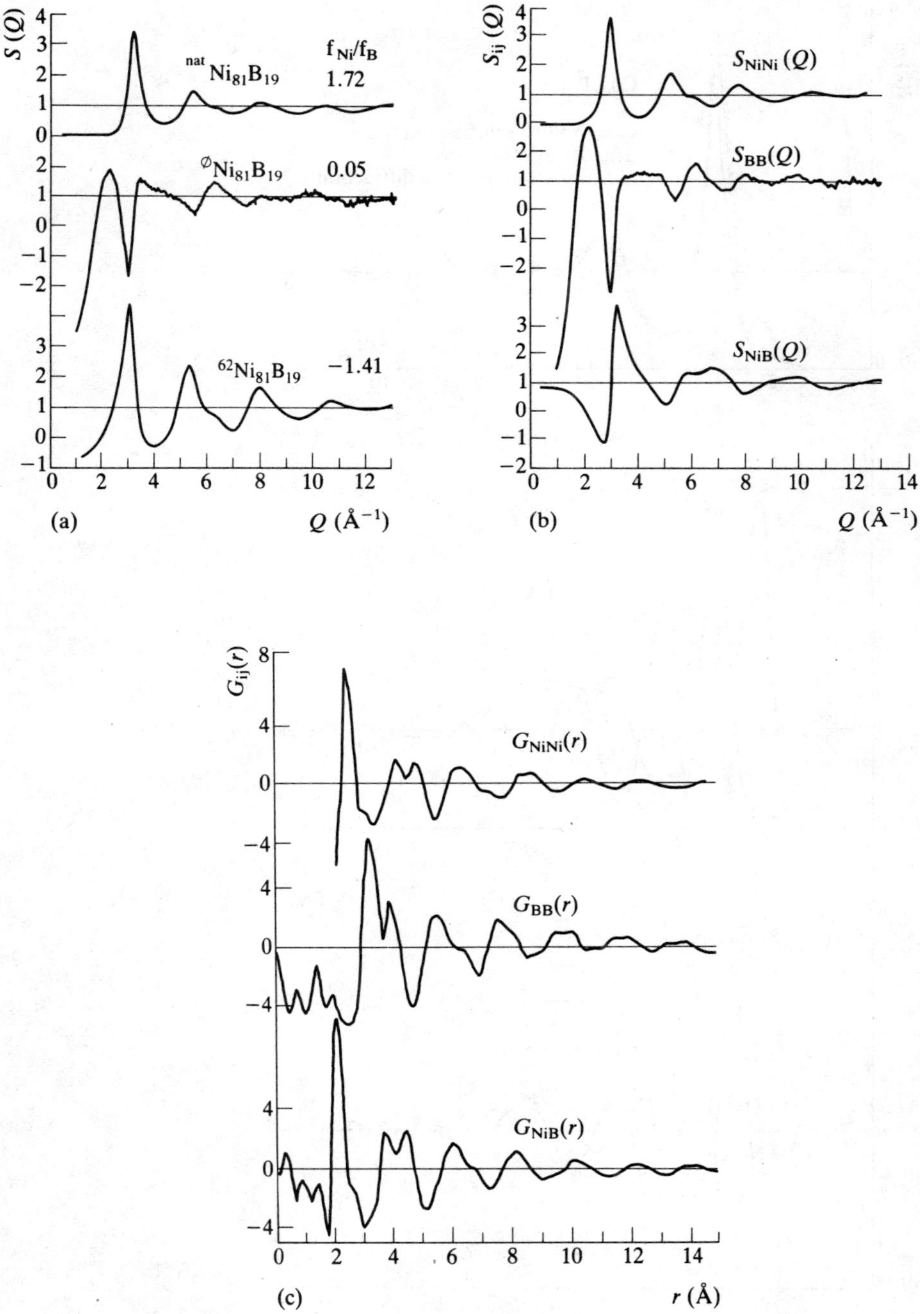

Fig. 3.11 Structural studies of glassy $Ni_{81}B_{19}$ using isotopic substitution and neutron diffraction (Lamparter *et al.* 1982).
(a) Structure factors for different isotopically substituted alloys, including the 'null element' isotopic mixture, $^{\varnothing}Ni$.
(b) Partial structure factors obtained from those shown in (a).
(c) Partial reduced RDFs obtained by Fourier transformation of the $S_{ij}(Q)$ shown in (b).

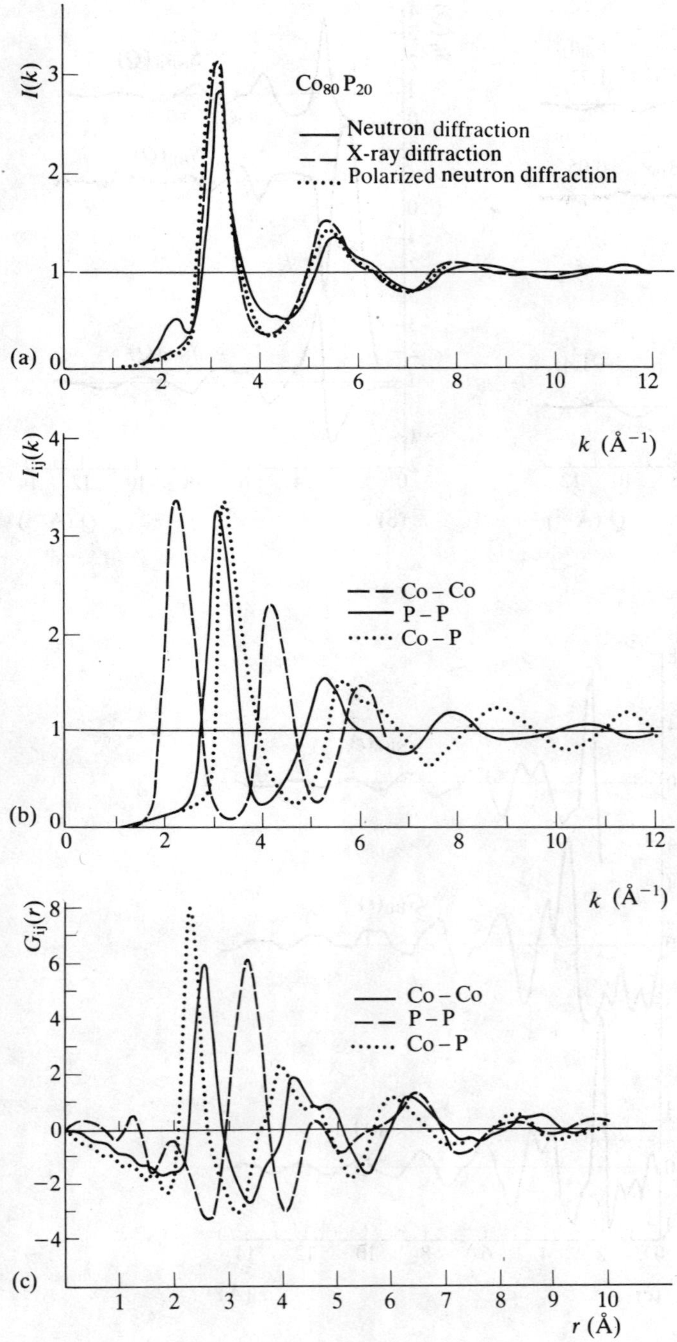

Fig. 3.12 Structural studies of glassy $Co_{80}P_{20}$ using polarized neutron scattering (Sadoc and Dixmier 1976):
(a) total scattering intensities $I(Q)$ for polarized, unpolarized neutron and X-ray diffraction;
(b) partial scattering intensities $I_{ij}(Q)$ obtained from the curves in (a);
(c) partial reduced RDFs obtained by Fourier transformation of curves in (b).

The third approach which can be used to obtain partial pair correlation functions is *isomorphous substitution.* This method is based on the supposition that if element B in a binary compound or alloy A_xB_{1-x} is substituted by element C having the *same size and chemical characteristics* as B, so the *structure* of the material $A_x(B_yC_{1-y})_{1-x}$ is expected to be the same as that of A_xB_{1-x}. Elements B and C are further chosen so as to maximize the difference in scattering factors between them. In this way, therefore, the weighting factors W_{ij} ([3.39]) can be varied significantly. Obviously, the method depends for success on whether true isomorphous substitution is possible; this is more likely to be the case for metallic systems with non-directional bonding than for covalent systems, where the degree of covalency (and hence the local structure) can vary significantly in passing down a given column of the periodic table. Thus, possible isomorphous partners for X-ray scattering have been found to be (Zr, Hf) and (Nb, Ta), and for neutron diffraction (Ni, Co). Wagner and Lee (1980) have used this technique to study the structure of a-$Ni_{35}Zr_{65-x}Hf_x$.

The fourth and final technique for systematically varying the atomic scattering factors in order to obtain partial structural correlation information lies in the phenomenon of *anomalous scattering* (or dispersion). In general, the scattering factor is a function of both k, the change in scattering wavevector and ω, the frequency of the radiation (X-rays or neutrons):

$$f(k,\omega)=f^{\circ}(k)+f'(k,\omega)+if''(k,\omega) \qquad [3.57]$$

The first term in this equation, $f^{\circ}(k)$, is the X-ray atomic scattering factor (i.e. the Fourier transform of the electronic charge density ([3.11])) or the normal neutron scattering length b, and is independent of ω; this is the quantity we have used previously. The other factors, $f'(k,\omega)$ and $f''(k,\omega)$, are the so-called 'dispersion corrections' and are only significant when ω lies very close to an absorption threshold of the atom. For X-rays, the absorption threshold corresponds to a characteristic electronic transition energy in the atom (e.g. the ejection of a 1s electron at the K-edge). For neutrons, the absorption threshold is associated with the (rare) occurrence of the nucleus undergoing a nuclear reaction upon absorption of a thermal neutron; an example of an isotope which shows this effect is ^{149}Sm. The form of the dispersion corrections to the X-ray scattering factor of Ge near its K-edge at $\simeq 11.1$ keV is shown in Fig. 3.13(a); that for the neutron scattering length of ^{149}Sm is shown in Fig. 3.13(b). The imaginary part of the X-ray scattering factor, $f''(k,\omega)$, is related to the X-ray absorption coefficient $\alpha(\omega)$ by

$$f''(\omega)=\frac{\omega\alpha(\omega)}{4\pi c} \qquad [3.58]$$

where c is the speed of light, and f' and f'' are related by a Kramers–Krönig transformation

$$f'(\omega)=\frac{2}{\pi}\int_0^{\infty}\frac{\omega f''(\omega)\,d\omega}{(\omega^2-\omega_o^2)} \qquad [3.59]$$

In principle, therefore, it should be possible to extract partial correlation functions for, say, a binary alloy, by measuring the scattering intensity for three

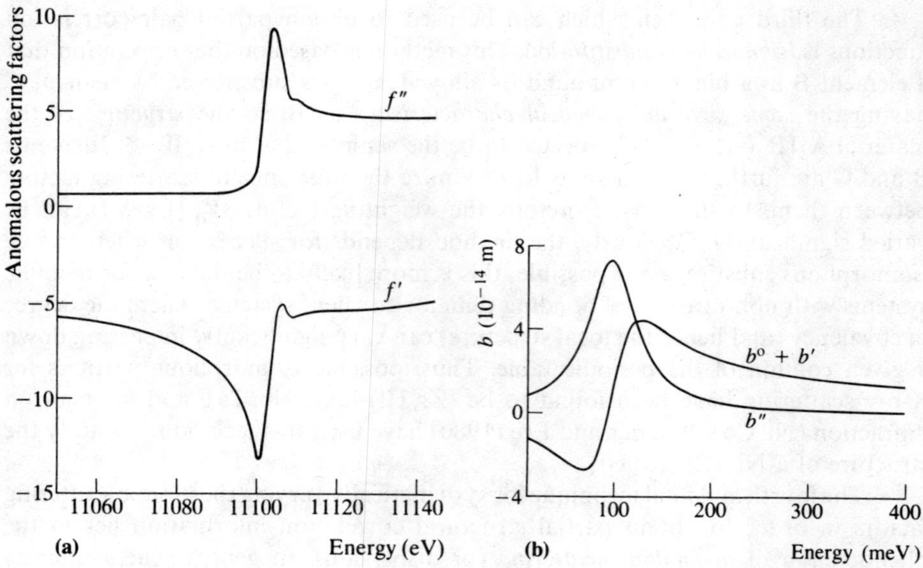

Fig. 3.13 (a) The real and imaginary parts of the X-ray scattering factor, f' and f'', for Ge near the K-edge (Fuoss 1980).
(b) The corresponding quantities, b' and b'', for the neutron scattering length for ^{149}Sm (Wright *et al.* 1982)

X-ray or neutron wavelengths including the regions near an absorption edge such that f', f'' and hence f vary widely between each experiment. The weighting factors $W_{ij}(k)$ ([3.39]) can then be written as

$$W_{ij}(k) = x_i x_j \frac{[(f_i^o + f_i')(f_j^o + f_j') - f_i'' f_j'']}{\langle f \rangle^2} \qquad [3.60]$$

where $\langle f \rangle^2 = [\Sigma x_i(f_i^o + f_i')]^2 + [\Sigma x_i f_i'']^2$.

In general, however, the changes in f' and f'' are perhaps only several tens of per cent of f^o even in the most favourable situation, giving rise to large errors in the derived partial correlation functions.

A more promising approach is to make use of *differential* anomalous scattering, since the changes in the *derivatives* of f' and f'' with energy are much larger than is the case for f' or f'' alone (Shevchik 1977). The derivative of the X-ray scattering intensity for a binary alloy AB with respect to X-ray photon energy near the absorption edge of atom A is:

$$\left[\frac{\delta I(k)}{\delta E}\right]_k = x_A \frac{\delta}{\delta E}[f_A(k) f_B^*(k) + f_A^*(k) f_B(k)] F_{AB}(k) \qquad [3.61]$$

where $x_A F_{AB} = x_B F_{BA}$ has been used, and F_{AB} is related to the partial real-space correlations function by:

$$F_{AB}(k) = \int_o^\infty 4\pi r[\rho_{AB}(r) - \rho_B^o] \sin kr \, dr \qquad [3.62]$$

Thus Fourier transformation of the derivative of the scattering intensity ([3.61]) gives information about the environment about atom A (i.e. $\rho_{AB}(r)$), and in the general case a weighted sum of the partial functions $\rho_{A\beta}(r)$ over all atom types β other than A in the multicomponent system, just as for EXAFS (sect. 3.2.3). In practice $I(k)$ is measured for several wavelengths in the region of each edge , and the difference taken between various wavelengths.

The technique is most widely applicable for X-ray diffraction, since every element has a series of X-ray absorption edges, whereas only very few elements have isotopes which exhibit anomalous neutron dispersion behaviour, and it obviously depends for success on having available a wide range of (X-ray) wavelengths to encompass all absorption edges of interest. For this purpose, synchrotron radiation sources which produce broad-band, high-intensity X-ray radiation are well suited. The advantages of the (differential) anomalous scattering technique in extracting structural information for multicomponent systems lie in the simplicity of the approach; once an appropriate radiation source is available no special procedures in sample preparation are needed (such as isotopic or isomorphous substitution) and large volumes of sample are not required as for neutron diffraction. Furthermore, data can be measured over a range of k values extending to $k = 0$; contrast this with the case for EXAFS (see sect. 3.2.3) in which data at low values of k are not easily interpretable using simple theories, and thus structural information at large separations cannot readily be ascertained. The disadvantages lie in the fact that the dispersion corrections, f' and f'', must be known with considerable accuracy if appreciable errors in the final partial RDFs are not to be incurred; errors are inherent in the differential technique in any case since to form the differential two quantities, comparable in size, are subtracted from each other. A further drawback is that the maximum value of k achievable is determined by the lowest energy absorption edge used, and this automatically limits the spatial resolution in the transformation of the diffracted intensity because of the small value of k_{max}. In addition, an experimental complication is the large fluorescent background for photon energies near an absorption edge, which can be 10–100 times the desired coherent intensity, and which must be subtracted before the data can be manipulated. Nevertheless, like EXAFS to be discussed next, differential anomalous scattering is chemically specific in that with a tunable radiation source, absorption edges for different elements in the sample can be selected and hence the structure about that atom investigated.

An example of the application of the technique is for amorphous GeSe (Fuoss *et al.* 1981). There has been a considerable controversy over the coordination in this material; both 3:3 coordination, characteristic of *crystalline* GeSe (which need not involve 'wrong' or homopolar bonds), and 4:2 coordination as found in amorphous or crystalline $GeSe_2$ (i.e. silica-like, and which must involve Ge—Ge bonds to satisfy stoichiometry in GeSe) have been proposed. The scattering intensity was measured at several photon energies within 100 eV of the Ge K-edge ($\simeq$11.1 keV) and the Se K-edge ($\simeq$12.65 keV), and by means of the differential method discussed previously, the distribution of atoms around Ge and Se, respectively were determined. These are shown in Fig. 3.14(a), from which it can be seen that the areas of the first peaks are nearly identical (3.1 atoms for Ge and 2.5 atoms for Se) and the second peaks are coincident in position, lending support

Table 3.3 Structural parameters for some covalent amorphous materials

Material	*Preparation method*	r_1 (Å)	r_2 (Å)	θ	ρ (g cm^{-3})	*Radiation type*	*Comments*	*Reference*
Elements								
C	TD	1.425	2.45	119	0.923	n	Prepared from polyfurfuryl alcohol resin	1
P	TD	2.18	3.48	106	2.16	X	Prepared from white P	2
As	VT	2.49	3.76	98	4.7–5.1	X	Hydrogen transport	3
Se	LQ	2.33	3.70	105	4.25	X	—	4
Si	E	2.35	3.86	109	2.10	e	—	5
Ge	E	2.463	3.997	109	4.79	n	Electron beam evaporated	6
Tetrahedral materials								
SiO_2	TD	1.61	2.62/3.13	109(Si)/152(O)	2.20	n	Spectrosil	7
GeO_2	LQ	1.74	3.18	109(Si)/132(O)	3.65	X		8
BeF_2	LQ	1.54	2.56	109(Be)/156(F)	2.00	n		9
$ZnCl_2$	LQ	2.29	3.72	109(Zn)/110(Cl)	2.71	n		10
Chalcogenide materials								
As_2S_3	LQ	2.28	3.48	102(As)/99.5(S)	3.17	X		11
As_2Se_3	LQ	2.43	3.70	100(As)/95(Se)	4.56	X		11
As_2Te_3	LQ	2.65	4.00	98(As)/95–98(Te)	5.53	X		11
Sb_2S_3	LQ	2.50	3.85	90(Sb)/100.7(S)	4.21	X		11
Sb_2O_3	LQ	1.99	2.80	92(Sb)	5.07	X	5% B_2O_3 added to stop crystallization	12
B_2O_3	LQ	1.37	2.38	120(B)/130(O)	1.84	n		13

Key: TD = thermal decomposition; VT = vapour transport; LQ = liquid quenched; E = evaporated; X = X-rays; n = neutrons; e = electrons.

References: (1) D. F. R. Mildner and J. M. Carpenter (1982), *J. Non-Cryst. Sol.* **47**, 391; (2) H. Krebs and H. U. Grüber (1967), *Z. Naturf.* **22a**, 96; (3) H. Krebs and R. Steffen (1964), *Z. anorg. allg. Chem.* **327**, 224; (4) P. Andonov (1982), *J. Non-Cryst. Sol.* **47**, 297; (5) S. C. Moss and J. F. Graczyk (1969), *Phys. Rev. Lett.* **23**, 1167; (6) G. Etherington, A. C. Wright, J. T. Wenzel, J. C. Dore, J. H. Clarke and R. N. Sinclair (1982), *J. Non-Cryst. Sol.* **48**, 265; (7) A. C. Wright, G. Etherington, J. A. Erwin Desa, R. N. Sinclair, G. A. N. Connell and J. C. Mikkelsen (1982), *J. Non-Cryst. Sol.* **49**, 63; (8) A. J. Leadbetter and A. C. Wright (1972), *J. Non-Cryst. Sol.* **7**, 37; (9) A. J. Leadbetter and A. C. Wright (1972), *J. Non-Cryst. Sol.* **7**, 156; (10) J. A. Erwin Desa, A. C. Wright, J. Wong and R. N. Sinclair (1982), *J. Non-Cryst. Sol.* **51**, 57; (11) L. Cervinka and A. Hruby (1982), *J. Non-Cryst. Sol.*, **48**, 231; (12) H. Hasegawa, M. Sone and M. Imaoka (1978), *J. Phys. Chem. Glasses* **19**, 28; (13) P. A. V. Johnson, A. C. Wright and R. N. Sinclair (1982), *J. Non-Cryst. Sol.* **50**, 281.

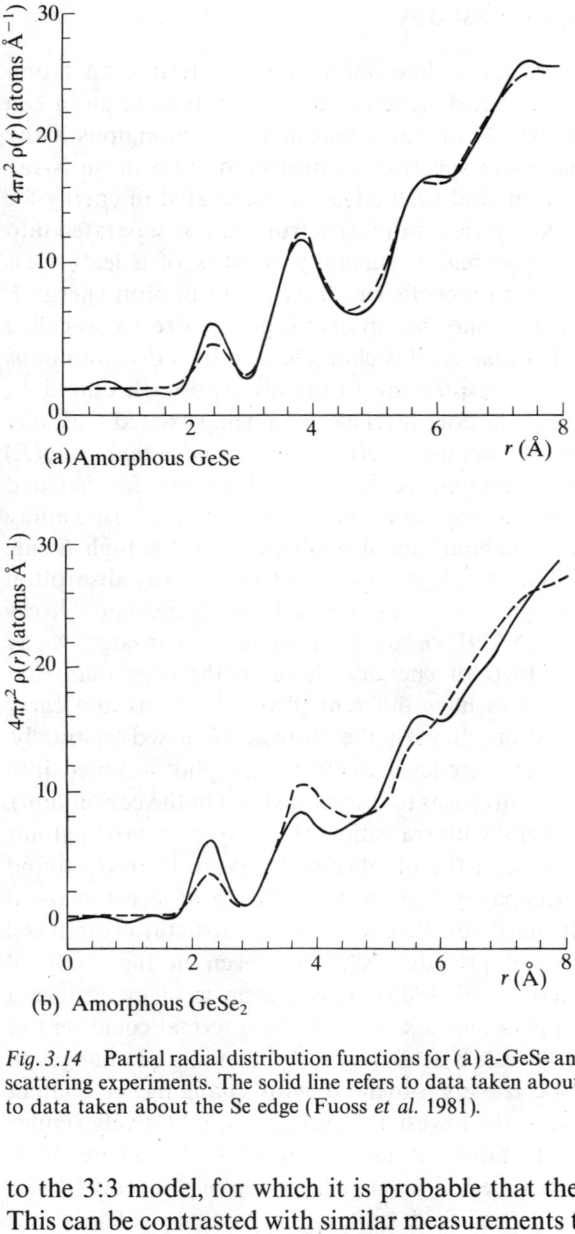

Fig. 3.14 Partial radial distribution functions for (a) a-GeSe and (b) a-$GeSe_2$ deduced from anomalous scattering experiments. The solid line refers to data taken about the Ge edge and the dashed line refers to data taken about the Se edge (Fuoss *et al.* 1981).

to the 3:3 model, for which it is probable that the bond angles would be similar. This can be contrasted with similar measurements taken on a-$GeSe_2$ (Fig. 3.14(b)), which yield very different coordinations of nearest neighbours (3.8 for Ge and 2.3 for Se) and the bond angle about the Ge atom is larger than about the Se atom, consistent with the notion of sp^3 hybridization of the fourfold coordinated Ge ($\theta \simeq 109°$) and p-bonding of the twofold coordinated Se ($\theta \simeq 90°$). Another application of the anomalous X-ray scattering technique has been to the amorphous metallic system $Mo_{50}Ni_{50}$ (Aur *et al.* 1983).

A collection of structural parameters obtained from diffraction experiments is given in Table 3.3.

3.2.3 X-ray absorption spectroscopy

X-ray absorption spectroscopy (XAS) is, like anomalous scattering, an atom-specific structural probe; that is, the local structure of a given type of atom can be probed. It is atom-specific for exactly the same reason as for anomalous X-ray scattering, namely an energy-dispersive spectrum is measured close to an X-ray absorption edge of a particular atom, and such edges are separated in energy for different elements. In general, an X-ray absorption spectrum can be separated into three regions in energy: the *pre-edge* region generally consists of a featureless monotonic decrease of X-ray absorption coefficient $\mu_X(E)$ with photon energy E (although occasionally certain peaks may be observed, giving rise to so-called 'pre-edge structure'): the *absorption edge* itself is characterized by a discontinuous increase in $\mu_X(E)$ at an energy E corresponding to the absorption threshold E_o (i.e. the difference in energy between the core-level electron being excited – usually a 1s electron at the K-edge – and the vacuum level); in the *post-edge* region $\mu_X(E)$ is again a featureless, monotonic decreasing function of energy for isolated (gas-phase) atoms but for atoms in any form of condensed state (including gas-phase molecular species) $\mu_X(E)$ exhibits small oscillations on the high-E side of the absorption edge. These oscillations are known as 'extended X-ray absorption fine structure' (EXAFS) for energies $\simeq 100$ eV beyond the edge, and 'X-ray absorption near-edge structure' (XANES), or equivalently 'near-edge X-ray absorption fine structure' (NEXAFS), for energies closer to the edge than this. The pre-edge and post-edge structures have different physical origins and carry different types of structural information; they will therefore be discussed separately.

Pre-edge structure arises when core-level electrons are photo-excited into low-lying bound states of ionized atoms (or into resonant states in the continuum). This form of XAS is particularly useful with transition-metal (or rare-earth) atoms incorporated into glassy matrices, since the number and type of ligands around such atoms affects the relative occupancy and energy splitting of levels in the d (or f) manifold; XAS can be particularly sensitive to such local structurally induced changes. An example of the use of pre-edge XAS has been in the study of $(TiO_2)_x(SiO_2)_{1-x}$ glasses (Greegor *et al.* 1983). This system is of interest as a potential zero thermal-expansion glass; for $x \leqslant 3$ at. %, the thermal coefficient of expansion, α_T is positive, whilst for $x \geqslant 8$ at. %, $\alpha_T < 0$. Pronounced changes in the pre-edge Ti K-edge XAS spectra are observed with changing Ti content (Fig. 3.15). The pre-edge structure at the lowest Ti contents is qualitatively similar to that observed in the rutile crystalline modification of TiO_2, where Ti is sixfold-coordinated in a regular octahedral arrangement, namely several (three) *weak* peaks characteristic of the ligand-field split t_{1g}, t_{2g} and e_g d-levels (see sect. 5.6.3). These peaks are weak because the centrosymmetry characteristic of perfect octahedral coordination precludes the orbital mixing necessary for the electronic transition involved to be strongly allowed. However, for Ti atoms in tetrahedral coordination such orbital mixing does occur, with the result that a characteristic *intense* single peak comprises the pre-edge structure; such a peak is clearly seen in the glasses with higher Ti contents (Fig. 3.15).

Thus, in certain favourable cases, XAS is a sensitive probe of the local coordination of atoms. In particular, it can distinguish between tetrahedral and

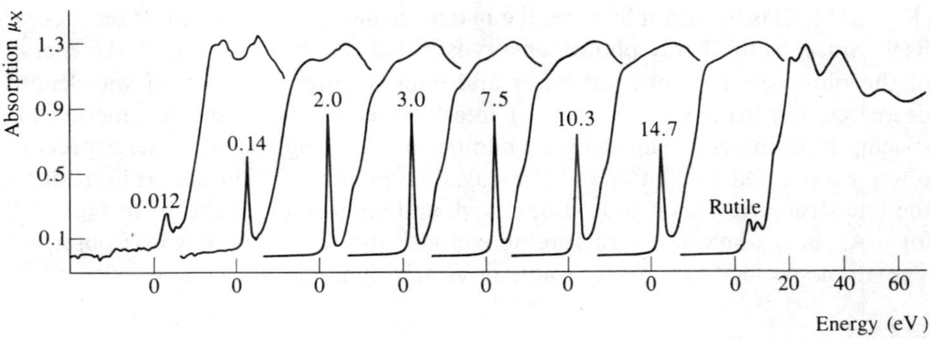

Fig. 3.15 X-ray absorption spectra for a series of TiO_2–SiO_2 glasses (with the TiO_2 content labelled as wt.%), and crystalline rutile, showing both the pre-edge structure and the near-edge (XANES) region (Greegor *et al.* 1983).

octahedral coordinations, although some other structural information (e.g. of bond lengths) it cannot reveal. In this sense, pre-edge XAS is able to provide a type of local, atom-specific structural information similar to that provided by nuclear magnetic resonance, NMR (see sect. 3.2.4), although in practice the two techniques tend to be complementary, since the types of atoms which are amenable to study by pre-edge XAS often turn out to be unsuitable for investigation by NMR, and vice versa.

The *post-edge* fine structure, either EXAFS or XANES, arises not from direct absorption processes, as does the pre-edge structure discussed above, but from what can be viewed as an 'internal diffraction' phenomenon. Consider what happens when a photon of energy slightly higher than threshold is absorbed by an atom, and a photo-electron is ejected. The photo-electron wave propagates outwards to infinity if the absorbing atom is isolated (e.g. in a rare gas); in this case the absorption coefficient decreases smoothly from the absorption edge. However, if other atoms surround the absorbing atom, as in a gaseous molecule or in a condensed phase, the outgoing photo-electron wave will be back-scattered (diffracted), and the back-scattered waves will interfere with the outgoing waves

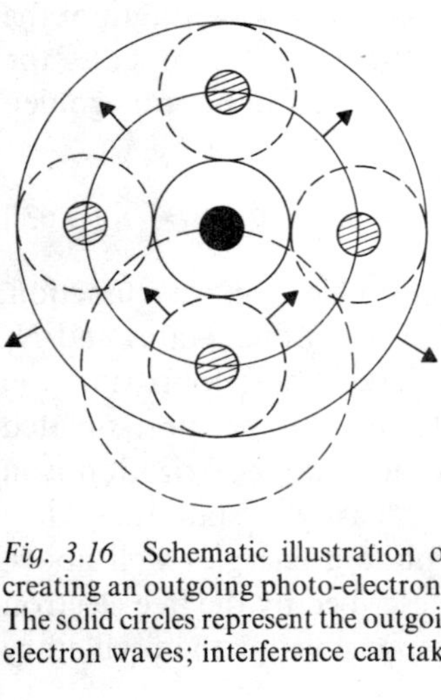

Fig. 3.16 Schematic illustration of the EXAFS process. An atom (filled circle) absorbs an X-ray creating an outgoing photo-electron wave which back-scatters off neighbouring atoms (hatched circles). The solid circles represent the outgoing electron wave and the dashed circles represent the back-scattered electron waves; interference can take place where these overlap.

(Fig. 3.16). This in turn influences the matrix elements for the absorption process itself. Since, as the X-ray photon energy is increased above threshold, the energy of the photo-electron also increases and consequently the electron wavelength decreases, constructive (destructive) interference occurs when the interatomic spacing is an integral (half-integral) number of wavelengths. Thus one expects to observe a periodic oscillation of the X-ray absorption above the absorption edge – the fine structure. This is indeed observed, and an example is shown in Fig. 3.17, for a-As_2Se_3, where fine structure above both the As and Se edges is observed. The atom-specificity of this technique is clearly apparent.

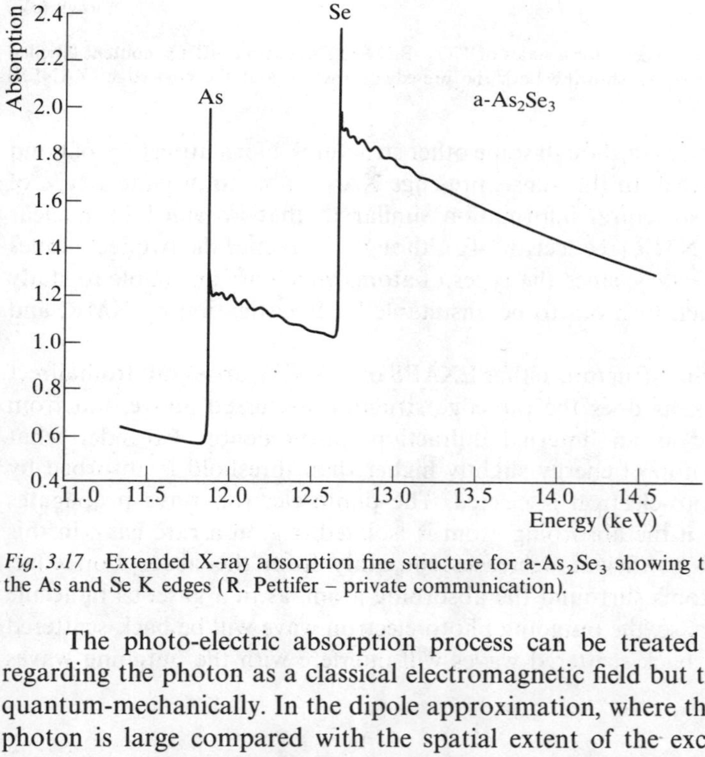

Fig. 3.17 Extended X-ray absorption fine structure for a-As_2Se_3 showing the fine structure beyond the As and Se K edges (R. Pettifer – private communication).

The photo-electric absorption process can be treated semi-classically by regarding the photon as a classical electromagnetic field but treating the electron quantum-mechanically. In the dipole approximation, where the wavelength of the photon is large compared with the spatial extent of the excited core state, the absorption can be treated to first order in perturbation theory and Fermi's golden rule then yields for the absorption coefficient:

$$\mu_X = 4N_0\pi^2 e^2 \frac{\omega}{c} |\langle f|z|i\rangle|^2 \rho(E^F) \qquad [3.63]$$

where $|i\rangle$ and $|f\rangle$ are the initial core state and final photo-electron wavefunction, respectively, ω is the X-ray frequency, $\rho(E^F)$ is the density of final states and N_o is the number of atoms of one type in the sample. Since in the dipole approximation the photon field can be regarded as being spatially uniform, it can be approximated by a scalar potential proportional to the distance z if the X-ray polarization is in the z-direction. Hence, only the matrix element or the density of states in [3.63] could give rise to oscillatory behaviour of μ_X. For photon energies well above threshold, $\rho(E^F)$ varies monotonically since it is well described by the free electron value. Therefore, the matrix element alone must be responsible for the oscillatory behaviour, and this is because the final state wavefunction $|f\rangle$ is made up of contributions from both the outgoing wave and the back-scattered wave, and interference between the two modulates the matrix element and hence μ_X (Stern 1978) – see Fig. 3.16.

EXAFS or XANES is conventionally measured using a transmission geometry in which the incident (I_o) and transmitted (I) X-ray intensities passing through a thin foil of thickness d are measured as a function of photon energy (Fig. 3.18(a)); they are related to the absorption coefficient μ_X by $I = I_o \exp(-\mu_X d)$. Two sources of broad-band X-rays may be used. Either the *bremsstrahlung* spectrum from conventional X-ray tubes or the radiation emitted by electron synchrotrons may be used, synchrotron radiation offering the advantage of a factor of $\sim 10^4$ increase in photon flux and concomitant reduction in counting time; in both cases the beam is monochromatized by a crystal. For dilute samples, the desired EXAFS is swamped by background intensity in a transmission mode, and the more sensitive technique of fluorescence detection must be employed. In this, the K-shell hole left

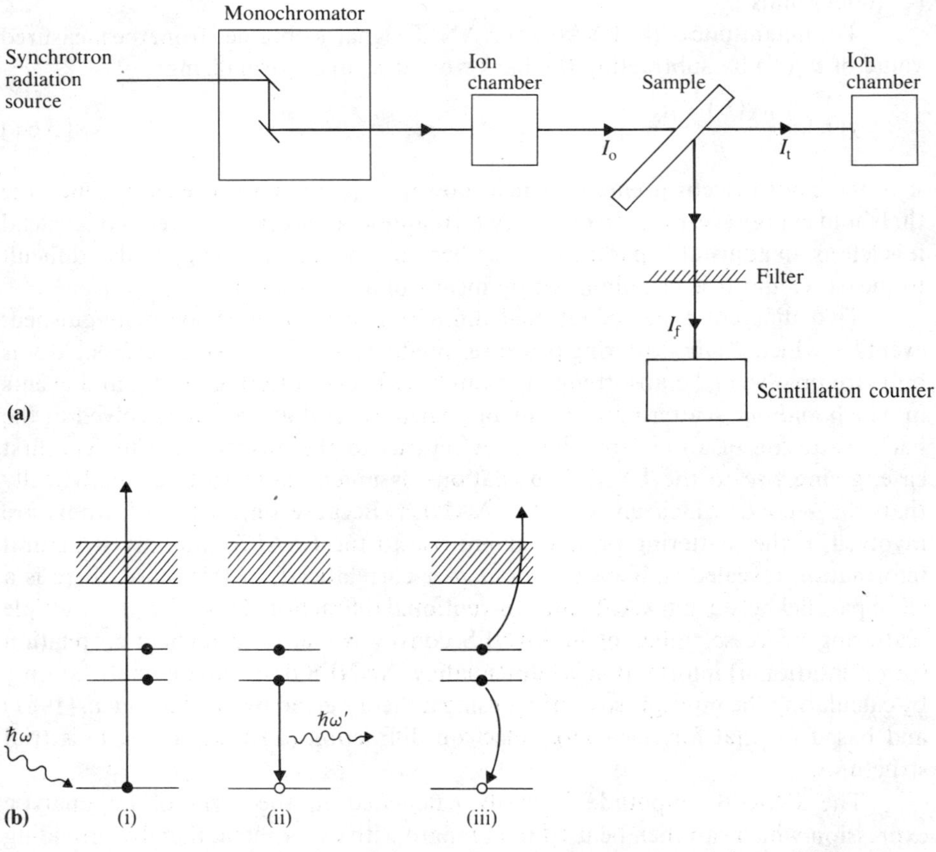

Fig. 3.18 (a) Schematic illustration of the experimental arrangement used for performing X-ray absorption spectroscopy measurements in both transmission and fluorescence modes.
(b) Electronic transitions in the initial X-ray absorption (i) and subsequent fluorescent radiative recombination (ii) or Auger non-radiative recombination (iii) processes.

by the photo-electron is filled by a p-electron from the L-shell, emitting an X-ray photon of characteristic energy less than that of the exciting X-rays (Fig. 3.18(b)). Thus by measuring the fluorescent X-ray emission, only those atoms which are excited are monitored, with a consequent dramatic increase in sensitivity. Radiative (fluorescent) recombination is the predominant recombination pathway for heavy elements. For lighter elements, Auger (secondary) electron emission is the preferred recombination mechanism (Fig. 3.18(b)), and monitoring of the Auger electron emission can also be used as a measure of $\mu_X(E)$; since the mean free path of the emitted Auger electrons is so short ($\simeq 20$ Å), this technique is particularly surface-sensitive. Although X-rays are usually used for photo-excitation, high-energy ($\simeq 100$ keV) electrons can also be used to excite indirectly the core electrons. These electronically excited electrons subsequently experience exactly the same internal diffraction processes as do those excited by X-rays; this results in the so-called 'extended energy-loss fine-structure' (EXELFS). Such experiments can be carried out in a conventional transmission electron microscope equipped with an energy-analysing detector, and have the advantage that very light elements can be studied; their disadvantage is that the samples must be in the form of very thin ($\simeq 100$ Å) films.

The magnitude of the EXAFS or XANES signal is obtained from the measured value of $\mu_X(E)$ by subtracting the background μ_o and normalizing:

$$\chi(E) = \frac{\mu_X(E) - \mu_o}{\mu_o} \qquad [3.64]$$

E is the photo-electron energy which, however, is not known exactly since the threshold energy (i.e. the zero of energy E_0) cannot be ascertained precisely; instead it is left as an adjustable parameter. The background absorption μ_o is also difficult to measure, and it is usually fitted by means of a polynomial.

Two different types of internal diffraction process must be distinguished: events in which *single* scattering processes predominate, i.e. in which *pairs* of atoms only are involved (the absorbing atom and the back-scattering atom), and events in which *multiple* scattering is dominant, where several atoms are involved in the back-scattering of an electron before it returns to the absorption site. The first case, giving rise to the EXAFS oscillations, is much easier to treat analytically than the second, which gives rise to XANES. Because only pairs of atoms are involved in the scattering process giving rise to the EXAFS, the only structural information revealed is, likewise, about pair correlations; in this sense, there is a close parallel between EXAFS and conventional diffraction. However, the multiple scattering processes inherent in XANES convey triplet (and higher) correlation (or orientational) information. Unfortunately, XANES data can be analysed only by calculating the multiple scattering, using a theory given by Durham *et al.* (1981) and based on that for low-energy electron diffraction (LEED), for various trial structures.

The EXAFS amplitude is easily calculated in the form of an analytic expression which can then be used to compare with experiment, thereby providing structural information *without* the need for trial structures. If one assumes single scattering, and moreover that the outgoing electron wavefronts are planar at the back-scattering shells of atoms (valid for energies $\simeq 80$ eV from the edge), then

[3.63] becomes (Sayers *et al.* 1971):

$$\chi(k) = -\sum_{j} \frac{N_j |f_j(\pi)|}{R_j^2 \quad k} \exp\left(-\frac{2R_j}{\lambda_e}\right) \exp(-2\sigma_j^2 k^2) \sin(2kR_j + 2\delta(k) + \eta_j(k)) \qquad [3.65]$$

where a sum over all shells of backscattering atoms, j, is taken, each containing N_j atoms at a distance R_j from the absorbing atom. The magnitude of the EXAFS is proportional to N_j, inversely proportional to R_j^2 (since both the outgoing and back-scattered waves are assumed to be spherical, decreasing in amplitude as $1/R$) and proportional to the back-scattering amplitude $|f_j(\pi)|$ from the atoms in the jth shell. The amplitude is attenuated because of the finite mean free path λ_e of the electrons in the material and by the Debye–Waller term involving r.m.s. displacements σ_j (static or thermal) about the equilibrium bond length along the line joining absorbing and backscattering atoms. Finally, the amplitude is sinusoidally modulated by a function involving the phase shift of the electron; the additional phase shifts $\delta(k)$ and $\eta(k)$ arise because the photo-electron is emitted and back-scattered, respectively from atomic potentials. The wavenumber of the photo-electron, k, is given by

$$k = \frac{2\pi}{\lambda} = \left[\frac{2m_e(E - E_o)}{\hbar^2}\right]^{1/2} \qquad [3.66]$$

where m_e is the electron rest mass. Note that this is *half* the normal *diffraction* wavevector Q or k ([3.9]). Thus, from [3.65] we expect the EXAFS to be a complicated oscillatory function, each shell contributing a sinusoid of a different period which mix together; note, however, that it is only the first few shells which contribute strongly because of the limited mean free path of the photo-electrons, and the large widths, σ, of higher-lying shells.

In the energy range $40 < E < 80$ eV from the edge, the curvature of the waves cannot be neglected and the simple [3.65] is no longer appropriate; therefore the full curved wave formula due to Lee and Pendry (1975) must be used (which is however still a single scattering theory).

The structural information available from EXAFS experiments can be seen by reference to [3.65]. In principle, the coordination number, N_j, the interatomic spacing, R_j, and the mean square deviation, σ_j^2, for each shell, j, of atoms surrounding the absorbing atom are obtainable; in practice, only the first and second neighbour shells are significant. Two methods of extracting the structural information are commonly employed. One approach is to treat all the structural parameters in [3.65] as adjustable variables and to vary these (perhaps with crystalline values as input parameters) until the best fit between the calculated EXAFS spectrum and the experimental data is achieved. The major difficulty in this, as in most analyses of EXAFS, is what values to take for the phase shifts δ and η. There are two possible remedies to this problem: either values may be calculated theoretically by solving Schrödinger's equation taking account of the perturbations to the potential caused by neighbouring atoms, or else comparison is made with a standard (usually a crystal) for which the structure is known and by fitting parameters to the EXAFS data of the standard, values of δ and η are thereby obtained which, it is assumed, may be transferred to the unknown system under study.

The second method of analysis of EXAFS data is to take the Fourier transform:

$$\phi(r)=\frac{1}{(2\pi)^{1/2}}\int_{k_{min}}^{k_{max}}\chi(k)M(k)k^{n}\exp(2ikr)\,\mathrm{d}k \qquad [3.67]$$

where $M(k)$ is a window function, and n can be 1, 2 or 3, the larger values weighting the data more at high k-values. This procedure gives rise to a series of peaks in real space corresponding to the first, second and perhaps higher shells: since EXAFS retains phase information, in contrast to X-ray diffraction, the Fourier transform of $\chi(k)$ consists of positive and negative excursions. The peaks do *not* lie at the correct distances if, as in [3.67], the Fourier transform is taken with respect to the argument $2ikr$ only, ignoring the phase-shift term $2\delta(k)+\eta_j(k)=\alpha_j(k)$. This is because the k-dependence of $\alpha_j(k)$ also contributes to the frequency of oscillation of $\chi(k)$; if $\alpha_j(k)$ is expanded in powers of k as $\alpha_j(k)=\alpha_o+(\partial\alpha/\partial k)k$, then the frequency of $\chi(k)$ would be proportional to $R_j+\partial\alpha/\partial k$ and conversely the positions of the peaks in the simple Fourier transform, [3.67], would appear at distances $R_j-\partial\alpha/\partial k$. Both techniques depend for success on the assumption of transferability of phase shifts and back-scattering amplitudes; i.e. δ, η and f are invariant between systems which are chemically similar. Although interatomic spacings obtained in this fashion can be very accurate (± 0.02 Å), the values of the coordination number are not so precise and can be $\simeq 50\%$ low due to changes in the emitter potential on ejection of the photo-electron, but the errors can be corrected for in comparative studies employing standards.

An example of the usefulness of the EXAFS technique in the study of amorphous materials is in the case of hydrogenated amorphous silicon (a-Si:H) and its alloys with C, N or Ge. These materials are of considerable technological interest in view of their photoelectric properties and the consequent applications in solar-cell devices (see sect. 7.2.4). The introduction of the alloying elements allows the optical band gap to be varied systematically, thereby optimizing the efficiency of the solar cell; however, the manner in which such atoms are structurally incorporated into a-Si:H network will control the variation in the band gap.

Figure 3.19(a) shows the Fourier-transformed Si K-edge EXAFS data for a series of amorphous silicon–nitrogen alloys, a-Si_xN_{1-x}:H (Mobilio and Filipponi 1987). The Si–Si and Si–N nearest-neighbour coordination shells are clearly distinguishable; however, only first-shell information can be reliably obtained from the EXAFS data for the amorphous samples; in contrast, the second peak is clearly distinguishable for crystalline Si_3N_4. Analysis of the data by means of curve-fitting $\chi(k)$ leads to the conclusion that the Si–Si and Si–N bond lengths are essentially independent of x (a circumstance also found for the other alloys containing C and N). However, there *is* a difference in the way in which N and Ge are incorporated, as is revealed by the EXAFS data. The average relative coordination number $C=N_A/(N_{Si}+N_A)$ of atom A around each Si atom is shown as a function of x for Ge and N alloys in Fig. 3.19(b). It is clear that the data for the N alloy can be understood only in terms of a *chemically ordered* model for the short-range order, one in which there are no N–N bonds. In contrast, Ge appears to alloy in

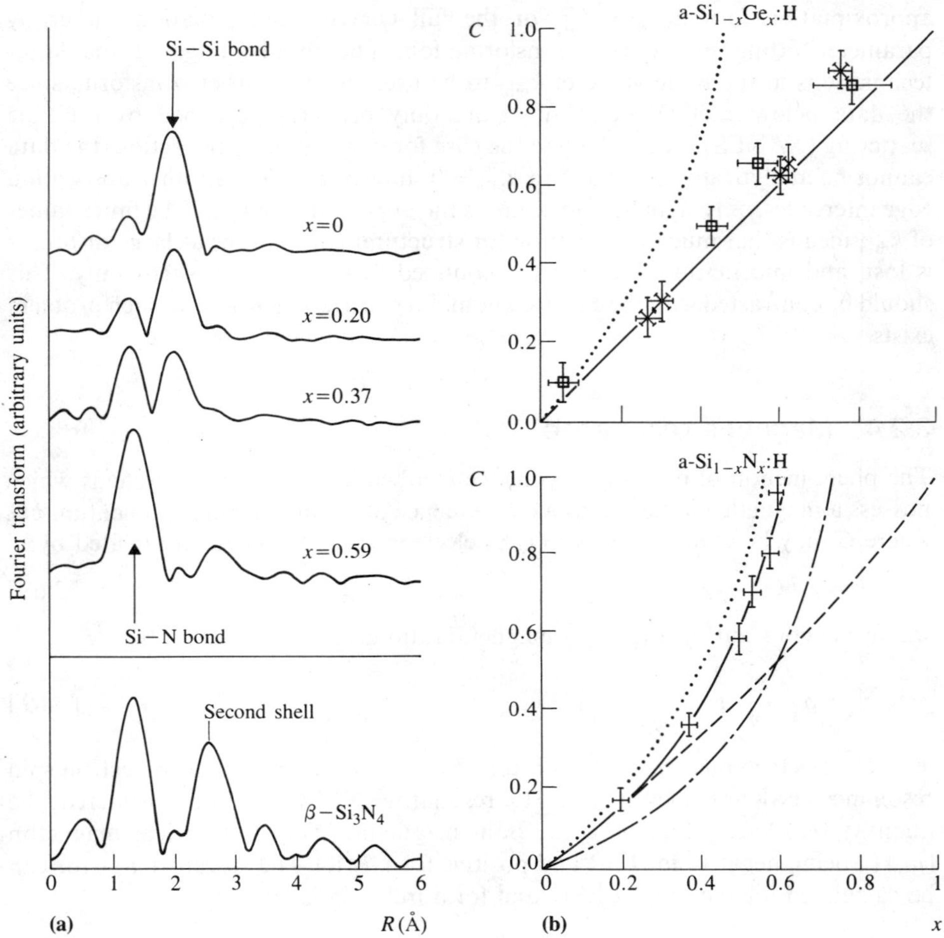

Fig. 3.19 (a) Silicon K-edge EXAFS spectra for a-$Si_{1-x}N_x$:H films, together with that for c-Si_3N_4 for comparison (Mobilio and Filipponi 1987).
(b) Average relative first-shell Si coordination numbers, $C = N_A/(N_{Si} + N_A)$, obtained from EXAFS data, as a function of composition for amorphous silicon alloys (Mobilio and Filipponi 1987). For a-$Si_{1-x}Ge_x$:H, the solid line represents a random-alloy model and the dotted line a chemically ordered model, whereas for a-$Si_{1-x}N_x$:H, the solid line represents a chemically ordered model including H, the dotted line a simple chemically ordered model, the dashed line a random-alloy model, and the dash-dotted line a chemically ordered model with no wrong bonds.

a *random* substitutional manner (so that for any intermediate value of x, Si–Si, Si–Ge and Ge–Ge bonds simultaneously coexist).

The advantages of EXAFS lie principally in the fact that it is an atom-specific, local-structure probe which is very suitable for the study of complex multicomponent systems, such as comprise many amorphous materials of interest, and for which conventional scattering techniques only give very limited information. It is an easy experiment to perform, and with the high photon fluxes available from synchrotron radiation sources, spectra can be taken in a matter of minutes. Fine structure above about 50 eV from the edge can be simply analysed using either the plane wave approximation formula ([3.65]) or the full curved wave equation, either by parameter fitting or Fourier transformation. The disadvantage of the latter technique is in the finite value of k_{min} to be used in the Fourier transform, since the data below $\simeq 50\,\mathrm{eV}$ ($k \lesssim 3\,\text{Å}^{-1}$) are only properly described by multiple scattering (XANES) theory. Unlike the case for conventional diffraction, the data cannot be analytically continued to $k_{\mathrm{min}} = 0$; however, unless another absorption edge intercedes, only signal-to-noise limits the upper value of k_{max}. The finite values of k_{min} means that much information for structural correlations at large distances is lost, and information is essentially confined to the first two shells only. This should be contrasted with the case for anomalous scattering, where no such problem exists.

3.2.4 Magnetic resonance

The phenomenon of magnetic resonance can be detected in those systems which possess a magnetic moment $\boldsymbol{\mu}$ as a consequence of having angular momentum $\hbar\boldsymbol{G}$, where $\boldsymbol{G}$ may be either a nuclear ($\boldsymbol{I}$) or electron spin ($\boldsymbol{S}$). These are related by:

$$\mu = \gamma_{\mathrm{g}} \hbar G \qquad [3.68]$$

where the constant γ_{g} is the gyromagnetic ratio given by:

$$\gamma_{\mathrm{g}} = g\left(\frac{e}{2m_{\mathrm{o}}}\right) \qquad [3.69]$$

m_{o} is the electron mass m_{e} or the proton mass m_{p} depending whether electron spin resonance (ESR) or nuclear magnetic resonance (NMR) is being considered; the quantity $(e\hbar/2m_{\mathrm{o}})$ is known as the Bohr magneton (β) or the nuclear magneton (μ_{N}) (e being negative for ESR and positive for NMR). The so-called g-factor can be calculated for the case of ESR, and for a free spin is given by:

$$g_{\mathrm{s}} \simeq 2\left(1 + \frac{\mu_{\mathrm{o}} c e^2}{4\pi h}\right) = 2.0023 \qquad [3.70]$$

When both orbital and spin moments are present, g becomes a function of the coupling between them and can also be calculated; for the case of NMR, g must be obtained from experiment through [3.71] below.

In the presence of a strong magnetic field, $\boldsymbol{H}$, the Zeeman interaction between the field and the magnetic dipole moment $\boldsymbol{\mu}$ is given by:

$$E = -\boldsymbol{\mu}\cdot\boldsymbol{H} = -g\left(\frac{e\hbar}{2m_{\mathrm{o}}}\right)\boldsymbol{G}\cdot\boldsymbol{H} = -g\left(\frac{e\hbar}{2m_{\mathrm{o}}}\right)mH \qquad [3.71]$$

and m is the appropriate magnetic quantum number. Resonance occurs when an oscillating magnetic field of an electromagnetic field at right angles to the static field and of (circular) frequency:

$$\omega = g\left(\frac{e\hbar}{2m_{\mathrm{o}}}\right)\frac{H}{\hbar} \qquad [3.72]$$

induces transitions between the Zeeman levels. For ESR and with $g = 2$ and

$H = 0.33$ T, the resonance frequency is $\nu = 9.3$ GHz, which lies in the so-called X-band range (9.2–12.4 GHz) of microwaves; in contrast, resonant frequencies for NMR lie in the radio-frequency range (MHz) because of the ratio of masses m_p/m_e ($= 1836$) and hence the gyromagnetic ratios.

Magnetic resonance is often measured by detection of the derivative signal obtained by the application of a small amplitude, oscillating magnetic field in addition to the static Zeeman field. The resonance condition may be probed by either sweeping the frequency or the magnetic field (cf. [3.72]); in practice for ESR and older NMR spectrometers, it is generally the field that is varied while the frequency remains constant (continuous wave, c.w.). However, modern NMR spectrometers operate with a fixed magnetic field and apply a pulse or series of pulses of r.f. power to the samples. Since each (square-wave) pulse contains very many Fourier components of frequency, all resonances are excited simultaneously; the application of repeated pulses and subsequent signal averaging therefore leads to a marked improvement in the signal-to-noise ratio compared with c.w. measurements. The magnetization response of the nuclei in the sample following the application of the r.f. pulse(s) is monitored as a function of time, giving the so-called 'free-induction decay' (FID). Fourier transformation of the FID then gives the absorption spectrum in the frequency domain.

Magnetic resonance is only observed in systems which have a non-zero angular momentum. For NMR, this is limited to those isotopes which have non-zero nuclear spin, and the following examples have been studied in the amorphous state (I given in units of $\hbar$ in parentheses): ^{1}H(1/2), ^{2}H(1), ^{10}B(3), ^{11}B(3/2), ^{17}0(5/2), ^{23}Na(3/2), ^{27}Al(5/2), ^{29}Si(1/2), ^{31}P(1/2), ^{51}V(7/2), ^{75}As(3/2), ^{77}Se(1/2), ^{125}Te(1/2) and ^{205}Tl(1/2). (On the other hand, ESR results when unpaired electrons exist, and this can be either when transition metals or rare earth elements (having an incomplete d or f shell) are present, or when unpaired electrons exist at defects, such as broken bonds; we will consider the latter in the chapter on defects (Ch. 6).)

The structural information that can be obtained from magnetic resonance spectra arises from the perturbations due to neighbouring atoms on the energy levels of the resonant species, whether nucleus or electron; the Zeeman interaction by itself ([3.71]) carries no structural information. For the case of ESR, values of the g-factor as measured using [3.72] often differ from the free spin value g_s ([3.70]). For spin–orbit coupling between the electron spin and the magnetic field resulting from the motion of other charges (for the case of nuclei with even spin), second-order perturbation theory gives for the g tensor:

$$g_{ij} = g_s(\delta_{ij} - \lambda \mathbf{\Lambda}_{ij}) \qquad [3.73]$$

where δ_{ij} is the Kronecker delta, λ is the spin–orbit coupling constant, and $\mathbf{\Lambda}$ is the tensor formed from a combination of the ground-state wavefunction $|0\rangle$ and excited states $|n\rangle$ which are mixed by the spin–orbit interaction, and the energy difference $(E_n - E_o)$ between the states:

$$\mathbf{\Lambda}_{ij} = \sum_{n \neq 0} \frac{\langle 0|L_i|n\rangle\langle n|L_j|0\rangle}{(E_n - E_o)} \qquad [3.74]$$

where L_i is the angular momentum operator. The g-shift is simply given by

$(g_{ij} - g_s) = -g_s \lambda \Lambda_{ij}$. It is always possible to find a coordinate system whereby the g-tensor is rendered diagonal and so the principal components g_{11}, g_{22} and g_{33} are generally recorded. Amorphous solids are generally isotropic, and so all orientations are taken by the paramagnetic centres; therefore resonance no longer occurs at the three discrete frequencies determined by g_{11}, g_{22} and g_{33} (if all different) but instead the absorption spectrum is broadened out into a 'powder pattern' (Taylor *et al.* 1975) in which the frequencies associated with the principal values of the g-tensor occur at shoulders or peaks in the spectrum. The directly measured derivative spectrum may therefore exhibit a split peak in both the positive and negative going parts, but if the (dipolar) broadening is too large, this resolved structure disappears and a broad, asymmetric line results (Fig. 3.20). The structural information obtainable from the g-tensor devolves on [3.74]. It is often possible to determine the coordination of a transition-metal ion in a glass (tetrahedral or octahedral, say) from a consideration of crystal–field splittings due to the ligands and the effect these have on the g-shifts; in addition, distortions from symmetry can often be distinguished. Wong and Angell (1976) give a large number of examples of the use of ESR in the study of paramagnetic ions in amorphous solids.

In general, the atomic nucleus (or nuclei) may have a non-integral spin, and the so-called 'hyperfine' interaction between the electron and nuclear spin occurs, resulting in hyperfine structure in the ESR line. The effective spin Hamiltonian representing ESR can thus be written:

$$\mathscr{H}_{\mathrm{ESR}} = \beta \boldsymbol{S}\cdot\mathbf{g}\cdot\boldsymbol{H} + \boldsymbol{S}\cdot\mathbf{D}\cdot\boldsymbol{S} + \boldsymbol{S}\cdot\mathbf{A}\cdot\boldsymbol{I} + \boldsymbol{I}\cdot\mathbf{Q}\cdot\boldsymbol{I} - \gamma_g \hbar \boldsymbol{I}\cdot\boldsymbol{H} \qquad [3.75]$$

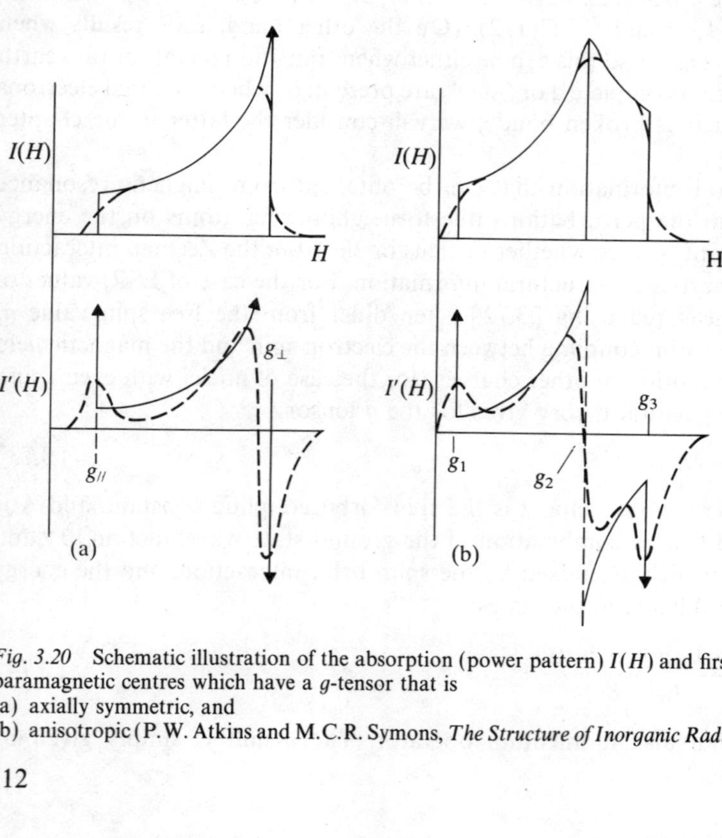

Fig. 3.20 Schematic illustration of the absorption (power pattern) $I(H)$ and first derivative $I'(H)$ for paramagnetic centres which have a g-tensor that is
(a) axially symmetric, and
(b) anisotropic (P.W. Atkins and M.C.R. Symons, *The Structure of Inorganic Radicals* (Elsevier: 1967)).

The first term describes the Zeeman splitting of the lowest energy levels, the second is the fine structure interaction produced by ligand field or spin–spin interactions in zero field, the third is the hyperfine interaction between the electron and the nucleus mentioned previously, the fourth is the nuclear quadrupolar interaction and the final term is the direct interaction between the nucleus and the magnetic field $\boldsymbol{H}$. The quantities **g**, **D**, **A** and **Q** are all tensors.

The Hamiltonian describing NMR can be written as

$$\mathscr{H}_{\text{NMR}} = \gamma_g \hbar \boldsymbol{I} \cdot \boldsymbol{\sigma} \cdot \boldsymbol{H} - \gamma_g \hbar \boldsymbol{I} \cdot \boldsymbol{H} + \boldsymbol{I} \cdot \mathbf{Q} \cdot \boldsymbol{I} + \boldsymbol{I} \cdot \mathbf{D} \cdot \boldsymbol{I} \qquad [3.76]$$

where $\boldsymbol{\sigma}$ is the magnetic shift tensor, and represents a 'chemical shift' for diamagnetic or slightly paramagnetic solids (or a 'Knight shift' for metals) resulting from the shielding of the external magnetic field from the resonant nucleus by the bonding electrons. Comparison of [3.75] and [3.76] shows that a formal similarity between ESR and NMR exists in that the electronic Zeeman term ($\beta \boldsymbol{S} \cdot \mathbf{g} \cdot \boldsymbol{H}$) has the same form as the nuclear Zeeman plus the magnetic shift term ($\gamma_g \hbar \boldsymbol{I} \cdot (\boldsymbol{\sigma} - \mathbf{1}) \cdot \boldsymbol{H}$) and the ESR fine-structure interaction ($\boldsymbol{S} \cdot \mathbf{D} \cdot \boldsymbol{S}$) is formally the same as the quadrupolar interaction in NMR ($\boldsymbol{I} \cdot \mathbf{Q} \cdot \boldsymbol{I}$); there the analogy ends, however, since there are no terms in the NMR Hamiltonian corresponding to the nuclear terms in ESR.

A general characteristic of NMR spectra obtained from solids is the broadness of the lines, particularly compared with liquids, and the consequent lack of resolution of the spectra. This broadening is a result of any 'anisotropic' terms (i.e. tensor components associated with various directions in the coordinate frame having different values) in the interactions comprising the NMR Hamiltonian ([3.76]) and arises as shown schematically in Fig. 3.20. It can be dramatically reduced to give high-resolution NMR in the solid state by adopting one or both of two techniques, 'magic-angle spinning' (MAS) or 'multiple-pulse sequencing'. The dipolar interaction is often the principal cause of line broadening and the Hamiltonian $\boldsymbol{I} \cdot \mathbf{D} \cdot \boldsymbol{I}$ can be rewritten in the form of a sum over all pairs of nuclei i, j (Andrew 1981):

$$\mathscr{H}_{\text{D}} = \sum_{i<j} \tfrac{1}{2} \gamma_i \gamma_j \hbar^2 r_{ij}^{-3} (\boldsymbol{I}_i \cdot \boldsymbol{I}_j - 3 I_{iz} I_{jz})(3 \cos^2 \theta_{ij} - 1) \qquad [3.77]$$

where $\boldsymbol{r}_{ij}$ is the internuclear displacement and θ_{ij} the angle between $\boldsymbol{r}_{ij}$ and the Zeeman field $\boldsymbol{H}_o$ which is in the z-direction in the laboratory frame. It is immediately apparent from [3.77] that the thermal motion of atoms that occurs in fluids reduces the dipolar interaction to zero, since the isotropic average of $\cos^2 \theta_{ij} = 1/3$. But what of solids? If the sample is rotated at an angle ϕ relative to the magnetic field direction, each vector $\boldsymbol{r}_{ij}$ rotates about the axis of rotation with the result that θ_{ij} becomes time dependent. In fact, the dipolar Hamiltonian can be separated into a constant term and a time-dependent part which has zero mean value; the constant term involves the factor $f(\phi) = (3 \cos^2 \phi - 1)$ and so if the sample is rotated at an angle of $\phi = \cos^{-1}(1/\sqrt{3}) = 54°44'$ (the 'magic' angle) to the magnetic field direction, $f(\phi) = 0$ and the dipolar interaction is completely removed (Andrew 1981). It is the chemical-shift interaction that contains most of the structural information of interest since $\boldsymbol{\sigma}$ is determined by the local structural environment around the resonant nucleus. Rapid sample spinning causes a change in the chemical shift anisotropy (CSA) Hamiltonian, similar to that suffered by the dipolar term, and it can be written in the form

$$\mathscr{H}^{\mathrm{CSA}} = f(\phi) \times (\text{other terms}) + \tfrac{3}{2}\boldsymbol{\sigma}_{\mathrm{i}} \boldsymbol{I}_z \boldsymbol{H}_{\mathrm{o}} \sin^2 \phi \qquad [3.78]$$

where $\boldsymbol{\sigma}_{\mathrm{i}}$ is the 'isotropic' value of the CSA tensor given by the trace of $\boldsymbol{\sigma}$:

$$\boldsymbol{\sigma}_{\mathrm{i}} = \tfrac{1}{3} \operatorname{tr} \boldsymbol{\sigma} \qquad [3.79]$$

and is the value of chemical shift observed in liquid-state spectra. At the magic angle, $f(\phi) = 0$ in [3.78] and $\sin^2 \phi = \frac{2}{3}$, i.e. the CSA Hamiltonian becomes equal to

$$\mathscr{H}^{\mathrm{CSA}} = \boldsymbol{\sigma}_{\mathrm{i}} \boldsymbol{I}_z \boldsymbol{H}_{\mathrm{o}} \qquad [3.80]$$

Thus, MAS causes the lines broadened by CSA interactions to collapse to narrow lines having values of chemical shift $\boldsymbol{\sigma}_{\mathrm{i}}$, exactly as if they were obtained using high-resolution liquid NMR; consequently, local differences in the structure reflected in subtle changes in values of chemical shift can in principle be detected in solid-state samples. This technique is particularly useful for (poly)crystalline samples where site disorder is minimal; in amorphous materials, though, the technique is really useful only where the degree of short-range topological disorder (e.g. bond-angle fluctuations) is small enough for the line broadening not to obscure the differences in chemical shift being sought, e.g. those caused by local coordination differences. Nevertheless, MASNMR has provided a degree of structural information for certain amorphous systems unattainable by any other technique.

Although line broadening due to CSA and dipolar interactions can be removed in principle by MAS, for the narrowing to be effective the sample-rotation frequency ω_r must be greater than the strength of the anisotropic interaction which it is desired to remove; if not, the anisotropic contributions relegated to 'spinning side-bands' are not sufficiently separated from the isotropic line for narrowing to be effective. This is possible for *heteronuclear* dipolar interactions ($1 < \omega_r < 10$ kHz), but *homonuclear* dipolar interactions (particularly ^1H–^1H) are much stronger than this, and MAS is ineffective. In this case, an alternative method must be employed which effectively rotates the nuclear spins in spin-space, rather than in the laboratory frame, by means of particular r.f. pulse sequences. Details of this multiple-pulse sequence technique are beyond the scope of this book and the interested reader should refer to the article by Mansfield (1981).

Line-narrowing resulting from MAS is not completely effective for quadrupolar interactions, however. Quadrupolar interactions occur in those nuclei (with $I \geqslant 1$) having an electric quadrupole moment (eQ) which is able to interact with the electric-field gradient (EFG) at the nucleus resulting from the distribution of electrons around the nuclei (in bonds, etc.); this distribution is obviously a sensitive function of the local structure and bonding. The quadrupolar interaction has several effects: it shifts the Zeeman levels, splitting the resonance line into $2I$ components and broadens the central line (e.g. $m = \frac{1}{2} \rightarrow m = -\frac{1}{2}$ for an $I = \frac{3}{2}$ nucleus like ^{11}B) by an amount that depends on the quadrupolar coupling constant

$$Q_{\mathrm{cc}} = e^2 qQ/h \qquad [3.81]$$

where eq is the largest component of the EFG tensor $\mathbf{V}$, defined as $eq = V_{zz} = \partial^2 V/\partial z^2$. The other quantity characterizing the EFG tensor is the 'asymmetry parameter'

η_a, describing the departure from axial symmetry:

$$\eta_a = (V_{xx} - V_{yy})/V_{zz} \quad [3.82]$$

The perturbation of the Zeeman frequency resulting from quadrupolar interactions can be written as the sum of two terms, one of which is first-order in Q_{cc} and the other which is second-order. It turns out that the first-order term *can* be removed by MAS, as for the dipolar interaction, since the Hamiltonian contains the factor $f(\phi)$ under spinning conditions; the second-order term, however, *cannot* be removed in this way since the spinning Hamiltonian involves a more complicated function of ϕ. (It can be minimized, nevertheless, by operating at very high magnetic fields, since the magnitude of the second term is inversely proportional to $\boldsymbol{H}_o$.) The very recent technical advance of double sample rotation, (Samoson *et al.* 1988, Chmelka *et al.* 1989), where the sample is rotated about *two* axes simultaneously, one being at the magic angle, offers considerable promise for the removal of second-order quadrupolar broadening effects as well. At the time of writing, however, this technique has not yet been applied to the study of amorphous materials.

An example of the application of broad-line (c.w.) NMR to the study of the structure of amorphous solids concerns glassy B_2O_3, and the structural changes that occur upon the addition of alkali oxides. There is much evidence, from e.g. X-ray and neutron diffraction studies, that in pure glassy B_2O_3, each B atom is coordinated by three O atoms in the form of a planar triangle, as in the hexagonal crystalline modification, but how the triangles themselves are connected together is still a matter of some dispute. C.W. NMR studies are of use principally in this system through the quadrupolar interaction. The ^{11}B NMR spectra for both glassy and hexagonal crystalline B_2O_3 are very similar, and both are characterized by a large quadrupolar coupling constant of 2.76 MHz, a value which is theoretically predicted for a planar BO_3 configuration. However, on addition of up to about 33% of Na_2O, a very sharp central line grows continuously in the spectrum (Fig. 3.21(a)). This has been ascribed to the creation of tetrahedral BO_4 units on the addition of alkali oxide. (Certain anomalies in the UV absorption spectra are also observed in glassy B_2O_3 upon the addition of alkali oxides and this effect has been termed the 'borate anomaly'. Originally it was thought that the anomalies in the optical spectra coincided exactly with the creation of tetrahedral BO_4 units, but this correlation does not appear so clear when quantitative NMR data are examined.) Q_{cc} for ^{11}B in a near-tetrahedral configuration is very small ($\simeq 50$ kHz) because of symmetry, and hence the line is narrower. Analysis of the area under the NMR absorption curves confirmed the validity of the theoretical relation linking the number of tetrahedral B atoms (N_4) to the molar percentage (x) of modifier (Fig. 3.21(b)): $N_4 = x/(100 - x)$ (each alkali ion added converts one B atom to tetrahedral coordination). This would be extremely difficult to detect using scattering techniques, and conventional EXAFS would be unsuitable since very soft X-rays are required to probe the B K-edge. A review of the application of c.w. (broad-line) NMR to the study of borate glasses has been given by Bray (1978).

Although broad-line NMR studies have been extremely useful in the investigation of a limited number of amorphous solids, particularly borates, the

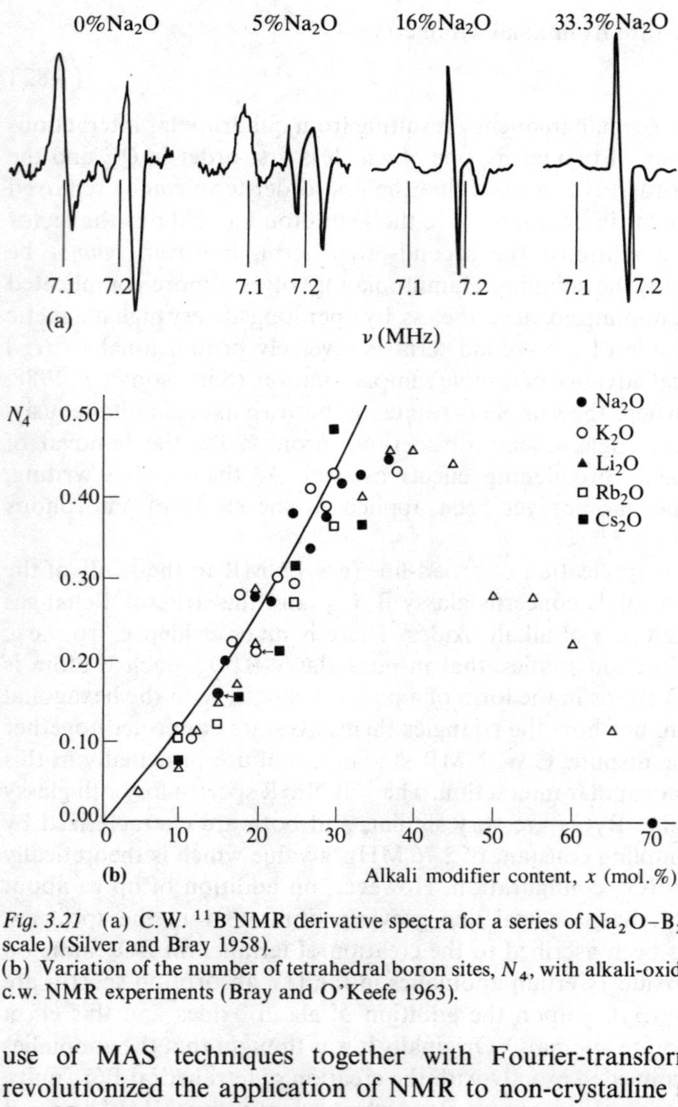

Fig. 3.21 (a) C.W. ^{11}B NMR derivative spectra for a series of $Na_2O–B_2O$ glasses (not to the same scale) (Silver and Bray 1958).
(b) Variation of the number of tetrahedral boron sites, N_4, with alkali-oxide content, as obtained from c.w. NMR experiments (Bray and O'Keefe 1963).

use of MAS techniques together with Fourier-transform spectrometers has revolutionized the application of NMR to non-crystalline solids. Studies of ^{29}Si and ^{27}Al have been particularly fruitful, and some examples are given below.

It has been found that NMR is particularly sensitive to changes in local coordination number about the resonant nucleus, since in this case the changes in chemical shift values can be considerable. One example concerns aluminium-containing glasses, e.g. aluminosilicates, in which the Al can occupy both tetrahedral sites (i.e. substitutional and network-forming), and octahedral sites (i.e. interstitial and network modifying):

$$2Al_2O_3 \rightarrow [Al^{3+}]_o + 3[AlO_{4/2}^-]_t$$

The addition of P_2O_5 increases the solubility of Al_2O_3 in SiO_2, and furthermore forces the Al into tetrahedral coordination:

$$Al_2O_3 + P_2O_5 \rightarrow 2[AlO_{4/2}^-]_t + 2[PO_{4/2}^+]_t$$

Thus the ratio of octahedral to tetrahedral Al sites can be changed. However, conventional spectroscopic techniques have proved unable to distinguish between (AlO_6) and (AlO_4) sites: X-ray photoemission spectroscopy, for example, provides no evidence for the change in the Al 2p core states expected if the ratio of octahedral to tetrahedral sites alters, and broad-line ^{27}Al NMR cannot distinguish between them either, because the lines are too broadened by the effect of anisotropic interactions. However, MASNMR produces such a dramatic increase in resolution that sites with the two types of coordination can readily be distinguished: octahedral (AlO_6) sites (with Al second neighbours) are found to give peaks with chemical shifts in the range 0 to +22 p.p.m. (with respect to a standard $AlCl_3$ solution in which the Al is octahedrally coordinated, i.e. $Al(H_2O)_6^{3+}$), shifting to −15 to −20 p.p.m. if P atoms are the second neighbours, since they are more electronegative than Al atoms, whereas tetrahedrally coordinated (AlO_4) sites (with P second neighbours) have chemical shifts near +40 p.p.m. The ^{27}Al MASNMR spectrum of a binary Al_2O_3–P_2O_5 glass shown in Fig. 3.22(a) confirms these findings (Müller *et al.* 1983).

The influence of the incorporation of P_2O_5 on the coordination of glass-

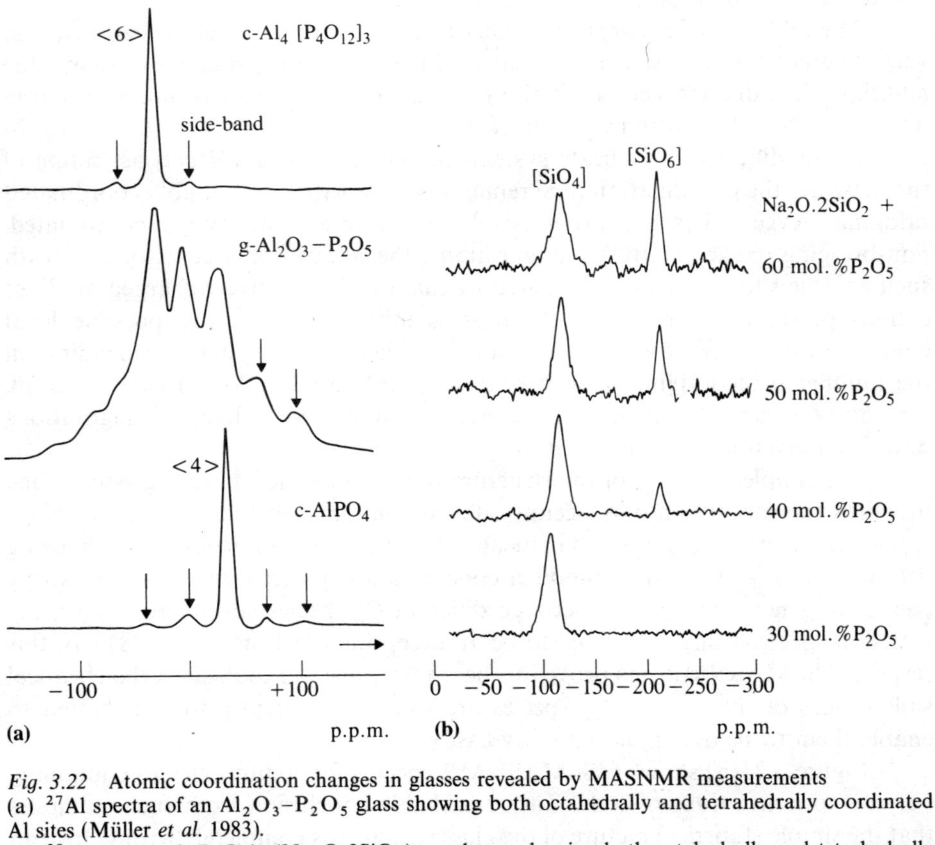

Fig. 3.22 Atomic coordination changes in glasses revealed by MASNMR measurements
(a) ^{27}Al spectra of an Al_2O_3–P_2O_5 glass showing both octahedrally and tetrahedrally coordinated Al sites (Müller *et al.* 1983).
(b) ^{29}Si spectra of $(P_2O_5)_x(Na_2O.2SiO_2)_{1-x}$ glasses showing both octahedrally and tetrahedrally coordinated Si sites (Dupree *et al.* 1987).

forming atoms has also been clarified by the use of MASNMR with other systems, notably silicates. Under normal thermodynamic conditions, Si atoms are tetrahedrally coordinated by oxygen atoms in crystalline and amorphous silicate materials. Preparation at high pressures, however, can result in octahedrally coordinated Si becoming energetically favourable, as for example in the stishovite modification of crystalline SiO_2, and in glasses quenched from melts subjected to high pressures (>60 kbar) (Ohtani *et al.* 1985); octahedral coordination is also found in crystalline SiP_2O_7. The chemical shift for (SiO_4) sites is $\simeq -110$ p.p.m. (with respect to tetramethylsilane), whereas that for (SiO_6) is very different at $\simeq -200$ p.p.m. Normally, the addition of alkali oxides to silicate glasses preserves the tetrahedral coordination whilst depolymerizing the network (see later). It is very interesting to note, therefore, that the addition of $\gtrsim 40$ mol.% of P_2O_5 to $Na_2O.2SiO_2$ results in glasses exhibiting clear evidence from ^{29}Si MASNMR of the presence of octahedrally coordinated (SiO_6) sites (Fig. 3.22(b)); the strongly polarizing electronegative P atoms can be regarded as providing a chemical 'internal' pressure driving the increase in coordination of the Si. (It is interesting to note, however, that attempts to vitrify pure SiO_2–P_2O_5, without the presence of alkali modifiers, invariably produced only tetrahedral Si sites (Weeding *et al.* 1985), although the crystallization products of such glasses *did* show evidence for octahedral coordination.)

The addition of network modifiers to glass-forming materials can result in very different types of structural change. In borate and germanate systems, the addition of alkali up to a certain limiting amount causes an *increase* in coordination number of the glass-forming cation (from 3 to 4 for B, from 4 to 6 for Ge). In contrast, modification of silicate systems in general causes a *depolymerization* of the network, the coordination of Si remaining tetrahedral, but doubly coordinated bridging oxygen sites are progressively transformed into singly coordinated, non-bridging oxygen (NBO) configurations; the negative charge associated with such anionic sites is then compensated by that on the positively charged modifier cations presumed to reside in the near vicinity. There are *five* possible local configurations for Si atoms in such modified silicate glass systems, depending on the number n of bridging oxygen nearest neighbours bonded to each Si atom. Obviously, n can take values in the range 0–4, and these different configurations have been given in the nomenclature Q_n in the literature.

The simplest picture of the structure of such modified silicate glasses is one in which the modifier atoms occupy sites at random and the distribution of Q_n species is determined purely statistically, the relative proportion of each being determined only by the total modifier concentration. Other forms of spectroscopy cannot in general distinguish between different Q_n species, although Si K_β X-ray emission spectroscopy is claimed to be an exception (de Jong *et al.* 1981). In this respect ^{29}Si MASNMR has proved to be a very powerful tool, since the chemical shift ranges of the various Q_n species are usually sufficiently well separated to enable them to be distinguished fairly easily.

Figure 3.23(a) shows ^{29}Si MASNMR spectra for sodium glasses containing varying concentrations of soda. The remarkable conclusion from such studies is that the simple statistical picture of the glass structure is completely wrong. Instead of all five Q_n species being detected at any given composition, only a *single* Q_n

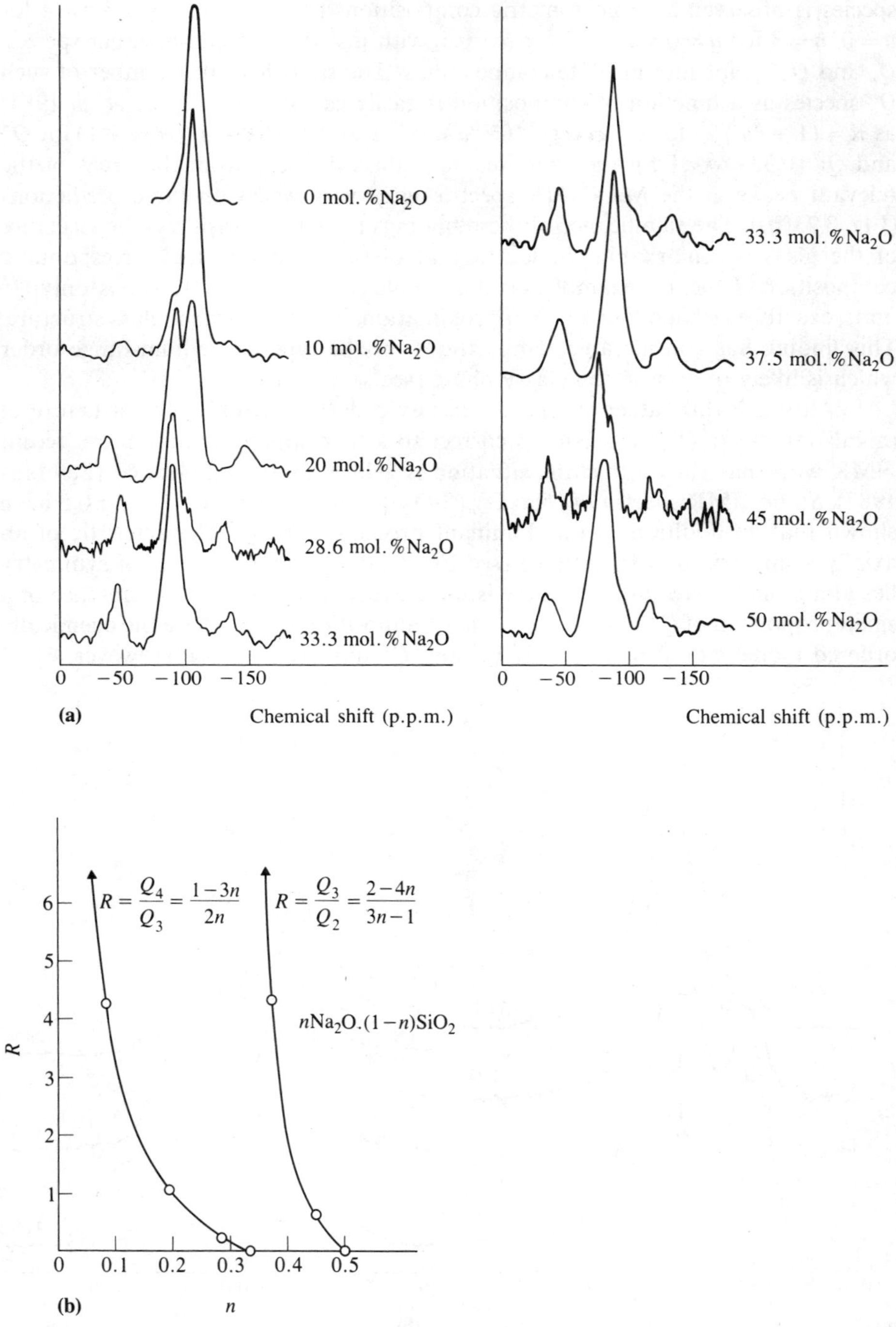

Fig. 3.23 NMR evidence for local structural ordering in soda silicate glasses (Dupree *et al.* 1984).
(a) ^{29}Si MASNMR spectra as a function of alkali content showing evidence for a single Si site (Q^n species) at stoichiometric compositions.
(b) Calculated and observed ratios for the proportion of Q^n species as a function of alkali content.

species is observed at stoichiometric compositions $(Na_2O)_a(SiO_2)_{1-a}$: $n=4$ for $a=0$, $n=3$ for $a=0.33$, $n=2$ for $a=0.5$, with just the two end-member species, Q_n and Q_{n-1}, for intermediate compositions. The ratio R of the number of such Q_n species as a function of composition is easily calculated (Dupree *et al.* 1984) as $R=(1-3a)/2a$ for Q^4 and Q^3 $(0<a<0.33)$, and $R=(2-4a)/(3a-1)$ for Q^3 and Q^2 $(0.33<a<0.5)$: the experimental values deduced from the areas of the relevant peaks in the MASNMR spectra agree very well with these predictions (Fig. 3.23(b)). The significance of these findings is that, surprisingly, the structure of the glass is remarkably similar to that of the crystal of the corresponding composition. Thus, for example, crystalline $Na_2O.2SiO_2$ $(a=0.33)$ has only Q^3 units, exactly as is found, to a first approximation, in the equivalent glass structure. This finding has considerable ramifications for the type of medium-range order which is likely to exist in the glassy phase (see sect. 3.4).

Although this rather simple chemically ordered picture of the structure of (alkali) modified silicate glasses is correct to a first approximation, more recent NMR work has shown that the situation is a little more complicated (Stebbins 1987). *Static* NMR spectra of $(Na_2O)_a(SiO_2)_{1-a}$ glasses with $0.33<a<0.5$ have shown that, in addition to the dominant profile due to Q^3 characteristic of an axially symmetric powder pattern (see Fig. 3.20(a), where the axis of symmetry lies along the Si–NBO bond), there is unequivocal evidence for the presence of a small proportion of Q^4 units (2–7%) in a composition range where the chemically ordered model would predict only Q^3 and Q^2 units (Fig. 3.24). However, MAS

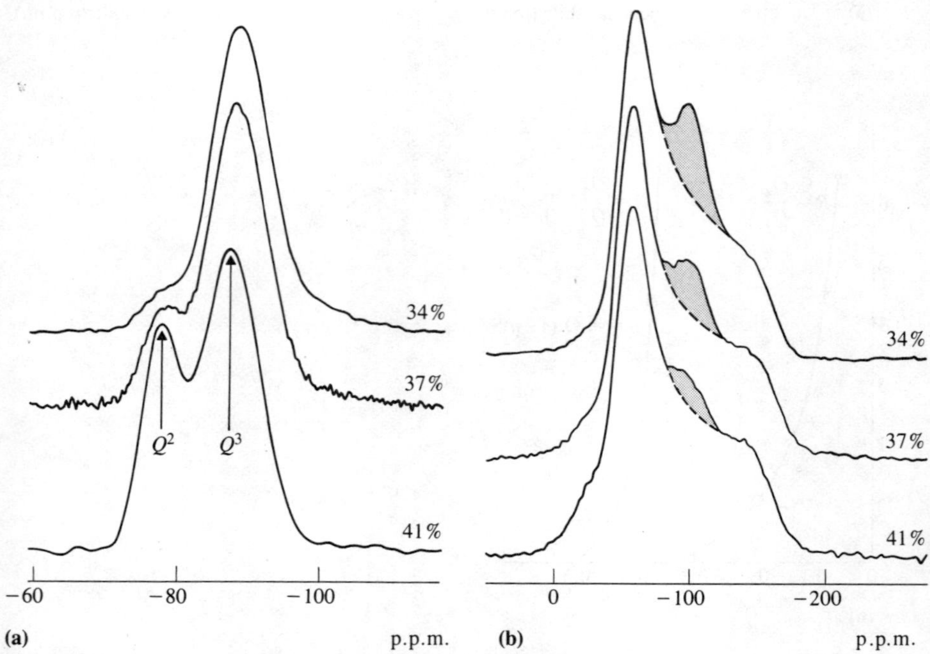

Fig. 3.24 (a) MASNMR spectra for soda silicate glasses showing the existence of Q^2 and Q^3 species, with the latter predominating near the stoichiometric composition of 33 at.% Na_2O.
(b) Static NMR spectra of the same glasses, showing clear evidence for Q^4 species (shaded peak) superimposed on the powder-pattern profile of Q^3 species.

spectra of the same glasses show no evidence for the presence of such Q^4 units (Stebbins 1987), so it must be concluded that MASNMR data by themselves do not always provide completely quantitative information about the relative proportion of different types of site.

In certain circumstances, MASNMR studies can provide structural information other than just local site symmetries, as discussed above. For example, the ^{29}Si MAS spectrum of pure vitreous SiO_2 is a symmetric line centred at a chemical shift of $\simeq -110$ p.p.m., with a width some ten times that of crystalline polymorphs (Gladden *et al.* 1986). The greater width of the resonance line for the glass reflects the static disorder in the Si–O–Si bond angle θ_O characteristic of the amorphous state. The chemical shift δ of the ^{29}Si resonance would be expected to depend subtly on θ_O (even though the number of oxygen nearest neighbours remains unchanged) because a change in θ_O would be associated with a change in Si–O bond charge due to a change in orbital overlap. An empirical relationship between δ and θ_O determined crystallographically has been found to hold for a variety of crystalline polymorphs of SiO_2 (Smith and Blackwell 1983), namely

$$\delta = a + b \sec \theta_O \qquad [3.83]$$

where a and b are constants. Dupree and Pettifer (1984) suggested that [3.83] could be used to convert the dispersion in values of chemical shift $g(\delta)$ observed experimentally for a-SiO_2 into a distribution of oxygen bond angles $P(\theta_O)$. Figure 3.25 shows the distribution $P(\theta_O)$ obtained in this way using an analytical method (Gladden *et al.* 1986), compared with an estimate for $P(\theta_O)$ obtained in an X-ray diffraction study by Mozzi and Warren (1969). It can be seen that the two methods predict an asymmetric distribution for $P(\theta_O)$ with the same value for the most probable oxygen bond angle, $\theta_O \simeq 143°$, although the predicted widths of the distributions differ.

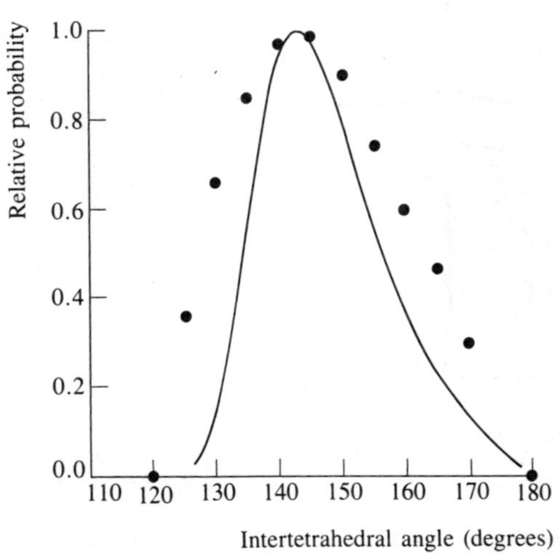

Fig. 3.25 Oxygen bond-angle distribution deduced from a ^{29}Si MASNMR spectrum of v-SiO_2 (solid line) compared with that deduced by Mozzi and Warren (1969) from X-ray diffraction data (Gladden *et al.* 1986).

The final example of the application of NMR to the study of the structure of amorphous materials that we wish to discuss here concerns a-Si:H, a material of considerable technological importance. The incorporation of approximately 10 at.% H is essential for the material to have useful electronic properties; obviously a proper understanding of the electronic behaviour of this material relies on a knowledge of the structural environment of the H. (A detailed review of the structure of a-Si:H has been given by Elliott (1989).) For this material, though, ^{29}Si MASNMR studies have been uninformative, yielding broad and featureless spectra whose width is most likely determined by static bond-angle fluctuations of the Si-based sub-lattice (as for a-SiO_2 above) rather than by the chemical-shift differences due to different hydride groupings SiH_n ($n = 1, 2, 3$). Proton NMR studies, on the other hand, have been much more informative, and evidence has been obtained for (at least) two types of site for the H. Figure 3.26 shows clearly that the ^1H line has two components: a narrow line and a wide line some ten times broader (Reimer *et al.* 1981). The linewidth in such cases is inversely proportional to the spin–spin relaxation time T_2, which varies with the spin–spin (^1H–H) distance R_{ij} as R_{ij}^{-6} (if homonuclear dipolar interactions are responsible for the relaxation). The narrow component has been ascribed to isolated *monohydride* groups (SiH), and the broad line to *clusters* of hydrogenous species, e.g. dihydride SiH_2 units, polysilane regions $(SiH_2)_n$, or monohydride units covering the surfaces of voids, the extra width in these cases coming from the ^1H–^1H dipolar broadening which is not removed by MAS. Heat treatment of H-rich a-Si:H films causing H effusion, or deposition on high-temperature substrates leading to a minimum of H incorporation and electronic-grade material,

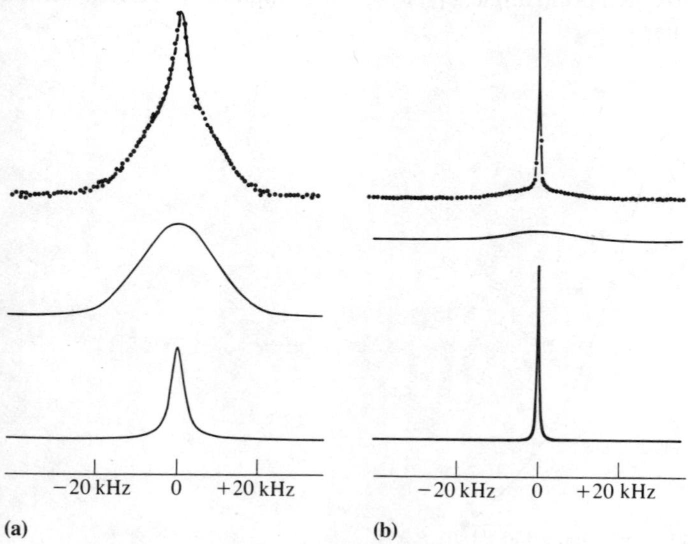

Fig. 3.26 ^1H NMR spectra for a-Si:H for:
(a) a sample deposited at a low substrate temperature ($T_s = 25°C$), and
(b) after annealing at 400°C during which effusion occurred.
Each spectrum can be deconvoluted into two components: a broad line due to clustered hydrogen species, and a narrow line due to isolated monohydride species (Reimer *et al.* 1981).

results in NMR spectra in which the narrow line becomes predominant at the expense of the broad line, indicating that the effusing H originates in the clustered hydrogen regions.

In addition to the information on H bonded directly to Si in different configurations, as outlined above, ^{1}H NMR studies have also produced unequivocal evidence for the existence of *molecular* H_2 trapped in microvoids in the matrix of a-Si:H. At very low temperatures ($T \leqslant 5$ K) a third component of the ^{1}H line, consisting of two additional lines situated symmetrically either side of the main peak shown in Fig. 3.26, is observed which is practically identical to the so-called Pake doublet powder pattern characteristic of solid H_2 (Boyce and Stutzmann 1985). The conclusion that H_2 molecules in the ortho state ($I=1$) are trapped in a-Si:H is reinforced by studies of the ^{1}H NMR spin-lattice relaxation time, T_1. Ortho-H_2 relaxes very efficiently by means of interactions between the rotational angular momentum of the molecule and phonons of the host lattice, whereas para-H_2 ($I=0$) does not. At room temperature, a thermodynamic equilibrium is established so that *o*-H_2 and *p*-H_2 molecules occur in the ratio 3:1, but at low temperatures an *o–p* transition takes place. Measurements of samples of a-Si:H held at $T = 4$ K for many days show a progressive increase in T_1 (Carlos and Taylor 1982), a finding only understandable in terms of an *o–p* transition of H_2 molecules entrained in the amorphous solid.

We conclude this section with a brief description of a new NMR technique which has been applied to a-Si:H with interesting results. Normally, the selection rule operating in NMR is $\Delta M = +1$; however, by means of particular r.f. pulse sequences, a system of N nuclear spins (each with $I = \frac{1}{2}$) can be forced to undergo what is essentially a cooperative ('multiple-quantum') transition in which the *total* change in magnetic quantum number can be $\Delta M \leqslant N$ (Baum *et al.* 1977). The coupling between the spins which is employed to produce the multiple-quantum transitions is the dipolar interaction, the magnitude of which varies as R_{ij}^{-3}, where R_{ij} is the inter-spin distance. Thus, for completely isolated clusters containing N spins each, the maximum quantum transition possible is $\Delta M = N$, because there is effectively no coupling between different clusters. However, for close clusters, or otherwise connected networks of spins, ΔM progressively increases as the 'preparation' time in the r.f. pulse sequence is increased to allow time for more distantly separated spins to interact; in fact, ΔM can increase without limit for an infinitely large, completely connected spin network.

The results of this multiple-quantum NMR technique applied to various samples of a-Si:H, prepared by GD decomposition and with different H contents (Baum *et al.* 1986, Gleason *et al.* 1987) are most interesting (Fig. 3.27). At low H contents ($x \leqslant 8$ at.%), i.e. for electronic-grade material, a clear plateau behaviour is observed for N, the effective number of correlated spins, as a function of preparation time τ. This indicates that isolated, well-defined clusters of H atoms exist in a-Si:H, apparently of about six atoms each. As the H content increases, this plateau behaviour is progressively lost, until for $x = 50$ at.% H, when the material is a polysilane ($(SiH_2)_n$), N appears to increase without limit. Thus, it appears that even device-quality a-Si:H material contains clustered H, and moreover that the additional defect states which are introduced when $x \geqslant 10$ at.%, leading to a deterioration in electronic properties, appear to accompany or are

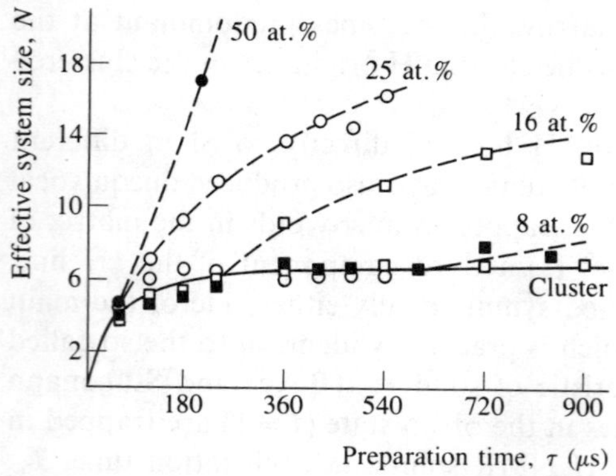

Fig. 3.27 Multiple-quantum ^1H NMR results for a-$Si_{1-x}H_x$ (Baum *et al.* 1986). The plateau behaviour indicates that isolated clusters of about six H atoms exist for $x \lesssim 10$ at.%, whereas for higher H contents the protons are in increasing spin contact.

caused by the extended connected clusters of H atoms that exist at higher H contents.

3.2.5 Mössbauer spectroscopy

Mössbauer spectroscopy, like NMR, is a nuclear spectroscopy, with the difference that γ-ray transitions between nuclear energy levels are involved. The technique relies for its sensitivity on the discovery by Mössbauer that γ-ray emission from atoms in a *solid* can be *recoil-free*, i.e. without loss of energy. Typically, an experimental set-up includes a source of the γ-rays, an absorber (i.e. a sample containing the resonant nucleus), a detector (e.g. a scintillation counter) and a drive system capable of imparting a relative linear velocity between source and absorber. The physical process underlying the technique may be understood by reference to the ^{57}Fe isotope, the most commonly used in Mössbauer studies. In this case the source contains radioactive ^{57}Co which converts to an excited state, $^{57}Fe^*$; this subsequently emits a recoil-less 14 keV γ-ray, which is correspondingly a very narrow emission. The absorber, containing naturally occurring or isotopically enriched ^{57}Fe in the sample, then absorbs the γ-ray in a recoil-less transition. The resonant γ-ray absorption in the sample is detected by moving source and absorber relative to each other with a variable small velocity v (a few mm s^{-1}). The Doppler shift in the frequencies and the extreme narrowness of the absorption line restrict the resonance to a particular value of v; otherwise the γ-ray transmission is high.

A number of interactions can perturb the resonance line which otherwise, if emitting and absorbing atoms had exactly the same structural environment, would occur at $v = 0$. These interactions are analogous to those which are important in NMR (sect. 3.2.4). The first is called the 'isomer' (or chemical) shift, and arises from the existence of a non-zero s-electron charge density at the nucleus; the effect of the interaction is to shift the relative energy of ground and excited nuclear levels (and hence that of the resonance line) without lifting the degeneracy (thereby splitting the line). The isomer shift can carry structural information since changes in the local structural environment of the resonant nucleus will produce changes in the electron charge density. A quadrupolar interaction between the electric field gradient (EFG) at the nucleus (due to a non-isotropic bonding electron charge density) and the nuclear energy levels can also occur. For ^{57}Fe, for example, the ground state ($I = \frac{1}{2}$) is unaffected, but the degeneracy of the excited state ($^{57}Fe^*$ with $I = \frac{3}{2}$) is partially lifted and the level is split into two, corresponding to $I = \pm\frac{1}{2}$ and $\pm\frac{3}{2}$ separated by an energy

$$E = \tfrac{1}{2}Q_{cc}(1 + \tfrac{1}{3}\eta_a^2)^{1/2} \qquad [3.84]$$

Thus, the resonance line becomes split into a 'quadrupolar doublet'. Finally, if a magnetic field (either external or internal) acts at the nucleus, the degeneracy of both ground and excited nuclear energy levels is completely lifted, yielding a characteristic six-line 'magnetic hyperfine' spectrum in the case of ^{57}Fe; six transitions only are allowed because of the $\Delta m = 0, \pm 1$ selection rules. The intensities of the hyperfine lines exhibit a characteristic 3:2:1:1:2:3 ratio if the (internal) fields are distributed isotropically.

Although ^{57}Fe Mössbauer studies are by far the most numerous, other nuclei (including ^{119}Sn, ^{121}Sb, ^{125}Te, ^{129}I, ^{151}Eu and ^{169}Tm) have also been studied in the amorphous state. It is apparent from this list that Mössbauer spectroscopy is therefore most useful in the study of amorphous metals, although there are examples of its use for non-metallic systems as well, such as Fe-doped silicate glasses and Te-containing chalcogenide glasses. This technique is useful not only as a structural probe, however; considerable information concerning the magnetic properties of metallic systems can also be obtained from Mössbauer spectra (see Gonser and Preston (1983) for a review).

An example of the Mössbauer spectrum of the ferromagnetic transition-metal–metalloid glass $Fe_{80}B_{20}$ (Gonser and Preston 1983) is shown in Fig. 3.28(a). The basic six-line hyperfine pattern is evident, and is denoted by the labels α–η.

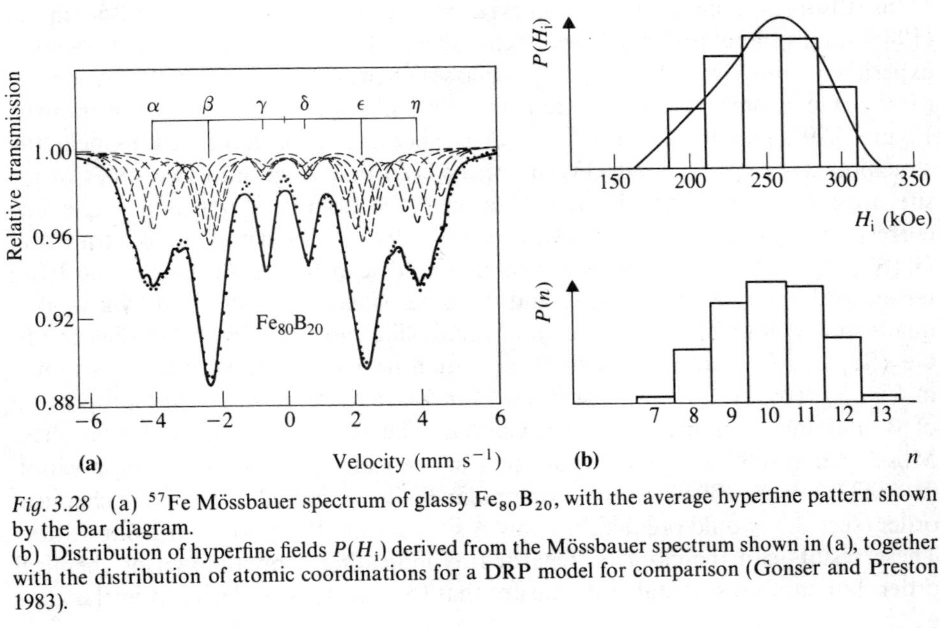

Fig. 3.28 (a) ^{57}Fe Mössbauer spectrum of glassy $Fe_{80}B_{20}$, with the average hyperfine pattern shown by the bar diagram.
(b) Distribution of hyperfine fields $P(H_i)$ derived from the Mössbauer spectrum shown in (a), together with the distribution of atomic coordinations for a DRP model for comparison (Gonser and Preston 1983).

It is apparent, however, that a single hyperfine spectrum does not fit the broad lines of the data; rather, a distribution of internal magnetic hyperfine fields $P(H_i)$ appears to be present. Such a distribution could result from a number of structural causes such as topological or compositional fluctuations. With the assumption that just five discrete values of H_i can be used to represent $P(H_i)$, this histogram distribution can be varied to give the best overall fit to the experimental spectrum. The result is shown in Fig. 3.28(b) (Gonser and Preston 1983); it is interesting that the shape of $P(H_i)$ inferred in this way closely resembles the distribution of nearest-neighbour coordination numbers $P(n)$ characteristic of a model of a dense random packing of spheres for the metallic glass structure (see sect. 3.5.4). However, more detailed work by Le Caer and Dubois (1987) has shown that the hyperfine field experienced at the Fe nucleus is sensitive mainly to the number of metalloid nearest neighbours.

Le Caer and Dubois (1987) have also discussed EFG interactions in the absence of hyperfine fields for non-magnetic metallic systems. Although in principle it should be possible to extract structural information from the degree of quadrupolar splitting, in practice this depends on the availability of a universal theory which can be used to calculate the EFG tensor, and, as Le Caer and Dubois (1987) stress, no such theory is available. Thus, unambiguous information about the local structural environment of probe atoms (e.g. Fe) in metallic glasses cannot yet be obtained solely from Mössbauer measurements.

In non-metallic systems, Mössbauer spectroscopy has often been used to explore the structural environment of Fe^{2+} and Fe^{3+} ions in oxide glasses (see e.g. Wong and Angell 1976), where it is found, for example, that Fe^{3+} ions apparently have a preference for tetrahedral coordination by oxygen ligands in alkali silicate glasses, but octahedral coordination is preferred in phosphate glasses.

Boolchand and co-workers have used a number of variants of Mössbauer spectroscopy to investigate the structure of chalcogenide glasses, using for example ^{119}Sn absorption and ^{129}I emission experiments to study a-$GeSe_2$ (see Boolchand (1985) for a general review of these techniques). They performed ^{119}Sn Mössbauer experiments on $(Ge_{0.99}Sn_{0.01})_xSe_{1-x}$ glasses to probe the structural environment of the Ge atoms, assuming that the Sn substitutes for it isomorphously. Figure 3.29(a) shows the results of such experiments for compositions near the stoichiometry $GeSe_2$ ($x = 0.333$). It can be seen that in general two types of Ge site appear to be probed by the Sn atoms: the dominant line is a single line corresponding to a so-called A-site, which is ascribed to a symmetrical tetrahedral $Ge(Se_{1/2})_4$ unit (the symmetry of which would lead to a vanishing of the EFG tensor at the tetrahedron centre), and the other site, B, characterized by a weaker quadrupolar doublet, ascribed to an 'ethane-like' grouping with Ge–Ge bonds, $Ge_2(Se_{1/2})_6$. The variation of the B-site intensity with composition (x) is shown in Fig. 3.29(b), and it can be seen that for Ge-rich compositions the proportion of B sites increases markedly, as expected. The surprising feature is that these Mössbauer results seem to indicate that at the exact stoichiometric composition ($x = 0.333$) there still exist some homopolar B sites; a model of complete chemical order (sect. 3.3) would predict that only A sites exist at this particular composition. These results can therefore be regarded as indicating a breakdown of chemical order, but care must be taken to ensure that (micro)phase separation in the glass

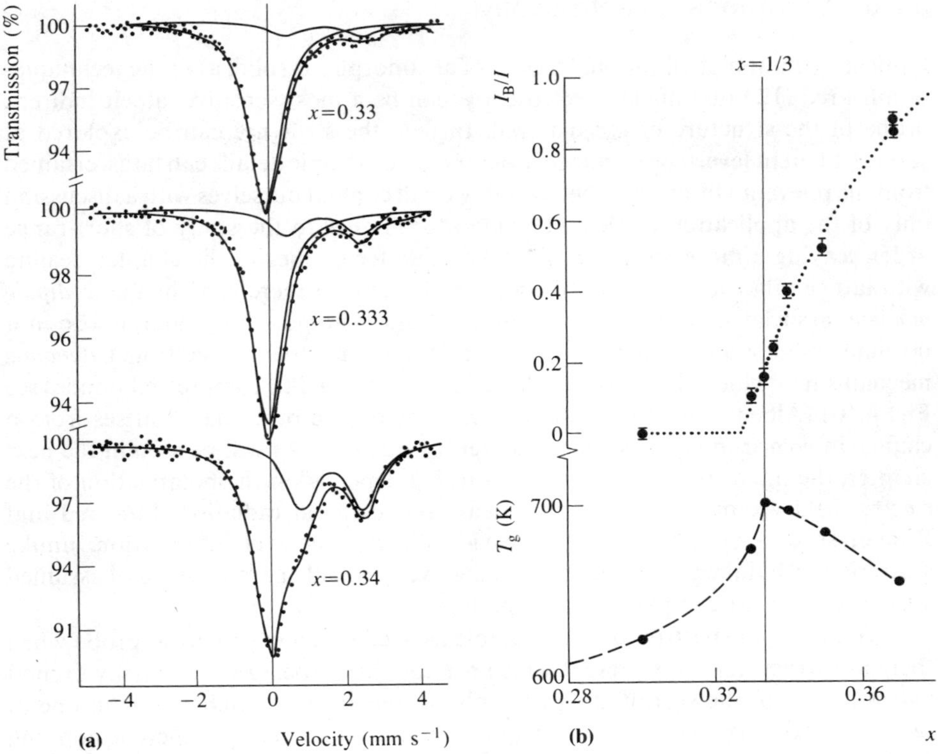

Fig. 3.29 (a) ^{119}Sn Mössbauer spectra of $(Ge_{0.99}Sn_{0.01})_xSe_{1-x}$ glasses showing evidence for two Sn(Ge) sites: site A is a singlet at $v = 0$, and site B is a doublet (Boolchand 1985).
(b) Compositional dependence of the intensity ratio for the B site in $(Ge_{0.99}Sn_{0.01})_xSe_{1-x}$ glasses, showing the abrupt change near the maximum in T_g (Boolchand 1985).

does not occur and that perfect isomorphous substitution by the probe atom takes place before such an inference can be conclusive.

The ^{129}I emission Mössbauer experiments are a little more complicated (Boolchand 1985). In these, radioactive $^{129}Te^m$ atoms, incorporated into the amorphous host (e.g. $GeSe_{2-x}Te_x$), decay by β^- emission into $^{129}I^*$ atoms which subsequently emit recoil-less γ-rays; the spectrum is detected using a mono-energetic ^{129}I absorber, e.g. crystalline NaI. Thus, the parent Te atoms (assumed to substitute isomorphously for Se) transmute to I atoms with a consequent change in chemistry; twofold-coordinated Te atoms become onefold-coordinated I atoms, and so on. Analysis of the experimental Mössbauer spectra provides evidence for two types of site originally occupied by the Te (Se) atoms: the A sites correspond originally to chemically ordered $(Te(Ge_{1/4})_2)$ sites (resulting in I–Ge bonds after the Mössbauer transitions), and the B sites correspond to I–chalcogen bonds which must originate from some of the sites in which the chemical order is broken, namely Ge–Te–Te configurations. As for the Ge-based experiments, it is found that the chemical order centred on the chalcogen atoms appears to be broken even for the stoichiometric composition.

3.2.6 Vibrational spectroscopy

Optical excitation of vibrational modes of an amorphous solid, as in the techniques of infra-red (IR) or Raman spectroscopy, can be a most sensitive, albeit indirect, probe of the structure of the material. In fact, the structure can be explored at several different levels; both microscopic and macroscopic details can be ascertained from the relevant vibrational spectra, but we will content ourselves with a discussion only of the application of IR and Raman techniques to the study of short-range order, leaving a more general account of these techniques to the chapter dealing with lattice vibrations (Ch. 4). Modes are IR-active if there is a change in *dipole moment* associated with the vibration; covalent homopolar materials do not normally possess a static dipole moment, and so their activity arises from a *dynamic* mechanism in which charge is transferred from extended to compressed bonds (see Fig. 4.10) (Alben *et al.* 1975). Raman activity, on the other hand, arises from a change in *polarizability*, a second rank tensor quantity. As will be seen in the next chapter, the intensity of the Raman scattering depends on the polarization of the electric field vector for both the incident and scattered radiation. Infra-red and Raman spectra do not generally provide direct structural information, unlike scattering techniques, but instead calculated vibrational characteristics of assumed structures may be compared with experiment.

Vibrational spectroscopy is valuable as a *short-range* structural probe when there are structural units present which are capable of being vibrationally excited *independently* of the surrounding amorphous matrix. This might occur in one of two ways: (i) 'terminator' atoms may be present which have a different bonding coordination to the majority of atoms forming the amorphous network and where, if the terminator atoms are considerably lighter than the other atoms, high frequency localized 'impurity' modes will result (an example of this is the hydrogen found in glow-discharge deposited amorphous silicon), (ii) if the structural units are vibrationally 'decoupled' from each other, which can happen when the units are connected together by atoms at which the bond angle is close to 90° (Sen and Thorpe 1977). In cases other than these, the vibrations are not confined to a small local region and can only properly be described in terms of network dynamics in which all the atoms present contribute to the modes, as will be discussed in the next chapter.

As an example of the application of IR spectroscopy as a structural probe, we will first consider the hydrogenated amorphous silicon system. As indicated in Chapter 1, the material may be prepared in a variety of ways, including the glow-discharge (GD) decomposition of SiH_4, reactive sputtering in Ar/H_2 and chemical vapour deposition from SiH_4 (possibly with post-hydrogenation in an H-plasma). The presence of (5–10 at.%) H is vital for the desirable electronic properties of this material (e.g. its dopability), yet preparation even by the same technique under different conditions can produce material with widely differing hydrogen contents, which for the GD method can even include polysilanes, a-$(SiH_2)_n$. A technique which is capable of measuring the concentration of H in a sample and which can also determine the bonding *configuration* of the H is obviously of great value, and in certain circumstances IR absorption can fulfil such a need. Samples can be deposited on single crystal Si wafers (polished on both sides and wedged in thickness to eliminate interference fringes), and the IR transmittance measured as a function of wavelength in a double-beam spectrophotometer with an identical uncoated Si substrate in the reference beam to correct for absorption, if any, of the substrate. Crystalline Si wafers are suitable as substrates for a-Si:H films because of the good match in refractive indices which minimizes interference fringes from multiple reflections within the film. The absorption coefficient α can be obtained from the equation

$$T=(1-R)^2 e^{\alpha d}/(1-R^2 e^{-2\alpha d}) \qquad [3.85]$$

where T is the percentage transmission, R the reflectivity and d the thickness of the film – [3.85] is strictly valid only for a freely supported film, but is a sufficiently good approximation in this case.

Figure. 3.30(a) shows the IR transmission of a series of films prepared by GD decomposition for different applied r.f. powers, for which the H content increases with r.f. power from $\simeq 9$ at.% to $\simeq 15$ at.% (see Lucovsky and Hayes 1979, Lucovsky *et al.* 1979). The spectra can be analysed by noting that since $m_H/m_{Si} \simeq 0.03$, the vibrational modes can be regarded as arising solely from the vibration of the H atoms. Three distinct H-bonded configurations can in principle exist in a-Si:H, namely monohydride (SiH), dihydride (SiH_2) and trihydride (SiH_3) groups; in addition, there is evidence that samples with high H contents should be regarded as being polysilanes, $(SiH_2)_n$. Each configuration will have its own characteristic vibrational modes, and these are indicated in Fig. 3.30(b). Stretching modes will occur at the highest frequencies ($\simeq 2000\ cm^{-1}$), well separated from bond-bending or wagging modes, which occur in the range 600–900 cm^{-1}. From a comparison with the IR mode frequencies characteristic of molecules containing the various Si–H groupings, it has proved possible to assign peak positions in the IR spectrum of a-Si:H samples fairly unambiguously (Fig. 3.30(c)). Monohydride groupings are the most straightforward to identify, with just two IR bands, at 2000 and 630 cm^{-1} (Fig. 3.30(b, c)); if IR peaks at $\simeq 900\ cm^{-1}$ are absent, only monohydride groups can be present. Isolated dihydride groups can also be readily identified: the stretching frequency is shifted up to 2090 cm^{-1} and a single scissors-bend band appears at $\simeq 875\ cm^{-1}$ (Fig. 3.30(b, c)).

There has been considerable controversy, however, over the origin of the doublet peak found at $\simeq$850–900 cm^{-1}. Originally it was believed that trihydride groups were responsible, but the present consensus (Lucovsky 1985) seems to be that there is no compelling evidence for the existence of such groups, but that polysilane groupings are responsible instead.

A second example of the application of vibrational spectroscopy to the study of short-range order is the series of covalent amorphous materials having the general formula AX_2, where A is the cation and X is the anion. Examples of these are SiO_2 and the Ge-chalcogenides (S and Se); in all cases the structure is believed to be comprised of tetrahedral AX_4 units linked either through the apices of the chalcogen atom or through shared edges. While the tetrahedral nature of these units is well established from a variety of experiments (e.g. diffraction), the angle subtended at the linking chalcogen atom is not in general known precisely, for the reasons referred to in section 3.2.2. However, the oxide glasses are believed to have higher A—X—A bond angles than other chalcogenide materials (e.g. for SiO_2,

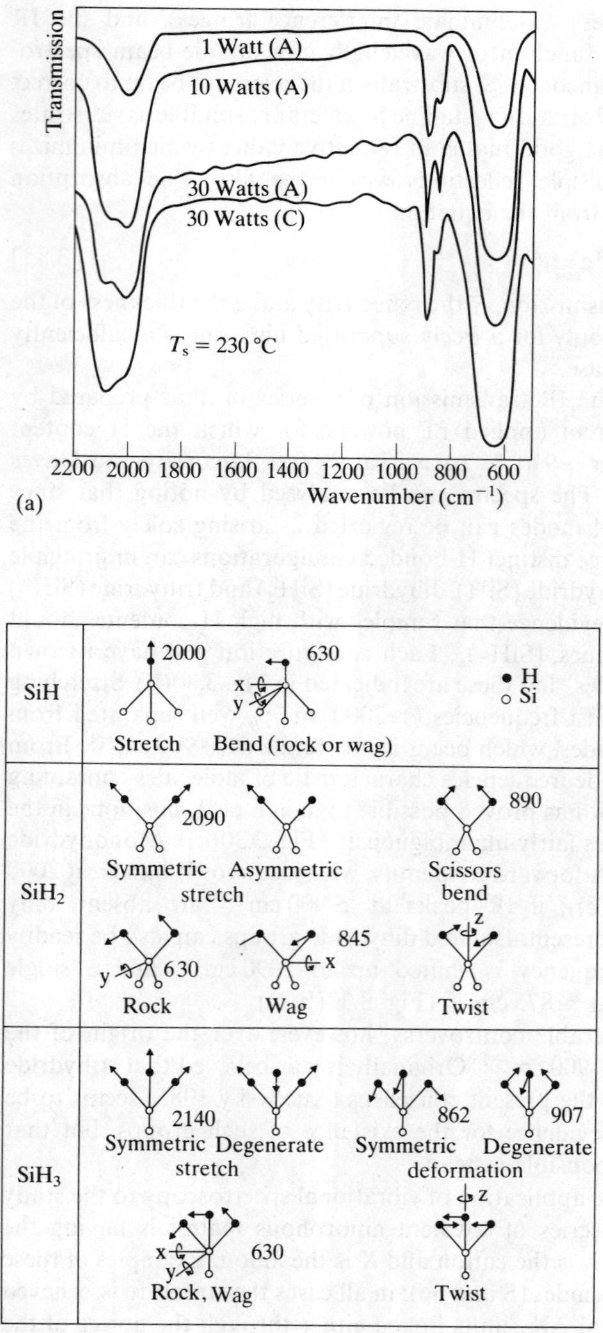

Fig. 3.30 (a) Infra-red transmission of films of a-Si:H deposited on to substrates at 230°C as a function of r.f. power in the discharge. (A refers to material deposited on the anode, C to material deposited on the cathode.) (Lucovsky *et al.* 1979).
(b) Local Si:H vibrations for SiH, SiH_2 and SiH_3 groups (Lucovsky *et al.* 1979).

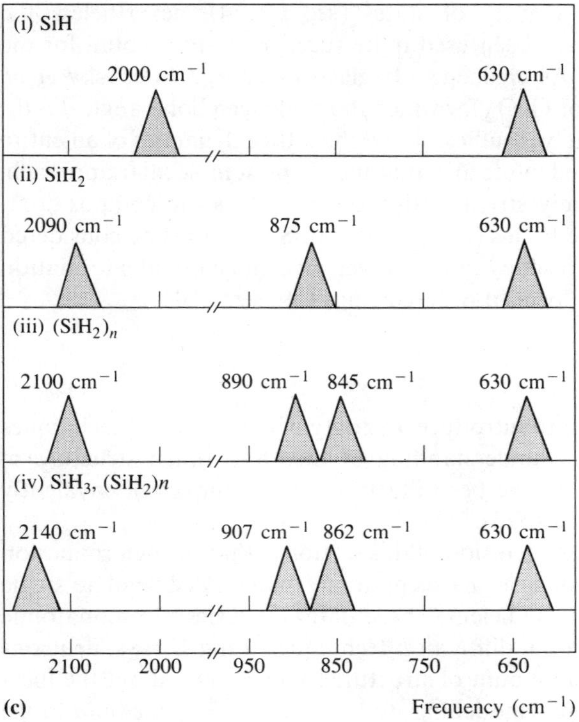

Fig. 3.30 (c) Schematic representation of the assignment of features in the IR absorption spectrum of a-Si:H to various H-related structural groupings (Lucovsky 1985).

$\theta \sim 140°$ and for GeS_2, $\theta \sim 105°$), and this feature plays a crucial role in the interpretation of the vibrational spectra, and indeed in the appearance of such spectra.

Sen and Thorpe (1977) have shown that for a simple model employing central (bond-stretching) and much weaker non-central (bond-bending) forces, the vibrational modes change from being 'band-like' (i.e. phonon modes involving all the atoms in the network in a cooperative motion) to 'molecular-like' (i.e. motion of isolated tetrahedra) as the chalcogen bond angle is reduced. Molecular modes are characterized by sharp, narrow features in the vibrational density of states, and therefore in the IR and Raman spectra, whereas band-like modes are typically broad and featureless. Thus, as the A—X—A bond angle is reduced, the coupling between molecular modes of different tetrahedra diminishes, and in fact is zero for an angle of 90° for a model in which there are central forces only (see Problem 4.1). This is the basis of this aspect of the use of vibrational spectroscopy as a structural probe, for if the signature of molecular-like modes is observed in the spectrum, the implication is that the chalcogen bond angle is near 90° or, as the calculations of Sen and Thorpe (1977) show, less than the critical angle $\theta_c = \cos^{-1}(-2m_X/3m_A)$, where m is the atomic mass; for SiO_2, θ_c is 117° and for GeS_2 it is 107°. Although detailed interpretation of IR and Raman spectra is complicated by matrix element effects, which mean that the vibrational spectra do not necessarily reflect accurately the shape of the vibrational density of states (see Ch. 4), nevertheless the independent molecular model has been used quite successfully to account for the vibrational spectra of Ge and As chalcogenide glasses (see e.g., Lucovsky *et al.* 1974). For materials like SiO_2 or GeO_2 for which the chalcogen bond angle $\theta > \theta_c$, the molecular model is completely insufficient and the lattice dynamics of an entire (CRN) cluster must be calculated. Note also that the simplest molecular approach, i.e. using AX_4 tetrahedra, is only strictly valid for stoichiometric compositions, and for other compositions, like-bonds (e.g. Ge—Ge or S—S) must be considered (see sect. 3.3). Vibrational spectroscopy can, however, give much useful information on medium-range order in the form of ordered rings (see sect. 3.4).

3.2.7 Summary

The aim of this section has been to introduce a variety of experimental techniques which may be used to gain an understanding of the short-range structure of amorphous materials, and these have been illustrated by examples of covalently bonded and metallic amorphous solids.

The somewhat depressing conclusion of this section is that the determination of the structure of amorphous solids is an exceptionally difficult task, and no single experimental technique is wholly sufficient. The situation is easiest for monatomic systems and for these, conventional diffraction techniques using X-rays, neutrons or electrons yield the maximum amount of structural information about the local order and possibly, with the help of modelling studies (see sect. 3.5), medium-range order. For multicomponent systems which, after all, form the majority of amorphous materials, the situation is much worse. Diffraction techniques and their variants (anomalous scattering, isomorphous and isotopic substitution or magnetic neutron scattering) are really only suitable for binary compounds which are describable by three partial pair correlation functions (and hence need three separate diffraction experiments to determine them). For systems containing more than two components, only EXAFS is really suitable as a direct structural probe. It has the virtue of being chemically specific, so the average local order around the atomic species which absorbs the X-ray photon is determined. Its disadvantages are that it is insensitive to atomic correlations much beyond the second coordination shell and it is difficult to deduce coordination numbers quantitatively. Other techniques such as magnetic resonance, Mössbauer and vibrational spectroscopies can supply indirect structural information, usually as an adjunct to more direct structural probes.

3.3 Short-range order

In the previous section we have discussed a variety of experimental techniques which can, to a greater or lesser extent, provide quantitative structural information for amorphous solids, particularly at short length scales. This is perhaps the place, then, to take stock and to discuss in a little more detail precisely what is meant by short-range order (SRO) in both covalent and metallic amorphous systems as a prelude to discussing medium-range order (MRO), in sect. 3.4.

We start with a discussion of covalent materials since the directed nature of the bonds characteristic of these systems considerably simplifies the picture. In such cases, the SRO is relatively easy to define in terms of well-defined local coordination polyhedra (see Fig. 3.31). Thus, the parameters which are sufficient to describe topological SRO are the number N_j of nearest neighbours j around an origin atom of type i, the nearest-neighbour bond length r_{ij}, the bond angle subtended at atom i, θ_i, and the corresponding quantities when atom j is regarded as the origin, N_i and θ_j. Note that in this description of SRO, nothing is said about the *connectivity* of the coordination polyhedra (e.g. corner-, edge- or face-sharing); the form of connection of the polyhedra essentially determines the type and extent of the MRO, as will be seen in the next section.

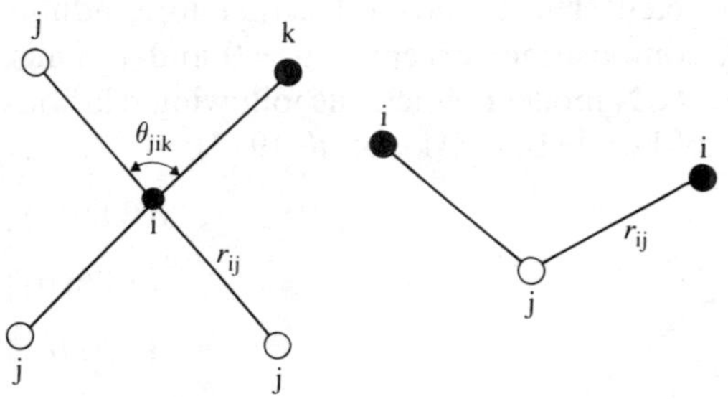

Fig. 3.31 Schematic illustration of the parameters describing short-range order in covalent amorphous materials.

An additional parameter is required if the degree of *chemical* SRO needs to be described as well, e.g. when *different* types of atoms constitute the coordination polyhedron around a given origin atom. Thus, for non-stoichiometric compositions, for example, excess atoms must be accommodated by the introduction of 'wrong' (i.e. homopolar) bonds (in the absence of coordination or valence changes), and the chemical order which might otherwise occur at the stoichiometric composition is thereby broken; the parameter of relevance here would then be the proportion of wrong bonds. Such considerations are appropriate for stoichiometric compositions too, since even then chemical ordering may not exist (e.g. Mössbauer spectroscopy provides some evidence for such a breakdown of chemical order in a-$GeSe_2$ – see sect. 3.2.5).

A related type of chemical order is when the different types of atomic species in the coordination shell around a given origin atom are in fact the *same* element, but each atom has a different charge state, bonding connectivity, etc. An example of this is the non-bridging oxygens (NBOs) introduced in, for example, silicate glasses by the incorporation of alkali modifier atoms, i.e. the Q_n species, where $n\,(=0\text{–}4)$ is the number of *bridging* oxygens around a central Si atom. As we have seen in sect. 3.2.4, MASNMR studies have demonstrated that chemical ordering is strong in such systems, since at the stoichiometric compositions only a *single* type of Q_n unit is detected, to a first approximation, whereas for compositions between two stoichiometric compositions, the two species Q_{n+1} and Q_n, characteristic of the end-member compositions, coexist.

The question of chemical ordering in covalent systems is most simply discussed for binary alloy systems such as A_xB_{1-x} where, if the elements A and B are in columns a and b of the periodic table, coordinations of $8-a$ and $8-b$, respectively, will occur, thereby satisfying their normal valence and the so-called '$8-N$' rule (strictly valid only for groups IV–VII). We neglect for the time being the effect of any coordination defects such as dangling bonds. In general, A—A, A—B and B—B bonds can coexist in an alloy of arbitrary composition, and two models can describe the distribution of such bond types (Lucovsky *et al.* 1977).

The first is the *random covalent network* (RCN) model which, as its name implies, treats the distribution of bond types as being purely statistical, determined only by the local coordinations $N_a = 8-a$ and $N_b = 8-b$ and the concentration variable x. Thus, any effects leading to preferential ordering (e.g. arising from differences in bond energies) are neglected. The RCN model, therefore, admits A—A, A—B and B—B bonds at *all* compositions except at $x=0$ and 1. Thus, for a 4:2 alloy (e.g. A = Ge, B = Se), the RCN model predicts the following relations for the numbers of the different types of bond (Lucovsky *et al.* 1977):

$$N_{AA} = 4(1-x)^2/(2-x) \qquad [3.86(a)]$$

$$N_{AB} = 4x(1-x)/(2-x) \qquad [3.86(b)]$$

$$N_{BB} = x^2/(2-x) \qquad [3.86(c)]$$

The predicted bond-counting statistics are shown in Fig. 3.32, together with those for a 3:2 system (e.g. As_xS_{1-x}).

The other model is the *chemically ordered network* (CON) model, which assumes that heteropolar A—B bonds are favoured; a completely chemically ordered phase thus occurs at the stoichiometric composition $x_c = N_a/(N_a+N_b)$ (e.g. as in As_2Se_3, $GeSe_2$). In contrast to the RCN model, only A—A and A—B bonds are allowed for A-rich compositions ($0<x<x_c$), and conversely only B—B and A—B bonds for $1>x>x_c$. Thus the bond statistics for the CON model for a 4:2 alloy are, for the A-rich region ($0<x<0.67$):

$$N_{AA} = 2-3x \qquad [3.87(a)]$$

$$N_{AB} = 2x \qquad [3.87(b)]$$

$$N_{BB} = 0 \qquad [3.87(c)]$$

and for the B-rich region ($1>x>0.67$):

$$N_{AA} = 0 \qquad [3.88(a)]$$

$$N_{AB} = 4(1-x) \qquad [3.88(b)]$$

$$N_{AA} = 3x-2 \qquad [3.88(c)]$$

These bond-counting statistics are also shown in Fig. 3.32 for comparison. There is considerable evidence from vibrational spectra and other results that the CON model is obeyed for the most part in most chalcogenide glasses, although some departures from chemical order can sometimes be detected (see e.g. sect. 3.2.5).

Obviously the RCN and CON approaches are valid only for covalent systems with well-defined directed bonds, i.e. not for metallic or ionic systems. Furthermore,

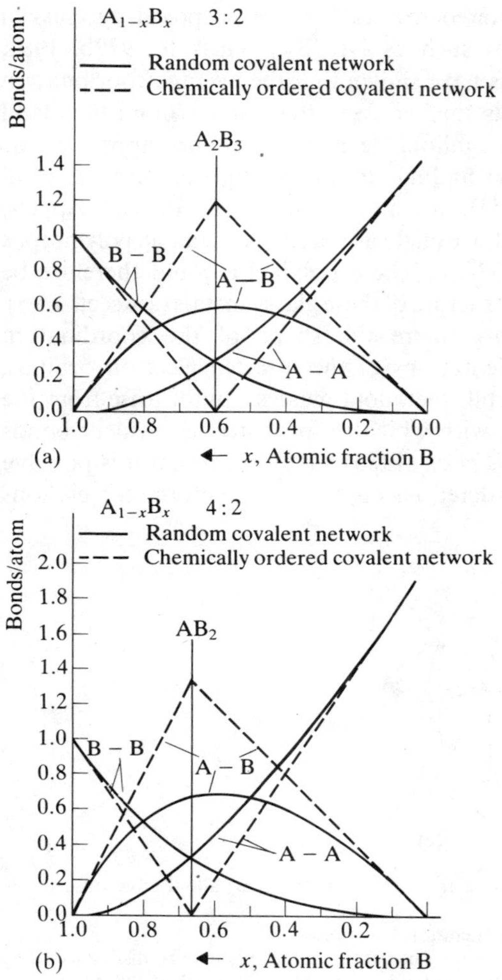

Fig. 3.32 Bond-counting statistics (after Lucovsky and Hayes 1979) for
(a) 3:2 networks (e.g. As_2Se_3) and
(b) 4:2 networks (e.g. SiO_2).

they are appropriate only for glass-forming systems, i.e. excluding 'modified' glasses in which network modifiers have either depolymerized the network by introducing non-bridging anionic sites or changed the coordination of the network-forming cation.

The description of SRO outlined above for covalently bonded systems is given in terms of rather well-defined coordination polyhedra. Such a description is obviously inappropriate for those systems in which the bonding is non-directional, e.g. for metals or ionic systems in which the structure can be regarded simply in terms of a random packing of spheres (see sect. 3.5.2). In such cases, a description of the local order is more difficult, as will be discussed later in this section. However, there are several structural models for such amorphous systems which do retain a significant degree of SRO.

One example is the so-called 'stereochemical' model proposed by Gaskell for transition-metal–metalloid glasses such as $Ni_{80}B_{20}$ (Gaskell 1979b, 1981, Dubois *et al.* 1985). Diffraction studies have shown that the average coordination number around the metalloid atoms is smaller ($\simeq 9$) than that around the metal atoms (12–14) and that metalloid–metalloid nearest neighbours appear to be absent. Gaskell has rationalized these findings by proposing that there is local chemical and topological order around the metalloid atoms in the form of (capped) trigonal prismatic coordination polyhedra, exactly as found in crystalline polymorphs such as cementite, Fe_3C (see Fig. 3.33(a)). These polyhedra could therefore be regarded as the building blocks of the structure of these glassy metals (see sect. 3.5.3). However, unlike in covalent systems where the shape of the coordination polyhedron is determined by the orientations of the directed electron orbitals, trigonal prismatic coordination in metal–metalloid glasses cannot arise from the same cause (e.g. B has no d-electrons with which to form strong covalent bonds to its nine nearest neighbours). Dubois *et al.* (1985) have shown that it is possible that this SRO is a consequence of ordered *medium-range* structural correlations (see sect. 3.4.3).

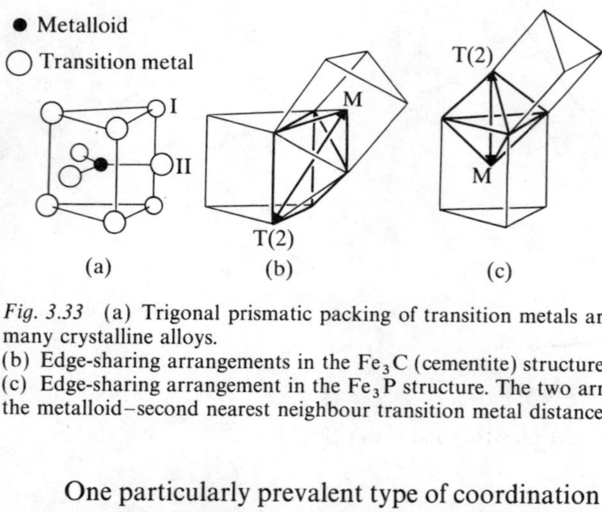

Fig. 3.33 (a) Trigonal prismatic packing of transition metals around a metalloid atom observed in many crystalline alloys.
(b) Edge-sharing arrangements in the Fe_3C (cementite) structure.
(c) Edge-sharing arrangement in the Fe_3P structure. The two arrangements may be distinguished by the metalloid–second nearest neighbour transition metal distance, M–T(2) (Gaskell 1981).

One particularly prevalent type of coordination polyhedron which can occur in randomly close-packed arrangements of spheres is the *icosahedron* (see e.g. Widom 1988 for a review). Thus, small neutral inert gas clusters appear to exist with 'magic numbers' of atoms (13, 19, 23, 26, 29, etc.) comprising such clusters; all these can be obtained by the serial addition of atoms to an initial 13-atom icosahedral cluster seed, giving so-called Mackay icosahedra (Mackay 1982) – see Fig. 3.34. Moreover, it has long been asserted (Frank 1952) that supercooled liquid metals should contain icosahedral clusters (which cannot therefore subsequently act as nucleation centres for crystallization). Thus it is natural to consider the extent of local icosahedral order in solid material rapidly quenched from the melt, namely metallic glasses and quasicrystals. The reason for the ubiquity of 13- and 55-atom Mackay icosahedra is, of course, that these structures are the lowest-energy configurations for small clusters, lower in energy even than crystalline f.c.c. arrangements. A large number of molecular-dynamics and other simulation studies

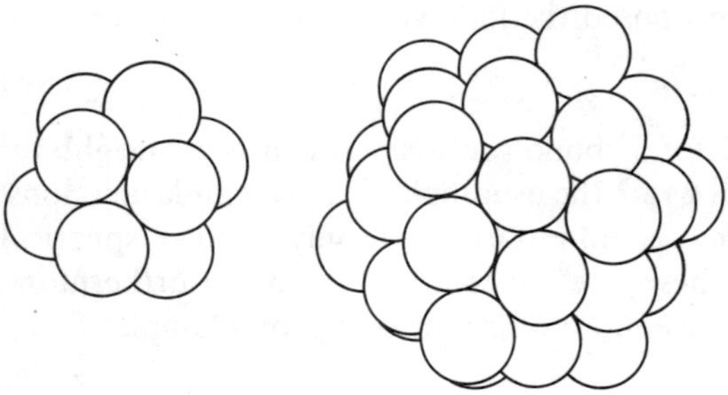

Fig. 3.34 13- and 55-atom Mackay icosahedra (or multiply twinned f.c.c. microcrystals).

have been performed to investigate the structure of the liquid and glassy states of rare-gas systems (modelled using Lennard-Jones potentials) and metals (see sect. 3.3.6). In order to determine whether local icosahedral order is present in such calculated structures, some kind of order parameter is required.

One such concerns the so-called 'Voronoi polyhedra' (or equivalently Wigner–Seitz cells) associated with such clusters. The Voronoi construction is illustrated in Fig. 3.35 using a two-dimensional analogue: in three dimensions, the polyhedron is defined as the smallest convex polyhedron formed by the planes which bisect perpendicularly the 'bonds' from a given origin atom to its nearest neighbours. It can be demonstrated that the number of edges of a particular face of a Voronoi polyhedron is equal to the number of tetrahedra packed around the 'bond' bisected by that face (Nelson 1983), e.g. pentagonal faces are a signature of icosahedral order.

Another indicator of local icosahedral order is associated with its *orientational* properties. Thus, in an investigation of bond-orientational order in a variety of

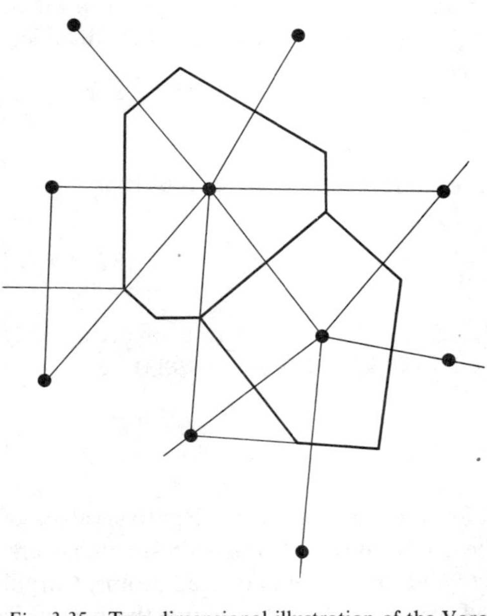

Fig. 3.35 Two-dimensional illustration of the Voronoi construction.

structures, Steinhardt *et al.* (1983) proposed the following order parameter:

$$Q_{lm}(\boldsymbol{r}) = [Y_{lm}(\theta(\boldsymbol{r}), \phi(\boldsymbol{r}))] \quad [3.89]$$

where θ and ϕ are the polar angles of a 'bond' between two nearest-neighbour atoms whose centre is at $\boldsymbol{r}$, and $Y_{lm}(\theta, \phi)$ are the usual spherical harmonic functions. The particular direction of a bond is unimportant if only even–l spherical harmonics are considered, since these are inversion-invariant. Furthermore, rotationally invariant combinations are more discriminatory, for example

$$Q_l = \left[\frac{4\pi}{2l+1} \sum_{m=-l}^{l} |\bar{Q}_{lm}|^2 \right]^{1/2} \quad [3.90]$$

where $\bar{Q}_{lm}$ represents an average over a number of bonds in the structure. Clusters with icosahedral symmetry have non-zero spherical harmonics only for $l = 6, 10, 12$, etc., whereas various forms of cubic close packing, for example, have non-zero values of Q_l for $l = 4$ as well. A dominant contribution to Q_l for $l = 6$ can therefore signify local icosahedral order.

In the absence of reasonably well-defined coordination polyhedra, as discussed above, SRO can be discussed only in terms of averaged quantities; this is the case, for example, with glassy metal–metal alloys. For this purpose, the Bhatia–Thornton formalism for the scattering identity ([3.44]–[3.46]) is useful. Thus, the number–number correlation function $\rho_{NN}(r)$ ([3.49]) describes the topological SRO of such alloys. (For a substitutional alloy with atoms of *equal* size, $\rho_1(r) = \rho_2(r)$ and the number–concentration function $\rho_{NC}(r)$ becomes zero – see [3.54].) The concentration–concentration correlation function $\rho_{CC}(r)$ ([3.52]) is more interesting, however, since it reflects the *chemical* SRO present in the alloy. In the approximation that all atomic species have the *same* size, [3.52] can be rewritten as $\rho_{CC}(r) = \rho_{NN}(r) - (\rho_{12}(r))/x_2$; for strong chemical ordering, so that atom 1 is surrounded by atom 2 and vice versa, the second term is larger and $\rho_{CC}(r)$ is *negative.* Thus, a negative peak in $\rho_{CC}(r)$ is a sign of chemical ordering in the shell of atoms at that distance.

Equation [3.52] can be rewritten as

$$\rho_{CC}(r) = [x_2 \rho_1(r) + x_1 \rho_2(r)] \alpha_{WC}(r)$$

where $\alpha_{WC}(r)$ is the Warren–Cowley order parameter (Warren 1969):

$$\alpha_{WC}(r) = 1 - \frac{\rho_{12}(r)}{x_2(x_2 \rho_1(r) + x_1 \rho_2(r))} \quad [3.91]$$

This order parameter can be further transformed into a function involving only the coordination numbers of the first shell (Sadoc and Wagner 1983):

$$\alpha_1 = 1 - \frac{N_{12}}{x_2(x_2 N_1 + x_1 N_2)} \quad [3.92]$$

where the N_i and N_{ij} have the same definitions as in [3.50]. Negative values of α_1 also indicate chemical ordering, i.e. the occurrence of unlike nearest neighbours. Equations [3.91] and [3.92] are strictly valid only for equal-sized atoms, Cargill and Spaepen (1981) have also proposed an expression describing the extent of

SRO in terms of nearest-neighbour coordination numbers:

$$\eta_{ij} = \frac{N_{ij}\bar{N}}{x_j N_i N_j} - 1 \qquad [3.93]$$

where $\bar{N}$ is the average coordination number, given by $\bar{N} = x_i N_i + x_j N_j$. For complete chemical disorder, $\eta_{ij} = 0$; otherwise, if chemical ordering exists (AB nearest-neighbours favoured) $\eta_{ij} > 0$, and in the other limit of chemical clustering (i.e. AA, BB pairs favoured), $\eta_{ij} < 0$. Ruppersberg (1988) has also discussed the case with atoms of unequal size.

3.4 Medium-range order

Medium-range order (MRO) is less obvious to define than SRO, and consequently several definitions exist in the literature. We will therefore commence this section by giving the definition of MRO which will be followed here (Elliott 1987b): MRO can be regarded simply as the next highest level of structural organization beyond SRO, existing on a length scale of, say, 5–20 Å. It is convenient in practice, particularly for covalent materials where MRO is more prevalent, to divide MRO into three categories corresponding to progressively increasing length scales; and these are now discussed in turn.

3.4.1 Short-range MRO

If we define SRO in terms of well-defined coordination polyhedra, as in sect. 3.3, short-range MRO (SRMRO) is therefore concerned with the type of connection of such polyhedra as well as their relative orientation. Thus, corner-, edge- or face-sharing of polyhedra lead to very different ordering schemes on the relatively local scale, as well as to pronounced differences at larger distances in terms of the network dimensionality (see sect 3.4.3). As an example consider the AX_2 family of glasses, where A = Si, Ge and X = S, Se; in all cases, the SRO is characterized by AX_4 tetrahedra, but the type of connection between polyhedra can differ in each case. Thus, SiO_2, in both glassy and crystalline states, is characterized by corner-sharing connections between neighbouring tetrahedra, leading to a 3D (giant molecule) structure. In contrast, both $GeSe_2$ and GeS_2 have a proportion of *edge*-sharing connections in the crystalline state, and there is considerable evidence that this situation exists in the glassy state also; such edge-sharing of tetrahedra can lead to (locally) 2D structural characteristics (see sect. 3.4.3). Finally, $SiSe_2$ and SiS_2 have crystalline polymorphs in which *only* edge-sharing of tetrahedra occurs; there is evidence that this type of connection predominates also in the amorphous phases, in this case leading to locally 1D structures (sect. 3.4.3). Thus, it can be seen that there is often a strong link between MRO at the various length scales.

Obviously, MRO is most pronounced when the degrees of freedom associated with the relative orientation of neighbouring polyhedra are restricted in some way, e.g. by the occurrence of edge-sharing connections. However, even with corner-sharing units, where free rotation about the common bonds could occur in principle, certain orientations may be more probable than others. A structural parameter which is a measure of such orientational correlations for covalent systems (with well-defined bonds) is the 'dihedral' (or torsion) angle, ϕ, defined as in Fig. 3.36; it is the angle of rotation about a common bond required to bring into coincidence the projections onto the plane perpendicular to this bond of the other bonds in the connected units. Departures from a uniform (i.e. random) distribution, $P(\phi)$, are correspondingly a hallmark of MRO. Consequently, preferred values of ϕ are associated with features of MRO such as (small) rings (see sect. 3.4.2). However, this need not always be the case: for example, it appears from structural modelling studies (Corb *et al.* 1982) and considerations of features in the electronic density of states that the structure of a-Se is not representative of a freely rotating chain (with constant $P(\phi)$), but instead ϕ takes more-or-less fixed values ($\simeq 102°$, as in the crystal), with the sign of ϕ varying either alternately or randomly down a chain.

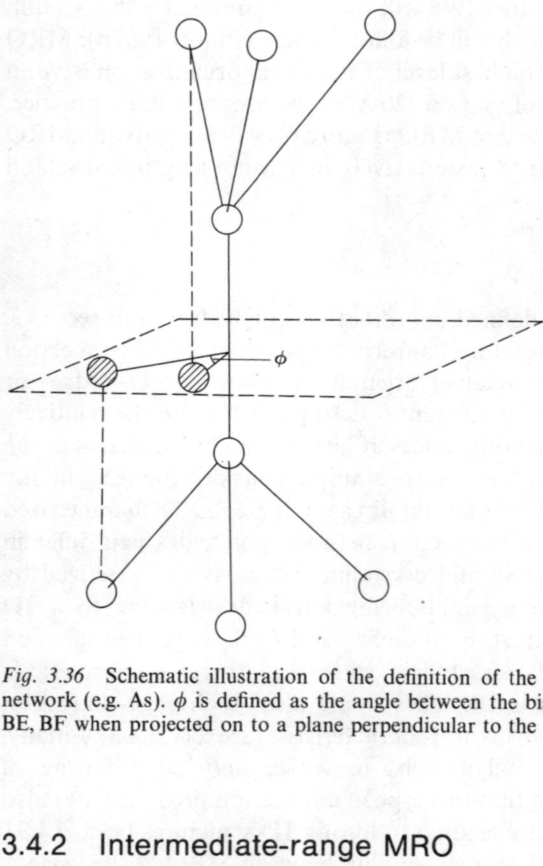

Fig. 3.36 Schematic illustration of the definition of the dihedral angle ϕ in a triply-coordinated network (e.g. As). ϕ is defined as the angle between the bisectors of the pairs of bonds AC, AD and BE, BF when projected on to a plane perpendicular to the bond AB.

3.4.2 Intermediate-range MRO

At the next largest length scale, say $\geqslant 5$ Å, MRO is obviously determined by the type and extent of SRMRO present. However, we make the distinction that intermediate-range MRO (IRMRO) is characterized by the presence of 'super-structural' units (consisting of several basic SRO units), e.g. rings or clusters of atoms of particular shapes or sizes, which occur in a considerably higher proportion than would be expected on a purely statistical (random) basis. In certain cases, such superstructural units can be regarded as forming the essential building blocks of the amorphous structure.

One example of regular rings acting as superstructural units is associated with the edge-sharing form of connection of tetrahedra discussed in sect 3.4.1. Such edge-shared units can equally be regarded as fourfold (e.g. Ge_2Se_2) rings. These four-membered rings may be the structural origin of the MRO responsible for the so-called 'companion line' in the Raman spectrum of glassy $GeSe_2$ observed at 220 cm^{-1}, next to the main A_1 breathing mode of the $GeSe_4$ tetrahedra at $\simeq 200$ cm^{-1} (Sugai 1987) – Fig. 3.37(a) (see sect. 4.4.3 for a discussion of Raman scattering). The companion line is unusually narrow and is strongly polarized, and moreover shows an unusual compositional dependence of the intensity varying as $\sim x^5$ in the alloys Ge_xSe_{1-x}; the intensity of the normal tetrahedral mode varies linearly with x, as would be expected for a systematic breaking-up of $GeSe_4$ tetrahedra as the composition becomes non-stoichiometric. A number of structural models have been suggested to account for the companion line, including large

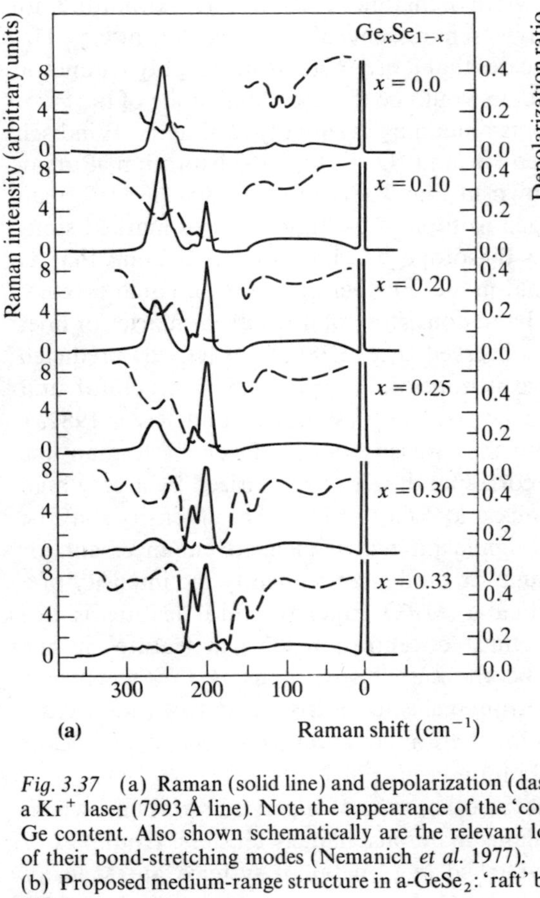

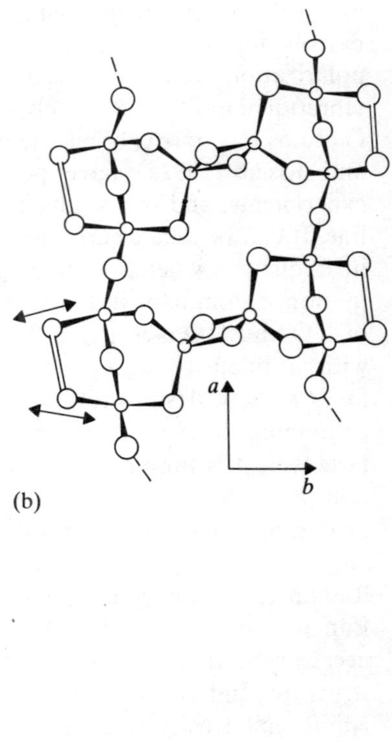

Fig. 3.37 (a) Raman (solid line) and depolarization (dashed line) spectra of a-Ge_xSe_{1-x} excited with a Kr^+ laser (7993 Å line). Note the appearance of the 'companion' line at $\simeq 220$ cm^{-1} with increasing Ge content. Also shown schematically are the relevant local atomic clusters and the frequency range of their bond-stretching modes (Nemanich *et al.* 1977).
(b) Proposed medium-range structure in a-$GeSe_2$: 'raft' bordered by Se–Se wrong bonds (J.C. Phillips 1981).

12-membered symmetric rings (Nemanich *et al.* 1977) and a yet larger quasi-ordered structural arrangement termed a 'raft' (Fig. 3.37(b)), consisting of 12 Ge atoms and 30 Se atoms (similar to the structure of the 2D β-crystalline form), but which is bordered by the Se—Se homopolar ('wrong') bonds, the vibrations of which are purported to be responsible for the companion line (J. C. Phillips 1981). It is perhaps significant that the 2D character of the raft arises from the presence of *edge*-shared tetrahedra running down the middle. It is still not clear whether the anomalous compositional dependence of the 220 cm^{-1} Raman line is due to a compositional variation of a feature in the vibrational density of states or of the Raman matrix elements. The highest-resolution inelastic neutron scattering experiments reported to date have not been able to resolve the companion line from the A_1 line to decide between these possibilities (Gladden *et al.* 1988).

Another classic example of a regular ring acting as a superstructural unit in the structure of a glass is that of the so-called boroxol ring in vitreous B_2O_3; this B_3O_3 ring has the form of a planar hexagon, and can equally be regarded as consisting of three corner-linked planar BO_3 triangles connected together in the form of an equilateral triangle (see Fig. 3.38). The Raman spectrum of glassy B_2O_3 is remarkable in that it is dominated by a single, very narrow, strongly polarized line at $\simeq 808$ cm^{-1} (Fig. 3.38(a)); the width of this line ($\simeq 10$ cm^{-1}) is extraordinarily narrow for a glassy material, and is reminiscent of a molecular mode. The polarization character of the line indicates that it originates from a highly symmetric vibrational mode, and a likely candidate would be the breathing mode of the three O atoms in a boroxol ring, the B atoms remaining essentially stationary. Windisch and Risen (1982a) have performed B and O isotopic substitution Raman experiments, and the results are shown in Fig. 3.38(b) for the 808 cm^{-1} Raman line. It can be seen clearly that oxygen isotopic substitution causes marked shifts in frequency, whereas changing the B isotope has no effect, indicating that O motion is dominant in the vibrational mode. Furthermore, for an equal mixture of ^{16}O and ^{18}O isotopes, a Raman band consisting of a resolved quartet of lines with an intensity ratio of 1:3:3:1 is observed (Fig. 3.38(b)), exactly as predicted for a statistical substitution of these isotopes in a well-defined structural unit containing three O atoms, namely a boroxol ring (Windisch and Risen 1982a). However, it is unsafe to assume from the intensity of the 808 cm^{-1} Raman line that the structure of glassy B_2O_3 is correspondingly characterized by a very high proportion of boroxol rings (Williams and Elliott 1982). The intensity may be anomalously large as a result of a significant matrix-element enhancement for Raman scattering from a highly symmetric mode; consequently the intensity of a Raman line associated with a particular MRO superstructural feature is not necessary a reliable guide to the actual concentration of such features in the structure. Indeed, this appears to be the case for vitreous B_2O_3, where the vibrational density of states obtained from inelastic neutron scattering (admittedly with six times worse resolution than for the Raman spectra) shows no significant feature at $\simeq 808$ cm^{-1} (100 meV) (Hannon *et al.* 1988).

We now turn to a discussion of the other type of superstructural unit which can contribute to IRMRO in amorphous materials, namely clusters. Examples of such superstructural units for close-packed (e.g. metallic) systems are Mackay icosahedra (see sect. 3.3). In covalently bonded systems, clusters can be regarded

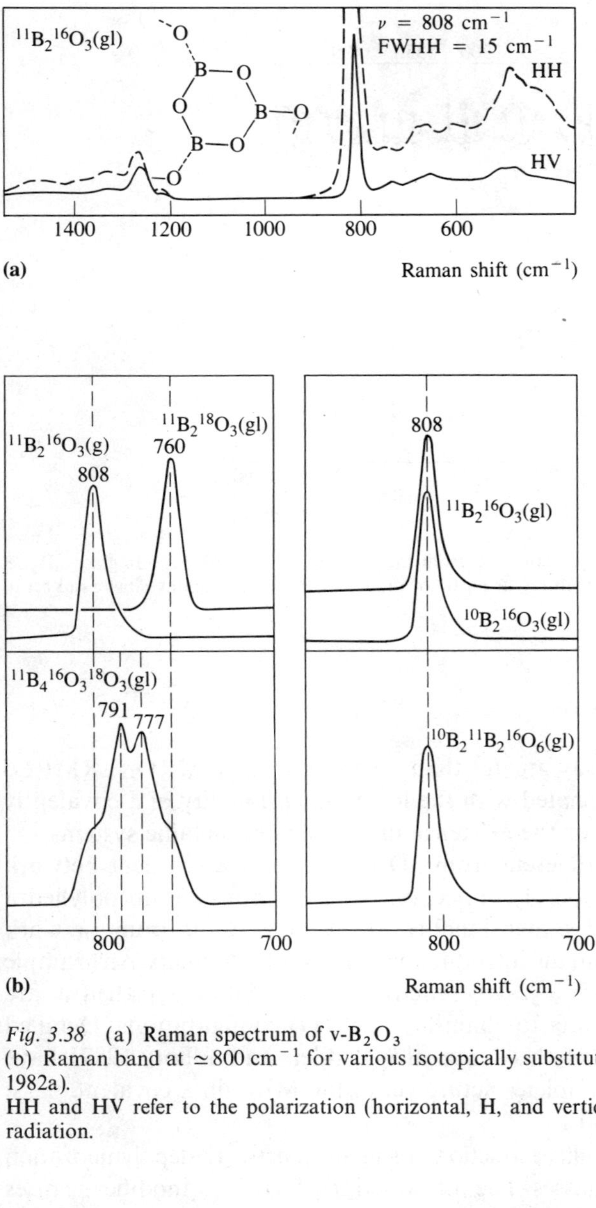

Fig. 3.38 (a) Raman spectrum of v-B_2O_3
(b) Raman band at $\simeq 800\ cm^{-1}$ for various isotopically substitute B_2O_3 glasses (Windisch and Risen 1982a).
HH and HV refer to the polarization (horizontal, H, and vertical, V) of the incident and scattered radiation.

as aggregates of regular rings. In P–Se glasses, for example, in the P-rich glass-forming region centred around P_2Se, the main structural unit appears to be a P_4Se_3 cluster (see Fig. 3.39) (as in the molecular crystal of the same composition); an inelastic neutron scattering measurement of the vibrational density of states (sect. 4.4.4) of glassy P_2Se (Fig. 3.39) reveals a close correspondence with the vibrational modes expected for isolated P_4Se_3 molecules (Verrall and Elliott 1988). The structure factor can also be well fitted by an arrangement of P_4Se_3 molecules bonded together by Van der Waals' forces; the material is therefore an inorganic molecular glass.

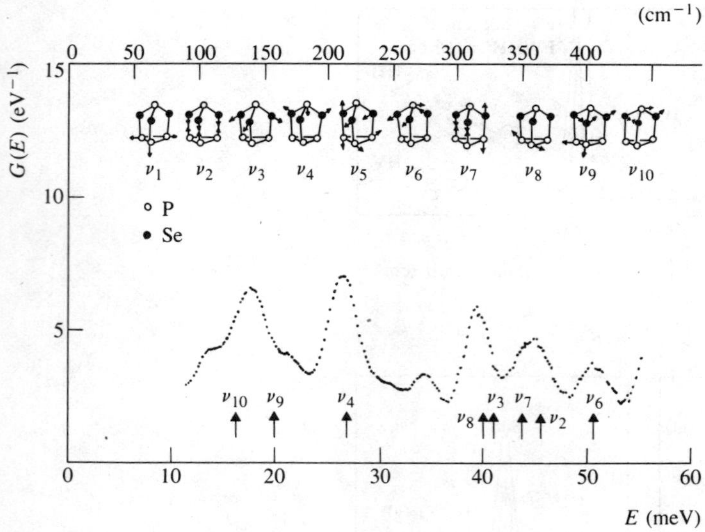

Fig. 3.39 Vibrational density of states obtained from inelastic neutron scattering for glassy P_2Se, together with the assignment of the features in terms of modes of P_4Se_3 molecular clusters (Verrall and Elliott 1988).

3.4.3 Long-range MRO

On a yet larger length scale, say greater than 10 Å, long-range MRO (LRMRO) can be regarded as being associated with the local dimensionality of a covalently bonded amorphous network, or the existence of domains in metallic systems.

A local dimensionality different from 3D can arise in a covalent network from two causes: either the type of connection between coordination polyhedra may impose such a reduced dimensionality, or it may result from network depolymerization resulting from the introduction of network modifiers. An example of the former occurs in a-$SiSe_2$, where edge-sharing $SiSe_4$ tetrahedra are predominant (sect. 3.4.1), leading to chain-like (i.e. 1D) configurations. Detailed computer simulation studies of this material (Gladden and Elliott 1987) have provided perhaps the most complete picture yet of the MRO in a covalent glass; this is summarized in Fig. 3.40.

An example of dimensionality reduction arising from network depolymerization is provided by alkali silicate glasses. The introduction of network modifiers causes negatively charged non-bridging oxygen (NBO) centres to be formed at the expense of bridging oxygens (see sect. 3.2.4), thereby causing a depolymerization of the network. Furthermore, it appears from MASNMR studies that such NBO sites are introduced in a very ordered manner, only one or two different Q^n species being present for a given composition. The result of this network depolymerization is to cause the *local* dimensionality, defined by the dimension of the percolation paths connecting neighbouring NBO sites, to become less than three; thus, for compositions near the stoichiometry $M_2O.2SiO_2$ (M = alkali), 2D sheet-like internal surfaces form, bordered by NBOs, which can act as percolation channels

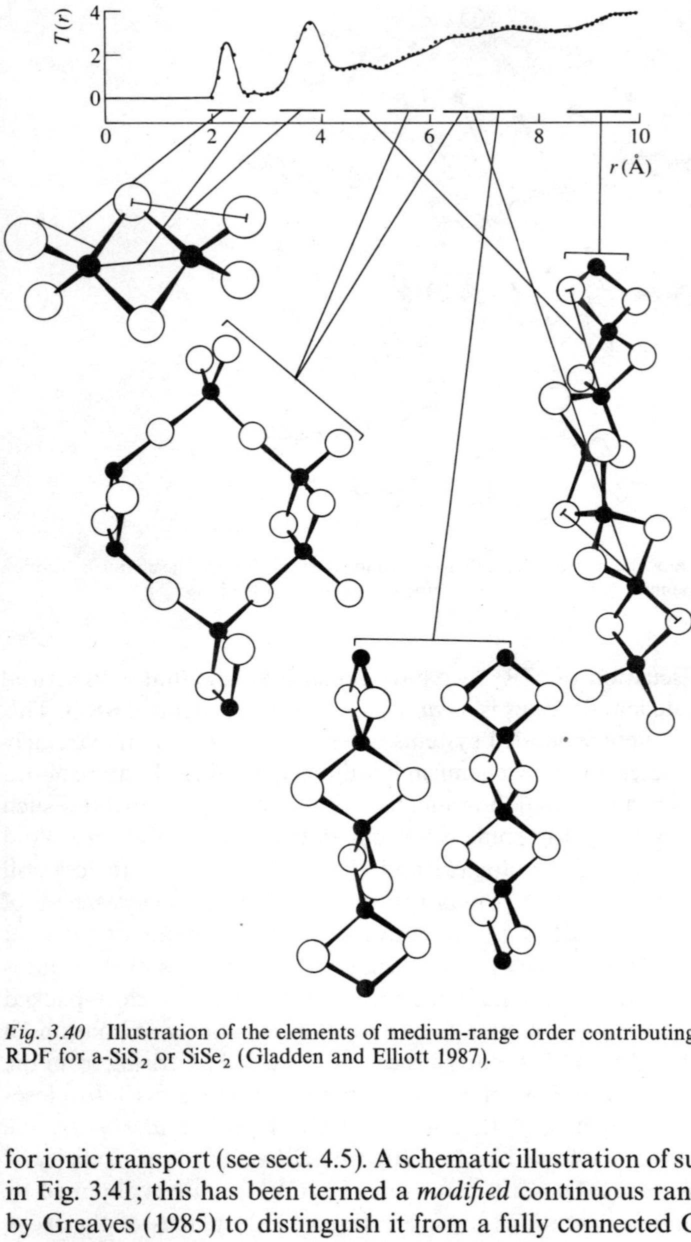

Fig. 3.40 Illustration of the elements of medium-range order contributing to various regions of the RDF for a-SiS_2 or $SiSe_2$ (Gladden and Elliott 1987).

for ionic transport (see sect. 4.5). A schematic illustration of such channels is shown in Fig. 3.41; this has been termed a *modified* continuous random network (CRN) by Greaves (1985) to distinguish it from a fully connected CRN (see sect. 3.5.3).

A final example of LRMRO determined by a superstructural unit is afforded by the boroxol ring discussed in sect. 3.4.2; this planar structural grouping automatically confers a considerable degree of two-dimensional character locally to the structure. This two-dimensional character has been confirmed by the detection of an orientation of the Raman scattering intensity in blown thin films of glassy B_2O_3, whereby the tensile stresses in the plane of the surface of the bubble exert an orienting influence on the 2D boroxol units (Windisch and Risen 1982b).

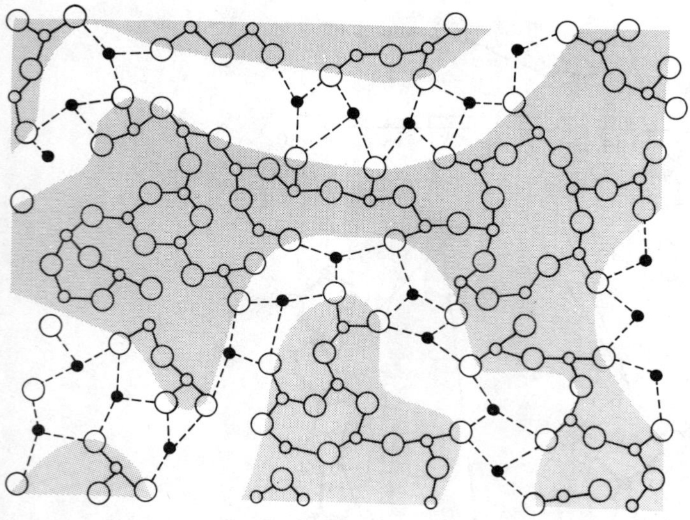

Fig. 3.41 Modified CRN model of network modifier-containing (oxide) glasses. The channels through the structure containing the modifier ions are clearly apparent (Greaves 1985).

So far in our discussion of MRO we have taken the view that a structural ordering at intermediate length scales is a *consequence* of a well-defined SRO. This is probably true for covalently bonded systems where the SRO is often extremely well-defined since it is determined by chemical bonding constraints. In amorphous materials in which non-directional bonding is predominant, e.g. metals, such constraints no longer apply; yet in some cases, e.g. in transition-metal–metalloid (TMM) glasses, a well-defined topological and chemical SRO nevertheless still appears to be present (see sect. 3.3). It may be that such SRO is a *consequence* of the existence of a well-defined MRO, i.e. the converse of what is found for covalent systems (Gaskell 1987). An example of this circumstance in the crystalline state is the occurrence of tetrahedra and octahedra arising from ordered close-packed arrays of atoms. In a similar manner, the trigonal prismatic coordination of metalloid atoms in crystalline TMM materials can be regarded as arising from the introduction of 'chemical twinning' planes containing the M atoms into close-packed crystalline arrangements of the metal atoms (Hyde *et al.* 1979); the composition of the TMM materials can be varied by changing the periodicity of the chemical twinning planes. Thus, the cementite (Fe_3C) structure is obtained by introducing such twinning operations for every three $\{112\}$ planes of the h.c.p. structure. Dubois *et al.* (1985) have proposed that similar considerations apply equally to the structure of *glassy* TMM alloys. In this model, the LRMRO is associated with domains of size 10–20 Å within each of which there are more-or-less ordered arrays of chemical twinning planes, and along which are therefore located the trigonal prismatic units; the orientation of the twinning planes is different for each domain, however, and considerable constraints exist at the domain boundaries in order to minimize strain and preserve the local coordination there (Dubois *et al.* 1985).

3.4.4 Experimental probes of MRO

We conclude this section with a brief discussion of the various experimental probes which may be used to obtain information about the type and extent of MRO in amorphous materials.

We have already seen (sect. 3.4.2) that Raman scattering can be particularly sensitive to the presence of IRMRO in the form of regular rings or ordered clusters, as a result of the matrix-element enhancement associated with the scattering from the breathing modes of such structural units. Examples already considered include the six-membered (boroxol) B_3O_3 planar rings in glassy B_2O_3, and four-membered (Ge_2X_2) rings in a-GeX_2 (X = S, Se); in addition, planar six- or eight-membered (Si_3O_3, Si_4O_4) rings in glassy SiO_2 have also been postulated to exist on the basis of Raman evidence (Galeener 1982). For clusters, Raman evidence supports the existence of P_8 or P_9 clusters in a-P (Fasol *et al.* 1984) and P_4Se_3 clusters in P–Se glasses (Price *et al.* 1984).

Conventional diffraction experiments can provide information about the extent of MRO in glasses, although this is often difficult to discern in real-space correlation functions (e.g. the RDF) because of overlapping peaks; in this case, systematic structural modelling studies are necessary in order to determine the degree of MRO that is present (e.g. as in glassy $SiSe_2$; Gladden and Elliott 1987). However, diffraction data in *reciprocal* space can often be more informative.

One particular feature which has long been regarded as a signature of the existence of MRO is the so-called 'first-sharp diffraction peak' (FSDP) or 'pre-peak' observed in a variety of mixed chalcogenide glasses (including oxides) and other covalent materials, such as $GeSe_2$, SiO_2, As_2S_3, P and As, but *not* in pure a-Se or Ge – see Fig. 3.42. This feature has been termed a pre-peak because Fourier transformation of the structure factor $S(Q)$, both including and omitting this first peak, produces indistinguishable real-space correlation functions, which indicates that it does not contain structural information about SRO, being

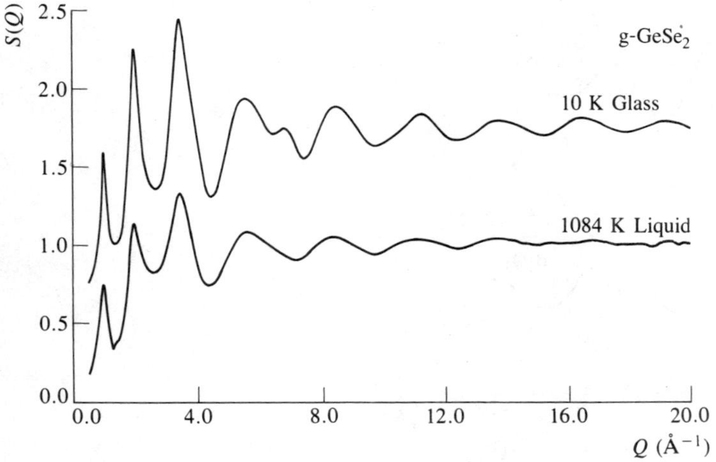

Fig. 3.42 Structure factor for liquid and glassy $GeSe_2$, showing the first sharp diffraction peak at $Q \simeq 1\ \text{Å}^{-1}$ present in both phases (Susman *et al.* 1988).

associated instead with subtle MRO structural arrangements. The following features connected with the FSDP in covalent systems have been found. For chalcogenide glasses (and P and As) it occurs at a value of scattering vector $Q_1 \simeq 1\ \text{Å}^{-1}$, although $Q_1 \simeq 1.5\ \text{Å}^{-1}$ for oxides such as SiO_2. However, when $S(Q)$ is plotted against the *reduced* parameter Qr_1 (where r_1 is the nearest-neighbour bond length), all curves are very similar (Wright *et al.* 1985, Price *et al.* 1987), and the FSDP lies at $Qr_1 \simeq 2.5$ (see Fig. 3.43).

The FSDP alone shows anomalous behaviour as a function of temperature and pressure. It is found that the intensity of the FSDP *increases* with increasing temperature in, for example, SiO_2 (Soklakov and Nechaeva 1967), As_2S_3 (Busse 1984), As_2Se_3 (Busse and Nagel 1981) and GeS_2 (Lin *et al.* 1984). (All other peaks in the structure factor *decrease* in intensity with increasing temperature, in accordance with the normal behaviour of the Debye–Waller factor.) Indeed, the FSDP is still pronounced, even in the *liquid* state (see Fig. 3.42). The application of pressure also produces an anomalous behaviour of the FSDP in chalcogenide glasses; its intensity decreases and the position of the peak shifts to higher Q with increasing pressure (Tanaka 1987).

The intensity of the FSDP increases as the atomic number of the chalcogen X decreases for a particular type of chalcogenide glass, e.g. GeX_2 or SiX_2 (Susman

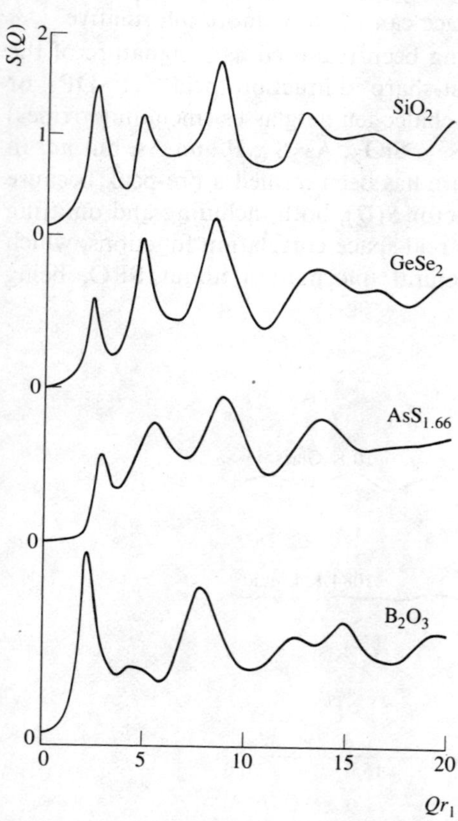

Fig. 3.43 Structure factor for oxide and chalcogenide glasses plotted against a reduced variable Qr_1, where r_1 is the nearest-neighbour bond length in each case (Wright *et al.* 1985).

et al. 1988). Furthermore, it appears that cation–cation correlations are primarily responsible for the FSDP; anomalous X-ray scattering results for glassy $GeSe_2$ indicate that Ge-centred correlations contribute predominantly to the FSDP intensity (Fuoss 1980).

There has been much controversy over the structural origin of the FSDP in covalent glasses, and a number of explanations have been proposed. These can be effectively divided into two categories, depending on how the peak in $S(Q)$ originates. In the first, the FSDP is regarded as being separate from the rest of $S(Q)$, i.e. it is a single Fourier component in reciprocal space arising, therefore, from a periodic function in real space characteristic of the MRO. The width of the peak in $S(Q)$ arises from the damping in the amplitude of the real-space oscillations, and the Debye–Scherrer equation relating diffraction line broadening and crystallite size for crystalline powder diffraction can be used to provide a relationship between the FSDP peak width ΔQ, and D, a characteristic coherence length over which the periodicity in real space is maintained:

$$D \simeq 2\pi/\Delta Q \qquad [3.94]$$

In addition, the position of the FSDP in this picture is determined by the periodicity in real space via

$$d = 2\pi/Q_1 \qquad [3.95]$$

where d is the spatial repeat distance. For the FSDPs observed in chalcogenide glasses where $Q_1 \simeq 1$ Å^{-1}, [3.95] would predict a repeat distance of $d \simeq 5$–6 Å, whilst the width of the peak leads to a correlation length $D \simeq 15$–20 Å. This explanation is obviously based on an essentially microcrystalline picture.

The second interpretation of the structural origin of the FSDP is in terms of a cluster picture which is, in Fourier terms, the converse of the above model. In the simplest case, in this picture there is a single (broad) peak in real space at r_m characterizing the MRO, which produces a strongly damped sinusoidal function in reciprocal space of which the observed FSDP is simply the first, most intense peak. Thus, in this picture the FSDP in itself has no special structural significance. This approach is similar to that used in the interpretation of the scattering from molecular liquids such as CCl_4. In this case, the overall measured structure factor can be written as the sum of two terms:

$$S(Q) = f_m(Q) + D_m(Q) \qquad [3.96]$$

where $D_m(Q)$ is the molecular-packing structure factor describing the *inter*-molecular interference resulting from the scattering of neutrons or X-rays from molecular entities separated by an average distance r_m, and $f_m(Q)$ is the molecular form factor describing the *intra*-molecular scattering. $D_m(Q)$ is dominant at low Q and is highly damped because of the large fluctuations in r_m, whereas $f_m(Q)$ is much less damped (because the fluctuations in intra-molecular covalent bond lengths are considerably smaller than those for r_m) and it therefore dominates at high Q.

Although the origin of the FSDP can be understood conceptually in terms of the two approaches described above, its actual structural origin remains controversial. The microcrystalline Bragg-peak picture has been supported (see e.g. Busse 1984), but the observation of the FSDP even in the liquid state, together with the fact that appreciable structural order over a correlation length of $D \simeq 20$ Å would introduce too many fine features into the RDF, makes it unlikely that this is the cause in all cases. Moreover, certain materials such as As_2S_3 which have layer-like crystal structures lend themselves to such a microcrystalline picture, but this is difficult to reconcile with the fact that the FSDP is observed irrespective of the network dimensionality (e.g. even in SiO_2, a 3D structure). Cervinka (1988) has proposed, in a cross between the microcrystalline and cluster approaches, that the FSDP (in e.g. GeS_2 or As_2S_3) arises from a parallel ordering in a particular orientation of pairs of configurations of between two and five coordination polyhedra, as found in the crystalline polymorphs. Although semi-quantitative agreement with the measured $S(Q)$ in the region of the FSDP is achieved, as well as a correct prediction of the pressure dependence, it is not clear how generally applicable this approach is (e.g. for SiO_2), or how many such ordered units are necessary in the overall structure in order to produce the correct intensity of the FSDP. On the other hand, in support of the cluster approach, Moss and Price (1985) have proposed that the FSDP results from the packing of coordination polyhedra, with Veprek and Beyeler (1981) stressing that whilst the FSDP arises from many Fourier components of the real-space pair correlation function, it may be associated with zones of low atomic occupancy characterized by a minimum in the RDF beyond the SRO correlations (itself related to the packing of polyhedra and the interstitial voids between them).

In general, it can be said that the FSDP is almost invariably observed in those covalent amorphous systems containing an atomic component having a low coordination number (e.g. a chalcogen); as a result, close packing of atoms in the vicinity of such low-coordination atoms is precluded. Consequently, pressure is a particularly sensitive probe because of the relatively compressible weak Van der Waals-type interactions between low-coordination atoms in non-covalently bonded directions. We have already seen that the FSDP is very sensitive to the application of pressure, but LRMRO in the form of network dimensionality can also be probed by monitoring the pressure-induced shift of the optical band gap E_o (sect. 5.6.1). In general, the application of pressure may give rise to two contributions to the shift in E_o:

$$\left(\frac{\partial E_o}{\partial p}\right)_T = \left(\frac{\partial E_g}{\partial p}\right)_T - \left(\frac{\partial J}{\partial p}\right)_T \qquad [3.97]$$

Here the first term represents the pressure shift of the 'average' energy separation E_g between valence and conduction bands, and since E_g depends on the nearest-neighbour covalent bond length r_1 approximately as $E_g \propto r_1^{-2.5}$, bond compression produces an increase in E_g. The second term reflects the pressure-induced band broadening caused by non-bonding (lone-pair) orbital overlap. In chalcogenide materials, pressure preferentially acts on the weaker Van der Waals' interactions between, say, chalcogen atoms, with the result that the second term in [3.97] is dominant and a red shift of E_o with pressure is expected. For materials like a-Si or Ge, though, the covalent bonds are preferentially compressed and the first term can be dominant, giving rise to a blue shift of E_o. Experimental data are shown

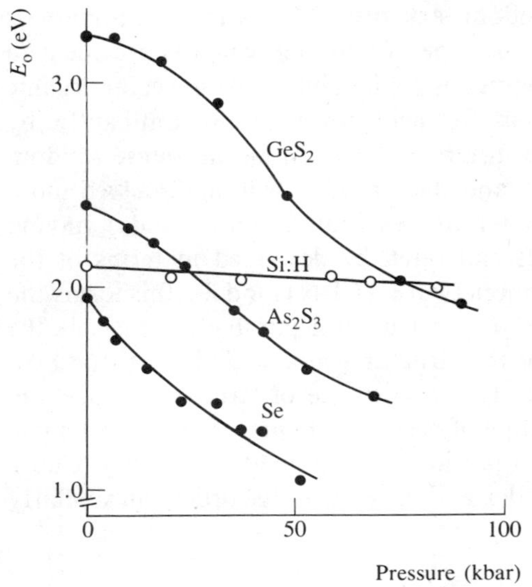

Fig. 3.44 Variation of the optical bandgap E_o with hydrostatic pressure for a number of amorphous solids (Tanaka 1987).

in Fig. 3.44, from which it can be seen that the locally low-dimensional chalcogenide glasses all exhibit appreciable pressure-induced red shifts, whereas E_o for a-Si:H is practically pressure-independent. (The fact that $\partial E_o/\partial p$ is not positive in the latter case may be due to the network-dimensionality-lowering effect of the incorporated hydrogen.)

3.5 Structural modelling

3.5.1 Introduction

We have seen in the last section that a variety of experimental techniques can give information about the microscopic structure of amorphous solids. In most cases, and for diffraction experiments in particular, this information is limited almost entirely to the first two coordination shells in the case of covalent materials; i.e. the bond lengths and angles of nearest-neighbour atoms comprising the basic structural unit (e.g. SiO_4 tetrahedra in a-SiO_2) can be determined, but the relative disposition of such units cannot be ascertained with certainty. For the case of RDFs derived from scattering experiments on monatomic systems, for example, the problem lies in the fact that peaks other than the first and second cannot be uniquely associated with a particular interatomic correlation, but are made up from a variety of contributions from higher-lying shells (see Fig. 3.1). Matters are complicated considerably for multicomponent materials for which a single diffraction experiment does not identify the origin of any peak in the RDF in terms of specific atomic pair correlations.

One solution to these difficulties is the construction of *models* which purport to simulate the structure. Any structural information that is known can be used as a basis for the generation of a random structure. The form of randomness involved depends to a great extent on the type of bonding which is prevalent in the material. The structure of amorphous metals, ionic solids, and molecular organic solids (and also some liquids), which are held together predominantly by *non-directional* forces, can be described in terms of the canonical 'dense random packing (DRP) of hard spheres' model, and these models will be discussed more fully in sect. 3.5.2). Non-metallic materials (excluding ionic solids) having predominantly covalent *directed* bonds can often be described in terms of the equivalent canonical 'continuous random network' (CRN) model – this forms the subject of much of sect. 3.5.3. In either case, structural parameters such as the RDF, density etc. may be computed for the structural model, and the theoretical predictions compared with experiment. The importance of structural modelling lies in the fact that a detailed understanding of the structure may thereby be gained. For instance, the structural origin of features in the (computed) RDF may readily be ascertained with the use of such a model, and medium-range order can similarly be investigated.

3.5.2 Dense random packing of spheres

Since many amorphous metals are quenched from the liquid state and have predominantly non-directional bonding, it is natural to consider the structure as being that of a frozen liquid. In this regard, the dense random packing (DRP) models of single-sized hard spheres pioneered by Bernal (1959, 1964) as descriptions of liquid structures have consequently been adopted as prototype structures for amorphous metal systems.

The first DRP models were physical models constructed from aggregates of equal-sized ball-bearings held in flexible bags; when a non-crystalline packing had been achieved and the density maximized by squeezing the bag, the balls were glued into position and the coordinates and connectivity measured. The most ambitious project of this kind was that of Finney (1970) which involved the construction and characterization of a model containing 7934 spheres. In addition, computer algorithms have been developed (e.g. Bennett 1972) which generate DRP models numerically starting from a given seed. Spheres are added sequentially such that they are in hard contact with three pre-existing spheres, choosing the particular site according to one of two criteria. Either the site which is closest to the centre of the cluster, or else that which satisfies a given 'tetrahedron perfection' condition (Ichikawa 1975), is chosen. Tetrahedron perfection is defined according to how nearly a group of four spheres approximates perfect tetrahedral symmetry, and may be quantified by considering the 'pocket' formed by three contacting surface atoms of the cluster (1, 2 and 3) and evaluating the expression (Cargill 1981)

$$k_{123} = \max_{\{12,13,23\}} \frac{r_{ij}}{(R_i + R_j)} \qquad [3.98]$$

where r_{ij} is the separation between centres of spheres i and j and R_i and R_j are their radii. Perfect tetrahedral arrangement is achieved if $k_{max} = 1$. In this manner, models containing four or five thousand atom sites have been constructed.

Characterization of both types of DRP model involves calculating the RDF and the packing fraction. The packing fractions, η_p, obtained for the physical models have been determined to be 0.6366 (cf. 0.74 for f.c.c. and h.c.p. structures); values for the computer-generated models tend to be slightly lower and Bennett found that η_p decreased with increasing radius, extrapolating to 0.61 for infinite radius. Shown in Fig. 3.45 are real-space distribution functions plotted as $\rho(r)/\rho_0$ (see sect. 3.2.2) for both the Finney and Bennett models. Points to note are the essential similarities of the two curves, a well-defined second peak following the singularity at the hard-sphere diameter, with progressively broader features at larger distances. Of particular interest is the clear splitting of the second peak in both cases, although the splitting appears to be more pronounced in physical ball-bearing than in computer-generated models and even more evident (with a reversal in the relative magnitudes of the sub-peaks of the second peak) in the much smaller models (200–1000 atoms) which have been reported. The reason for this difference is not clear, but may in some way be related to the drop-off in density that accompanies the construction of very large models. The two sub-peaks occur at 1.73 and 1.99 sphere diameters, in reasonable agreement with the values found experimentally for many TMM glasses, namely 1.67 and 1.9 respectively. The microscopic origin of this splitting can be ascribed to certain particular recurrent local configurations (see Fig. 3.45).

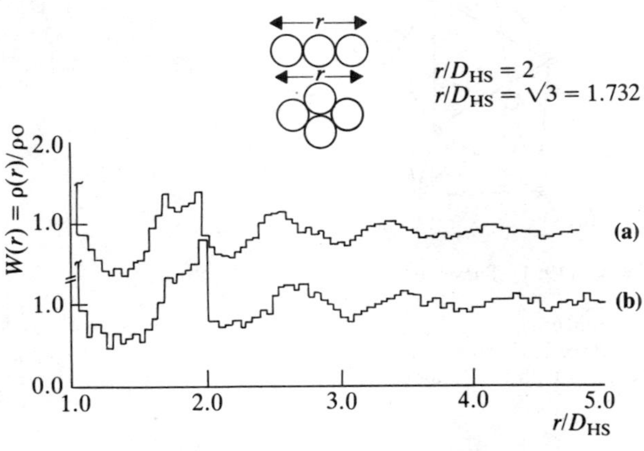

Fig. 3.45 Distribution functions for dense random packed models plotted versus hard-sphere diameter D_{HS}:
(a) DRP of single-sized balls (Finney 1970);
(b) computer generated single-sized sphere DRP (Bennett 1972). The inset shows the atomic configurations believed to be responsible for the split second peak.

Although these random packing models of equi-sized spheres are dense in the sense that no internal voids exist which are large enough to accept a normal-sized sphere, this does not mean that there are no reasonably large voids present in the structure; indeed the $\simeq 10\%$ deficit in packing density between DRP models and close-packed crystal structures immediately suggests that voids of appreciable size are present. The holes found in DRP models can be classified into one of five types of polyhedron (whose vertices are defined by the sphere centres), if allowances are made for distortions of up to 20% in the edges of the polyhedra; these were termed by Bernal 'canonical holes'. These polyhedral holes are illustrated in Fig. 3.46 and of the five, the smallest, the tetrahedron, occurs most often (2.9 per sphere) and this is some 25 times more frequent than the next most common polyhedra, the trigonal prism and the tetragonal dodecahedron (Cargill 1975). The topology of the DRP structures may also be described in terms of Voronoi polyhedra (or Wigner–Seitz cells – sect. 3.3). It is found that these have predominantly pentagonal faces, with an average number of 14.25 faces per Voronoi polyhedron (Finney 1970); for comparison, the comparable number for crystalline lattices are 12 (h.c.p.), 12 (f.c.c) and 14 (b.c.c).

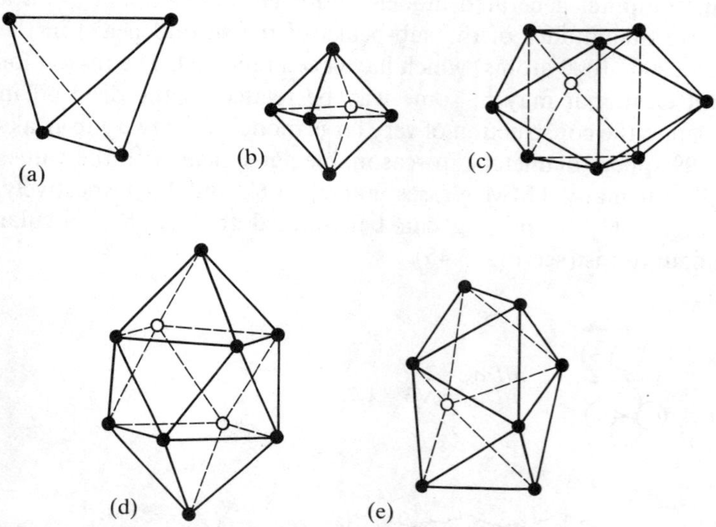

Fig. 3.46 The 'canonical holes' of Bernal (1964). They are
(a) tetrahedron;
(b) octahedron (often present as half-octahedra);
(c) trigonal prism (shown capped with three half-octahedra);
(d) Archimedean anti-prism (shown capped with two half-octahedra);
(e) tetragonal dodecahedron.

The observation that a random packing of spheres can be described equivalently as a packing of polyhedra is not new however. An amateur botanist, Stephen Hales, then Rector of Farrington and Minister of Teddington published in 1727 a work entitled *Vegetable staticks: an account of some statistical experiments on the sap in vegetables*, in which he described his experiments on the uptake of water by plants, particularly vegetables. In one such experiment, he studied the dilation of peas upon absorption of water by observing that they could raise the heavy lid of a pot in which they were contained. In an extension of this work he carried out the following experiment (Hales 1727):

Being desirous to try whether they would raise a much greater weight by means of a lever with weights at the end of it, I compressed several fresh parcels of Pease in the same Pot, with a force equal to 1600, 800 and 400 pounds; in which Experiments, tho' the Pease dilated, yet they did not raise the lever, because what they increased in bulk was, by the great incumbent weight, pressed into the interstices of the Pease, which they adequately filled up, being thereby formed into pretty, regular dodecahedrons.

What Hales had inadvertently found was that the DRP of 'Pease' in his experiments had deformed into Wigner–Seitz cells as they expanded in a container of constant volume. A dodecahedron is distinguished by the presence of pentagonal faces and it was probably the observation of these in his random packings which prompted Hales to describe the Voronoi polyhedra as dodecahedra.

Obviously, the simple DRP models described so far are suitable as representations of the structure of pure amorphous metals (Bi, Ga, etc.), but because of the difficulties of preparation and consequent impurity content, and the problems encountered in trying to extract quantitative data from low-temperature electron diffraction measurements on such films, there is unfortunately a dearth of reliable structural diffraction data on such systems. Instead a comparison of the experimental total reduced RDF obtained from X-ray scattering for a-$Ni_{76}P_{24}$ and $G(r)$ calculated for the single atom type Finney DRP model, both scaled to the same hard-sphere diameter, reveals good overall agreement between the two curves (Cargill 1975) but although the splitting of the second peak is reproduced by the DRP model, the relative weighting of the two sub-peaks is incorrectly predicted by the model compared with that observed generally in glassy TMM alloys. Thus, although the simple DRP model may be a reasonable approximation to the structure if only scattering from metal atoms is considered (or equally if both TM and M atoms have the same hard-sphere radius, e.g. 1.25 Å for Co and 1.28 Å for P), considerable problems arise once *partial* distribution functions are considered, since it is quite clear from experimental data that the metalloid atoms are *not* distributed at random (see Fig. 3.11), as would be the case if certain of the atomic sites in a single atom type DRP model were assigned as metalloids.

Instead, it has been suggested (Polk 1970, 1972) that the metalloid atoms could occupy the larger Bernal holes in a DRP of the transition metal atoms. While this simple idea has the virtue that it predicts correctly the amount of metalloid atoms ($\simeq 20\%$) apparently needed to stabilize the structure, and also the presence of local chemical ordering, i.e. metalloid–metalloid avoidance, inherent in such alloys, subsequent work (see e.g. Cargill 1975) has shown that there is in fact insufficient void volume available to accept any but the very smallest type of metalloid atoms. However, the attractions of the interstitial model for metalloid positioning have meant that further efforts to improve the simple DRP model have been made. Among these has been the computer generation of binary DRP models having two sizes of atom (see e.g. Boudreaux and Gregor 1977), and more recently the energy relaxation of DRP models, whether comprising spheres of the same size and interstitially stuffed, or binary packings of different sizes of spheres.

Energy relaxation of DRP models can be carried out using Lennard–Jones potentials to simulate the predominantly non-directional nature of the bonding inherent in amorphous metals:

$$E \propto \frac{1}{2}\sum_{i}\sum_{j\neq i}\left[\left(\frac{R_{ij}}{r_{ij}}\right)^{12} - 2\left(\frac{R_{ij}}{r_{ij}}\right)^{6}\right] \qquad [3.99]$$

where R_{ij} is the sum of hard-sphere radii of spheres i and j, and r_{ij} is the actual distance between their centres, and only the interactions between nearest-neighbours are included. Potentials of the form of [3.99] are obviously an approximation to the true potential, particularly so for TMM alloys where some d-electron character to the bonding between metal and metalloid may exist. It should be noted also that energy relaxation using purely non-directional potentials does not necessarily preserve the topology of the model, i.e. considerable atomic rearrangement may occur during the relaxation. Thus, the relatively high degree of porosity characteristic of packings having near-ideal tetrahedral symmetry is considerably reduced upon energy relaxation. In addition, because of the 'soft' nature of potentials such as given by [3.99], energy relaxation causes the previously hard non-overlapping spheres to overlap to a certain extent; such models have been termed 'dense random packing of soft spheres', DRPSS.

As an example of this approach, we shown in Fig. 3.47 a comparison of experimental scattering curves for a-$Gd_{32}Co_{68}$ with computed curves for a binary DRP model ($k_{max} = 1.1$) in both unrelaxed and relaxed states (Cargill 1981). Note that the overall fit to experiment is significantly improved for the relaxed model, and that the density of the model also approaches the experimental value after relaxation (as can be seen from the slope of $G(r)$ at small r). Note too that the

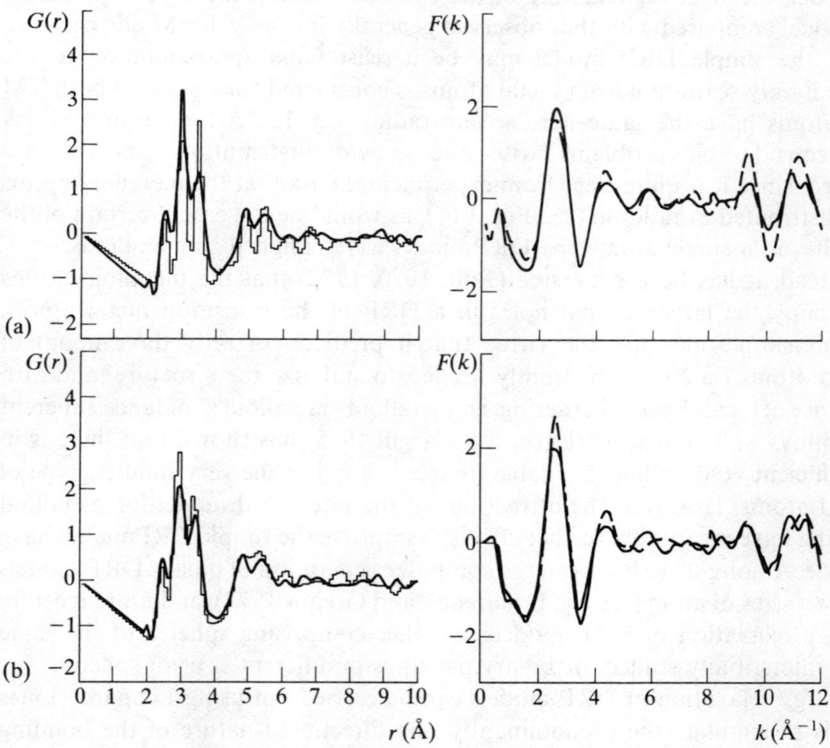

Fig. 3.47 Reduced RDF and scattering factor of a-$Gd_{32}Co_{68}$ compared with calculated curves for DRP models (Cargill 1981). In all cases, the experimental data are shown by the continuous solid lines:
(a) unrelaxed model;
(b) energy relaxed model.

first peak in the reduced RDF is resolved into a triplet, corresponding to the correlations, in order of increasing distance, Co–Co, Co–Gd and Gd–Gd, respectively.

3.5.3 Continuous random networks

A continuous random network (CRN) model treats the structure of a covalently bonded amorphous material as a giant macromolecule in which perfect connectivity is maintained (except for the surface of the model), with all the atoms retaining their normal valence (Zachariasen 1932). Therefore the CRN can be considered to be a hypothetical idealized structure of an amorphous covalent solid in the same sense that a single crystal is for the crystalline state; in both cases, the structures can be regarded as perfect because bonding defects such as dangling bonds or voids are excluded. However, the term 'random network' is somewhat of a misnomer in that such networks are not truly random in a statistical sense, but have a considerable degree of local order conferred upon them by the presence of directed bonding (see sect. 3.3). Thus, covalently bonded amorphous solids generally have well-defined bond lengths, bond angles and nearest-neighbour coordinations, often very similar to those found in corresponding crystalline polymorphs. The fundamental polyhedra formed by the nearest neighbours (e.g. SiO_4 in SiO_2, BO_3 in B_2O_3 or As_4 in As) are connected together in the network, and their relative orientation is characterized by the dihedral angle, ϕ (see Fig. 3.36). If the units were completely randomly arranged, the dihedral angle distribution would be constant; in practice, it is found for most CRNs that although all values of dihedral angle are represented, certain values are more probable than others as a result of ring-closure constraints (see sect. 3.4). Non-crystallinity is achieved by allowing some degree of bond-angle distortion and more or less complete freedom for values of the dihedral angle, subject to steric constraints and the imposition of a degree of MRO (e.g. in the form of favoured ring sizes). In fact, the influence of steric constraints alone on the structure can be extremely important, as stressed by Thathachari and Tiller (1982) for the case of vitreous SiO_2. Galeener (1985) has actually computed analytically the oxygen bond-angle distribution $P(\theta_o)$ for a-SiO_2, assuming only steric effects to be operative (i.e. neglecting ring-closure constraints). The steric hindrance factor relates to the minimum oxygen–oxygen distance R allowed (see Fig. 3.48(a)); Galeener (1985) correspondingly obtained the following analytic expression relating R to θ_O^M, the value at which $P(\theta_O)$ is a maximum:

$$(R/r)^2 = 4(1 - \cos\psi)[1 + \cos(\theta_O^M + \psi)] \qquad [3.100]$$

where r is the Si—O bond length and ψ the (tetrahedral) bond angle subtended at the Si atom (see Fig. 3.48(a)). For glassy SiO_2, both X-ray diffraction (Mozzi and Warren 1969) and MASNMR (Gladden *et al.* 1986) indicate that $\theta_O^M \simeq 144°$ (see Fig. 3.25); [3.100] therefore requires that $R/r = 1.95$. The distribution $P(\theta_O)$ can be calculated using this value of R/r, and by assuming that *all* dihedral angles ϕ are allowed for $\theta_O > \theta_O^M$, whereas the allowed values of ϕ are progressively reduced for $\theta_O < \theta_O^M$; the result shown in Fig. 3.48(b) is in remarkably good agreement with the curve obtained by X-ray diffraction.

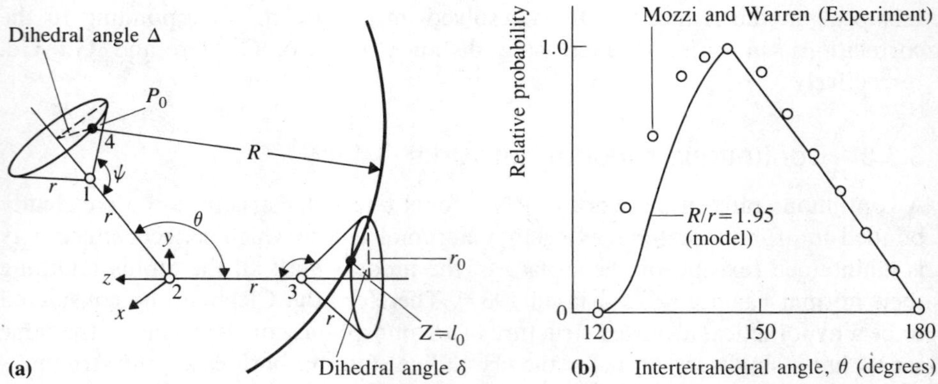

Fig. 3.48 (a) Steric constraints controlling the minimum O–O distance in v-SiO_2. Atoms 1 and 3 are Si; 2, 4 and 5 are O.
(b) The oxygen bond-angle distribution in v-SiO_2 derived solely from steric considerations and compared with that derived by Mozzi and Warren (1969) from X-ray diffraction data (Galeener 1985).

It should be noted that in the definition of a CRN given above, nothing is said about the local dimensionality associated with the covalently bonded backbone of the network. Consequently, we can regard the 'random-coil' model for the structure of amorphous polymers (Flory 1953) as being simply a special case of a CRN in which the local network dimensionality is two (in the absence of cross-links). Zallen (1983) discusses the random-coil model in some detail, in particular its relationship to random walks.

There are two ways of constructing such structural models. The first involves the physical building of a model consisting of a 'ball-and-stick' arrangement, in which tubes of the appropriate length represent covalent bonds, and atomic sites are represented by either balls with holes drilled at the appropriate orientations or units with spokes at the correct relative angles (Fig. 3.49 shows two CRNs of these types for the oxide glasses SiO_2 and B_2O_3). Each atom is numbered and a connectivity table constructed; the spatial coordinates of all the atoms are measured either by stereo-photography or by some form of surveying. The other way of constructing structural models is by using a computer and this technique in itself can take several forms. The method which comes closest to physical construction is the serial addition of atoms, adatom by adatom, to a starting seed, perhaps chosen to be a polyhedron not found in any crystalline polymorphs (e.g. containing fivefold rings for the case of Ge or Si) so as to preclude possible periodic growth of the cluster. The algorithm used must search among all the possible empty bonding sites to find those most favourable, i.e. those which minimize strain and the number of non-bonded atoms (dangling bonds) in the case of covalent materials. (An example of this approach is for the case of a-$SiSe_2$ – see Fig. 3.40 (Gladden and Elliott 1987).)

An alternative way of using computational methods to generate structural models is by the *modification* of existing models, for example by altering the energy of the structure by appropriate algorithms; for the case of CRN models this means minimizing the local strain energy resulting from bond-length and bond-angle

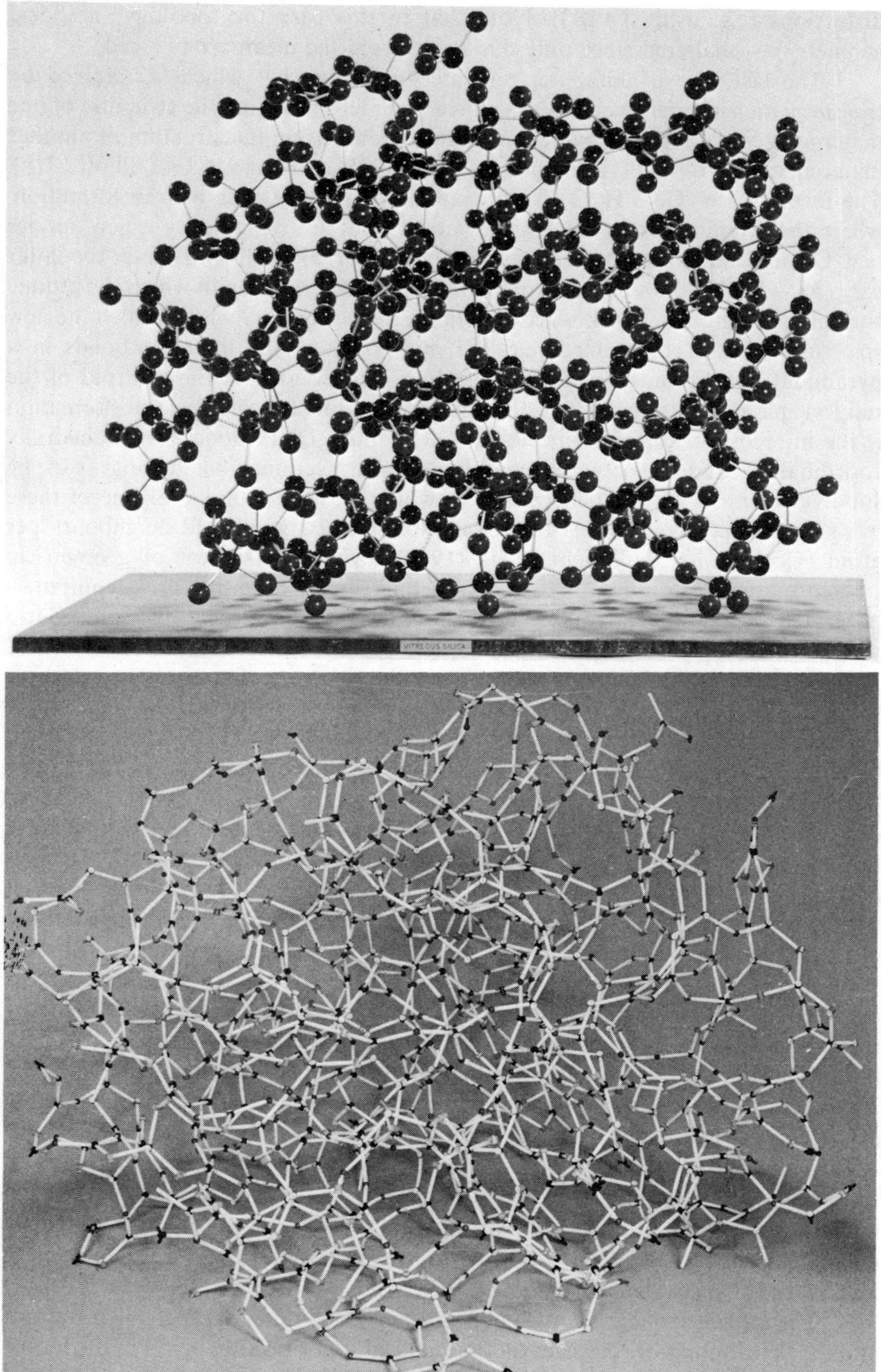

Fig. 3.49 (a) Continuous random network model simulating the structure of amorphous SiO_2 (Bell and Dean 1966) (Crown copyright).
(b) Continuous random network model simulating the structure of vitreous B_2O_3 (no boroxol rings) (Williams and Elliott 1982).

distortions (e.g. using [4.18]). Note that in this case the topology (i.e. local geometry) is unaltered since only deviations from the mean are reduced.

The last type of computer-generated model is that which is obtained by *transformation* of existing models in which a model simulating the structure of one material is converted into a different network simulating the structure of another material, e.g. Ge to H_2O (Boutron and Alben 1975) and As to B_2O_3 (Elliott 1978). The former case (Ge → H_2O) is an example of a direct type of transformation, where the atomic sites in the original model (Ge) are replaced by a new species (e.g. O) and the other atomic species (H) are inserted relative to these; the latter case (As → B_2O_3) is an example of an indirect transformation in which the atomic sites of the original network serve to define a local geometry within which the new structural unit is constructed (e.g. the mid-points of the three As bonds in a pyramidal unit become O sites and B atoms are inserted at the centroid of the resulting planar O_3 triangle). A further interesting form of topological transformation is the interconversion of a tetrahedral CRN and a DRP model. A tetrahedrally coordinated random network containing only even-membered rings can be subdivided into two interpenetrating networks and the removal of one of these leads to a structure which is randomly close-packed with $\simeq 12$ neighbours per atom (Connell 1975). Wright *et al.* (1980) review the subject of geometrical transformations of random networks in detail. Finally note that gross computer-generated structural changes (e.g. crystal → CRN) can be effected by Monte Carlo methods; for convenience, these techniques are discussed separately (sect. 3.5.5).

The principal advantage of simulating the structure of amorphous materials using physical models is that many structural features may be identified directly by inspection. As an example may be cited the case of the $\simeq 20\%$ density deficit between the crystalline and amorphous forms of arsenic which was not explained before the construction of a threefold coordinated CRN demonstrated that voids of sufficient size to account for the density difference arose naturally as a result of the asymmetric As_4 building unit forming the basis of a random network (Greaves and Davis 1974). In addition, it is a relatively straightforward matter to construct a CRN model of sufficient size (say $\simeq 500$ atoms) to give a reasonably small amount of statistical noise in the calculated RDFs. The disadvantage of physical model building lies mainly in the tedium of measuring the coordinates of large numbers of atomic sites (although it is not necessary to measure them precisely if energy relaxation procedures are to be applied subsequently), and also in the possibility that a bias may be introduced into the network by the model builder.

The advantage of computer generation of models is that large numbers of atoms can be generated in this way without bias and that the model can be energy relaxed as it is formed. Transformation of networks would be virtually impossible to achieve without using a computer. The disadvantage of computer-generated models is that they cannot easily be examined visually, and that it is extremely difficult to produce algorithms that ensure perfect connectivity in the interior of a network for a covalent material.

The influence of factors such as bond-angle and dihedral-angle distributions and the distribution of ring sizes on the shape and positions of peaks in the model RDF can be investigated which, if the model RDF agrees with experiment, are important since these quantities cannot be obtained directly from experimental data but only inferred from such modelling studies. The distribution of ring sizes is a significant parameter since it is believed to influence the shape of features in the electronic density of states in the valence band for certain amorphous semiconductors (e.g. Ge or Si) (see sect. 5.2.1); Greaves (1976) and Temkin (1978) have shown how the RDF can be synthesized from a knowledge of the ring statistics and the bond-length and dihedral-angle distributions for threefold (e.g. As) and fourfold (e.g. Ge) coordinated networks, respectively.

We show in Fig. 3.50 an experimental RDF derived from a high-resolution neutron scattering study of a-Ge, compared with a calculated RDF for a CRN model; also shown is another structural parameter ($P(\phi)$) for the network. It can be seen that the agreement is reasonable, but this worsens if the relative proportion of fivefold to sixfold rings is changed from the value of 0.34:1.5. It must be admitted, however, that the position of the first two peaks is fixed for a tetrahedral material such as a-Ge, since the first peak of the model RDF is scaled to the experimental value for the nearest-neighbour distance, and the average bond angle (and hence the position of the second peak) is determined by symmetry to be the tetrahedral angle, $109°28'$. It is the shape and position of higher-lying peaks which are characteristic of a particular model (through $P(\phi)$, etc.) and whose agreement with experiment is necessary for the model to be valid.

For the case of models containing more than one component, all the structural parameters discussed above can of course be obtained, but in addition and most importantly, the individual partial correlation functions can be computed and the relative contribution of each to the total weighted RDF ($\Sigma W_{ij}\rho_{ij}(r)$) can be ascertained. As an example of the power of structural modelling applied to multicomponent systems, we show in Fig. 3.51 data for the archetypal glassy system, SiO_2, and the structural information about the individual pair correlations that can be obtained; again agreement between theory and experiment is reasonably good. We have already shown in Fig. 3.40 the corresponding curves for the chemically similar but non-isomorphous material $SiSe_2$.

So far we have considered CRN models only as representations of the structure of *covalent*, amorphous materials. However, these form just one example

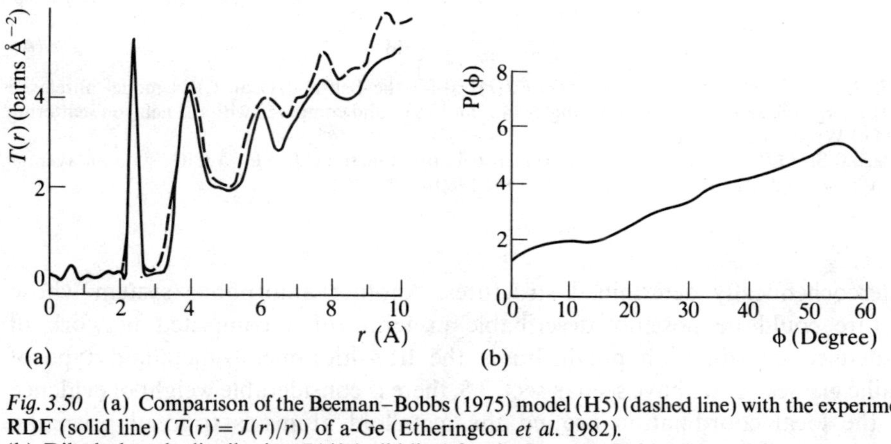

Fig. 3.50 (a) Comparison of the Beeman–Bobbs (1975) model (H5) (dashed line) with the experimental RDF (solid line) ($T(r) = J(r)/r$)) of a-Ge (Etherington *et al.* 1982).
(b) Dihedral angle distribution $P(\phi)$ (solid line) for the Beeman–Bobbs H5 model.

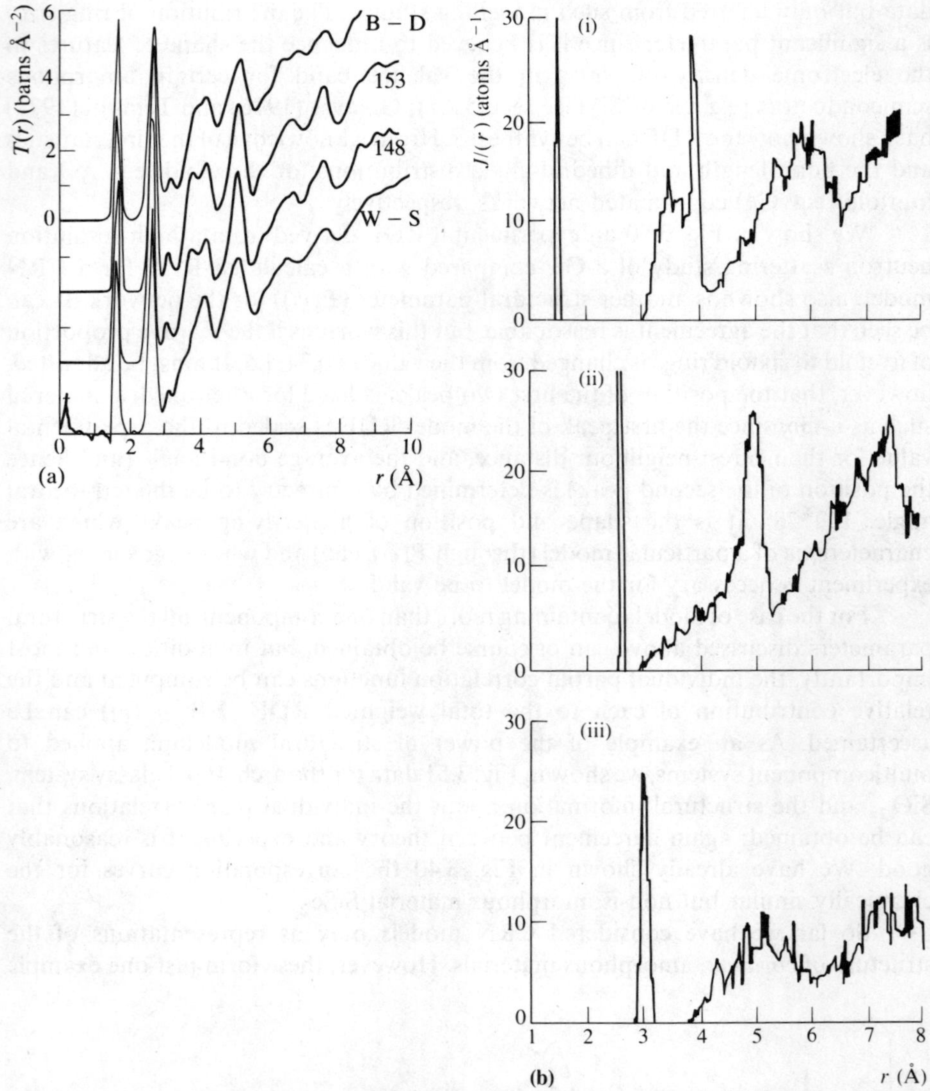

Fig. 3.51 (a) Correlation functions ($T(r) = J(r)/r$)) for the Bell and Dean CRN model unrelaxed (B–D), energy relaxed to oxygen bond angles 148° and 153°, and compared with the neutron scattering results of Wright and Sinclar (W–S).
(b) Partial Si–O (i), O–O (ii) and Si–Si (iii) distribution functions $J(r)$ for a CRN with an average oxygen bond angle of 142° (Gaskell and Tarrant 1980).

of stereochemically determined structures. Another amorphous system whose structure could be possibly describable in terms of a connected network of well-defined coordination polyhedra is the transition-metal–metalloid type of metallic glasses. As we have seen in sect. 3.3, there is considerable weight of evidence that the local coordination around the metalloid atoms is trigonal prismatic (Fig. 3.33), and Gaskell (1979b, 1981) has suggested that the overall structure of

TMM glasses could be regarded as a form of CRN of such connected trigonal prismatic units. However, later considerations by Dubois *et al.* (1985) suggest that the observed SRO may instead result from the existence of a particular type of MRO in the form of regularly spaced chemical twinning planes (sect. 3.4.3).

At this point it must be emphasized that a model is an idealization of the real amorphous structure; ideal in that most CRN models are perfectly connected in the interior of the model, admitting no dangling bonds except at the surface, even though it is known from other studies (e.g. ESR) that appreciable densities ($\simeq 1\%$) of bonding defects can occur, particularly in vapour-deposited materials. Furthermore, CRN (and DRP) models are idealistic in that they simulate the amorphous structure in a dense state; that is voids, or other such growth defects, are not included unless they arise naturally in the construction of the model (e.g. when low-dimensional structural units are involved, as in the As or B_2O_3 CRNs). Thus, a tetrahedral CRN simulating a-Ge or a-Si and containing no voids has a density 99% of the diamond cubic form (Polk and Boudreaux 1973). This demonstrates the important point that model amorphous structures can be built, without the supposed stabilizing influences of the presence of voids or dangling bonds, such that the inbuilt strain (e.g. in bond-angle distributions) does not increase without limit as the size of the model increases. This empirical observation seems to be at variance with the claim by Phillips (1979) that 'mechanically overconstrained' covalent systems (see sect. 2.5.4), such as a-Si or Ge, cannot exist with homogeneous structures as represented by a CRN model; instead, it is claimed that strain must be relieved discontinuously at interfacial boundaries between clusters.

3.5.4 Other structural models

A number of other models for the structure of covalently bonded amorphous solids have been proposed. These should be mentiond both for completeness, and also because the recent realization that ideal CRN models are often inadequate in accounting for features observed in e.g. medium-range order and growth morphology may mean that other models need to be considered.

The model most at variance with the CRN hypothesis and yet which is the simplest conceptually is the 'microcrystalline' approach. In this, it is assumed that the amorphous solid is composed of very small crystals of a polymorph or polymorphs which are so small that the diffraction rings, characteristic of a polycrystalline sample, are broadened into the diffuse haloes characteristic of amorphous materials. The problem with this approach is that to achieve even moderate agreement with scattering data, the microcrystallites must be made so small that they comprise only some 50% of the structure, the other half being composed of an unspecified random connecting tissue (presumably CRN-like). At these dimensions it becomes almost a matter of semantics as to whether microcrystallites exist or not. Furthermore, the fit between model and experiment is not good, the microcrystallites predicting too large oscillations at large r in the RDF, or conversely at small k in $F(k)$ (Wright *et al.* 1987). As an example, Fig. 3.52 shows the reduced scattering intensity calculated for microcrystals of diamond- and wurtzite-like polymorphs of Ge compared with experimental data

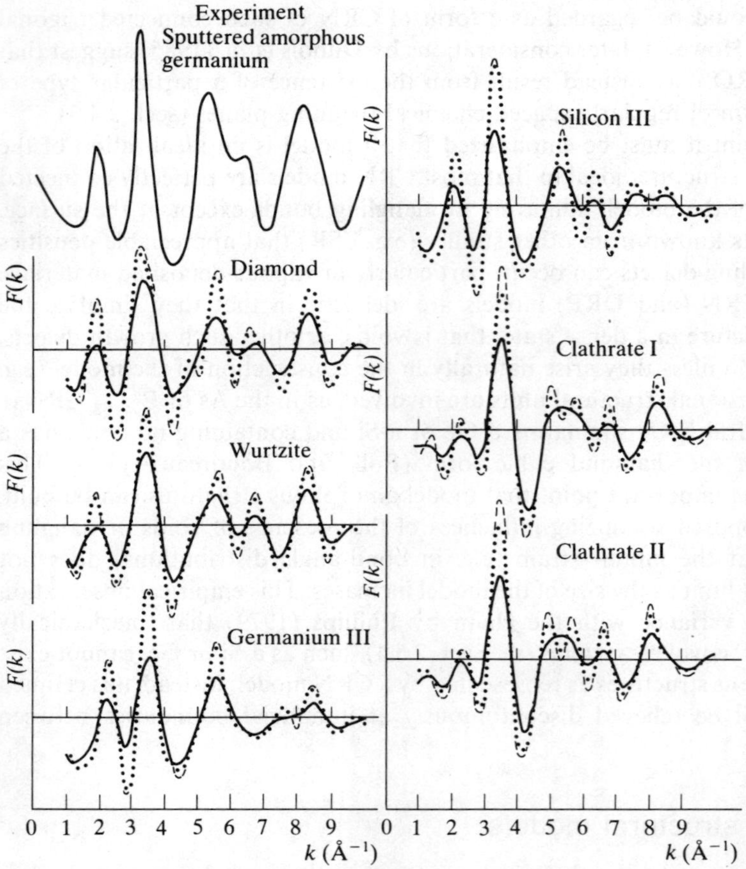

Fig. 3.52 Reduced scattering intensity ($F(k)$) for a-Ge compared with calculated curves for microcrystallites of various structures. In each case, the solid line refers to crystallites of small radii ($\simeq 4$ Å), whereas the dotted and dashed lines refer to crystallites of medium and large ($\simeq 6$ Å) radii (Weinstein and Davis 1973/4).

(Weinstein and Davis 1973/4) where these shortcomings are clearly visible. Figure 3.53 shows a comparison of the experimental RDFs of amorphous and crystalline Ge (Temkin *et al.* 1973); points to note are the strong oscillations present in the crystalline RDF and, particularly, the presence of a peak at 4.7 Å in the crystalline RDF but its complete absence in the amorphous RDF. This feature arises from correlations across the sixfold rings which occur in the diamond cubic structure when all bonds are staggered and the dihedral angle is 60° (see inset to Fig. 3.53(a)). The RDF calculated from a collection of microcrystallites of such polymorphs, incorporating such rings, has a peak at ~4.7 Å, in disagreement with experiment. Attempts have been made to improve the fit with experiment for a variety of materials by using microcrystals of varying shapes and sizes, but in all cases agreement is significantly worse than that of the corresponding CRN.

In between the continuous random network and microcrystalline approaches lie those models built using non-crystalline structural units. Retaining the

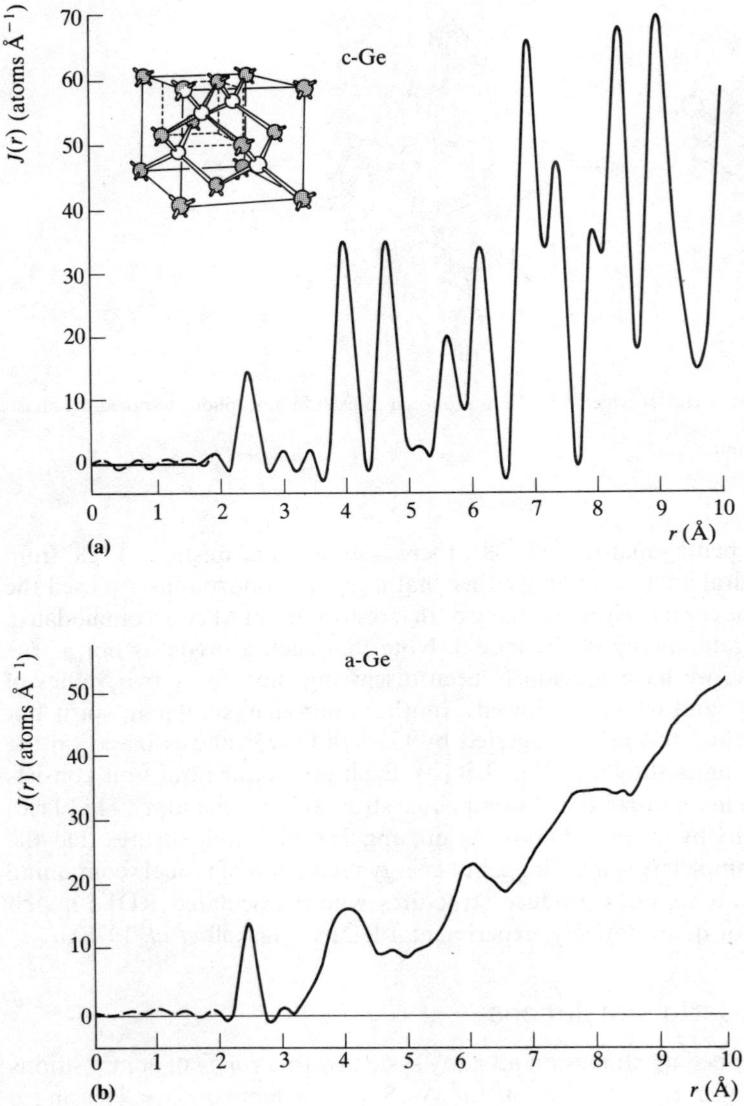

Fig. 3.53 Radial distribution functions for:
(a) crystalline Ge (with inset of diamond cubic lattice, and
(b) amorphous Ge (sputtered on to a substrate at 150°C) (Temkin *et al.* 1973).

tetrahedrally bonded amorphous semiconductors, Ge and Si, as examples for the moment, the first such study was undertaken by Grigorovici and Manaila (1969) who devised a random network model consisting essentially of a mixture of diamond-like units and so-called 'amorphons', regular dodecahedra consisting of planar pentagonal rings (Fig. 3.54(a)). The fivefold rings result from the presence of eclipsed bonds (i.e. $\phi = 0°$) in contrast with the staggered configuration ($\phi = 60°$) characterizing the diamond cubic structure. However, since the interior angle of

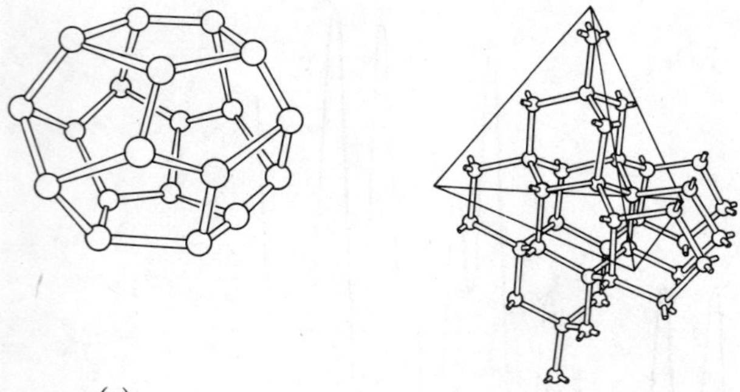
(a) (b)

Fig. 3.54 Two forms of cluster which have been proposed to exist in amorphous tetrahedral solids:
(a) amorphon;
(b) polytetrahedral unit.

a regular planar pentagonal ring is 108°, there is an angular misfit of 1°28′ from the ideal tetrahedral angle, implying either that if regular amorphons are used the structure cannot be completely space filling or that distortions must be accommodated, increasing the strain energy of the model. Note that such a model is not a true CRN in the sense we have previously been discussing, since only two values of dihedral angle, 0° and 60°, are allowed. Another approach, similar in spirit but difference in practice, has been suggested by Gaskell (1975) and is based on the 'polytetrahedral' units shown in Fig. 3.54(b). Each polytetrahedral unit consists of five tetrahedra in an ordered, diamond cubic structure and the four {111} faces contact other units by eclipsed bonds. Again, angular mismatch ensures that the models are not completely space filling, but energy relaxation of models containing several hundred atoms does produce structures whose calculated RDFs match qualitatively, if not quantitatively, experimental RDFs (Gaskell *et al.* 1977).

3.5.5 Monte Carlo simulations

One of the few modelling studies on an alloy system with a *range* of compositions is that by Renninger *et al.* (1974) on the As_xSe_{1-x} system ($0 < x < \frac{1}{2}$) using a Monte Carlo approach. Random starting structures, containing clusters of about 150 atoms, were generated using structural data known from experiment, e.g. bond length, coordination number (in accord with the $(8-N)$ rule) and bond angle. The Monte Carlo approach consisted of choosing atoms at random and moving them in an arbitrary direction and only if the move resulted in an improvement to the fit of the calculated RDF with the experimental curve was the move accepted; this process was continued until the r.m.s. deviation was less than a certain amount. As the percentage of As was increased it was found that the tendency to form large clusters of atoms increased, the Se chains becoming progressively cross-linked; a trend from chains to rings was therefore predicted.

Another application of Monte Carlo methods has been the conversion of the even-ring Connell–Temkin (1974) tetrahedrally coordinated CRN into ones containing varying fractions of odd-membered rings by the random repositioning of atoms with subsequent energy relaxation (Beeman and Bobbs 1975).

Finally, a type of Monte-Carlo process has been used by Wooten and Weaire (1984) and Wooten *et al.* (1985) to construct models of a-Si. Starting from an f.c.c. crystalline lattice, bonds are 'switched' randomly (i.e. broken and reformed between different pairs of atoms); after a sufficiently large number of such operations, an amorphous structure is produced with no residual crystalline order. Such models, subsequently energy-relaxed using bond-stretching and bond-bending potentials, give good agreement with experimental scattering data for a-Si (Wooten *et al.* 1985).

3.5.6 Molecular dynamical simulations

The final method of computer simulation differs in spirit from all the preceding approaches. Molecular dynamics, as its name implies, involves the study of the time evolution of the classical motion of a collection of atoms (or molecules), interacting by means of a given interatomic potential, as a function of the change in temperature (or some other thermodynamic variable) of the system. In this manner, the course of solidification on reducing the temperature of the computer 'liquid' may be followed, and importantly, the thermal history (trajectories, momenta) of all the atoms in the cluster are known. In practice, a small cluster of atoms, 100–1000 in number, is confined to a cube subjected to periodic boundary conditions whose dimensions accord with the known density of the liquid or solid under investigation, and each atom is assigned to a site (usually random). The atoms are then given a random Boltzmann distribution of velocities corresponding, typically, to a temperature of several thousand Kelvin. The Newtonian equations of motion are then solved for a series of time steps each of typical length $\simeq 10^{-15}$ s. Cooling of the system is achieved by a step-wise reduction in temperature followed by an isothermal molecular dynamics run for say $\simeq 10^{-10}$ s in total to ensure thermal equilibration. This is possible in these time-scales for liquids, but not for *glasses* where the time taken to equilibrate with respect to time-averaged atomic positions becomes much longer than the 'experimental' time-scale; the transition between these two situations of course marks the glass transition (see Ch. 2). The structure (e.g. the RDF) of the system can be calculated at the end of each temperature decrement.

The advantages of the molecular dynamics (MD) approach include a detailed knowledge of the time evolution of the atomic positions and momenta as the liquid forms a glass, and the ability to impose instantaneous changes of thermodynamic variables, such as pressure and temperature, on the system and to monitor the concomitant changes, if any, on the glass transition and the final structure of the glass. The disadvantages of the method, which are considerable, are that the effective rate of cooling in such computer simulations is very high, of the order of 10^{11} K s^{-1} or higher, and the loss of internal equilibrium during cooling, marking the transition to the glass (i.e. the glass-transition temperature), occurs often at temperatures much higher than those encountered in the laboratory. The very high rate of cooling is simply a result of computer economics; an isothermal run can seldom exceed $\simeq 10^{-10}$ s (i.e. $\simeq 10^5$ time steps) since it takes several hours of computer time, and if 10 K temperature decrements are chosen, the minimum cooling rate is consequently $\simeq 10^{11}$ K s^{-1}. Such rates are well in excess of any achievable in the laboratory, exceeding probably even the effective cooling rates encountered in thermal evaporation (Table 1.1). The other major drawback of the MD approach is the fact that in most cases only non-directional (e.g. ionic) interatomic potentials have been used; the attractive part of the potential is often taken to be simply of coulombic form for ionic systems or Van der Waals for rare gas systems, whereas the repulsive part may be either an exponential function or an inverse power ($1/r^n$) of the interatomic separation. There are appreciable difficulties with the implementation of the purely coulombic interaction in MD simulations. The long-range effect must be taken into account by using a full Ewald sum, using a truncation factor, or making the box size larger than the effective range of the interaction so as to avoid atoms interacting with their own images as a result of the imposition of boundary conditions.

A recent significant advance has been the use of interatomic potentials for simple metal glass systems, e.g. Ca–Al or Mg–Zn, derived from optimized first-principles pseudopotentials (Hafner *et al.* 1988). In this way *ab initio* calculations of the atomic structure of these glassy alloys have been made using MD techniques in which the only input information required is the composition, atomic numbers and atomic weights of the constituent metals.

Glasses have been simulated by MD techniques for a variety of systems, in addition to the metals mentioned above, namely Lennard-Jones systems (e.g. Ar), simple ionic glasses (e.g. KCl, which cannot normally be vitrified in the laboratory), BeF_2 and $ZnCl_2$ (which are among the few halides to be relatively easy glass-formers) and also predominantly covalently bonded systems such as SiO_2 or B_2O_3. However, the use of spherically symmetric interactions in MD simulations cannot alone reproduce the stereochemistry characteristic of covalent systems. In the cases where such potentials have been used, the tetrahedral coordination characteristic of $(SiO)_4$ tetrahedra in SiO_2 and the planar trigonal coordination characteristic of BO_3 units in B_2O_3 can be reproduced only by particular (unphysical) choices of the atomic radius ratios (Angell *et al.* 1977, Soules 1980, Amini *et al.* 1981). This deficiency has recently been partly remedied by the empirical inclusion of three-body force terms in the potential (Feuston and Garofalini 1988, Newell *et al.* 1989).

In conclusion, MD simulations can be very informative about the *dynamics* of liquids and the behaviour on cooling to form a glass. However, the cooling rates used are far higher than those encountered in the laboratory, and as we have seen in Chapter 2, the value of T_g observed is a sensitive function of the cooling rate. Molecular dynamics simulations of the structure of glasses, furthermore, suffer from the limitation that often only central forces acting between atoms are employed, restricting the scope of such simulations strictly to rare-gas or ionic solids. If covalent, non-central interactions could be introduced into the interatomic potential, the method is a potentially powerful modelling technique. The very recent work by Car and Parinello (1985) points the way in this regard. In these simulations (of the structure of a-Si), the potential that is used is obtained quantum mechanically in a self-consistent way using first-principles methods. It is only with the use of such accurate interatomic potentials that the MD technique can be expected to produce realistic structures of amorphous materials.

3.5.7 Summary

Structural modelling can be a valuable aid to determining the structure of amorphous solids. A model built using knowledge about the chemistry (i.e. local order) of the system, and whose computed scattering properties (e.g. structure factor, RDF etc.) fit well those derived from experiment, can give considerable additional insight into the structure, which otherwise cannot be obtained. An example might be the relationship for a CRN between the ring-size and the dihedral-angle distributions and features in the RDF, such as the peaks corresponding to the third and fourth coordination shells.

Many structural models have been *physical* models, either of the 'ball-and-stick' type (for CRNs and other models of covalent materials) or random packings of spheres (for metallic solids), because visualization of the model is most important. Nevertheless, for convenience many models have been constructed using a computer, particularly DRP models, often adatom-by-adatom to simulate real growth. Computer methods play an important role in the modification (e.g. strain-energy relaxation) or transformation of such models, as well as in Monte Carlo simulations in which models are constructed by the random positioning of atoms starting from either a crystalline or another amorphous structure. Computer techniques also figure prominantly in *molecular dynamics* simulations, in which the dynamics of a fluid subject to certain interatomic potentials are followed as a function of time. By lowering the 'temperature' during a simulation run, the liquid can be vitrified. The considerable disadvantages of this technique for the study of the structure of glasses (as opposed to that of liquids) are that the simulated rates of cooling are much higher than those encountered in the laboratory, and non-central (i.e. non-covalent) interatomic potentials have mostly been used.

It should be stressed finally that a structural model, of whatever type, is always an idealization of the true structure of an amorphous solid. Unlike a crystal, there is no unique structure of an amorphous material, and furthermore, most of the structural models built so far have not taken account of the growth processes and the resulting morphologies which occur in real amorphous systems.

II MACROSCOPIC STRUCTURE

In our discussion of the structure of amorphous solids so far, we have implicitly assumed the structure to be homogeneous on a macroscopic scale and concerned ourselves solely with a consideration of the fine details of short- and medium-range order. However, many amorphous solids exhibit gross inhomogeneities on the macroscopic scale, such as voids in vapour-deposited material or phase separation for certain compositions of multicomponent melt-quenched glasses. A discussion of the experimental techniques sensitive to such coarse features is therefore desirable, and we follow this with a consideration of the factors that can given rise to macroscopic structure in amorphous materials.

3.6 Experimental techniques

3.6.1 Microscopy

The most direct evidence for gross structural inhomogeneities comes from microscopy. If the different regions are sufficiently large (say greater than 10 μm), optical microscopy can be used to study the macroscopic structure. In most cases, however, macroscopic features occur on considerably smaller length scales and the greater resolving power of electron microscopy must be used. In certain cases where there are *density* fluctuations (as, for example, in the void network of evaporated a-Ge), these may be observed directly by electron microscopy. In the other major type of macroscopic structure, that involving *compositional* fluctuations, the phase-separated regions may be revealed by etching, thus transforming the compositional variation into density variations which are observable by electron microscopy. For example, glasses in the $Na_2O–B_2O_3–SiO_2$ system containing more than 20 wt.% of B_2O_3 separate into two interconnecting phases, one being silica rich; the other is preferentially attacked by hydrochloric acid, leaving behind the siliceous skeleton which is very apparent in electron microscopic images. This is the basis of the Vycor glass process.

3.6.2 Small-angle scattering

In addition to the scattering of radiation by atoms as discussed in section 3.2.2, larger scale composition or density fluctuations can also coherently scatter radiation but in a different region of k-space to that for conventional (sometimes called 'wide-angle') scattering. By the reciprocity relation between real and reciprocal space, the scattering from large-scale features will be confined to small values of k, i.e. 'small-angle scattering'. For example, the scattering from a collection of spheres of diameter D will be confined between the angles $\theta = 0$, where all the centres scatter in phase, and $\theta \sim \lambda/2D$ where destructive interference occurs. Thus, the presence of macroscopic inhomogeneities is signalled by a large increase in the scattered intensity as k tends to zero, and furthermore the functional form of the intensity can yield information on the size, shape and number of such inhomogeneities.

The small-angle scattering intensity is calculated in a similar manner to that for wide-angle scattering (sect. 3.2.2), except that (e.g. for X-ray scattering) an assumed constant density deficit, $\Delta\rho$, is introduced which corresponds to the difference in scattering factors between the two phases of material (whether between void and matrix, or two compositionally different phases). The small-angle X-ray scattering (SAXS) can be calculated for an inhomogeneous system modelled by a collection of particles of one type of material embedded in a homogeneous matrix of the other. Since the structure factor $S_h(k)$, and hence the X-ray scattering intensity, for a *homogeneous* medium tends to zero as $k \to 0$, the SAXS intensity (in electron units) can be taken to a very good approximation to be due solely to the dispersed particles

$$I_{SA}(k) = \Delta f^2 |S_p(k)|^2 \qquad [3.101]$$

where Δf is the difference in scattering amplitude for the two phases, and $S_p(k)$ is the structure factor for the particles only. This can be written in the form

$$I_{SA}(k) = (\Delta\rho)^2 \left| \int \rho_c(r) e^{ik\cdot r} \, d^3r \right|^2 \quad [3.102]$$

where $\rho_c(r)$ is defined such that for r inside the particle, $\rho_c = 1$ and for r outside the particle, $\rho_c = 0$. The final expression for the SAXS intensity for an assembly for equally but arbitrarily shaped particles of maximum dimensions, a, is (Letcher and Schmidt 1966):

$$I_{SA}(k) = (\Delta\rho)^2 V(a) N(a) \int_0^\infty 4\pi r^2 \gamma_0(r) \frac{\sin kr}{kr} \, dr \quad [3.103]$$

where $V(a)$ is the volume of a particle, $N(a)$ is the concentration of the particles and $\gamma_0(r)$ is defined by:

$$\gamma_0(r) = \int \rho_c(R) \rho_c(R+r) \, d^3R \quad [3.104]$$

which is the probability that for an arbitrary point R within a particle, a vector of unit length drawn in an arbitrary direction also lies within the particle. Therefore, $\gamma_0(r=0) = 1$ and $\gamma_0(r \geqslant a) = 0$. For a set of particles which do not have the same dimensions (i.e. 'polydisperse'), the factor $\gamma_0(r)$ in [3.103] is replaced by:

$$\gamma(r) = \int_0^\infty N(a) V(a) \gamma_0(r) \, da \quad [3.105]$$

Comparison of [3.103] for SAXS with the corresponding expression for conventional (wide-angle) scattering ([3.17]) shows that in the latter case, it is the local variation in (atomic) density which determines the scattering, whereas for SAXS it is variations in *shape* which governs the scattering intensity. In a similar fashion to the Fourier inversion of the wide-angle scattering data to extract the reduced RDF ([3.21]), the SAXS data may also be inverted to isolate the shape factor $\gamma(r)$:

$$\gamma(r) = \frac{1}{2\pi^2 (\Delta\rho)^2 r} \int_0^\infty k \sin kr \, I_{SA}(k) \, dk \quad [3.106]$$

While this gives some information on the dimensions of the dispersed particles (from the value of r at which $\gamma(r) \to 0$) a more useful quantity is $N(r)$, which is much more difficult to obtain in the general case. The expression for the SAXS intensity of a dilute dispersion of particles of volume v in the limit of very small k ($kR \ll 1$) may be approximated as (Guinier and Fournet 1955):

$$I_{SA}(k) = N v^2 (\Delta\rho)^2 \exp\left(-\frac{R^2 k^2}{3} \right) \quad [3.107]$$

where R is the 'electronic radius of gyration' given by:

$$R^2 = \frac{\int r^2 \rho(r)\,dv_r}{\int \rho(r)\,dv_r} = \frac{\int r^2\,dv_r}{\int dv_r} \qquad [3.108]$$

for a homogeneous particle of density $\rho(r) = \rho_0$. For example, the radius of gyration of a spherical particle of diameter L is given by $R = (3/5)^{1/2} L/2$. Thus, examination of [3.107] shows that a plot of $\ln I_{SA}(k)$ vs. k^2 should be linear, of slope $-R^2/3$; this is the so-called 'Guinier plot'. In practice, the Guinier plot often deviates from linearity which is an indication of a polydisperse system of particles of varying shapes and sizes. The slope as $k \to 0$ gives a value for the weighted radius of gyration R_G defined by:

$$R_G^2 = \frac{\Sigma_i R_i^2 v_i^2 P_i}{\Sigma_i v_i^2 P_i} \qquad [3.109]$$

where R_i, v_i and P_i are the gyration radius, volume and fraction of particles of species i. The determination of particle shapes from Guinier plots is not straightforward, and the interested reader is referred to the book by Guinier and Fournet (1955). An expression involving the concentration and volume of particles can be obtained from the zero intercept of the Guinier plot:

$$I_{SA}(0) = (\Delta\rho)^2 N v^2 \qquad [3.110]$$

or from the integrated intensity given by:

$$Z = \frac{1}{2\pi^2} \int_0^\infty k^2 I_{SA}(k)\,dk \qquad [3.111]$$

since it can be shown that:

$$Z = (\Delta\rho)^2 N v \qquad [3.112]$$

It should be stressed that the Guinier approximation only holds for an inhomogeneous system composed of a matrix containing *widely* separated particles which each scatter independently; i.e. interparticle interference is neglected.

Other forms of radiation can also be used in small-angle scattering experiments, and in recent years with the advent of nuclear reactor sources of neutrons, small-angle neutron scattering (SANS) has been much employed. This technique has the advantage that neutrons of long wavelength (e.g. $\simeq 10$ Å) can be used, thereby still satisfying the condition for small k-values but considerably easing the problems of collimation encountered for small angles.

Note that a finite amount of small-angle scattering is to be expected in certain cases even if no obvious gross macroscopic inhomogeneities, such as voids or phase separation, in present. Variations in density can remain in melt-quenched glasses as configurational (but not vibrational) fluctuations frozen-in on cooling the melt through the glass-transition temperature, T_g. Thermodynamic fluctuation theory predicts that the density fluctuations in a single component liquid are related to the isothermal compressibility κ_T by:

$$\frac{V\langle(\Delta\rho)^2\rangle}{\rho^2} = k_B T \kappa_T \qquad [3.113]$$

where V is the volume. However, the density fluctuations in a glass are essentially the same as those existing in the liquid at T_g and so the structure factor becomes (see problem 3.3):

$$S(0) = \rho k_B T_g \kappa_T(T_g) \qquad [3.114]$$

Thus, the small-angle scattering for a glass does not in fact tend to zero as k tends to zero, but instead approaches the 'compressibility limit' given by [3.114]. Values for the density fluctuation $\langle(\Delta\rho)^2\rangle$ obtained from experiment and calculated using [3.114] for glassy SiO_2 and GeO_2 agree to within a factor of 1.5 (Pierre *et al.* 1972).

The above discussion assumes that the scattering objects giving rise to the small-angle scattering are regular and have smooth shapes and surfaces, i.e. the mass per unit volume of such objects scales as r^3, or the mass per unit area scales as r^2, where r is a characteristic linear dimension. In many cases this assumption is valid, but there can be instances where this normal behaviour no longer applies, e.g. for rough surfaces (of voids) or for certain types of aggregates of particles. If these types of structure exhibit 'self-similarity' over a range of length scales they are termed 'fractals' (Mandelbrot 1982); some examples of their occurrence in amorphous systems are given in sect. 3.7.1. Small-angle scattering (SAS), using X-rays or neutrons or light scattering, is particularly sensitive to fractal geometries (if the size of the features therein is comparable to the wavelengths of the scattering radiation) since the SAS intensity then has a characteristic *power-law* dependence on the scattering vector.

Thus, *bulk* fractals, such as aggregates of particles, can be described by a correlation function $g(r)$ which remains invariant under a scale change ε, so that

$$g(\varepsilon r) = \varepsilon^{-\alpha} g(r) \qquad [3.115]$$

This implies that $g(r)$ has a power-law dependence $g(r) \sim r^{-\alpha}$, where $\alpha = 3 - D_H$ in three-dimensional space, and D_H is the Hausdorff or fractal dimensionality ($1 \leqslant D_H < 3$). In this case, the SAS intensity obeys the scattering law

$$I \propto k^{-D_H} \qquad [3.116]$$

For fractally rough *surfaces*, where the mass per area scales as r^{D_s} (where D_s is the surface fractal dimensionality, $2 \leqslant D_s < 3$), the SAS intensity obeys a different power-law dependence (Bale and Schmidt 1984):

$$I \propto k^{D_s - 6} \qquad [3.117]$$

For smooth regular surfaces where $D = 2$, [3.117] reduces to the Porod law, $I \propto k^{-4}$. Thus, the scattering power-law exponent varies between -1 and -3 for bulk fractals, and -3 and -4 for surface fractals. It should be noted that in both [3.116] and [3.117] non-k-dependent terms have been omitted.

There are four important length scales which define different SAS behaviour: η, the size of a cluster of particles or pores; a, the size of a pore itself; l, the size of the smallest asperity on the surface of the pore; and r, the atomic spacing,

Consequently three regions of scattering vector values can be defined, within each of which different dimensionalities are probed by SAS: (i) $\eta^{-1} < k < a^{-1}$ (dimensionality of the pore cluster); (ii) $a^{-1} < k < l^{-1}$ (dimensionality of the pore interface); and (iii) $l^{-1} < k < r^{-1}$ (the Porod region, where the surface is atomically smooth). Usually either region (i) or (ii) can be distinguished at small k-values, together with the Porod region at large k-values. In other words, bulk fractals can consist of smooth (non-fractal) particles, or the surfaces of voids can be fractally rough even though macroscopically the material may be non-fractal.

Finally, it should be noted that explanations other than that involving fractals can account for the power-law behaviour of SAS. In particular, a particle-size distribution ('polydispersity') can also give a power-law dependence: if the mass distribution function for a collection of uniform particles can be represented by $N(M) \propto M^{-\alpha}$, then the SAS intensity behaves as $I \propto k^{-\beta}$, where $\beta = 3(3-\alpha)$ (Schmidt 1982).

Another possible cause of the power-law SAS behaviour is scattering from domains within which the scattering factor is not constant (the special case where it is constant gives rise to Porod behaviour, $I \propto k^{-4}$). Boucher *et al.* (1983) have discussed a number of different scattering-factor profiles which predict power-law SAS behaviour, with exponents in the range $2 < \beta < 6$.

3.7 Examples of macroscopic structure

3.7.1 Defects in growth morphology

Amorphous materials, particularly if prepared by vapour deposition on cold substrates, are often not macroscopically homogeneous because of growth morphology defects. The existence of small voids in evaporated a-Ge, for example, is strikingly demonstrated by SAXS, as shown in Fig. 3.55. The small-angle scattering increases dramatically as k tends to zero, and can be ascribed to voids with radii between 5 and 40 Å which occupy about 5% of the total volume; the density deficit relative to the crystal obtained by weighing is in reasonable agreement at $\simeq 11\%$. Samples obtained by sputtering or electrolytic deposition exhibit a progressive decrease in the magnitude of the SAXS (although the density deficit measured by weighing is comparable), suggesting that the majority of the void space in these two cases lie in voids less than $\simeq 6$ Å in diameter, for which the SAXS intensity extends under the first peak in $I(k)$ and hence is obscured. Notice particularly from Fig. 3.55 that the SAXS from chalcogenide glasses (and vapour-deposited thin films) shows no evidence for the presence of voids, but instead tends to the compressibility limit. An increase in the SAXS is however observed in vitreous SiO_2 after irradiation by high-energy neutrons. The increase in the SAXS is uniform with k and the very sharp rise at small angles as observed in evaporated a-Ge is not seen. Interestingly, the density of the glass increases by $\simeq 3\%$ after irradiation; the density of the crystalline form (quartz), on the other hand, decreases by $\simeq 14\%$ to the *same* density on amorphization for high doses. The SAXS and the increase in density of the neutron-irradiated glass both disappear on annealing to near T_g. The origin of this effect is believed to be the 'thermal spikes' induced

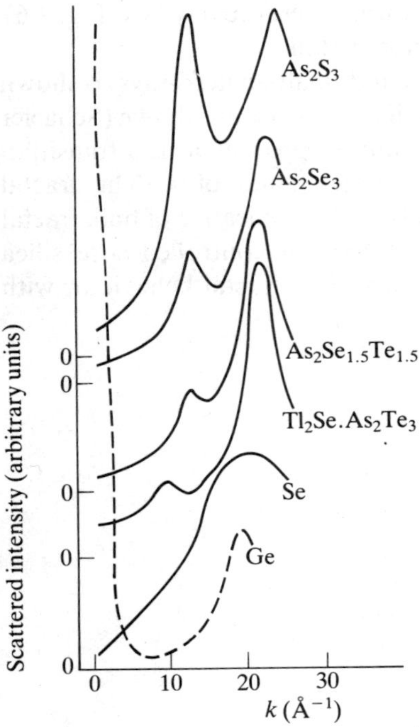

Fig. 3.55 Small-angle X-ray scattering for chalcogenide glasses and amorphous germanium (Bishop and Shevchik 1974). Note the absence of SAXS for the chalcogenide glasses, but the large intensity at small k for a-Ge arising from the presence of small voids.

by the neutron bombardment which cause local volumes of damage. The paramagnetic nature of the defects induced by neutron bombardment will be discussed in Chapter 6.

Columnar growth has been observed in many amorphous and crystalline vapour-deposited thin films (Leamy *et al.* 1980). Small-angle scattering again is a sensitive and quantitative probe for such macroscopic structural features. The orientation of such columns with respect to the substrate (and consequently with respect to the original vapour beam direction) can be ascertained by tilting the film relative to the X-ray beam, or the electron beam in an electron microscope. The small-angle scattering intensity approximately follows the shape transform of the assembly of particles ([3.101]), which is a torus for the case of parallel, uniformly separated columns. The column orientation may thus be ascertained by the position at which the diffraction plane of the tilted plane bisects the toroid. The orientation and dimensions of the columns may be measured in this single experiment, since the separation between columns is proportional to the reciprocal of the radius of the torus. Intriguingly, the column orientation is found *not* to be parallel to the angle of incidence of the vapour, but rather the following relation is generally obeyed:

$$2 \tan \beta = \tan \alpha \qquad [3.118]$$

where α and β are the angles between the vapour beam and the substrate normal, and the column orientation and the substrate normal, respectively (see Fig. 1.6). Leamy *et al.* (1980) discuss this behaviour in more detail.

An example of the use of scattering (by light and small-angle X-rays) is shown in Fig. 3.56, where data for colloidal amorphous silica particles are shown (Schaefer *et al.* 1984). Clear evidence for power-law behaviour is apparent, as is a transition to Porod-like behaviour (with slope -4) at large values of k. The fractal dimensionality inferred from the low-k region is $D_H = 2.1$, indicative of bulk fractal characteristics. Similar SAS studies by Höhr *et al.* (1988) on 'controlled-pore' silica glasses (e.g. Vycor – see sect. 3.7.2) have revealed surface-fractal behaviour, with fractal dimensionalities $D_s = 2.2$–2.4.

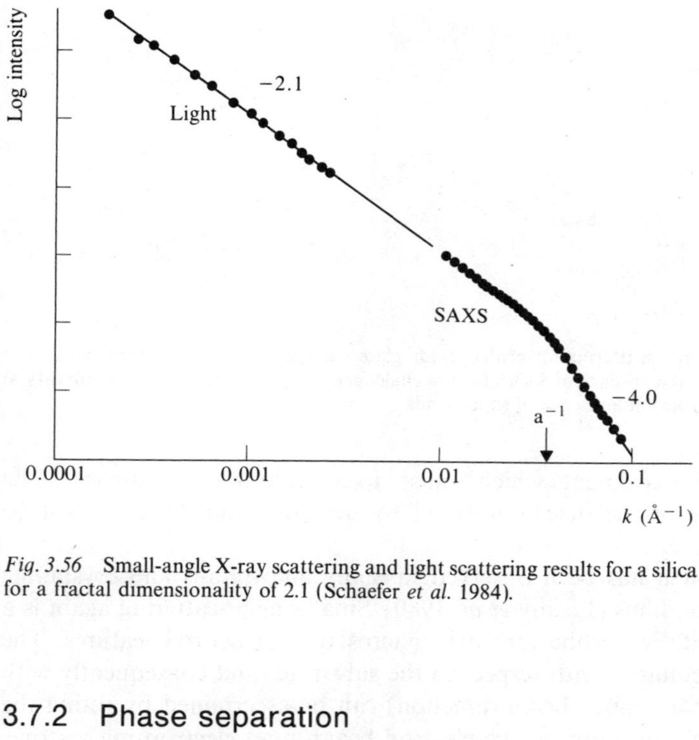

Fig. 3.56 Small-angle X-ray scattering and light scattering results for a silica aerogel showing evidence for a fractal dimensionality of 2.1 (Schaefer *et al.* 1984).

3.7.2 Phase separation

An obvious example of a macroscopic inhomogeneity produced by phase separation is the nucleation and growth of crystallites within an amorphous solid. The presence of these can certainly be detected by the techniques mentioned in section 3.6, particularly since the crystallites would produce sharp Bragg peaks in any scattering experiment if they were of sufficient size. However, we will not consider this situation further, but restrict our attention to the compositional (i.e. glassy → glassy) phase segregations that can occur because of immiscibility of the melt in certain systems.

To discuss this problem in detail, we need to consider phase separation (or immiscibility) of liquids. The simplest example concerns a binary system, AB, which can exist in the liquid phase either as a single phase or as a two-component mixture, depending on the temperature. Phase separation will occur when the free energy of the phase-separated mixture is less than that of the single homogeneous phase.

The Helmholtz free energy, F, is related to the internal energy, U, and the entropy, S, at any temperature, T, by:

$$F = U - TS \qquad [3.119]$$

The entropy term can be regarded as arising from the entropy of mixing at any composition c:

$$S_{\text{mix}} = -Nk_B[c \ln c + (1-c)\ln(1-c)] \qquad [3.120]$$

If U for the single phase is greater than for a mixture of the components, the free energy curve will exhibit minima at *two* compositions in general, since the $-TS$ curve is a symmetric function of composition having a minimum at $c = \frac{1}{2}$, whereas U generally is an asymmetric function of composition having a maximum, but the end-points for the pure components are not equal. Thus the free energy–composition curve in this case is as shown in Fig. 3.57(b). Between the compositions marked

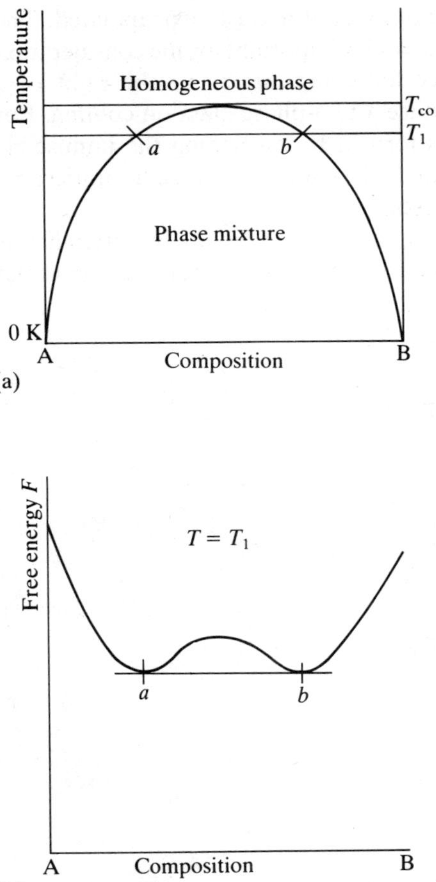

Fig. 3.57 Phase diagram (a) and composition dependence of the free energy (b) for a simple binary alloy AB which can exist either as a single phase or a two-component mixture in the liquid phase depending on the temperature. The temperature marked T_{co} in (a) is the upper consolute temperature, and the composition range a–b in (b) marks the miscibility gap (after Rawson 1967).

a and b, the free energy of the phase-separated mixture at c is lower than that of the single homogeneous phase (given by the solid curve), and hence phase separation takes place in the melt. The composition range a–b at that temperature marks the 'miscibility gap'. The phase diagram for the system is thus as shown in Fig. 3.57(a). At a certain critical temperature, T_{CO}, the two minima in the F vs c curve coalesce and a single homogeneous phase is now stable for all $T \geqslant T_{CO}$; this temperature is known as the 'upper consolute' temperature.

Complications can arise if crystalline phases are formed below a certain temperature. The phase diagram for such a case is shown in Fig. 3.58(a). At temperature T_1 the liquid has the lower F for all c, and since it exhibits a single minimum, a homogeneous liquid phase is formed. At T_2 two minima have developed and so between compositions a and b two liquid immiscible phases form, but for compositions very close to pure B the solid phase has the lower energy. At T_3 the solid phase and a single component phase are stable for compositions along the tangent line. For a supercooled liquid, such as might form a glass, the immiscibility limits extend below the liquidus and are simply continuously extrapolated. The composition q is called the 'immiscibility limit' and is important for the consideration of glass formation in such a system, since for compositions to the right of q a two-phase melt is formed which will produce a two-phase glass on cooling. For both this and the former case, if a glass is formed by quenching the immiscible melt, a very coarse-grained phase separated glass ensues, which is noticeably inhomogeneous to the naked eye (i.e. opalescent).

The final case we will consider which can lead to liquid immiscibility is indicated in Fig. 3.59, in which the two-liquid immiscibility region (i.e. the upper

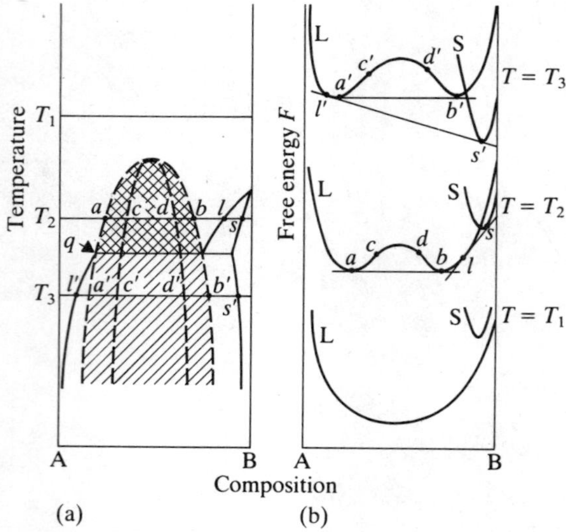

Fig. 3.58 (a) Schematic phase diagram showing the phase boundary and spinodal (dotted line) of a two-liquid immiscibility region (cross-hatched) with a sub-liquidus extension (hatched), together with the solid solubility behaviour of an unrelated third crystalline phase. The composition q is termed the immiscibility limit.
(b) Free energy-composition curves of the liquid (L) and solid (S) phases for the temperatures shown in (a) (after Rawson 1967).

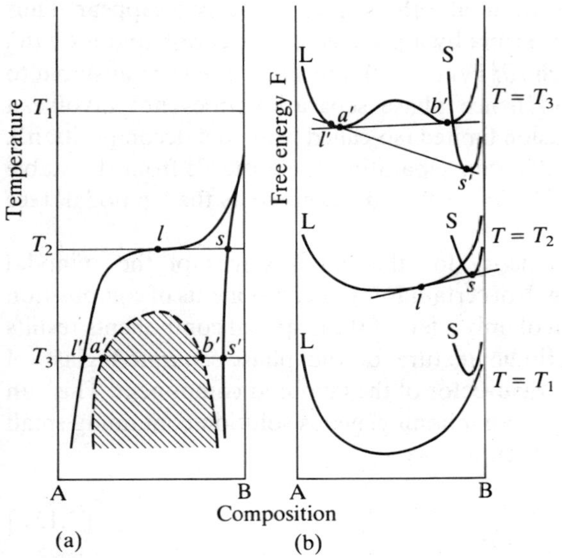

Fig. 3.59 (a) Schematic phase diagram showing an entirely sub-liquidus two-liquid immiscibility region (hatched) and an unrelated third crystalline phase.
(b) Free energy-composition curves of the liquid (L) and solid (S) phases for the temperatures shown in (a) (after Rawson 1967).

consolute temperature) is only encountered for temperatures *below* the liquidus. For temperatures below T_2, a solid crystalline phase is coexistent with a single liquid phase for the phase diagram given, but if the liquid is supercooled until the immiscibility region is encountered (e.g. at T_3), the liquid (or the glass) will show a propensity to separate into two phases. This 'metastable immiscibility' gives rise to phase separation on a very small scale and such glasses appear homogeneous (transparent) to the naked eye, but not, of course, in electron microscopy or small-angle scattering. Its likelihood is signalled by the liquidus having an approximately flat section or a point of inflection. It is this 'microphase separation' extending over some hundreds of ångströms, and the forms of separation, with which we concern ourselves for the remaining part of this chapter. It is of some considerable commercial and technological importance, since the Vycor process for the manufacture of non-crystalline silica utilizes the phase separation into almost pure silica and sodium borate phases which occurs in Na_2O–B_2O_3–SiO_2 glasses; the borate regions are selectively removed by acid leaching and the resulting porous silica framework compacted by subsequent heat treatment. In addition, glass–ceramic materials are made by controlled crystallization of one portion of a microphase-separated glass.

The texture of a microphase-separated glass is dependent to a certain degree, particularly during the initial states of segregation, on the mechanism and dynamics of decomposition. Two distinct mechanisms may be distinguished which differ in their dependence of free energy on composition. A phase is stable (or metastable) to an infinitesimal fluctuation in composition if the chemical potential of each component increases with increasing density of that component, or equivalently $(\partial^2 F/\partial c^2) > 0$. In this region, only fluctuations greater than a certain size are stable and hence can lead to phase separation; all others spontaneously disappear. Thus if $(\partial^2 F/\partial c^2) > 0$ phase separation occurs by a process of 'nucleation and growth'. Conversely, for a system for which $(\partial^2 F/\partial c^2) < 0$, the single phase is unstable to *all* fluctuations, of any size, and hence phase separation does not involve a nucleation step, being instead diffusion limited (so-called 'spinodal decomposition'). The boundary in a binary phase diagram separating the unstable from the stable region is given by the locus of $(\partial^2 F/\partial c^2) = 0$, and is known as the 'spinodal' (see Figs 3.58 and 3.59).

Cahn (1965) has given a theory for the initial stages of the spinodal decomposition in terms of the growth of certain Fourier components of composition fluctuations. The preferred growth of only a few of these spatial components results in the formation of a quasi-periodic texture of the phase-separated glass of wavelength corresponding to the wavevector of the favoured component. This can be seen as follows. The free energy of an inhomogeneous solution containing small concentration gradients can be written as:

$$F = \int [f(c) + \kappa(\nabla c)^2]\,dV \qquad [3.121]$$

where $f(c)$ is the free energy density for a homogeneous medium of concentration c, and $\kappa(\nabla c)^2$ represents the increase in free energy arising from a non-uniform concentration which, like surface tension, minimizes the extent of the interfacial region. The free energy density may be expanded about the average composition, c_o, by means of a Taylor series:

$$f(c) = f(c_o) + (c - c_o)\left(\frac{\partial f}{\partial c}\right) + \frac{(c - c_o)^2}{2}\left(\frac{\partial^2 f}{\partial c^2}\right) + \ldots \qquad [3.122]$$

Since by definition $\int (c - c_o)\,dV = 0$, the free energy difference between the homogeneous and inhomogeneous phases is:

$$\Delta F = \int \left[\frac{1}{2}\left(\frac{\partial^2 f}{\partial c^2}\right)(c - c_o)^2 + \kappa(\nabla c)^2\right] dV \qquad [3.123]$$

Consider the composition fluctuation to be made up from a series of Fourier components such as

$$(c - c_o) = A \cos \beta x \qquad [3.124]$$

The contribution of this component to the free energy difference is obtained by substitution in [3.123] and integrating

$$\Delta F_\beta = \tfrac{1}{4} V A^2 \left[\left(\frac{\partial^2 f}{\partial c^2}\right) + 2\kappa\beta^2\right] \qquad [3.125]$$

The critical wavenumber β_c below which spinodal growth occurs is given by:

$$\beta_c^2 = -\frac{1}{2\kappa}\left(\frac{\partial^2 f}{\partial c^2}\right) \qquad [3.126]$$

The kinetics of the initial stages of spinodal decomposition can be obtained by solving the diffusion equation in a linearized form:

$$\frac{\partial c}{\partial t} = M\left(\frac{\partial^2 f}{\partial c^2}\right)\nabla^2 c - 2M\kappa\nabla^4 c \qquad [3.127]$$

where M is an atomic mobility ($M > 0$). The solution to this equation is:

$$(c - c_o) = \exp[R(\beta)t] \cos(\boldsymbol{\beta}\cdot\boldsymbol{r}) \qquad [3.128]$$

where

$$R(\beta) = -M\left(\frac{\partial^2 f}{\partial c^2}\right)\beta^2 - 2M\kappa\beta^4 \qquad [3.129]$$

The amplification factor, $R(\beta)$, is positive in the spinodal region and since $M = D/(\partial^2 f/\partial c^2)$, it can be seen from [3.129] that in this region the diffusion constant, D, is *negative* (i.e. 'uphill' diffusion). This is because the attraction between like species is so great that the flux of atoms opposes the concentration gradient. Substitution of $M = D/(\partial^2 f/\partial c^2)$ into [3.129] yields:

$$R(\beta) = -D\beta^2(1 - \beta^2/\beta_c^2) \qquad [3.130]$$

which is sharply peaked at the 'spinodal wavenumber' of the system

$$\beta_m = \beta_c/\sqrt{2} \qquad [3.131]$$

The spatial components of concentration fluctuations of this wavenumber receive maximum amplification in the early stages of the spinodal growth and hence give rise to a phase-separated glass having a texture of characteristic wavelength $\Lambda_m = 2\pi/\beta_m$. The type of texture expected is shown schematically in Fig. 3.60, which is a section through a computed two-phase structure (the minor phase occupying 24 vol.%) calculated by the addition of a number of random sine waves all having the same wavelength Λ (Cahn and Charles 1965).

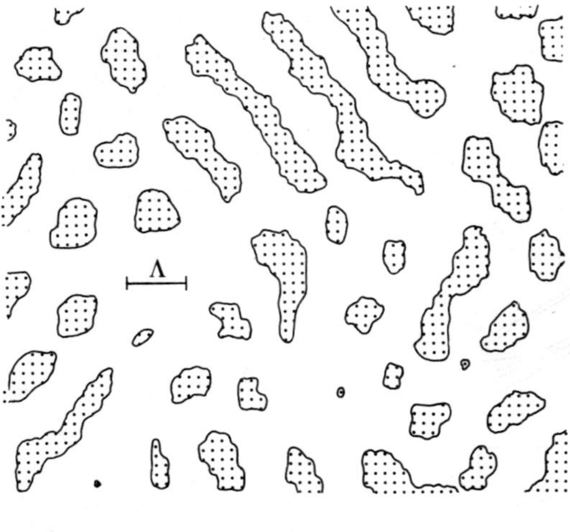

Fig. 3.60 Computed two-phase structure characteristic of spinodal decomposition (Cahn and Charles 1965).

The attraction of Cahn's theory is that it is formulated in reciprocal space, and can therefore be tested directly by small-angle scattering experiments. The scattered intensity at wavevector k is proportional to $|A(\beta, t)|^2$ for the concentration fluctuation, and furthermore, since $k = \beta$

$$\frac{4\pi \sin\theta}{\lambda} = \frac{2\pi}{\Lambda} \quad [3.132]$$

where 2θ and λ are the scattering angle and wavelength of the probing radiation, respectively. Thus, a glass which is metastably homogeneous originally and which, after heat treatment at a given temperature, for a given time, decomposes spinodally should exhibit a diffraction intensity which from [3.128] should obey the equation:

$$I(k,t) = I(k,0)\exp[2R(k)t] \quad [3.133]$$

Furthermore, since $R(\beta)$ is sharply peaked at β_m, an intensity maximum is expected in the small-angle scattering spectrum at $k = \beta_m$, and this peak should not shift in position with increasing time of heat treatment. Figure 3.61 shows the SAXS spectra for a B_2O_3–PbO–Al_2O_3 glass quenched from 1150°C and heat treated to 450°C for varying lengths of time (Zarzycki and Naudin 1969). The spinodal mechanism is clearly indicated for the initial heat treatments. A plot of $R(k)/k^2$ vs. k^2 should be linear ([3.130]); from such a plot, the value for the spinodal wavelength $\Lambda_m = 130$ Å was deduced. At longer times, the peak in the SAXS shifts to smaller values of k, indicative of a coarsening process where Cahn's theory is no longer appropriate. Such a process might be the 'Ostwald ripening mechanism' where differences in solute concentration due to various particle sizes (or local curvature of the interface for a fully connected biphasic system) set up concentration gradients which cause the growth of large particles at the expense of smaller ones.

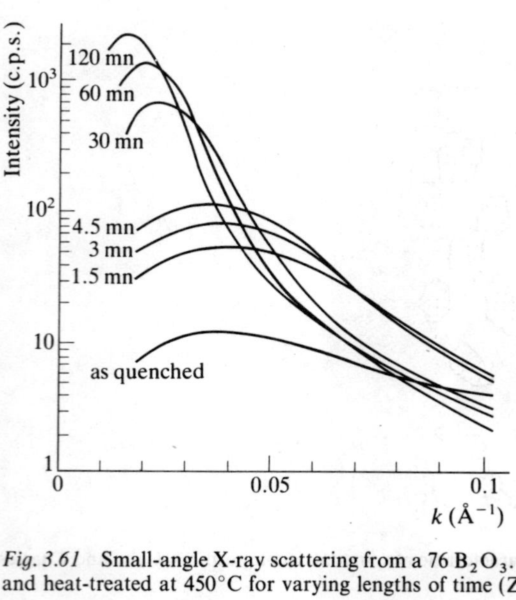

Fig. 3.61 Small-angle X-ray scattering from a 76 B_2O_3.19PbO.5Al_2O_3 glass, quenched from 1150°C and heat-treated at 450°C for varying lengths of time (Zarzycki and Naudin 1969).

3.7.3 Summary

Amorphous materials in general are *not* as homogeneous as structural models would indicate, and such gross structural imperfections may be detected using optical or electron microscopy or small-angle (X-ray or neutron) scattering.

Amorphous thin films prepared by vapour deposition are almost invariably structurally inhomogeneous to some degree. Voids are often present, and in certain extreme cases can coalesce, resulting in a columnar growth morphology for the film. Glasses prepared by melt-quenching are not macroscopically homogeneous either, but at the very least possess density fluctuations, characteristic of a liquid, and frozen-in at T_g. These density fluctuations are responsible for the non-zero intercept of the structure factor at $k=0$ (the 'compressibility limit').

In certain cases, melt-quenched glasses can undergo varying degrees of phase separation, depending on the miscibility of the components in the melt. If the melt is immiscible (two-phase), the resulting phase-separated glass will be very coarse-grained. If the immiscibility region occurs instead at temperatures below the liquidus, supercooling the melt below this temperature to form the glass will result in a very fine-grained 'microphase' separation. In this case, phase separation can occur either by nucleation and growth, or by a diffusion-limited process called spinodal decomposition.

Problems

3.1 Show that by suitable isotopic substitution (e.g. in the Ni–Ti system – see sect. 3.2.2 and Table 3.2) the following correlation functions are obtainable *directly* from neutron scattering experiments (in which $f \equiv b$):

(a) $b_1 = b_2$ gives $S_{NN}(k)$
(b) $x_1 b_1 + x_2 b_2 = 0$ ('zero alloy') gives $S_{CC}(k)$
(c) $b_1 = 0$ or $b_2 = 0$ ('null elements') give $I_{22}(k)$ and $I_{11}(k)$, respectively.

3.2 Obtain the equations describing the bond statistics for AX_2- and A_2X_3-type glasses for both the random covalent network and chemically ordered network models (see Fig. 3.32).

3.3 Derive the 'compressibility limit' of the scattering intensity ([3.114]) by considering the density fluctuations in a liquid.

(a) Consider a small volume V_o in a liquid containing N_o molecules. Show that N_o can be written as:

$$N_o = \sum_{i=1}^{N} P(\mathbf{r}_i) \qquad P(r) = 1 \quad \mathbf{r} \text{ in } V_o \qquad [3.1P]$$

$$= 0 \quad \mathbf{r} \text{ not in } V_o$$

and hence that

$$\langle N_o \rangle = n V_o \qquad [3.2P]$$

where n is the number density and

$$\langle N_o^2 \rangle = \int_{V_o} \rho_1(\boldsymbol{r}_1)\,\mathrm{d}\boldsymbol{r}_1 + \iint_{V_o} \rho_2(\boldsymbol{r}_1, \boldsymbol{r}_2)\,\mathrm{d}\boldsymbol{r}_1\,\mathrm{d}\boldsymbol{r}_2 \qquad [3.3P]$$

where ρ_1 and ρ_2 are the single particle and pair distribution functions respectively. Thus show that the fluctuations in the number of molecules is:

$$\frac{\langle N_o^2 \rangle - \langle N_o \rangle^2}{\langle N_o \rangle} = 1 + \frac{4\pi}{n} \int_{V_o} [\rho_2(r) - n^2] r^2\,\mathrm{d}r \qquad [3.4P]$$

Deduce the relation of the right-hand side of [3.4P] with the X-ray scattering intensity ([3.17]).

(b) The grand partition function, for a system containing N particles is defined by

$$\Xi = \sum_N \lambda^N Z_n \qquad [3.5P]$$

where $\lambda = \mathrm{e}^{\beta\mu}$ (μ is the chemical potential and $\beta = 1/k_B T$) and Z_n is the partition function $Z_n = \Sigma_k \exp(-\beta E_{N,k})$. It is a standard result that the average number of particles in the system is

$$\langle N \rangle = \frac{\partial \log \Xi}{\partial(\beta\mu)} \qquad [3.6P]$$

Hence show that $\partial\langle N \rangle/\partial(\beta\mu) = \langle N^2 \rangle - \langle N \rangle^2$ and using the expression for the chemical potential

$$N\,\mathrm{d}\mu = -S\,\mathrm{d}T + V\,\mathrm{d}p \qquad [3.7P]$$

obtain finally [3.114], $S(0) = \rho k_B T \kappa_T$.

Bibliography

P. H. Gaskell (ed.) (1977), *Structure of Non-Crystalline Materials*, vol. I (Taylor and Francis).

P. H. Gaskell, J. M. Parker and E. A. Davis (eds) (1983), *Structure of Non-Crystalline Materials*, vol. II (Taylor and Francis).

Y. Waseda (1980), *Structure of Non-Crystalline Materials: Liquids and Amorphous Solids* (McGraw Hill).

J. Wong and C. A. Angell (1976), *Glass: Structure by Spectroscopy* (Dekker).

A. C. Wright (1974), *Adv. Struct. Res. Diffr. Meth.* **5**, 1.

A. C. Wright and A. J. Leadbetter (1976), *Phys. Chem. Glasses* **17**, 122.

R. Zallen (1983), *The Physics of Amorphous Solids* (Wiley).

4 Atomic dynamics

I VIBRATIONAL EXCITATIONS

4.1 Introduction

In the last chapter we discussed at length the static arrangement of atoms in an amorphous solid, i.e. its structure. In this chapter, we turn our attention to the dynamical behaviour of such atoms and discuss both the nature of localized (vibrational) excitations of the atoms from their equilibrium positions, as well as long-range (diffusional) behaviour.

Atomic vibrations in a solid, whether amorphous or crystalline, are manifested in many ways. In non-metallic systems, heat is transported via vibrational excitations and the specific heat capacity is determined by the spectrum of allowed vibrational modes. Under certain circumstances, electromagnetic radiation can interact with vibrational excitations, thereby affecting the transmission or scattering of light, and making possible the use of infra-red and Raman spectroscopies as vibrational probes. Furthermore, neutrons can be inelastically scattered from condensed matter, either losing or gaining energy to and from atomic vibrations during the scattering event, and thus are a direct probe of the density of vibrational states.

In many respects the vibrational behaviour of amorphous solids is very similar to that of the corresponding crystalline forms, with the exception that selection rules for transitions are relaxed and sharp features in the density of vibrational modes are smeared out due to the lack of periodicity. The next three sections of this chapter are therefore devoted to a discussion of the nature of vibrational excitations in amorphous materials and how the vibrational properties may be calculated, and of the experimental probes that may be employed to explore the vibrational characteristics of amorphous solids. Where amorphous materials differ markedly from their crystalline counterparts is in the realm of vibrational properties at very low temperatures, the heat capacity, thermal conductivity and other vibrationally dependent properties being singularly anomalous for the amorphous state. These low-temperature anomalies appear to be present in most amorphous solids and are characteristic of the presence of disorder; they form the basis of the fourth section of this chapter.

At the outset one must ask the question whether, in the absence of long-range order, propagating vibrational excitations exist. The answer of course is in the affirmative, since it is common experience that window glass transmits both sound and heat. The difference between amorphous and crystalline materials essentially lies in the fact that periodicity cannot be used to simplify the dynamical equations by the introduction of Bloch states.

The calculation of dynamical properties of three-dimensional solids is not a trivial problem (see e.g. Born and Huang 1954 and Maradudin *et al.* 1971). In general two approximations are widely made, the 'adiabatic' or Born–Oppenheimer approximation, and the 'harmonic' approximation. The former has the potential energy written only in terms of nuclear (not electronic) coordinates, since electronic motion is so much more rapid than nuclear motion (because the ratio of electronic to nuclear mass is of the order of $\simeq 10^{-3}$), or equivalently the energies of electronic excitations (in non-metals) are typically larger ($\simeq 1$ eV) than those characteristic of vibrational excitations ($\simeq 10^{-3}$ eV). The harmonic approximation is that in which the potential is expanded in powers of the displacement from equilibrium only up to the first non-vanishing, i.e. quadratic, term. The use of the harmonic approximation ensures, as we shall see later, that the elementary vibrational excitations are non-interacting and independent; if terms in the lattice potential higher than quadratic are included, interaction between excitations is introduced, leading to a shift in the frequencies and finite lifetimes of the excitations.

The Hamiltonian for lattice vibrations may be written generally as (see e.g. Madelung 1978):

$$\mathscr{H} = \sum_{n\alpha i} \frac{M_\alpha}{2} \dot{s}_{n\alpha i}^2 + \tfrac{1}{2} \sum_{\substack{n\alpha i \\ n'\alpha' i'}} \mathbf{\Phi}_{n\alpha i}^{n'\alpha' i'} s_{n\alpha i} s_{n'\alpha' i'} \qquad [4.1]$$

where n runs over N, the number of Wigner–Seitz cells in the volume of the material, α runs over r, the number of atoms in the basis, i is one of the three Cartesian coordinates, M_α is the mass of the αth atom, and $s_{n\alpha}$ is the instantaneous displacement from equilibrium. The force constant matrix $\mathbf{\Phi}_{n\alpha i}^{n'\alpha' i'}$ in [4.1] has $3Nr$ rows and columns and represents the force in the ith direction acting on the αth atom in the nth cell when the α'th atom in the n'th cell is displaced by unit distance in the i'th direction. Certain general 'symmetry relations' link various components of the force constants even in the absence of periodicity, such as the symmetric nature of the Φ's with respect to interchange of indices; furthermore the potential energy is invariant to infinitesimally small translations or rotations of the solid.

The equation of motion can be written as:

$$M_\alpha \ddot{s}_{n\alpha i} = -\sum_{n'\alpha' i'} \mathbf{\Phi}_{n\alpha i}^{n'\alpha' i'} s_{n'\alpha' i'} \qquad [4.2]$$

and if a solution to this is sought which is periodic in time, i.e.

$$s_{n\alpha i}(t) = M_\alpha^{-1/2} u_{n\alpha i} \exp(-i\omega t) \qquad [4.3]$$

where $u_{n\alpha i}$ are time independent, the equation of motion may be written as:

$$\omega^2 u_{n\alpha i} = \sum_{n'\alpha' i'} \mathbf{D}_{n\alpha i}^{n'\alpha' i'} u_{n'\alpha' i'} \qquad [4.4]$$

where the real symmetric matrix $\mathbf{D}$ ($= \mathbf{\Phi}/\sqrt{(M_\alpha M_{\alpha'})}$) has $3Nr$ real *eigenvalues* ω_{j}^2. The *eigenvectors*, u, in [4.4] are characterized by the index j, and for each eigenvalue ω_{j} there are $3Nr$ $u_{n\alpha i}^{(j)}$, which are termed the 'normal modes'. The introduction of 'normal coordinates', q_{j}, which are related to the displacements $s_{n\alpha i}(t)$, allows the reduction of the Hamiltonian into *independent* normal modes, and is only possible in the harmonic approximation in which the matrix $\mathbf{\Phi}$ (or $\mathbf{D}$) is able to be diagonalized (see e.g. Madelung 1978). Thus, the coupled individual oscillations of the atoms are replaced in a formal sense by *decoupled* collective excitations which are known as 'phonons' in quantum-mechanical language.

It is to be noted that nothing we have said so far about the vibrations of a solid has depended on the assumption of a periodic lattice, and hence the discussion is completely general, and can apply to amorphous solids as well as to crystalline materials. The vibrational problem thus reduces formally to the diagonalization of the matrix **D** to find the $3Nr$ eigenvalues.

The presence of lattice periodicity has two effects: the number of equations to be solved is reduced, and the introduction of a wavevector q, related to the size of the unit cell and characterizing the eigenvalues, is made possible. Thus, $\mathrm{D}_{n\alpha i}^{n'\alpha' i'}$ must be the same for all cells and is only a function of the difference in cell indices, $\mathrm{D}_{\alpha i}^{\alpha i'}(n'-n)$ for a periodic lattice. The eigenvectors may be written in the Bloch form as:

$$u_{n\alpha i} = c_{\alpha i} \exp(i\boldsymbol{q}\cdot\boldsymbol{R}_n) \qquad [4.5]$$

where $\boldsymbol{R}_n$ is a point in the Wigner–Seitz cell. These two consequences of periodicity cause [4.4] to become

$$\omega^2 c_{\alpha i} = \sum_{\alpha' i'} \mathrm{D}_{\alpha i}^{\alpha' i'}(\boldsymbol{q}) c_{\alpha' i'} \qquad [4.6]$$

where the matrix is defined by

$$\mathrm{D}_{\alpha i}^{\alpha' i'}(\boldsymbol{q}) = \sum_{n'} \frac{1}{\sqrt{(M_\alpha M_{\alpha'})}} \Phi_{\alpha i}^{\alpha' i'}(n'-n) \exp[i\boldsymbol{q}\cdot(\boldsymbol{R}_n - \boldsymbol{R}_{n'})] \qquad [4.7]$$

Hence, periodicity has reduced the set of equations from $3Nr$ to $3r$ in number, and consequently there are $3r$ eigenvalues, $\omega_j(\boldsymbol{q})$, at each value of $\boldsymbol{q}$. Since there are exactly N values of $\boldsymbol{q}$ in Brillouin zone because of the cyclic boundary conditions imposed on the system, there are $3Nr$ different $\omega_j(\boldsymbol{q})$ corresponding to the number of internal degrees of freedom of the crystal. The number of atoms in the basis, r, determines the number of 'branches' in the Brillouin zone, which thus number $3r$.

The type of vibrations in a given branch of the set of phonon dispersion curves can be distinguished by the behaviour of $\omega_j(\boldsymbol{q})$ as $\boldsymbol{q}$ tends to zero (i.e. in the infinite wavelength limit). In general, there are three branches whose frequencies go to zero as $\boldsymbol{q}$ tends to zero; these are known as 'acoustic' branches, and at $\boldsymbol{q}=0$ the vibrations of the basis atoms in each cell are obviously in the same direction. The remaining $3(r-1)$ branches are referred to as 'optic' branches, and the frequencies tend to a finite value as $\boldsymbol{q}$ tends to zero; at $\boldsymbol{q}=0$, the basis atoms vibrate against each other. In addition, at certain symmetry points, or along high symmetry directions in the Brillouin zone, the modes can be further subdivided into 'longitudinal' or 'transverse', depending on the direction of the eigenvector with respect to $\boldsymbol{q}$. At a general value of $\boldsymbol{q}$ in the Brillouin zone, however, the lattice vibrations are not strictly longitudinal or transverse, nor are acoustic branch vibrations completely in phase or optic branch modes completely out of phase.

How is this situation altered for amorphous solids, for which the simplifying consequences of periodicity are no longer appropriate? In the low-frequency, long-wavelength limit, the precise arrangement of atoms is of little consequence and the amorphous solid appears isotropic and acts as an elastic medium. The vibrational excitations can properly be described as acoustic phonons with the exception that the waves are more heavily damped (or equivalently the phonon lifetime is shorter). At higher frequencies and lower wavelengths, the modes in general will no longer be plane waves but instead they will be much more localized than the corresponding crystalline modes. Furthermore, and more importantly, the wavevector $\boldsymbol{q}$ is no longer a good quantum number, and is not useful in classifying vibrational modes for amorphous solids. This can be seen in a simple manner. For a single crystal, the size of the first zone in the Brillouin scheme is given by $2\pi/a$, where a is the size of the unit cell. An amorphous solid can be viewed as having a structure equivalent to a unit cell of infinite size, and hence the Brillouin zone reduces to the point $\boldsymbol{q}=0$ and so $\boldsymbol{q}$ becomes redundant as a label of the modes.

Thus, in amorphous solids the concepts of Brillouin zones and phonon dispersion curves lose their meaning (as is the case also for electron states, as will be seen in the next chapter). The dependence of the vibrational frequency ω on the scattering vector $\boldsymbol{q}$ *can* be ascertained from inelastic neutron scattering experiments (sect. 4.3.4), but the dispersion does not follow a single curve for a given branch as in a crystal. Instead, taking an acoustic branch as an example, the locus of points may fall close to a single curve at very low values of $\boldsymbol{q}$, but at higher values the curve becomes smeared out and ill-defined (see Fig. 4.1). This is a consequence of structural disorder; peaks in the structure factor $S(q)$ act in the scattering process like smeared-out reciprocal lattice vectors. The loci through the dispersion relation in Fig. 4.1 correspond to different forms of the scattering law $S(\boldsymbol{q}, \omega)$, and these are shown also in the figure.

We see therefore, that ω–q relations are not meaningful in describing vibrational excitations in amorphous solids. Is there a quantity which is used to

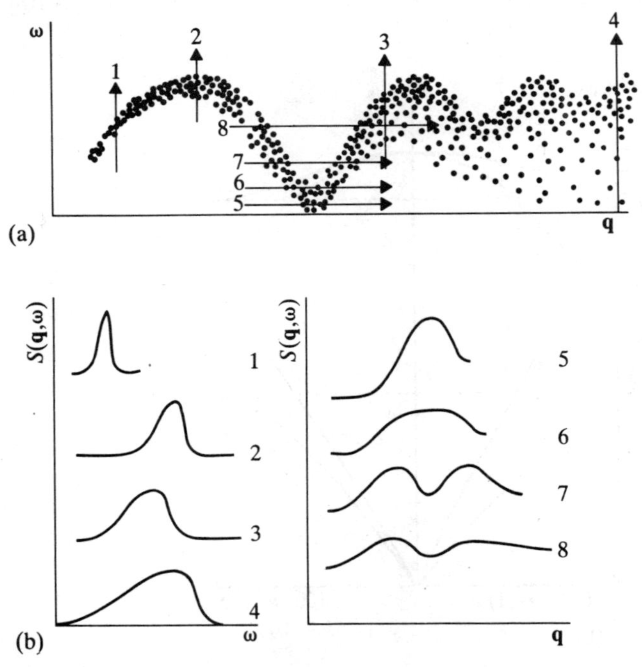

Fig. 4.1 (a) Schematic illustration of a dispersion relation $\omega(\boldsymbol{q})$ for an acoustic branch in a glass. The form of the scattering law, i.e. the dynamic structure factor $S(\boldsymbol{q}, \omega)$ is shown in (b) for the various experimental loci through the curve in (a) (Leadbetter 1973).

describe phonon modes in crystalline materials which can also be sensibly used for amorphous solids? The answer is yes, and the quantity is the 'density of states' $\rho(\omega)$, where $\rho(\omega)d\omega$ is the number of states with frequencies between ω and $\omega + d\omega$. This can be defined for a crystal as:

$$\rho(\omega) = \frac{V_g}{(2\pi)^3} \int_{S_\omega} \frac{dS_\omega}{|\nabla_k \omega|} \qquad [4.8]$$

where N/V_g is the reciprocal volume of a Wigner–Seitz cell (equal to the volume of a Brillouin zone divided by $(2\pi)^3$), and S_ω is a surface of constant ω. The density of states for a crystal exhibits 'Van Hove singularities', which are discontinuities in the slope of $\rho(\omega)$ in three dimensions whenever $\nabla\omega_k = 0$, i.e. where the dispersion curve is flat in ω–q space. The calculated density of states for diamond cubic Si (or Ge) is shown in Fig. 4.2, where the Van Hove singularities are clearly seen. Also shown are the phonon dispersion curves computed using a Born potential (Thorpe 1976):

$$V_{ij} = \tfrac{3}{4}\beta_B \sum_{i,j} [(\boldsymbol{u}_i - \boldsymbol{u}_j)\cdot\boldsymbol{r}_{ij}]^2 + \tfrac{1}{4}(\alpha_B - \beta_B)\sum_{i,j}(\boldsymbol{u}_i - \boldsymbol{u}_j)^2 \qquad [4.9]$$

where $\boldsymbol{u}_i$ are displacement vectors, $\boldsymbol{r}_{ij}$ is a unit vector joining sites i and j; the first term represents bond stretching and the second represents bond bending. The origins of some singularities in $\rho(\omega)$ with respect to features in the dispersion curves are indicated by dotted lines. An alternative way of defining the density of states is in terms of a Green function formalism (Thorpe 1976):

$$\rho(\omega) = \sum_k \delta(\omega - \omega_k) = -\frac{1}{\pi}\,\mathrm{Im}\sum_k \frac{1}{(\omega - \omega_k)} \qquad [4.10]$$

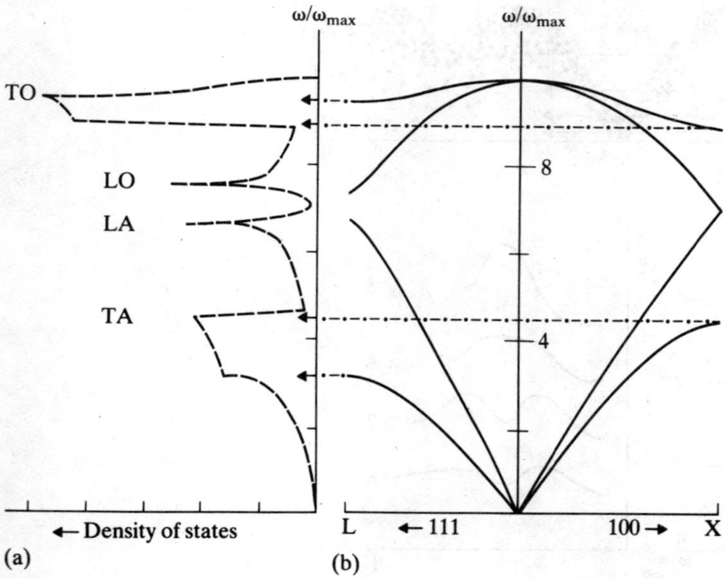

Fig. 4.2 (a) Density of states and (b) phonon dispersion curves for diamond cubic Si (or Ge) calculated using the Born potential. The dotted lines indicate the origin of some Van Hove singularities in k-space. The maximum frequency is given by $\omega_{max} = \sqrt{(8\alpha/M)}$ (Thorpe 1976).

For crystals, the sum over $\boldsymbol{k}$ is over all branches and covers the first Brillouin zone: [4.10] is also valid for an amorphous solid, but now $\boldsymbol{k}$ becomes merely a label for the vibrational eigenstates, is no longer a vector and has no other physical significance.

As an example of a vibrational density of states determined experimentally for an amorphous solid, we show in Fig. 4.3 $\rho(\omega)$ for amorphous Si, obtained by inelastic neutron scattering (see sect. 4.3.4) by Kamitakahara *et al.* (1984, 1987), compared with the VDOS calculated for c-Si and also a CRN model of a-Si. It can be seen immediately that the forms of the curves are rather similar, and this is a general feature of amorphous solids. The features in the crystal density of states resulting from the periodicity, namely the Van Hove singularities, are smeared out in the amorphous density of states, but otherwise the correspondence between peaks in $\rho(\omega)$ is very close, although the peaks are broadened for the amorphous case. This immediately suggests that it is *short-range order* which essentially determines the vibrational properties, and thus if this is the same for amorphous and crystalline polymorphs, as is the case for Si or Ge, one expects $\rho(\omega)$ to be similar. Indeed, we shall see in the next section that the vibrational characteristics of a *single* tetrahedral unit can account for many of the features in $\rho(\omega)$ for Si or Ge.

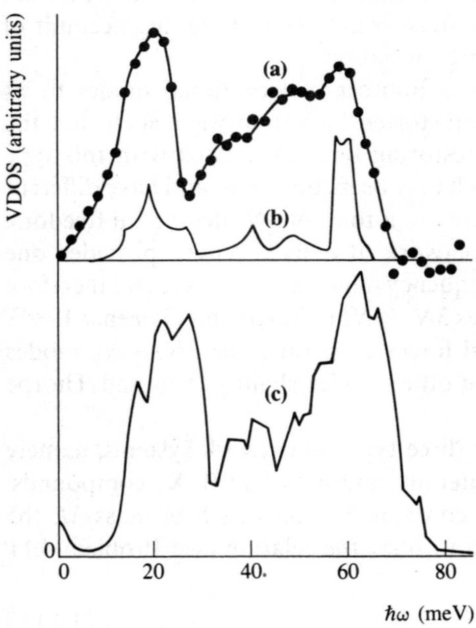

Fig. 4.3 Experimental vibrational density of states for a-Si (a), compared with the calculated VDOS for (b) c-Si and (c) a-Si (Kamitakahara *et al.* 1984).

4.2 Computational methods

4.2.1 Analytic methods

In general, numerical techniques must be used to obtain information on the vibrational characteristics of large structural models of amorphous materials, particularly if matrix diagonalization is to be used; such methods form the subject of the remaining part of this section. However, in certain special cases considerable physical insight can be achieved by the use of simple models which can be analysed *analytically*; this is particularly true for covalently bonded (network) amorphous materials such as tetrahedrally coordinated materials (e.g. Ge, Si), trigonally-bonded materials (e.g. As), and AX_2 or B_2X_3 compounds ($A = Ge, Si$; $B = As, B, P$; $X = O, S, Se$).

In a series of papers, Thorpe and co-workers have analysed the vibrational characteristics of such network amorphous materials in terms of a 'central-force' (i.e. bond-stretching) model, in which bond-bending (non-central) interatomic interactions are neglected; this approximation is valid since, for such materials, the ratio of bond-stretching to bond-bonding force constants is typically rather large ($\alpha:\beta = 5:1$). In this approximation, some features of the network dynamics can be obtained analytically either from simple equation-of-motion considerations (Sen and Thorpe 1977) or by using a Lagrangian formalism (Thorpe and Galeener 1980). In particular, criteria can be established which determine the character of the vibrational modes (whether molecular-like or band-like), and simple formulae can be obtained which describe the spectral limits of the vibrational bands. It should be noted that, in its simplest form, the theory is local, since the dynamics are calculated for small representative clusters in the network, taking account of the network connectivity but ignoring ring formation.

In the central-force approximation, a number of vibrational modes lie at *zero* frequency since the network can be distorted in certain ways such that the bond lengths are unaltered; there is no restoring force associated with this type of motion. If the network has N_s sites, which may be inequivalent and have different numbers of nearest-neighbour bonds, there are a total of $3N_s$ degrees of freedom. Keeping each of the N_b bonds in the network at a fixed length provides one constraint, making the number of finite-frequency modes equal to N_b, and therefore the number of zero-frequency modes equals $3N_s - N_b$ (Thorpe and Galeener 1980). The effect of the inclusion of non-central forces is to raise the $3N_s - N_b$ modes from zero frequency, and also to shift the other modes slightly (Sen and Thorpe 1977).

We can illustrate this approach for three types of network systems, namely monatomic tetrahedrally coordinated materials, and AX_2 and B_2X_3 compounds. For the first case, in which the network contains N atoms each of mass M, the allowed vibrational frequencies are found to obey the relation (see Problem 4.1)

$$\omega^2 = \frac{\alpha}{3M}(4 - \varepsilon) \qquad [4.11]$$

where $-4 \leqslant \varepsilon \leqslant 4$ (Thorpe and Galeener 1980). Thus, the vibrational frequencies are bounded by the inequality (Weaire and Alben 1972)

$$0 \leqslant \omega^2 \leqslant \frac{8\alpha}{3M} \qquad [4.12]$$

There are N modes at zero frequency and N modes at $\omega^2 = 8\alpha/3M$, giving rise to two delta-function contributions in the density of states; these peaks can be regarded as being TA- and TO-like respectively. The remaining N modes form a double-peak structure in the density of states between these bounds; these can be regarded as being LA- and LO-like. The inclusion of non-central forces broadens the delta-function peaks and forces the two LA and LO peaks closer together (see Fig. 4.2(a)).

More complex behaviour is shown by AX_2-like materials. In the spirit of the central-force model, the dynamics for AX_2 networks can be deduced in a similar manner as for monatomic tetrahedral networks (cf. [4.11]), and the edges of the two bands of vibrational states are given by (Sen and Thorpe 1977, Thorpe and Galeener 1980)

$$\begin{aligned} \omega_1^2 &= (\alpha/m)(1+\cos\theta), \qquad & \omega_2^2 &= (\alpha/m)(1-\cos\theta) \\ \omega_3^2 &= \omega_1^2 + 4\alpha/3M, & \omega_4^2 &= \omega_2^2 + 4\alpha/3M \end{aligned} \qquad [4.13]$$

where M and m are the masses of the fourfold-coordinated atoms (A) and twofold-coordinated atoms (X), respectively, and θ is the A—X—A bond angle; the X—A—X bond angle is the tetrahedral angle, 109° 28′, by symmetry. The regions of allowed frequencies are shown in Fig. 4.4(a) as a function of θ.

Several features are worthy of note. The first is that when the bond angle $\theta = 90°$, the displacements associated with each AX_4 tetrahedron become *vibrationally decoupled*, and consequently the vibrational modes become charac-

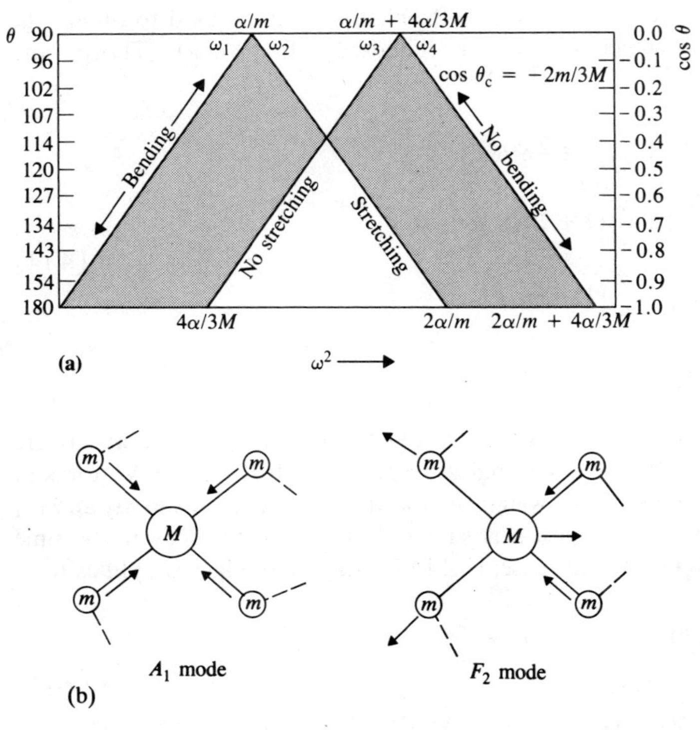

Fig. 4.4 (a) Allowed frequency regions as a function of the bond angle subtended at X atoms for vibrational excitations of an AX_2 network in the central-force approximation.
(b) Singlet A_1 and triplet F_2 vibrational modes of an isolated AX_4 tetrahedron (Sen and Thorpe 1977).

teristic of those of isolated molecules. The two non-zero-frequency eigenmodes of isolated AX_4 tetrahedra are shown in Fig. 4.4(b), and are the A_1 (singlet) breathing mode at $\omega^2 = \alpha/m$ and the F_2 (triplet) mode at $\omega^2 = \alpha/m + 4\alpha/3M$. There are also five zero-frequency modes, since non-central forces have been neglected. As θ increases from 90°, the coupling between molecular modes increases, leading to the formation of two vibrational bands, the higher-frequency of which has delta functions at each edge, ω_3 and ω_4 (each with unit weight), each band also having unit weight. Beyond a critical value $\theta_c = \cos^{-1}(-2m/3M)$, two bands each with a weight of 2 are found, the average separation of which is greater than that between the A_1 and F_2 molecular modes and increases further with increasing θ.

For the AX_2 materials $GeSe_2$, GeS_2, SiO_2 and GeO_2, the mass ratios m/M are 1.09, 0.44, 0.57 and 0.22 respectively, leading to critical values of the A—X—A bond angle $\theta_c = 136°, 107°, 117°$ and 98° respectively. Since the average Si—O—Si and Ge—O—Ge bond angles are $\simeq 144°$ and 130° respectively, $\theta > \theta_c$ in both cases, so one expects the vibrational modes to be dominated by strong coupling, i.e. solid-state, effects. For the chalcogenide glasses $GeSe_2$ and GeS_2, $\theta < \theta_c$, so molecular-like vibrational modes are expected. The observed IR and Raman bands of a-$GeSe_2$ (which are rather narrow) do correspond rather closely to the molecular-mode frequencies of $GeBr_4$ (Walter *et al.* 1988).

As our final example within the framework of the central-force model we consider B_2X_3 materials (B = As, P; X = O, S, Se) which are of interest since results for monatomic trigonally coordinated networks can be obtained as a special case. The Lagrangian method outlined in Problem 4.1 can be used to obtain the dynamical eigenvalues which give for the band-edge frequencies (Thorpe and Galeener 1980)

$$
\begin{aligned}
\omega_1^2 &= \frac{\alpha}{m}(1+\cos\theta) + \frac{\alpha}{M}(1+2\cos\psi) \\
\omega_2^2 &= \frac{\alpha}{m}(1-\cos\theta) + \frac{\alpha}{M}(1+2\cos\psi) \\
\omega_3^2 &= \frac{\alpha}{m}(1+\cos\theta) + \frac{\alpha}{M}(1-\cos\psi) \\
\omega_4^2 &= \frac{\alpha}{m}(1-\cos\theta) + \frac{\alpha}{M}(1-\cos\psi)
\end{aligned}
\qquad [4.14]
$$

where m and M are the masses of atoms X and B respectively, and θ and ψ are the B—X—B and X—B—X bond angles, respectively. If there are N atoms of type B, each of the two bands has weight N and the delta functions at ω_3 and ω_4 each have weight $N/2$. A special case arises if $\psi = 120°$, when the BX_3 units become *planar*, as in glassy B_2O_3. In this case, [4.14] for the band edges simplifies to

$$
\begin{aligned}
\omega_1^2 &= \frac{\alpha}{m}(1+\cos\theta) & \omega_2^2 &= \frac{\alpha}{m}(1-\cos\theta) \\
\omega_3^2 &= \frac{\alpha}{m}(1+\cos\theta) + \frac{3\alpha}{2M}, & \omega_4^2 &= \frac{\alpha}{m}(1-\cos\theta) + \frac{3\alpha}{2M}
\end{aligned}
\qquad [4.15]
$$

A further special case is obtained by putting $\theta = 180°$, letting $m \to 0$ and setting $\alpha = 2\alpha'$; in this case the trigonally coordinated As network is recovered with the As–As force constant α'. Equation [4.14] now becomes

$$\omega_1^2 = \frac{2\alpha'}{M}(1 - \cos\psi), \qquad \omega_2^2 = \frac{2\alpha'}{M}(1 + 2\cos\psi) \tag{4.16}$$

and the vibrational density of states within the central-force approximation consists of a single band, with weight N, and a delta function at ω_2 with weight $N/2$.

The central-force approximation outlined above has the advantage that analytic results can be obtained for the vibrational dynamics of covalent networks, but it is obviously simplistic in that non-central (bond-bending) forces are neglected, with the result that many modes appear unphysically at zero frequency. Furthermore, the effects of rings of atoms in the network are ignored, as are the effects of Coulomb forces.

Galeener and Thorpe (1983) have investigated the dynamics of B_2X_3 networks containing regular six-membered (B_3X_3) rings; an example of such a ring in a real glassy system is the planar boroxol ring in B_2O_3, already discussed in sect. 3.4.2 in connection with medium-range order. Within the central force approximation, it is found that the overall VDOS for B_2O_3-like materials becomes much more complex than when the network contains no rings (cf. [4.15] and Fig. 4.5(a)) when boroxol rings are incorporated (Fig. 4.5(b)). It can be seen that this simple model predicts rather well the essential features observed experimentally in the Raman spectrum of glassy B_2O_3 (Fig. 4.5(c)), despite the fact that the same B—O bond-stretching force constant is assumed to exist both within and outside the boroxol rings. The effect of the inclusion of non-central forces would be to raise modes from zero frequency to the lower frequency range of the spectrum.

We have already commented in sect. 3.4.2 that the 808 cm^{-1} line in the Raman spectrum of vitreous B_2O_3, which has been ascribed to the symmetric

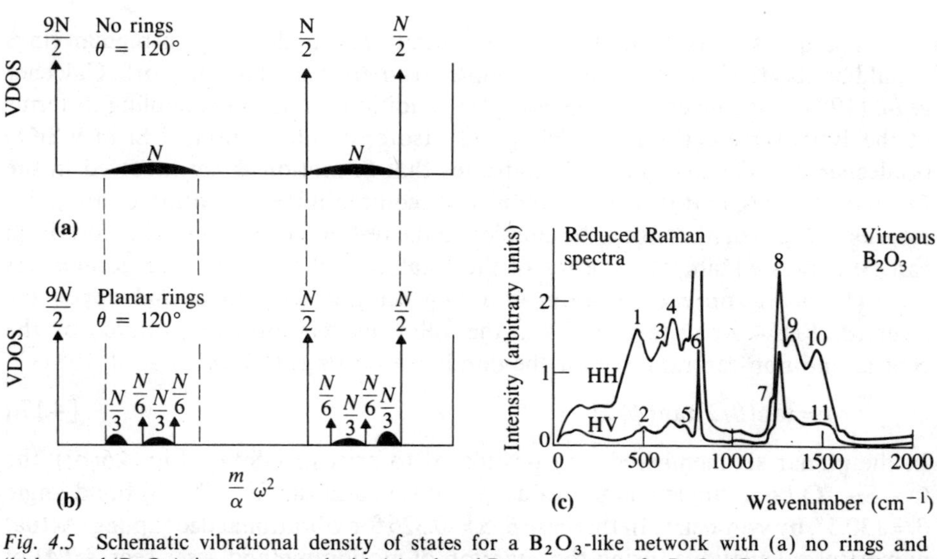

Fig. 4.5 Schematic vibrational density of states for a B_2O_3-like network with (a) no rings and (b) boroxol (B_3O_3) rings, compared with (c) reduced Raman spectra of v-B_2O_3 (Galeener and Thorpe 1983).

breathing motion of the oxygen atoms within boroxol rings, is remarkably narrow (see Fig. 3.38). Similar sharp lines have also been observed in the spectra of other glassy materials, e.g. $GeSe_2$ and SiO_2; the Raman spectrum for v-SiO_2 is shown in Fig. 4.6(a), and the two narrow 'defect' lines at 495 cm^{-1} (D_1) and 606 cm^{-1} (D_2) are clearly apparent. The origin of these two features has been the subject of much dispute, variously being ascribed to broken bonds, wrong bonds or regular rings. It has been argued that the last of these possibilities is the most likely on the basis of isotopic substitution experiments (Galeener 1982), the D_2 line being ascribed to the oxygen breathing mode of a planar six-membered ring, topologically similar to the boroxol ring in B_2O_3 (Fig. 4.6(b)), and the D_1 line has been ascribed to the same motion in a regular but slightly puckered eight-membered ring.

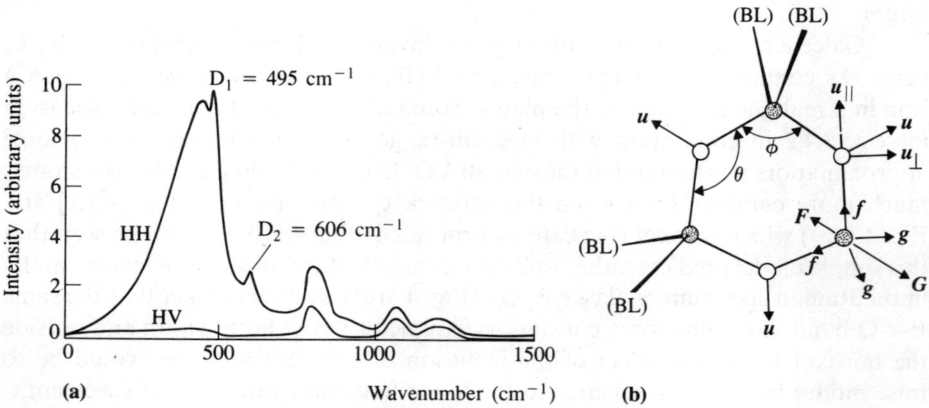

Fig. 4.6 (a) Polarized Raman spectrum of v-SiO_2 showing the 'defect' lines D_1 and D_2. (b) The symmetric breathing mode of a six-membered ring in SiO_2 assigned to the D_2 line (Galeener *et al.* 1984).

The question is, then, why certain vibrational modes of such regular rings should be effectively vibrationally decoupled from the rest of the network. Galeener *et al.* (1984) have given a simple analytic condition for such decoupling in terms of the Born force-constant model ([4.9]). Isotopic substitution of Si in v-SiO_2 (Galeener and Geissberger 1983) indicates that Si motion is *not* involved in the D_1 and D_2 lines, and this condition is also required if the O motion of a regular ring, as in Fig. 4.6(b), is not to be coupled to the rest of the network. This condition can be expressed simply in terms of the total central force (***F***) and non-central force (***G***) acting on each Si atom in a ring being exactly equal and oppositely directed (Fig. 4.6(b)); this results in the following relation being obeyed by the central and non-central force for the bonds within rings (Galeener *et al.* 1984):

$$\alpha/\beta = \tan(\theta/2)\tan(\phi/2) \qquad [4.17]$$

In the planar six-membered ring postulated to exist in v-SiO_2 (Fig. 4.6(b)), the O—Si—O bond angle has the value $\phi = 109.5°$ and the Si—O—Si bond angle $\theta = 130.5°$, by symmetry. In this case $\beta/\alpha = 0.326$ for vibrational decoupling. Actual numerical calculations (using the equation-of-motion method – see sect. 4.2.4) of the VDOS for a CRN for v-SiO_2 containing six- and eightfold rings have shown, however, that narrow bands are only obtained for extremely regular rings (de Leeuw *et al.* 1985).

4.2.2 Numerical methods: introduction

We have seen that the quantity describing vibrational excitations which is most appropriate for an amorphous solid and which can be compared more or less directly with experiment is the density of states. Thus, it is this, rather than say the eigenvectors, to which calculations are mostly directed. The problems in calculating vibrational (or electronic) properties of amorphous materials are principally twofold, both of which are due essentially to the lack of periodicity. The first problem is what to take as the structure; the equilibrium positions of all atoms are required in order to compute the force-constant matrix $\Phi_{n\alpha i}^{n'\alpha' i'}$ on which all calculations are based. The difficulties in ascertaining the structure of amorphous solids were explored in Chapter 3 and the most common starting point is to use the coordinates of a structural model (e.g. a CRN) whose scattering properties match experiment well. The second problem lies in the diagonalization of the force-constant matrix. We have seen that the presence of periodicity reduces the number of eigenvalues from $3Nr$ to $3r$ through the introduction of the Bloch formalism for the eigenvectors. Such a simplification is not available for calculations for amorphous solids. Furthermore, the number of atoms, r, in the 'unit cell' (e.g. the CRN model) of an amorphous solid is very large (~ 500) and so matrices containing several thousand elements are involved, leading to considerable computational difficulties. However, this problem may be eased to a certain degree by the use of only a small cluster (of several tens of atoms) in the interior of the model, since it is the *local* structure, rather than medium-range structure, which determines many of the features in $\rho(\omega)$; sophisticated matrix manipultion techniques also serve to speed diagonalization.

As might be expected, most of the theoretical techniques that have been used to explore vibrational properties of amorphous solids have been numerical in character. No general analytical methods exist which can handle topological disorder in three dimensions, although certain forms of disorder, e.g. 1-D disordered chains, or substitutional disorder, are amenable to such analysis.

All numerical approaches are formulated to solve the eigenvalue problem given by [4.4]. However, a variety of interatomic potentials have been used. Most have tended, for simplicity, to be restricted to nearest-neighbour interactions only, and some have been further restricted to include only stretching forces (see sect. 4.2.1). Generally, however, a non-central component is included for covalent materials to represent the bond-bending forces needed to stabilize the structure against shear. Several types of potential have been used for covalent systems. The first is the Born model ([4.9]) which we have already mentioned, and which is axially symmetric about the bond connecting two nearest-neighbour atoms because of the threefold symmetry of the other bonds in the perfectly tetrahedral case. The second type of potential, widely used, is due to Keating (1966):

$$V_{\mathrm{ij}} = \tfrac{3}{4}\alpha_{\mathrm{K}} \sum_{\mathrm{ij}} [(\boldsymbol{u}_{\mathrm{i}} - \boldsymbol{u}_{\mathrm{j}})\cdot \boldsymbol{r}_{\mathrm{ij}}]^2 + \tfrac{3}{16}\beta_{\mathrm{K}} \sum_{\mathrm{i(j,k)}} [(\boldsymbol{u}_{\mathrm{i}} - \boldsymbol{u}_{\mathrm{j}})\cdot \boldsymbol{r}_{\mathrm{ik}} + (\boldsymbol{u}_{\mathrm{i}} - \boldsymbol{u}_{\mathrm{k}})\cdot \boldsymbol{r}_{\mathrm{ij}}]^2 \qquad [4.18]$$

where the sums run over all atoms i and their distinct neighbours j and k and where, as in [4.9] $\boldsymbol{u}_i$ are displacement vectors and $\boldsymbol{r}_{ij}$ are unit vectors joining sites i and j. The bond-stretching and bond-bending force constants are written as α_K and β_K respectively, to differentiate them from the corresponding Born terms. This potential has the advantage that it is *rotationally* invariant because it only depends on the scalar product of nearest-neighbour vectors. Also rotationally invariant is the 'valence-force-field' model, in which the scalar product is replaced by the corresponding bond angle:

$$V_{ij} = \tfrac{1}{2}\alpha_{VFF}\sum_{ij}\Delta r_{ij}^2 + \tfrac{1}{2}\beta_{VFF}\sum_{i(jk)} r_{io}^2 \cos\theta_{jk}\Delta\theta_{jk} \qquad [4.19]$$

where r_{io} is the equilibrium bond length and θ_{jk} is the bond angle subtended at atom i by atoms j and k. Note that all these forms of interatomic potential only take into account topological disorder through the coordinates of the atomic positions. Quantitative disorder, manifested by differences in force constants between atoms due to local structural variations, is almost always neglected. For metallic systems, the pair potential may be obtained from pseudopotentials, at least for simple metallic systems without d-bands (Hafner 1983).

A further problem with vibrational computations for amorphous solids is the effect on the calculated modes of the surface of the model cluster used. Of course, there is no difficulty for perfectly crystalline materials for which periodic boundary conditions may be invoked. However, for cluster models simulating the structure of amorphous solids containing typically 500 atoms a significant number of atoms lie on or near the surface, and spurious surface modes can easily arise if calculations are performed on the entire cluster without proper regard for boundary conditions. Such problems can be reduced if a model with periodic boundary conditions is used; or, for cluster models, if only interior atoms are used in the calculation, if the density of states is projected out locally as in the 'recursion' method (sect. 4.2.5), or if a 'Bethe lattice' is grafted on to the surface unsatisfied bonds of a CRN. A Bethe lattice (also called a Cayley tree) has the same coordination as the network itself, except that there are *no* closed rings of bonds (Fig. 4.7). The use of such a mathematical device eliminates localized surface modes and does not introduce additional structure into the density of states since $\rho(\omega)$ for a Bethe lattice is smooth and featureless. An additional limitation imposed by the use of finite size models is that the low-frequency acoustic-type modes below, say, 50 cm^{-1} are not amenable to calculation for several reasons; principally this is because such modes have very long wavelengths which are comparable in extent to the size of the model, but also because the local potentials (e.g. [4.9], [4.18] and [4.19]) are insufficient to describe the long-range interactions acting in such modes. Furthermore, in general, Coulomb interactions are neglected.

As mentioned previously, the theoretical investigation of vibrational modes and calculation of the density of states reduces to the solution of the set of eigenvalue equations given by [4.4]. A variety of numerical techniques have been developed for this problem, and we briefly mention them in the following, starting with the oldest method.

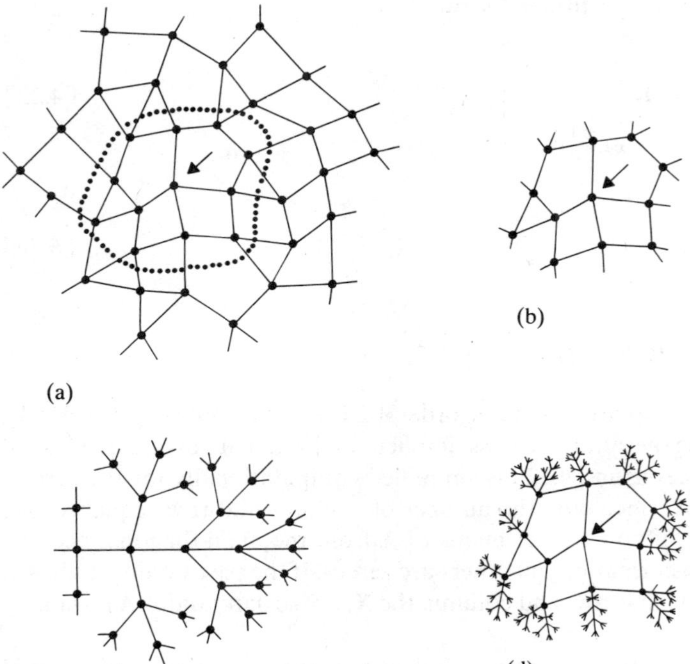

(a) (b) (c) (d)

Fig. 4.7 The cluster-Bethe-lattice method (Joannopoulos 1979) for calculating vibrational (or electronic) densities of states of amorphous structures. A given atom is chosen as a reference point (a) and a cluster removed from the network (b). A Cayley tree appropriate for the coordination of the network (c) is attached to each dangling bond of the cluster (d).

4.2.3 Negative eigenvalue method

This method utilizes a theorem introduced by Dean and Martin (1960) which offers an efficient method of computing the distribution of eigenvalues of a real symmetric matrix; extensive reviews of the method are given by Dean (1972) and Bell (1972).

The equation of motion ([4.4]) can be written in matrix terms as:

$$(\mathbf{D} - \mathbf{I}\omega^2)\mathbf{U} = 0 \qquad [4.20]$$

where as before $\mathbf{D}$ is the real symmetric force constant matrix, $\mathbf{U}$ is the unitary matrix, whose elements are the eigenvectors $u_{n\alpha i}$, which consequently obeys the relation $\mathbf{U}^{\mathrm{T}}\mathbf{U} = \mathbf{I}$ where $\mathbf{I}$ is the unit matrix. If we consider the matrix $\mathbf{A} = \mathbf{D} - \omega^2\mathbf{I}$, the negative eigenvalue theorem states that the number of negative eigenvalues (i.e. physical solutions) of the $n \times n$ matrix $\mathbf{A}$, denoted by $\eta(\mathbf{A})$ is given by

$$\eta(\mathbf{A}) = \sum_{i=1}^{m} \eta(\mathbf{X}_{\mathrm{i}}) \qquad [4.21]$$

if **A** can be written in the partitioned form:

$$\mathbf{A} = \begin{bmatrix} \mathbf{A} & \mathbf{B}_2 & & \\ \mathbf{B}_2^{\mathrm{T}} & \mathbf{A}_2 & \mathbf{B}_3 & \\ 0 & & & \mathbf{B}_{\mathrm{m}}^{\mathrm{T}}\mathbf{A}_{\mathrm{m}} \end{bmatrix} \qquad [4.22]$$

where

$$\mathbf{X}_1 = \mathbf{A}_1 - \omega^2\mathbf{I}_1 \qquad [4.23]$$

and

$$\mathbf{X}_{\mathrm{i}} = \mathbf{A}_{\mathrm{i}} - \omega^2\mathbf{I}_{\mathrm{i}} - \mathbf{B}_{\mathrm{i}}^{\mathrm{T}}\mathbf{X}_{\mathrm{i}-1}^{-1}\mathbf{B}_{\mathrm{i}} \qquad (\mathrm{i} = 2, 3, \ldots)$$

and $\mathbf{A}_{\mathrm{i}}$ is a symmetric square matrix of order 1_{i}, $\mathbf{I}_{\mathrm{i}}$ is a unit matrix of order 1_{i}, and $\Sigma_{i=1}^{m}\ 1_{\mathrm{i}} = n$. The iteration process implicit in [4.23] involving sequential partitioning of matrices as in [4.22] is simplified computationally for the case of a sparse matrix like **A**, since only the number of columns and rows equal to one less than the number of non-zero elements of **A** from the main diagonal need be considered in the transformation; all others are zeroes. If the partitioning is chosen to separate out just the first row and column, the $\mathbf{X}_{\mathrm{i}}$ are scalars, and $\eta(\mathbf{A})$ is simply the number of *negative* X_{i}.

The importance of this theorem is that $\eta(\mathbf{A})$ is just the number of squared frequencies of the vibrational system less than ω^2, i.e. the integrated density of states for squared frequencies, and is therefore related directly to the (integrated) density of states. Thus, $\rho(\omega)$ can be obtained in a considerably simpler computational manner than is necessary for diagonalization of the (large) matrix **D**. The negative eigenvalue method suffers from the disadvantage that only the frequency spectra are obtained directly, rather than individual eigenvalues or eigenvectors. However, in practice this is of little consequence since $\rho(\omega)$ is generally the desired quantity; however, if a knowledge of the character of the vibration corresponding to a certain feature in the density of states is required, the eigenvector may be calculated using the inverse of [4.20], having obtained a good approximation to the eigenvalue by using the iteration procedure for small intervals around the chosen frequency.

The method outlined above has been applied to the (4:2) glass-forming systems SiO_2, GeO_2 and BeF_2, which are expected to have the same local geometry, using as a basis the CRN models built to simulate the structure of a-SiO_2 (Bell 1972, Dean 1972). Figure 4.8(a) shows typical results of density of states calculations for SiO_2 using fixed-atom boundary conditions for the surface atoms. Three well-defined bands are observed, at about 400, 700 and 1100 cm^{-1} which agree in position fairly well with the features at 465, 800 and 1100 cm^{-1} observed in the IR spectrum (see also Fig. 4.6(a)). The various features in $\rho(\omega)$ have been identified with certain predominant types of atomic motion; *S* refers to stretching, B to bending, R to rocking of the oxygen atoms respectively, and C refers to those modes which involve a large amount of cation motion. The highest-frequency band is due predominantly to an *asymmetric* stretching motion of the oxygen atoms (i.e. along the line joining the two Si atoms bonded to each O atom), whilst the

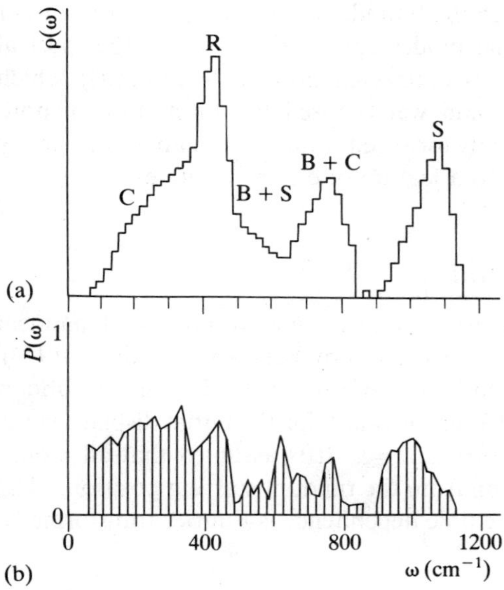

Fig. 4.8 Density of states (a) and participation ratio (b) of vibrational modes calculated for the Bell and Dean CRN of a-SiO_2 with fixed boundary conditions (Bell 1972).

lower-frequency stretching motion is a symmetric motion perpendicular to this direction.

The spatial extent of modes, which in an amorphous solid can span practically the entire range observed in crystals from the infinitely extended plane-wave-like excitations to highly localized vibrations associated with point defects, can be quantified to some extent by means of the 'participation ratio' $P(\omega)$. This describes the proportion of the total number N of atoms which contribute effectively to a given mode:

$$P(\omega) = \frac{M_1^2}{M_0 M_2} \qquad [4.24]$$

where for a mode at frequency ω_j, M_r is the rth moment of the kinetic energy of the mode

$$M_r(\omega_j) = \sum_{i\alpha} |u_{i\alpha}^{(j)}|^{2r} \qquad [4.25]$$

A mode involving only one atom (i.e. localized) has a value $P = 1/N$, whereas an extended mode in which all atoms participate equally gives $P = 1$; the sinusoidal waves characteristic of crystals give $P \simeq 0.5$. The participation ratios for the various features in $\rho(\omega)$ for SiO_2 are shown in Fig. 4.8(b), whence it can be seen that modes in the low-frequency acoustic region ($\omega < 500\,\text{cm}^{-1}$) are reasonably extended, with $P \simeq 0.5$. Higher frequency (optic) modes are much less extended, and there are strong indications of distinct localization ($P \to 0$) at the band edges. This feature of localization at band edges will re-emerge when we discuss the nature of *electron* states in amorphous materials (Ch. 5). Calculations by Bell and Dean

(1972) were also made for cases in which the boundary conditions were free rather than fixed. It was found that additional modes appeared in $\rho(\omega)$ in the spectral gap at $\simeq 900\ \mathrm{cm}^{-1}$ observed for the fully connected case (see Fig. 4.8(a)), which were extremely localized in character and were caused by the motion of non-bridging oxygens; the occurrence of highly localized states in the gap in the density of states for defect configurations is also a feature of electronic states.

4.2.4 Equation-of-motion method

This method, as its name implies, concentrates on the time evolution of dynamical quantities; a review of the technique has been given by Beeman and Alben (1977). The equation of motion ([4.2]) is solved by integration forward in time to obtain the displacements up to a time interval T, given values for the initial displacements $s^0_{i\alpha}$ and taking the velocities equal to zero at $t=0$. It is assumed that all modes contribute equally to the ensuing motion from the random starting positions. The function $R(\omega)$ is then computed (i.e. the time dependence is Fourier transformed):

$$R(\omega)=\frac{2}{\pi}\int_0^T \sum_{i\alpha} A_{i\alpha}s_{i\alpha}(t)(\cos \omega t)\exp(-\lambda t^2)\,\mathrm{d}t \qquad [4.26]$$

where $A_{i\alpha}$ and $s^o_{i\alpha}$ are chosen such that $R(\omega)$ represents the density of states, IR spectrum or Raman spectrum, as appropriate. For $R(\omega)$ to be the density of states, the following choice is made (Beeman and Alben 1977):

$$s^o_{i\alpha}=\sqrt{2}\cos\theta_{i\alpha} \qquad [4.27(a)]$$

$$A_{i\alpha}=\sqrt{2}\cos\theta_{i\alpha} \qquad [4.27(b)]$$

where $\theta_{i\alpha}$ are random angles distributed uniformly between 0 and 2π. The choices for $s^o_{i\alpha}$ and $A_{i\alpha}$ that cause $R(\omega)$ to represent IR and Raman spectra will be given in the next section. The upper limit T of the integral in [4.26] is taken to be about 30 periods of the fastest oscillation, and the damping factor λ is taken to equal $3/T^2$. This has the result that the computed spectra are broadened by the use of a finite time interval; the value of T is chosen to give a broadening of about 4% of the full spectral width. An average over several computed $R(\omega)$ for several sets of $\theta_{i\alpha}$ is necessary to achieve a reasonable spectral function.

The equation-of-motion method possesses several advantages, the most notable being that smoothed spectral functions ($\rho(\omega)$, etc.) are obtained directly from the coordinates of a model without the need for diagonalization of large matrices. It is also economical in computer time and storage. The method gives results which are similar to those produced by the 'recursion' method to be discussed next; a typical result is shown in Fig. 4.9 where $\rho(\omega)$ computed for the Greaves–Davis (1974) threefold coordinated CRN is compared with the density of states for a-As obtained from inelastic neutron scattering by Leadbetter *et al.* (1976). General features are reproduced reasonably well, but the sub-structure in the lower acoustic peak does not appear because of the simple (Born) force law used.

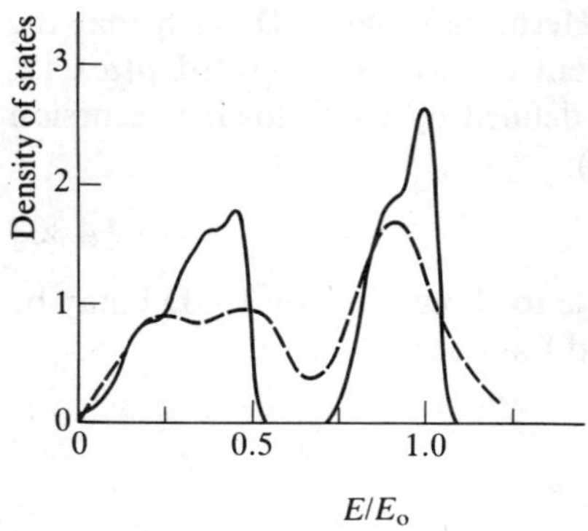

Fig. 4.9 Comparison of the density of states for a-As calculated using the equation-of-motion method applied to the Greaves–Davis CRN model (Beeman and Alben 1977) (solid line) with the density of states obtained experimentally by inelastic neutron scattering (Leadbetter *et al.* 1976) (dashed line).

4.2.5 Recursion method

We have already seen that it is generally the *local* structure which determines the gross features of the vibrational density of states; longer range structure only serves to introduce fine detail (cf. periodicity giving rise to Van Hove singularities). One computational technique that capitalizes on this feature for amorphous materials is the recursion method. This was originally developed for calculations of the *electronic* structure of disordered materials (see the review by Heine 1980), but was extended to the calculation of vibrational excitations by Meek (1976). The *local* density of modes, i.e. the density of states weighted by the amplitude squared of each mode at the atom considered, is calculated and an average taken over 10 or so central atoms in a model cluster to yield a representative spectrum.

The mathematical details of the method go beyond the level of this book, and so the interested reader is referred to the extensive discussion by Heine (1980). In brief, though, it offers an attractive computational technique for obtaining the required density of states for, say, the dynamical matrix **D** ([4.4]), without recourse to direct calculation of the eigenvectors or eigenvalues. The Hamiltonian for vibrational (or electronic) excitations of an amorphous cluster yields a discrete spectrum of eigenvalues, rather than a band, and so the density of states is given by a sum over such states, as in [4.10]. It can be shown that in terms of the Green function (or resolvent operator):

$$\mathbf{G}(\omega^2) = (\omega^2\mathbf{I} - \mathbf{D})^{-1} \qquad [4.28]$$

an equivalent way of writing the local density of modes is in terms of the spectral function:

$$\rho_{ox}(\omega^2) = \frac{-1}{\pi} \lim_{\varepsilon \to 0} \mathrm{Im}[G_{oo}(\omega^2 + i\varepsilon)] \qquad [4.29]$$

where the density of the modes is 'projected' on to the squared x-displacement of atom O (denoted by $|0\rangle$) and G_{oo} is the diagonal matrix element of **G**, i.e. $\langle 0|G|0\rangle$. The local density of states can now be calculated directly for **G** by choosing a new set of orthonormal basis vectors $|u_n\rangle$, of the Hermitian matrix $\mathbf{D}$, such that the first is chosen to be the particular displacement for which is wanted $\rho(\omega)$, i.e. $|u_o\rangle = |0\rangle$. The new basis of vectors $|u_n\rangle$ are defined by the following recursion relation (hence the name given to the method):

$$b_{n+1}|u_{n+1}\rangle = (\mathbf{D} - a_n)|u_n\rangle - b_n|u_{n-1}\rangle \qquad [4.30]$$

Expressing $\mathbf{D}$ in this basis set, it is then possible to show that $[\omega^2\mathbf{I} - \mathbf{D}]$ may be inverted *analytically* in the form of a continued fraction:

$$G_{oo}(\omega^2) = \cfrac{1}{\omega^2 - a_0 - \cfrac{b_1^2}{\omega^2 - a_1 - \cfrac{b_2^2}{\cdots}}} \qquad [4.31]$$

where the a_n, b_n are defined according to the recursion relation. The computation is very efficient since only three vectors are needed at any one time ([4.30]), and the continued fraction is stopped at some finite value of $n = N$ (say 15–25). Although the accuracy of the final density of states does increase with increasing N, the higher $|u_n\rangle$ represent displacements of atoms far removed from the origin atom, and therefore have little effect on the local density of states. It has been found from experience that the recursion method produces results very similar to those given by the equation-of-motion method, and there is therefore little to choose between them.

4.2.6 Summary

The presence of disorder in a solid affects the vibrational excitations in a variety of ways. The most important consequence of the lack of periodicity is that a reciprocal lattice can no longer be defined, and therefore any excitations of the solid (whether vibrational or electronic) cannot be described simply with respect to the underlying lattice by means of the Bloch formalism. This has the result that the phonon (or equivalently electron) wavevector is no longer a good quantum number, and therefore the selection rules obeyed by the wavevector in any interaction of the phonon (or electron) in a perfect crystal are relaxed in an amorphous solid. Furthermore, the description of vibrational excitations in terms of dispersion curves, ω vs. $\boldsymbol{q}$, (or equivalently band structures, E vs. $\boldsymbol{k}$, for electrons), valid for single crystals, becomes inappropriate for non-crystalline solids. Two additional consequences of the presence of disorder for excitations (both vibrational and electronic) are that disorder-scattering reduces the lifetime (or equivalently the mean free path), and if the amount of disorder is very great, the excitations may become spatially localized rather than extended; an example of this is an impurity vibrational mode.

Although excitations of an amorphous solid cannot be described in terms of a dispersion curve (or band structure), the *density of states* remains a valid description, albeit more broadened than in the crystalline case. Thus, calculations of vibrational properties are generally directed towards obtaining this quantity.

The 'brute-force' method is simply to diagonalize the dynamical matrix for all atoms in a model cluster, and the considerable computational difficulties which this involves can be simplified somewhat by using the negative eigenvalue method (section 4.2.3). A more elegant solution is to recognize that it is the *short-range* order which essentially determines the major features in the density of states (again true for both vibrations and electrons). This is the basis of the recursion method (sect. 4.2.5) (and the cluster-Bethe-lattice method – sects. 4.2.2 and 5.2.1), in which the *local* density of states is projected out onto the squared displacement of a given origin atom. Contributions from atoms far from the origin atom perturb only slightly the resulting density of states.

4.3 Experimental probes

4.3.1 Introduction

A variety of experimental techniques may be employed to probe the vibrational structure of solids. These range from various light absorption and scattering experiments (IR, Raman and Brillouin) which are available in well-equipped laboratories, to inelastic neutron scattering experiments which require a central facility (e.g. a nuclear reactor) as a source of neutrons. These techniques are equally applicable to crystalline or amorphous solids, but the lack of periodicity characteristic of non-crystalline materials changes somewhat the way photons or neutrons interact with the phonons, as will be discussed in the following.

4.3.2 Infra-red spectroscopy

Infra-red spectroscopy measurements usually employ either a grating (or prism) spectrometer together with a source (e.g. a tungsten lamp) and detector (e.g. a thermocouple or photo-cell), or an interferometric spectrometer which has a Michelson interferometer replacing the monochromator and more sophisticated data processing facilities to transform the data into a frequency spectrum. The measured transmission T is related to the IR absorption coefficient α and the reflectivity R by [3.85] for the case $\alpha d \gg 1$ where d is the sample thickness. Therefore two independent measurements of T and R are required to extract α uniquely.

As already mentioned in section 3.2.6, IR absorption takes place when there is a change in dipole moment associated with the vibrational mode excited. For non-polar, covalently bonded materials, there is no static dipole moment and so any induced dipole moment must arise from a *dynamic* effect involving atomic displacements and the resulting bond compressions and extensions. The dipole moment vector $\boldsymbol{M}$ is proportional to the displacement $\mathbf{u}_i$ (to first order) and according to one model can be written as (Alben *et al.* 1975):

$$\boldsymbol{M} = \sum_{i(j,k)} (\boldsymbol{r}_{ik} - \boldsymbol{r}_{ij})[(\boldsymbol{u}_i - \boldsymbol{u}_j)\cdot\boldsymbol{r}_{ij} - (\boldsymbol{u}_i - \boldsymbol{u}_k)\cdot\boldsymbol{r}_{ik}] \qquad [4.32]$$

or in terms of the bond compression $C_{ij} = (\boldsymbol{u}_i - \boldsymbol{u}_j)\cdot\boldsymbol{r}_{ij}$ as:

$$\boldsymbol{M} = 2\sum_i \left(\sum_j C_{ij}\right)\left(\sum_k \boldsymbol{r}_{ik}\right) \qquad [4.33]$$

where $\boldsymbol{u}_i$ and $\boldsymbol{r}_{ij}$, etc., have the same meaning as in [4.9]. The origin of the dynamic dipole moment can be seen with reference to Fig. 4.10; charge flows into a bond under compression and out under tension. Note from [4.32] or [4.33] that the induced dipole moment is zero for perfect tetrahedral symmetry since $\Sigma_{ij}\boldsymbol{r}_{ij}=0$, and hence crystalline Ge and Si should not be, and indeed are not, IR-active; in fact *all* elemental crystals with two atoms per unit cell exhibit a vanishing first-order dipole moment and are IR-inactive. Thus, we can see that one consequence of disorder is that modes become IR-active that otherwise would be inactive.

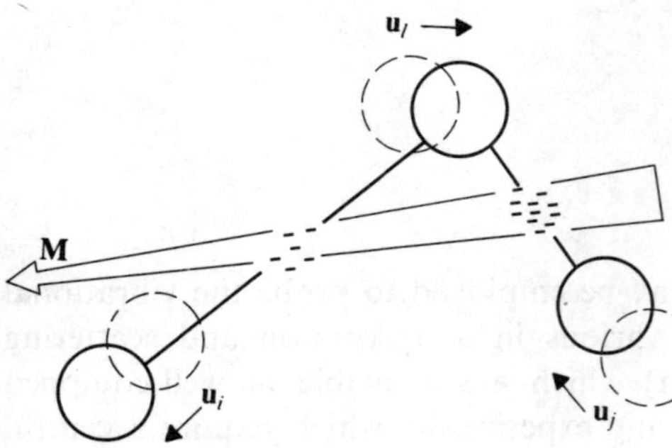

Fig. 4.10 Dynamic bond-charge model for the dipole moment which can give rise to IR activity (Alben *et al.* 1975). Bond charge moves from extended to compressed bonds, resulting in a local electric-dipole moment **M**. The local moments cancel for modes of the diamond cubic structure, but do not cancel for amorphous structures.

Another way of stating this consequence of the lack of long-range order is that the selection rules for the wavevector $\boldsymbol{k}$ that are obeyed in crystals are relaxed for amorphous solids. In crystals, as a consequence of periodicity, we have seen that $\boldsymbol{k}$ is a good quantum number. For this case, we must have conservation of momentum, or equivalently wavevector, i.e. $\boldsymbol{k}+\boldsymbol{q}=\boldsymbol{k}'$, where $\boldsymbol{q}$ corresponds to the photon momentum and $\boldsymbol{k}$ and $\boldsymbol{k}'$ are the incident and scattered phonon wavevector respectively. However, since $|\boldsymbol{q}| \ll |\boldsymbol{k}|$, only phonons very near $\boldsymbol{k}=0$ (i.e. at the centre of the Brillouin zone) are excited in crystals, giving sharp lines in the IR spectrum. For amorphous materials, in which $\boldsymbol{k}$ is *not* a good quantum number, the $\boldsymbol{k}$-selection rule breaks down, and *all* modes can contribute (not necessarily equally) to the IR (or Raman) activity. Thus, broad bands in the IR spectrum are expected for amorphous materials, qualitatively similar to the phonon density of states but weighted by the matrix element for the transition (proportional to the square of the dipole moment). The broad features in the spectrum of amorphous germanium and the degree of correspondence between experimental spectra and the density of states, both obtained experimentally from inelastic neutron scattering and calculated for a CRN model (Maley *et al.* 1988) are shown in Fig. 4.11.

Attempts have been made to calculate the IR spectra expected for various amorphous materials. The fine details in spectra produced by these calculations depend somewhat crucially on two aspects: the quality of the structural model; and the frequency dependence of the matrix elements (or dipole moment) that are used. Beeman and Alben (1977) have given values for $s^{\circ}_{i\alpha}$ and $A_{i\alpha}$ which, when used

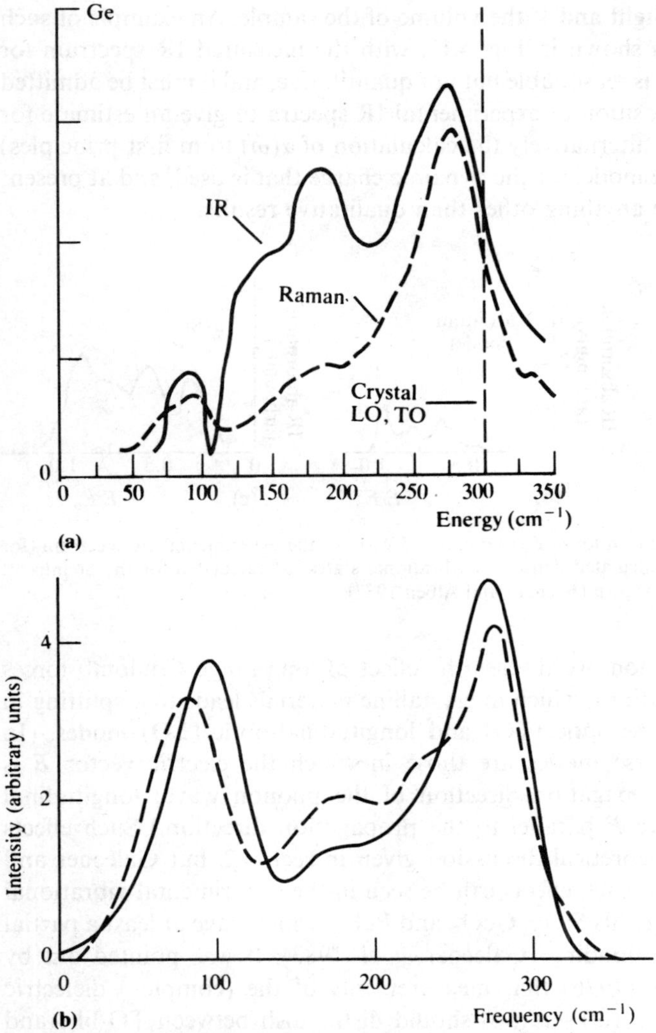

Fig. 4.11 Comparison of (a) IR and Raman spectra of a-Ge (Alben *et al.* 1975) with (b) the experimental density of states (dashed line) obtained from inelastic neutron scattering experiments and a broadened VDOS (solid line) calculated for a CRN (Maley *et al.* 1988).

with [4.26], yield $R(\omega)$ proportional to the IR absorption $\alpha(\omega)$:

$$s^{\mathrm{o}}_{\mathrm{i}\alpha} = \sum_{\gamma} \varepsilon^{\gamma} e^{\gamma}_{\mathrm{i}\alpha}/m_{\mathrm{i}} \qquad [4.34(\mathrm{a})]$$

$$A_{\mathrm{i}\alpha} = m_{\mathrm{i}} s^{\mathrm{o}}_{\mathrm{i}\alpha} \qquad [4.34(\mathrm{b})]$$

where ε^{γ} is the γ component of a unit vector in the direction of polarization of the incident light, and $e^{\gamma}_{\mathrm{i}\alpha}$ is the 'dynamic charge' which gives the dipole moment in the γ-direction associated with a displacement of atom i in direction α; the model used for $e^{\gamma}_{\mathrm{i}\alpha}$ was that given by [4.32]. Thus, the IR absorption is given by:

$$\alpha(\omega) = \frac{2\pi^2}{c} \frac{R(\omega)}{V} \qquad [4.35]$$

where c is the speed of light and V the volume of the sample. An example of such a calculation for a-Si is shown in Fig. 4.12, with the measured IR spectrum for comparison. Agreement is reasonable but not quantitative, and it must be admitted finally that the decomposition of experimental IR spectra to give an estimate for the density of states (or alternatively the calculation of $\alpha(\omega)$ from first principles) depends crucially on the model for the dynamic charge that is used, and at present this is too crude to give anything other than qualitative results.

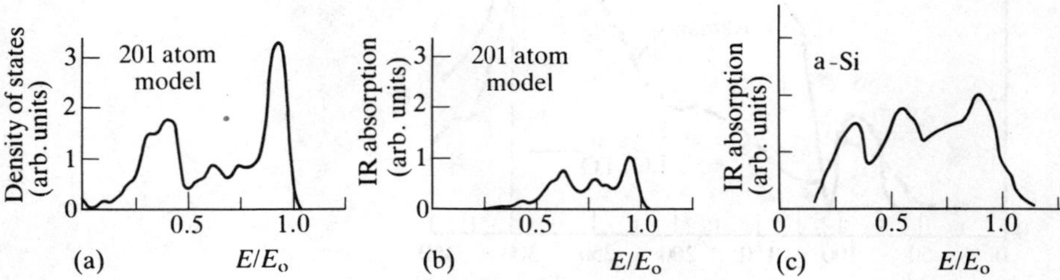

Fig. 4.12 Calculated IR spectrum for a-Si (b) compared with (c) the experimental IR spectrum (for which $E_0 = 521\ \text{cm}^{-1}$). The calculated density of vibrational states (a) calculated for the Steinhardt 201 atom CRN model is also shown (Beeman and Alben 1977).

Finally in this section, we discuss the effect of long-range Coulomb forces on the vibrational excitations, which in crystalline materials leads to a splitting in energy between transverse-optic (TO) and longitudinal-optic (LO) modes. (In isotropic media, transverse modes are those in which the electric vector $\boldsymbol{E}$ is perpendicular to the propagation direction of the phonon wave; longitudinal modes, by contrast, have $\boldsymbol{E}$ parallel to the propagation direction.) Such effects were neglected in the theoretical discussion given in sect. 4.2, but Galeener and co-workers maintain that such effects can be seen in the experimental vibrational spectra of the glassy materials SiO_2, GeO_2 and BeF_2, which have at least a partial ionic component to the bonding (Galeener *et al.* 1983). It was pointed out by Galeener and Lucovsky (1976) that measurements of the (complex) dielectric spectral function $\varepsilon(\omega) = \varepsilon_1(\omega) + i\varepsilon_2(\omega)$ should distinguish between TO-like and LO-like modes: peaks in $\varepsilon_2(\omega)$ are a signature of *transverse* modes, whilst peaks in the energy-loss function $\text{Im}(-\varepsilon^{-1}) = \varepsilon_2/(\varepsilon_1^2 + \varepsilon_2^2)$ reveal *longitudinal* modes. These quantities can be obtained from measurements of the IR reflectivity which, together with Kramers–Kronig transformations, lead to values for the optical constants n and k. The real and imaginary parts of the dielectric function are then given by

$$\varepsilon_1 = n^2 - k^2 \qquad [4.36(a)]$$

$$\varepsilon_2 = 2nk \qquad [4.36(b)]$$

For v-SiO_2, a comparison of the spectra for the IR dielectric constant ε_2 and the energy-loss function $\text{Im}(-\varepsilon^{-1})$, together with the reduced Raman spectra (sect. 4.3.3) and inelastic neutron scattering spectrum (sect. 4.3.4) is shown in Fig. 4.13 (Galeener *et al.* 1983). It can be seen that the major vibrational bands all exhibit LO–TO splittings; the splitting of the highest-frequency modes is

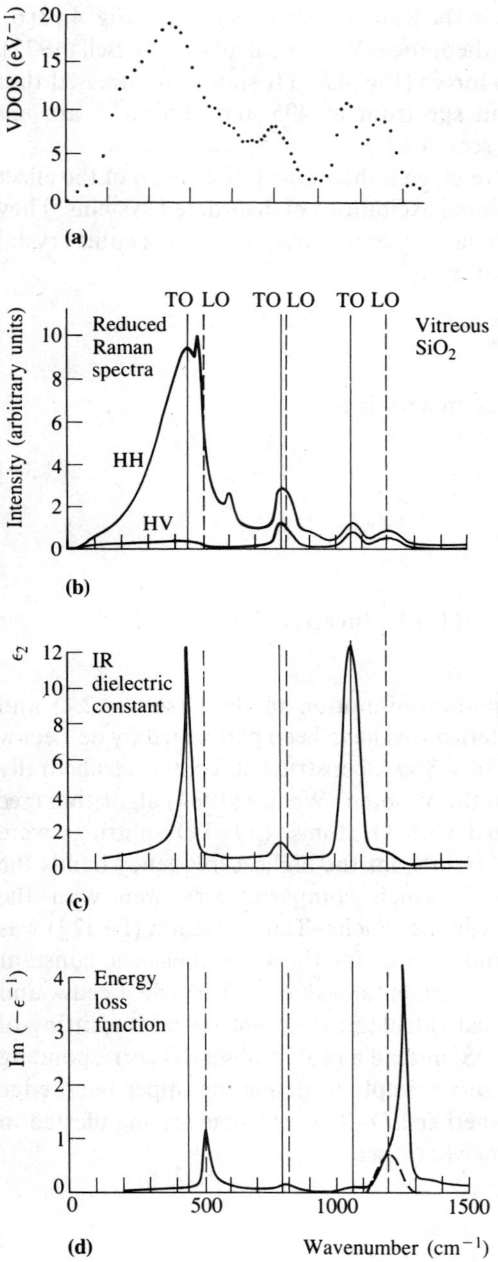

Fig. 4.13 (a) Vibrational density of states for v-SiO_2 (Price and Carpenter 1987), compared with (b) the reduced Raman spectra, (c) the IR dielectric constant and (d) the energy loss (Galeener *et al.* 1983).

$\Delta \simeq 135\ cm^{-1}$ for the case of v-SiO_2 (Galeener and Lucovsky 1976), whereas for the case of α-quartz it is $163\ cm^{-1}$ (Thorpe and de Leeuw 1986). It is significant, too, that these splittings are also observed in the vibrational density of states obtained experimentally from inelastic neutron scattering (Price and Carpenter

1987), shown in Fig. 4.13(a), as well as in the Raman spectra, shown in Fig. 4.13(b). However, they are *not* observed in the theoretical VDOS calculated by Bell (1972), who neglected the effects of Coulomb forces (Fig. 4.8). (It should be stressed that the sharp features in the HH Raman spectrum at 495 and 606 cm^{-1} are *not* associated with LO–TO effects – see sect. 4.2.1.)

Thorpe and de Leeuw (1986) have given a theoretical discussion of the effect of Coulomb interactions on the vibrational excitations of disordered systems. They show analytically that the Lyddane–Sachs–Teller relation, which for a cubic crystal with two atoms per primitive cell is given by

$$\frac{\omega_L^2}{\omega_T^2} = \frac{\varepsilon(0)}{\varepsilon(\infty)}$$

becomes generalized for non-crystalline materials to

$$\frac{\langle 1/\omega^2 \rangle_T}{\langle 1/\omega^2 \rangle_L} = \frac{\varepsilon(0)}{\varepsilon(\infty)} \qquad [4.37]$$

where the second moment is given by

$$\langle \omega^2 \rangle_x = (1/\pi) \int_0^\infty \omega^2 \, \mathrm{Im}\, \varepsilon_x(\omega) \, d\omega^2 / (1/\pi) \int_0^\infty \mathrm{Im}\, \varepsilon_x(\omega) \, d\omega^2, \quad x = \mathrm{L, T}$$

Numerical calculations using the equation-of-motion method (sect. 4.2.4) and incorporating long-range Coulomb interactions have been performed by de Leeuw and Thorpe (1986) for a CRN model of v-SiO_2, constructed from a tetrahedrally coordinated (Si) network according to the Wooten–Weaire (1984) algorithm (see sect. 3.5.5) and subsequently decorated with O atoms. LO–TO splittings were observed in plots of $\varepsilon_2(\omega)$ and $\mathrm{Im}(-1/\varepsilon(\omega))$; for the highest-frequency bands the calculated splitting was $\Delta = 141\ cm^{-1}$, which compares very well with the experimental value. The generalized Lyddane–Sachs–Teller relation ([4.37]) was verified using the numerical results, and a value for the static dielectric constant $\varepsilon(0) = 3.38$ was found, close to the experimental value of 3.30 (de Leeuw and Thorpe 1986). However, these numerical calculations do *not* reveal a splitting of the highest-frequency modes in the VDOS; instead a peak is observed corresponding to the TO modes, but the LO modes serve simply to define the upper band edge of the VDOS. Thus, the question of whether LO–TO splittings are manifested in the VDOS of glasses still remains somewhat open.

4.3.3 Raman spectroscopy

Raman and IR spectroscopies are usually discussed together, although the microscopic processes that are responsible are very different, making the two techniques (at least for crystals) complementary. This is because vibrational modes are Raman-active only if they are associated with a change in polarizability (rather than the change in dipole moment for IR transitions), and so modes in (centro-symmetric) crystals which are IR-inactive are Raman-active. This complementary relation is generally lost in amorphous materials since, as we have seen, *all* modes contribute (but not equally) to the IR activity and hence also to the Raman activity. An exception is for highly symmetric vibrational modes (e.g. the symmetric stretch of a boroxol ring in v-B_2O_3) for which the Raman response is dominant.

Raman spectroscopy has been revolutionized by the advent of lasers producing monochromatic, high-intensity light with which the weak Raman scattering may easily be detected. A typical Raman set-up consists of a laser, double grating monochromator and a detector (e.g. a photomultiplier), together with data processing facilities such as photon counting electronics. The double grating monochromator is required for spectral purity since the light is down-scattered to $\omega_L - \omega$, the 'Stokes' line, or up-scattered to $\omega_L + \omega$, the 'anti-Stokes' line, by some small frequency ω corresponding to a typical phonon frequency (of the order of a few hundred cm^{-1}) from the central laser line at ω_L. To achieve a further decrease in the interference from the laser line, the experiment is often carried out in the geometry shown in Fig. 4.14 for a scattering angle of 90°. The polarization of both the incident and scattered radiation affects the Raman intensity, and this dependence is made plain by labelling the scattered intensity as being vertically polarized (V) or horizontally polarized (H) with respect to the scattering plane; thus I_{VH} refers to the scattered intensity measured in the H direction for an incident beam polarized in the V direction. The polarization properties of the Raman spectrum are of utmost importance in the study of single crystals since it is then possible to determine the structure of the Raman tensor of each mode and hence its symmetry character. This is not possible for anything other than single crystals; for amorphous materials, the 'depolarization ratio', $p(\omega) = I_{HV}/I_{VV}$, is often used and can take values between 0 and 3/4. Raman lines for which $p = 3/4$ are referred to as depolarized, whereas those for which $p < 3/4$ are referred to as polarized; the dependence of the depolarization ratio on Raman shift can give information about the nature of the active vibrational modes. For instance, if the Raman spectrum is plotted as $(I_{VV} - 4I_{HV}/3)$ only the pure polarized component will contribute, since all depolarized modes give an identically zero contribution in such a plot.

Raman scattering is a complicated phenomenon and for more detail the interested reader is referred to Brodsky (1975). The induced polarizability is a second rank tensor (component $E_{i\alpha}^{\gamma\beta}$), which is a linear function of the atomic displacements $\boldsymbol{u}_i$. Beeman and Alben (1977) have considered a model involving three different mechanisms contributing to the Raman scattering for tetrahedrally bonded materials:

$$\mathbf{E}_1 = \sum_{ij} [\boldsymbol{r}_{ij}\boldsymbol{r}_{ij} - \tfrac{1}{3}\mathbf{I}]\boldsymbol{u}_i \cdot \boldsymbol{r}_{ij} \qquad [4.38(a)]$$

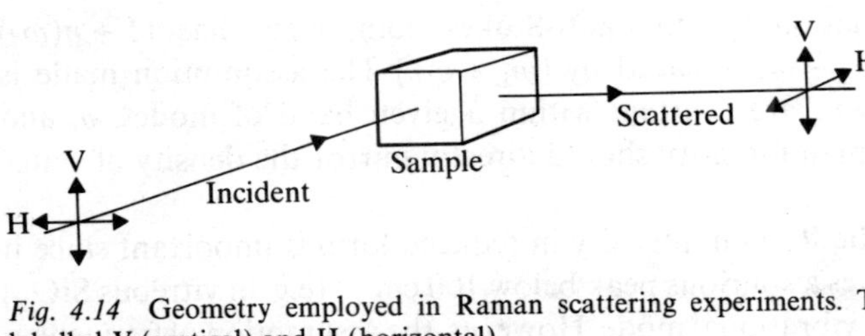

Fig. 4.14 Geometry employed in Raman scattering experiments. The polarization directions are labelled V (vertical) and H (horizontal).

$$\mathbf{E}_2 = \sum_{ij} \{\tfrac{1}{2}(\boldsymbol{r}_{ij}\boldsymbol{u}_i + \boldsymbol{u}_i\boldsymbol{r}_{ij}) - \tfrac{1}{3}\mathbf{I}\}\boldsymbol{u}_i \cdot \boldsymbol{r}_{ij} \qquad [4.38(b)]$$

$$\mathbf{E}_3 = \sum_{ij} \mathbf{I}\boldsymbol{u}_i \cdot \boldsymbol{r}_{ij} \qquad [4.38(c)]$$

where $\mathbf{I}$ is the unit dyadic. Note that [4.38(a)] and [4.38(c)] depend only on the bond compression C_{ij}, and the last two expressions vanish in the case of perfect tetrahedral symmetry ($\Sigma_{ij}\boldsymbol{r}_{ij} = 0$), although not necessarily for amorphous structures in which bond-angle distortion exists. The problem is that the relative contribution of each individual mechanism to the Raman activity is not known, although an estimate for the purposes of calculation may be made by noting that only [4.38(c)] gives a purely polarized spectrum. To let $R(\omega)$ represent Raman scattering in the equation-of-motion method, the choices for $s_{i\alpha}^{o}$ and $A_{i\alpha}$ are (Beeman and Alben (1977)):

$$s_{i\alpha}^{o} = \sum_{\gamma,\beta} \varepsilon_{\rm I}^{\beta}\varepsilon_{\rm s}^{\gamma} E_{i\alpha}^{\gamma\beta}/m_i \qquad [4.39(a)]$$

$$A_{i\alpha} = m_i s_{i\alpha}^{o} \qquad [4.39(b)]$$

where $\varepsilon_{\rm I}^{\beta}$ and $\varepsilon_{\rm s}^{\gamma}$ are components of the unit polarization vectors of the incident and scattered light respectively. The differential cross-section for Stokes-shifted Raman scattering of incident light of frequency $\omega_{\rm L}$ into scattered light of frequency $\omega_{\rm s} = \omega_{\rm L} - \omega$ is:

$$\frac{{\rm d}^2\sigma(\omega)}{{\rm d}\Omega\,{\rm d}\omega} = \frac{\omega_{\rm L}\omega_{\rm s}^3\hbar}{c^4}\frac{[n(\omega)+1]}{2\omega}R(\omega) \qquad [4.40]$$

where ${\rm d}\Omega$ is an element of solid angle, and $n(\omega)$ is the Bose occupation number:

$$n(\omega) = [\exp(\hbar\omega/k_{\rm B}T) - 1]^{-1} \qquad [4.41]$$

The use of [4.39]–[4.41], together with [4.26] enables the Raman scattering to be calculated for a given model and set of force constants.

In the reduction of *experimental* data, the scheme proposed by Shuker and Gammon (1970) is often used. They showed by consideration of the Fourier transform of the space–time autocorrelation function (related to fluctuations in the dielectric tensor) that the Stokes component of the *reduced* Raman intensity could be written as:

$$I_{\rm R}(\omega) = \frac{\omega\Sigma_{\rm b}c_{\rm b}\rho_{\rm b}(\omega)}{(1+n(\omega))(\omega_{\rm L}-\omega)^4} \qquad [4.42]$$

(The corresponding equation for the anti-Stokes component has $(1+n(\omega))$ replaced by $n(\omega)$ and $(\omega_{\rm L}-\omega)$ replaced by $(\omega_{\rm L}+\omega)$.) The assumption made is that the matrix elements $c_{\rm b}$ are *constant* within a given band of modes, b, and therefore the reduced Raman intensity should directly mirror the density of states $\rho(\omega)$.

The expression of the Raman intensity in reduced form is important since it thereby completely removes a spurious peak below 100 cm^{-1} (e.g. in vitreous SiO_2) originally interpreted as a vibrational mode. However, the assumption of frequency-independent matrix elements, and the division of the spectrum into separate bands due to different types of vibrational mode, is very poor for many materials. Different frequency dependences of the coupling constants $c_b(\omega)$ in [4.42] have been proposed to try and improve the fit to the experimental data compared with that given by the Shuker–Gammon formula (see Lannin 1977).

A connection with the structural use of Raman scattering (sect. 3.2.6) may be made for studies on glassy As_2S_3. We saw in sect. 4.2.1 that chalcogenide glasses, for which the chalcogen bond angle approaches 90°, exhibit vibrational modes characteristic of isolated molecular units. For these, there will be two bond-stretching modes, one antisymmetric and therefore IR-active, and the other symmetric and therefore Raman-active; these are shown in Fig. 4.15(a). For comparison, plotted in Fig. 4.15(b) is the frequency dependence of the depolarization ratio I_{HV}/I_{VV}; the higher frequency, more symmetric vibrations give more nearly polarized scattering as expected.

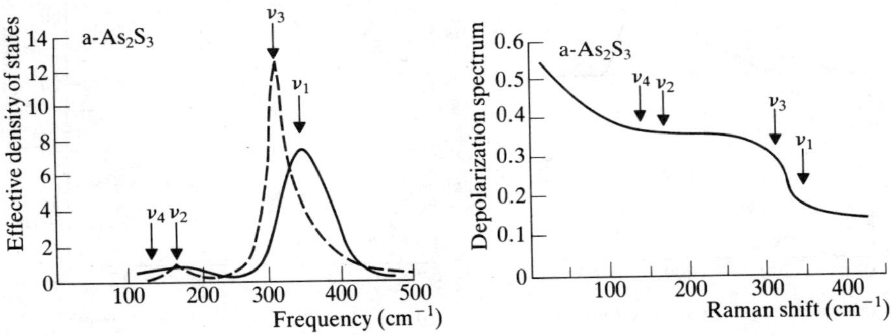

Fig. 4.15 (a) Comparison of the reduced Raman spectrum (solid line) of a-As_2S_3 with the IR absorption spectrum $\omega^2\varepsilon_2$ (dashed line) which also gives an estimate for the density of states. The arrows mark the normal mode frequencies of an AsS_3 molecule (Lucovsky 1972).
(b) Depolarization spectrum $p(\omega) = I_{HV}/I_{VV}$ of a-As_2S_3 (Kobliska and Solin 1973).

A further use of Raman spectra also has a structural connotation. With the tetrahedrally coordinated covalent materials a-Si and a-Ge, at least, the width (and, to a lesser extent, the position) of the highest frequency (TO-like) vibrational mode appears to be a function of the degree of (strain) disorder in the material. Both VH and fully polarized (HH−4VH/3) Raman spectra clearly show an increase in the width of the TO-like peak with increasing disorder (Maley *et al.* 1988) – see Fig. 4.16(a). Theoretical computations of the Raman spectra using the three bond-polarizability mechanisms given by [4.38] for a variety of CRN models for a-Si or Ge, each containing differing degrees of strain as monitored by the bond-angle fluctuations $\Delta\theta$, have revealed a linear relationship between the width of the TO peak and $\Delta\theta$ (Beeman *et al.* 1985, Maley *et al.* 1988):

$$\Gamma = a + b\,\Delta\theta \qquad [4.43]$$

where a and b are constants. The results of the most recent calculations by Maley *et al.* (1988) are shown in Fig. 4.16(b), and these indicate that $\Delta\theta \simeq 12°$ for the

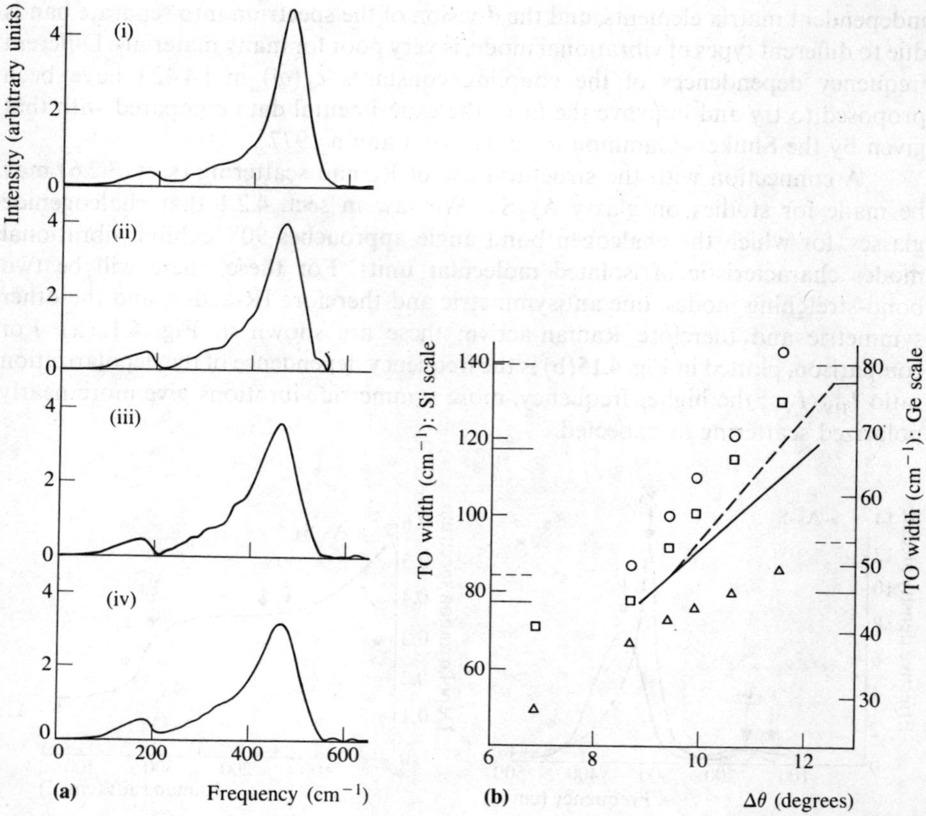

Fig. 4.16 (a) VH Raman spectra for a-Si samples, of which (i) is the most ordered and (iv) the most disordered.
(b) Theoretical TO Raman peak widths for Raman mechanisms 1 (Δ), 2 (○) and 3 (□), c.f. [4.38], as a function of bond-angle fluctuation $\Delta\theta$. Fits to the experimental data for a-Si and a-Ge are given by the solid and dashed lines, respectively, and the corresponding horizontal lines mark the minimum and maximum values of TO peak widths observed experimentally (Maley *et al.* 1988).

most disordered films of a-Ge or a-Si (i.e. sputtered onto low-temperature substrates), whereas the most highly ordered films (prepared on high-temperature substrates) are characterized by $\Delta\theta \simeq 9^\circ$.

One must be cautious when using features in the Raman spectrum to obtain quantitative information on, say, certain structural configurations giving rise to well-defined vibrational bands. In general, the matrix elements c_b are *not* frequency-independent or, in other words, matrix-element enhancement can occur for certain (symmetric) vibrational modes, which makes relative quantitative comparison of Raman intensities for different modes often rather uncertain. An extreme example of this effect is the case of v-B_2O_3, already mentioned in sect. 3.4.2. A very intense and narrow polarized Raman line is observed at 808 cm^{-1} (Fig. 3.38(a)), which is ascribed to the symmetric breathing mode of O atoms in regular six-membered (boroxol) rings. However, an inelastic neutron scattering determination of the vibrational density of states (see sect. 4.3.4) has revealed *no* pronounced feature at such a frequency (Hannon *et al.* 1988), indicating that the concentration of boroxol rings may be rather low.

4.3.4 Inelastic neutron scattering

Thermal neutrons typically have energies of the order 25 meV ($\simeq 200\ \mathrm{cm}^{-1}$) and can thus readily lose energy to or gain energy from phonons having frequencies in the range $\simeq 50$–$1500\ \mathrm{cm}^{-1}$. In contrast, X-rays have energies much greater than typical phonon energies, and inelastic events are therefore generally not important. An extensive discussion of the use of neutron scattering as a vibrational probe for glasses is given by Leadbetter (1973) and more recently by Sinclair (1985).

As we saw in the last chapter (sect. 3.2.2), the occurrence of inelastic scattering naturally means that the structure factor becomes frequency dependent, $S(\boldsymbol{Q}, \omega)$, and consequently the pair correlation function takes on a dynamic form, $G(\boldsymbol{r}, t)$ ([3.26] and [3.27]). The total correlation function can be divided into two parts, corresponding to self and distinct correlations, $G(\boldsymbol{r}, t) = G_s(\boldsymbol{r}, t) + G_D(\boldsymbol{r}, t)$. For experiments at high Q, the important contribution to $G_D(\boldsymbol{r}, t)$ is at small r, and here it tends to zero as only the origin atom is found at later times close to its position at $t = 0$, since all atoms vibrate about their quasi-equilibrium sites and do not approach closer than the interatomic distance r_o. The neglect of the distinct term in the expression for the neutron scattering cross-section (describing the interference of neutrons scattered by *different* atoms) is termed the 'incoherent approximation', and obtains when $Qr_o \gg 1$. The simplification that this approximation brings about is that the scattering becomes proportional simply to the phonon density of states; this is not true for coherent scattering in which only certain discrete energy (as well as momentum) transfers are allowed (analogous to Bragg scattering).

The scattering cross-section in the incoherent approximation may be written for neutron energy loss (Leadbetter 1973) as:

$$\frac{\mathrm{d}^2\sigma_{\mathrm{inc}}}{\mathrm{d}\Omega\,\mathrm{d}E'} = \frac{k'}{k}\sum_{\mathrm{j}} \delta(\omega - \omega_{\mathrm{j}})\frac{(n(\omega)+1)}{2\omega_{\mathrm{j}}}\left[\frac{\langle b^2\rangle}{M}|\boldsymbol{Q}\cdot\boldsymbol{u}_{\mathrm{j}}|^2 e^{-2W}\right] \qquad [4.44]$$

where primed and unprimed terms (E and k) refer to scattered and incident quantities, respectively, $n(\omega)$ is the Bose occupation number ([4.41]), M is the atomic mass, $\boldsymbol{u}_{\mathrm{j}}$ is the eigenvector of the jth vibrational mode and W is the Debye–Waller factor ([3.33]). The summation is over individual modes j, and [4.44] has been written for one atom type only. For multicomponent systems, all quantities within square brackets become atom-type dependent and a further sum must be taken over all atoms within a unit of composition (e.g. SiO_2). Equation 4.44 may be simplified further by assuming that chemically equivalent atoms are dynamically equivalent. Averaging over all orientations of $\boldsymbol{Q}$ with respect to $\boldsymbol{u}$ gives $\langle|\boldsymbol{Q}\cdot\boldsymbol{u}_{\mathrm{j}}|^2\rangle = \frac{1}{3}Q^2\langle u^2\rangle$, where the mean-square vibrational amplitude $\langle u^2\rangle$ is in fact frequency and atom-type dependent. The summation of the delta functions in [4.44] yields the one-phonon density of states,

$$\sum_{\mathrm{j}}\frac{(\omega - \omega_{\mathrm{j}})}{\omega_{\mathrm{j}}} = \frac{3N}{\omega}\rho(\omega) \qquad [4.45]$$

where $3N$ is the total number of phonon states. Thus, [4.44] becomes:

$$\frac{d^2\sigma_{inc}}{d\Omega\, dE'} = \frac{k'}{k} \frac{Q^2(n(\omega)+1)N\rho(\omega)}{2\omega}\left[\frac{\langle b^2\rangle}{M}\langle u^2\rangle e^{-2W}\right] \qquad [4.46]$$

A function $g(Q, \omega)$ may be defined in terms of experimentally known quantities:

$$g(Q, \omega) = \frac{d^2\sigma}{d\Omega\, dE'} \frac{k}{k'} \frac{2\omega}{N(n(\omega)+1)Q^2} \qquad [4.47]$$

Coherence effects are manifested by a Q-dependence of $g(Q, \omega)$, but if the incoherent approximation is valid, $g(Q, \omega)$ is equal to the density of states but weighted by the frequency dependent term $[(\langle b^2\rangle/M)(\langle u^2(\omega)\rangle e^{-2W})]$ (for each atom in the composition unit, if appropriate).

Obviously, interpretation is simplest for monatomic systems, and an early example of inelastic neutron results is shown in Fig. 4.17 for the case of amorphous arsenic and two of the crystalline modifications, orthorhombic and rhombohedral (Leadbetter *et al.* 1976). It can be seen that all forms of As have well-separated acoustic and optic bands, and in addition the amorphous and orthorhombic forms exhibit a distinct dip in the lower acoustic band, a feature not shared by the rhombohedral form; on the basis of recursion method calculations, Davis *et al.* (1979) ascribe the filling in of this dip to a predominance of bonds in the staggered configuration ($\phi = 180°$) in the rhombohedral polymorph.

More recently, a number of inelastic neutron scattering experiments probing the phonon density of states have been performed on a variety of amorphous systems, both covalent and inorganic. Some examples have already been mentioned

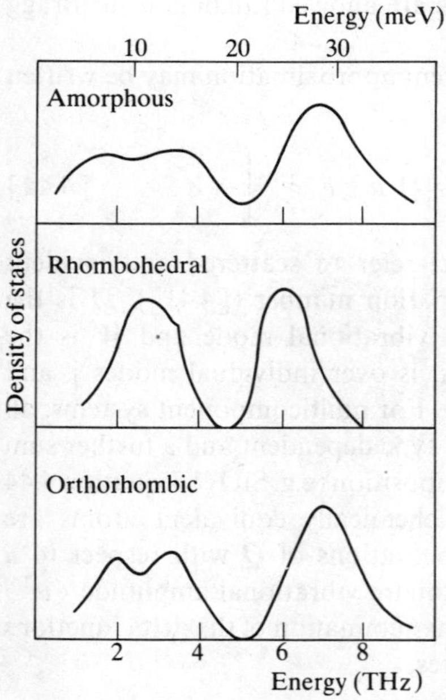

Fig. 4.17 Densities of states obtained from inelastic neutron scattering data for various structural modifications of arsenic (Leadbetter *et al.* 1976).

in passing in this chapter, namely a-Si shown in Fig. 4.3 (Kamitakahara *et al.* 1984, 1987), a-Ge shown in Fig. 4.11(b) (Maley *et al.* 1988) and v-SiO_2 shown in Fig. 4.13 (Price and Carpenter 1987). In addition, inelastic neutron scattering results have been obtained on other AX_2-type glasses, namely v-$GeSe_2$ (Walter *et al.* 1988, Gladden *et al.* 1988) and v-$SiSe_2$ (Arai *et al.* 1988), as well as B_2O_3 (Hannon *et al.* 1988) and P-Se glasses (Verrall and Elliott (1988), and several metallic glass materials, including simple-metal alloys such as $Ca_{70}Mg_{30}$, transition-metal alloys such as $Cu_{46}Zr_{54}$, and a transition-metal–metalloid alloy, $Pd_{80}Si_{20}$ (see Suck and Rudin (1983) for a review).

In many instances when just the VDOS $\rho(\omega)$ is required, the incoherent approximation is invoked, and $S(Q, \omega)$ is averaged over a range of Q values, leaving the quantity $Q^2 e^{-2W}$ as the only Q-dependent term in [4.46]. In such cases, features in $\rho(\omega)$ are often used to infer some information about the atomic-scale structure, in terms of either their shape or the position of peaks in $\rho(\omega)$. For instance, inelastic neutron scattering has been used in several instances to try to ascertain quantitatively the relative contribution of various vibrational modes to certain features in Raman spectra or, in other words, to investigate indirectly Raman matrix-element effects. A study has been made of the 808 cm^{-1} Raman band of v-B_2O_3 (Hannon *et al.* 1988), ascribed to the breathing mode of boroxol rings (see sect. 3.4.2), and also of the 'companion' line to the main A_1 breathing mode of AX_4 tetrahedra observed in AX_2-type chalcogenide glasses such as $GeSe_2$ (Gladden *et al.* 1988) – sect. 3.4.2. However, such studies are hampered by the fact that the energy resolution of Raman spectrometers is considerably greater than that currently available from inelastic neutron scattering spectrometers.

Considerably greater (structural) information is contained, of course, in the full dynamical structure factor, $S(Q, \omega)$, i.e. in the additional Q-dependence (resulting from coherence effects) which is averaged out in the incoherent approximation; some recent inelastic neutron scattering studies have started to address this point. We now consider these coherence effects, i.e. the influence of correlated atomic vibrational motions on the Q-dependence of inelastic neutron scattering spectra, for relatively large energies (frequencies) corresponding primarily to optic phonons, leaving the low-frequency (acoustic) behaviour for discussion mainly in sect. 4.5.

As examples of such coherence effects, we show in Fig. 4.18 $S(Q, \omega)$ for two representative covalent amorphous systems: v-SiO_2 (Price and Carpenter 1987) and a-Ge (Maley *et al.* 1988). The oscillations as a function of Q for a fixed value of energy can be clearly seen when the dominant elastic ($\omega = 0$) contribution is subtracted. In order to understand properly the origin of such oscillatory behaviour, it is necessary to calculate the 'matrix-element' terms ($\boldsymbol{Q} \cdot \boldsymbol{u}_j$) in [4.44], knowing the individual eigenvector displacements $\boldsymbol{u}_j$ for each vibrational mode, computed from, say, a structural model (see Carpenter and Pelizzari 1975 for mathematical details).

To illustrate such an approach, we consider v-SiO_2, whose dynamic structure factor is shown in Fig. 4.18(a). The simplest force law describing such a material is the 'central-force' model discussed in sect. 4.2.1. As mentioned there, this model predicts two delta-function modes for AX_2 materials corresponding to symmetric and asymmetric motions at frequencies ω_3 and ω_4, respectively; using the

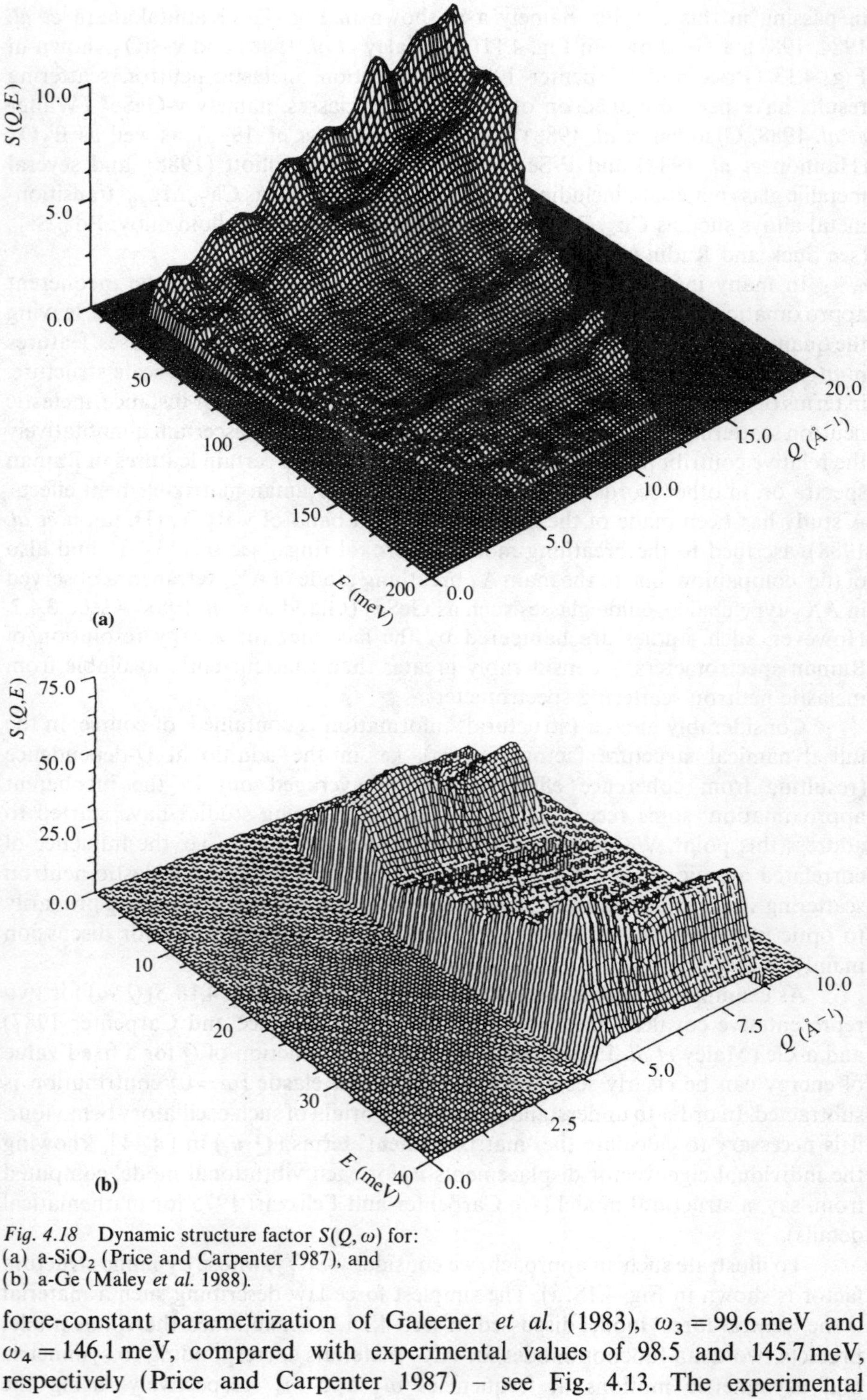

Fig. 4.18 Dynamic structure factor $S(Q, \omega)$ for:
(a) a-SiO_2 (Price and Carpenter 1987), and
(b) a-Ge (Maley *et al.* 1988).

force-constant parametrization of Galeener *et al.* (1983), $\omega_3 = 99.6$ meV and $\omega_4 = 146.1$ meV, compared with experimental values of 98.5 and 145.7 meV, respectively (Price and Carpenter 1987) – see Fig. 4.13. The experimental

Q-dependence obtained from $S(Q, \omega)$ (Fig. 4.18(a)) at these two vibrational frequencies is shown in Fig. 4.19(a), together with that for the main oxygen rocking band at 49.2 meV; the Q-dependences calculated using the Sen–Thorpe central-force model are shown in Fig. 4.19(b). It can be seen that there is a qualitative resemblance, with oscillations being superimposed on the smoothly varying $Q^2 e^{-2W}$ background. However, quantitatively, there are discrepancies. For instance, the oscillations in Q are predicted to be more pronounced for the lower-frequency mode, whereas the converse is observed experimentally. Furthermore, the experimental data indicate the existence of periods of oscillation extending down to 1.6–1.8 Å^{-1}, indicating that dynamic correlations extend beyond second-nearest-neighbour Si atoms, for which a period of 2.15 Å^{-1} ($= 2\pi/\Delta r_{Si}$) would be predicted (Price and Carpenter 1987).

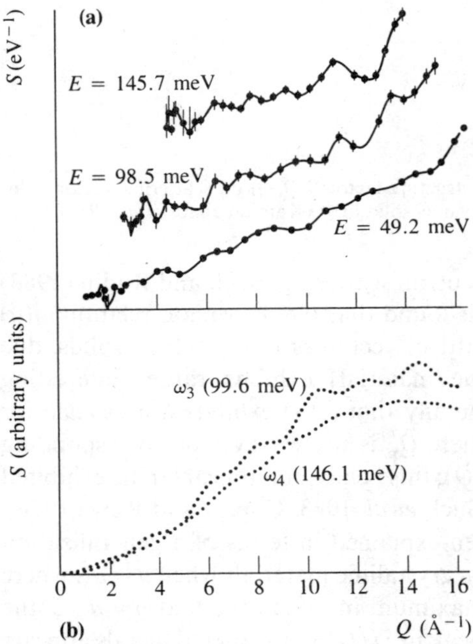

Fig. 4.19 Dependence of the dynamic structure factor $S(Q, \omega)$ on scattering vector Q for v-SiO_2 at various energies:
(a) measured experimentally;
(b) calculated using the Sen–Thorpe central-force model (Price and Carpenter 1987).

Somewhat more detailed calculations of the Q-dependence of the inelastic neutron scattering intensity have been performed for a-Si (Kamitakahara *et al.* 1987) using the CRN model of Wooten *et al.* (1985), together with the Keating potential. The calculated Q-dependences for several frequencies are shown in Fig. 4.20 with the experimental data for comparison; it can be seen that the agreement is reasonably satisfactory, and much better than when just *nearest-neighbour* dynamic correlations are considered (Maley *et al.* 1988).

Finally in this section we discuss briefly the behaviour of collective vibrational excitations (phonon-like modes) in glassy metals. Longitudinal (but not transverse)

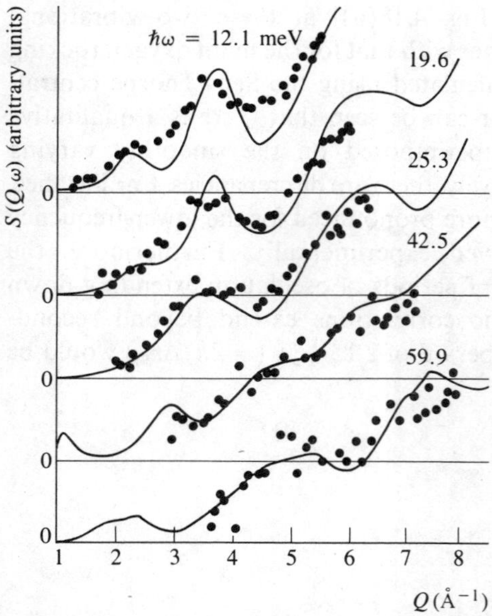

Fig. 4.20 Measured dependence of the dynamic structure factor $S(Q, \omega)$ on scattering vector Q for a-Si at various energies. Calculated curves are shown as solid lines (Kamitakahara *et al.* 1987).

excitations can be detected by inelastic neutron scattering; Suck and Rudin (1983) give a review of experimental results. It is found that the dispersion relation $\omega(q)$ (where q is a scalar, not a (reciprocal-lattice) vector as in crystalline solids, due to spatial isotropy in a non-crystalline material) exhibits rather interesting behaviour. In particular, it is found generally that $\omega(q)$ exhibits a *maximum* for $q \simeq Q_p/2$, and a *minimum* for $q \simeq Q_p$, where Q_p is the wavevector corresponding to the first peak in the structure factor, $S(Q)$; this behaviour appears to be exhibited by both amorphous and liquid metals (Suck *et al.* 1983, Copley and Rowe 1974). The minimum in $\omega(q)$ at $q \simeq Q_p$ has been explained in terms of a generalization of the 'umklapp minimum', well-known in crystalline materials when $q = q_o$, where q_o is a reciprocal lattice vector, and a maximum in $\omega(q)$ occurs at $q = q_o/2$, the Brillouin-zone boundary. Thus, the first peak in $S(Q)$ for disordered metals appears to act as a quasi-Bragg peak, Q_p corresponding to a reciprocal-lattice vector. Several theoretical treatments of this problem have been given: Hafner (1983a,b) has calculated the vibrational behaviour of the simple-metal glassy system $Ca_{70}Mg_{30}$ using the recursion method (sect. 4.2.5), and Wang and Overhauser (1988) have used a generalization of the method conventionally used for crystals in which they replace a sum over reciprocal space by an integration weighted by the static structure factor $S(Q)$. In both cases, the general observed behaviour of $\omega(q)$ is predicted for longitudinal excitations; the result for glassy $Ca_{70}Mg_{30}$ (Hafner 1983a) is shown in Fig. 4.21. Several features are worthy of note. There is very reasonable agreement between the experimental data, which clearly show a minimum in $\omega(q)$ at $q \simeq 2.1\ \text{Å}^{-1}$ (cf. the first peak in $S(Q)$, $Q_p = 2.13\ \text{Å}^{-1}$ for this material) and the theoretical points obtained from the position of the peak in $S(Q, \omega)$ (which can only be calculated at particular Q-values governed by the

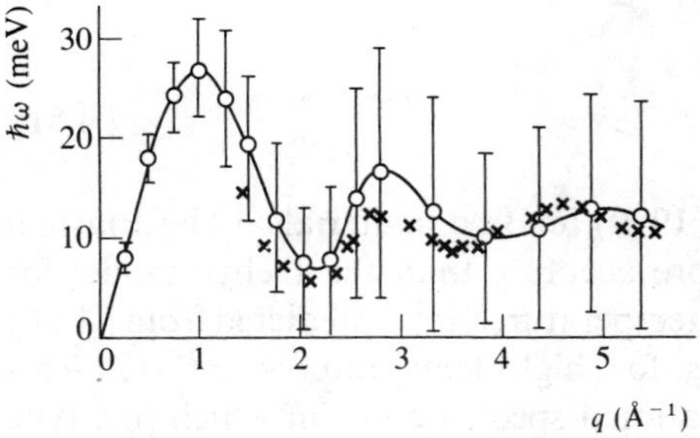

Fig. 4.21 Dispersion law, $\omega(q)$, for propagating collective excitations in a-$Ca_{70}Mg_{30}$ (Hafner 1983). The calculated dispersion is given by the solid line and the circle data points; the experimentally observed dispersion is given by the crosses.

periodic boundary conditions imposed on the DRP-like model used). Note also that the uncertainty in the $\omega(q)$ values, as measured by the width at half-maximum in the peak in $S(Q, \omega)$, increases markedly with increasing q, as shown schematically in Fig. 4.1.

4.3.5 Extended X-ray absorption fine structure

An interesting recent application of EXAFS has been in the investigation of *vibrational* characteristics of amorphous materials, in addition to its use in obtaining static structural characteristics as described in sect. 3.2.3. The vibrational information is obtained by measuring the temperature dependence of the Debye–Waller factor σ^2. In EXAFS, the Debye–Waller factor is given by the mean-square relative displacement of the absorbing atom j relative to its near neighbours k *along* the direction joining them:

$$\sigma^2 = \langle [(\boldsymbol{u}_j - \boldsymbol{u}_k) \cdot \hat{\boldsymbol{R}}_{jk}]^2 \rangle \qquad [4.48]$$

where $\boldsymbol{u}_i$ is the displacement vector of an atom at site i, $\hat{\boldsymbol{R}}_{jk}$ is a unit vector in the direction of the vector joining atoms j and k, and $\langle \ \rangle$ denotes a thermal average. The mean-square relative displacement is determined by the local projected vibrational density of states $\rho_p(\omega)$ describing only those modes with displacements along interatomic directions j-k via the relation (Lottici 1987):

$$\sigma^2 = \frac{\hbar}{2\mu} \int \frac{\rho_p(\omega)}{\omega} \coth(\hbar\omega/2k_B T)\, d\omega \qquad [4.49]$$

where μ is the reduced mass for the pair of atoms (j, k) and the projected VDOS is given by

$$\rho_p(\omega) = \sum_i (\hat{\boldsymbol{\varepsilon}}_i \cdot \hat{\boldsymbol{R}}_{jk})^2 \delta(\omega - \omega_i) \qquad [4.50]$$

where $\hat{\boldsymbol{\varepsilon}}_i$ is the eigenvector of the ith vibrational mode with eigenfrequency ω_i. In the Einstein approximation, where the vibrational behaviour is characterized by a single frequency ω_E, [4.49] becomes

$$\sigma^2 = \frac{\hbar}{2\mu\omega_E} \coth(\hbar\omega_E/2k_B T) \qquad [4.51]$$

a relation first given by Sevillano *et al.* (1979) for f.c.c. materials. (The Einstein approximation has been found to be more accurate than the Debye model for first-shell vibrational correlations.) At low temperatures, σ^2 is predicted from [4.51] to be temperature-independent, whereas for high temperatures, $\sigma^2 \propto T$. This technique is therefore a *site-selective* vibrational spectroscopy, in which one type of atom is selected as the X-ray absorbing central site, and its immediate nearest neighbours can be identified from the EXAFS data.

The temperature dependence of the Debye–Waller factor can be extracted from the EXAFS data either by fitting procedures as described in sect. 3.2.3, or by graphical means. In the plane-wave approximation, the EXAFS amplitude χ is given by [3.65], and when neither bond lengths nor coordination numbers vary, the ratio of χ in two given states (e.g. at two temperatures) is given by

$$\ln(\chi(T_1)/\chi(T_2)) = 2k^2(\sigma^2(T_2) - \sigma^2(T_1)) \qquad [4.52]$$

Thus, a plot of the logarithm of the ratio of EXAFS amplitudes against k^2 should yield a straight line whose slope gives the difference in Debye–Waller factors.

This technique has been used by Lottici (1987) and Yang *et al.* (1987) in the study of amorphous and crystalline As and As_2S_3, respectively. It was found that for temperatures below about 100 K, σ^2 was temperature-independent, and linearly dependent on T at higher temperatures, in agreement with [4.51]. The values of ω_E so obtained agree to within 5% with the values of the symmetrical stretching Raman frequencies.

4.4 Low-frequency modes and low-temperature properties

4.4.1 Introduction

It is a striking feature of glasses that, at low temperatures, they exhibit markedly different behaviour from their crystalline counterparts in phonon-related properties, such as specific heat capacity, thermal conductivity, acoustic and dielectric absorption. The purpose of this section is to describe these interesting phenomena under the one umbrella, and to show how far these low-temperature properties may be understood in terms of a simple phenomenological model.

The thermal properties of *crystalline* solids at low temperatures are well understood in terms of Debye's theory in which, assuming the distribution of phonons is cut off at some frequency with which a characteristic temperature θ_D may be associated, the famous T^3 temperature dependence of the heat capacity, as T tends to zero, results:

$$C_v = \frac{234Nk_B T^3}{\rho\theta_D^3} = \frac{2\pi^2}{5}\frac{k_B^4 T^3}{\hbar^3 \rho v_D^3} \qquad [4.53]$$

where ρ is the mass density, N is the atomic number density and v_D is given by a weighted average of longitudinal and transverse sound velocities:

$$\frac{3}{v_D^3} = \frac{1}{v_L^3} + \frac{2}{v_T^3} \qquad [4.54]$$

One would expect that at low temperatures, as the phonon mean free path and wavelength increased, the structural disorder characteristic of a glass would become unimportant, and in this continuum approximation the thermal properties should be similar to those of crystals, i.e. Debye-like. As Zeller and Pohl (1971) first found, this is far from the truth. Figures 4.22(a) and (b) show the specific heat and thermal conductivity of vitreous silica as a function of temperature compared with data for crystalline silica (α-quartz). The specific heat of the glass decreases much more slowly with temperature than the Debye T^3 prediction at low temperatures (for the temperature range $0.1\text{ K} < T < 1\text{ K}$) and can be expressed instead in the form:

$$C_v = aT + bT^3 \qquad [4.55]$$

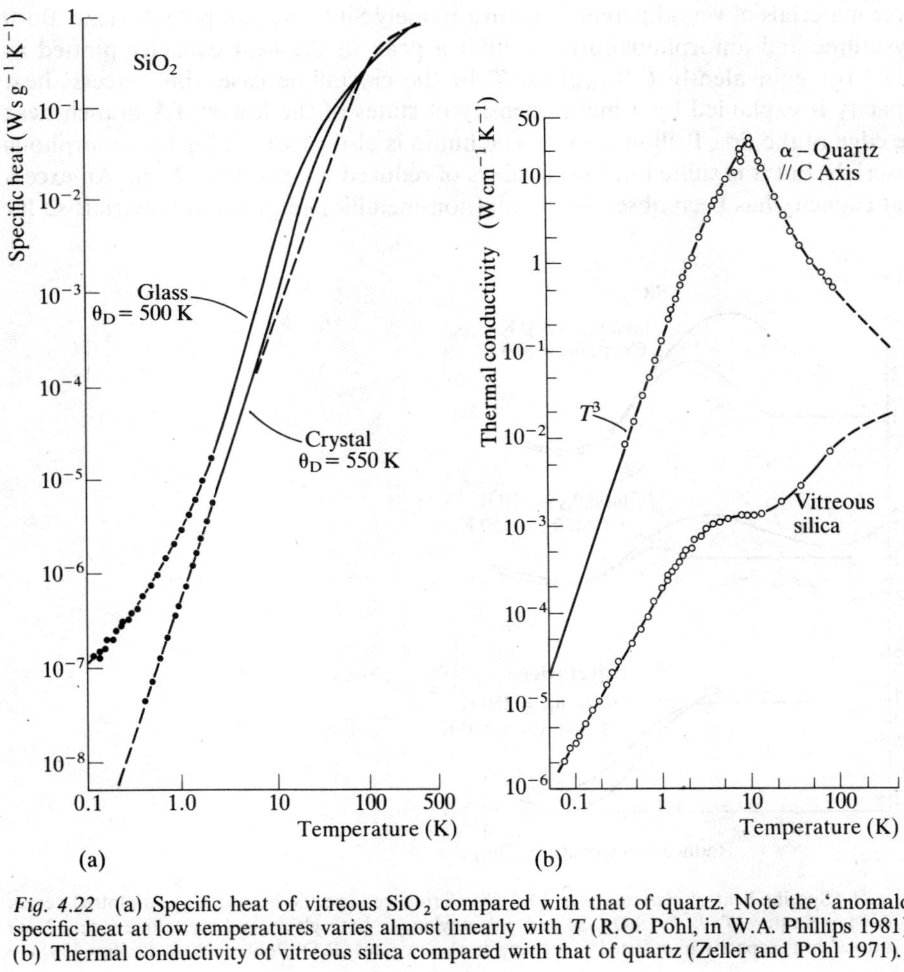

Fig. 4.22 (a) Specific heat of vitreous SiO_2 compared with that of quartz. Note the 'anomalous' specific heat at low temperatures varies almost linearly with T (R.O. Pohl, in W.A. Phillips 1981). (b) Thermal conductivity of vitreous silica compared with that of quartz (Zeller and Pohl 1971).

where b exceeds the Debye prediction ([4.53]). The thermal conductivity is also markedly different (Fig. 4.22(b)) being a monotonically decreasing function of temperature for the glass, in contrast to the peaked behaviour characteristic of crystals.

The remaining parts of this section discuss the low-temperature behaviour of individual phonon-related properties in detail, as well as the study of low-frequency modes by inelastic neutron scattering, finishing with a discussion of a model which goes a long way to explaining these anomalies for glasses. An extensive review of these phenomena is given in the book edited by W. A. Phillips (1981).

4.4.2 Specific heat

The behaviour of the specific heat, in common with other thermal properties, can conveniently be separated into two temperature regimes, above and below 2 K, since theories for the former are often extensions of existing (crystalline) models whereas those describing the latter case are distinct and novel (Phillips 1978).

Dealing first with the case of $T > 2$ K, Fig. 4.23 shows typical behaviour for three materials of very different structure, namely SiO_2, Se and polyethylene. Both crystalline and amorphous forms exhibit a peak in the heat capacity plotted as C/T^3 (or equivalently C/C_{Deb}) vs. T. In the crystalline case, this 'excess' heat capacity is explained by a higher density of states of the lowest TA branch near the edge of the first Brillouin zone. The hump is also observed for the amorphous materials, but it is shifted to lower values of reduced temperature T/θ_D. An excess heat capacity has been observed for all (non-metallic) amorphous materials so far

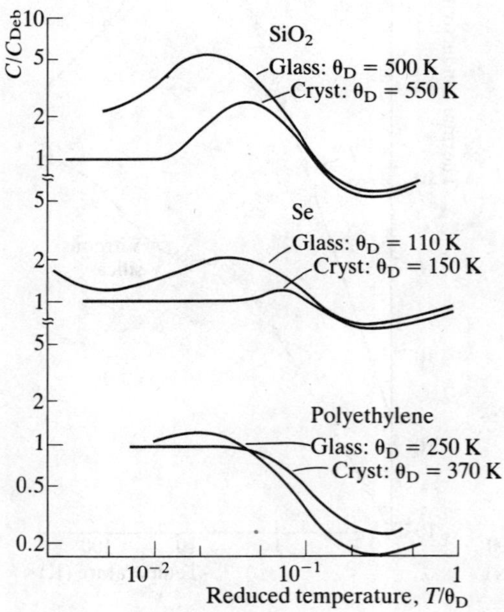

Fig. 4.23 Specific heat of glasses above 2 K. The normalized specific heat C/C_{Deb} is plotted versus reduced temperature T/θ_D for SiO_2, Se and polyethylene in both glassy and crystalline forms. Note the 'excess' heat capacity manifested as a hump in these curves (R.O. Pohl, in W.A. Phillips 1981).

studied, and the peak generally occurs at a different value of T/θ_D to that of the corresponding crystal. The cause of the shift is not generally known, but for v-SiO_2 it has been ascribed to the coupled rotational motions of SiO_4 tetrahedra (see later).

Another feature of note in Fig. 4.23 is the observation that the curves of C/C_{Deb} vs. T do not tend to unity, the Debye limit, for the amorphous materials for $T<2$ K. Instead, as was seen in Fig. 4.22(a), the heat capacity decreases more slowly than the Debye T^3 law, and can be approximated well by [4.55]; a plot of C/T vs. T^2 should plot linearly according to this equation, giving an intercept C_1 at $T=0$. The 'anomalous linear specific heat' has been observed in a wide variety of amorphous materials, silica and silicates, chalcogenide glasses, various organic polymers and even epoxy resins. An example of a non-crystalline insulator in which the linear term is very small is bulk amorphous arsenic (Jones *et al.* 1978). This, and amorphous Ge films which behave similarly, are interesting since they do not possess any flexible structural units, such as the bridging oxygen in SiO_2. The linear term in the specific heat appears to be predominant in disordered materials containing low coordinated atoms. Note that we have said *disordered* rather than amorphous; disordered (e.g. neutron irradiated) *crystals* have also been found to have an anomalous specific heat, although perfect crystals do not. We might add at this juncture that [4.55] is often only approximately obeyed; in practice, deviations from strict linearity of the anomalous term are often found (Fig. 4.24).

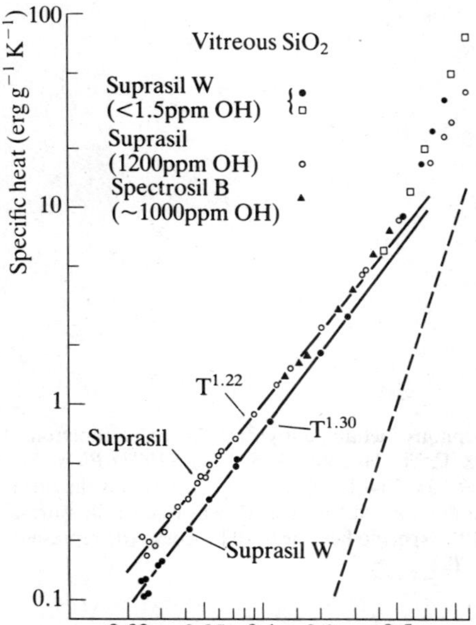

Fig. 4.24 Specific heat of a-SiO_2 above 25 mK. Suprasil and Spectrosil B have large OH concentrations but small metal concentrations. Suprasil W has small OH and metal concentrations, but 230 p.p.m. Cl and 290 p.p.m. F. The dashed line is the Debye specific heat, $C_{Deb}=8T^3 \text{erg g}^{-1} \text{K}^{-4}$ (R.O. Pohl, in W.A. Phillips 1981).

So far we have addressed the case of amorphous insulators – what of glassy metals? The problem here is that the lattice vibrational contribution to the specific heat is masked by the electronic contribution which also depends linearly on T. The results of an elegant experiment to eliminate the electronic effect in order to reveal any possible anomalous specific heat are shown in Fig. 4.25. Glassy $Zr_{70}Pd_{30}$, prepared in ribbon form by melt extraction, is a superconductor below $T_c = 2.53$ K. Thus, below T_c no unpaired normal electrons exist, and hence the electronic contribution to the specific heat is removed. It is found that below T_c this material *does* appear to exhibit an anomalous linear specific heat as given by [4.55], with the value of the constant 'a' very similar to those found in insulating glasses.

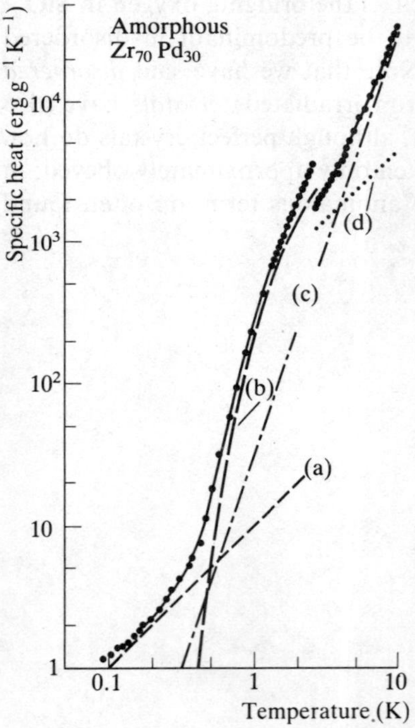

Fig. 4.25 Specific heat of the superconducting amorphous metallic alloy $Zr_{70}Pd_{30}$ as a function of temperature (J.E. Graebner, B. Golding, R.J. Schutz, F.S.L. Hsu and H.S. Chen (1977) *Phys. Rev. Lett.* **39**, 1480). The superconducting transition at $T_c = 2.53$ K is clearly evident. Curve (a) is the linear term in the specific heat; curve (b) is the exponential decrease of the specific heat due to the normal electrons below T_c; curve (c) is a T^3 extrapolation of the specific heat near 10 K; curve (d) represents the electronic contribution to the specific heat above T_c.

In summary, it appears that an anomalous linear term in the specific heat is a characteristic feature of most glass-forming solids, with the constant 'a' being in the range 1–5 J g^{-1} K^{-1}. However, at very low temperatures, strict linearity is often not observed and the magnitude of the specific heat is often sample dependent (see Fig. 4.24).

4.4.3 Thermal conductivity

Thermal transport in crystalline non-metallic materials is by means of propagating phonons; the thermal conductivity may be written as:

$$\kappa = \tfrac{1}{3} \sum_i \int_0^{\omega_D} C_i(\omega) v_i(\omega) \ell_i(\omega)\, d\omega \qquad [4.56]$$

where the sum is over the three phonon modes, v and ℓ are the phonon velocity and mean free path, and $C(\omega)$ is the contribution to the lattice specific heat by phonons of frequency ω. The temperature dependence of κ for a crystal is typically peaked (Fig. 4.22(b)). Above 10 K, phonons are scattered by intrinsic (phonon–phonon umklapp) processes, and hence ℓ decreases as T increases since the number of phonons increases. At lower temperatures, the mean free path ℓ eventually becomes comparable to the sample dimensions and is hence constant, and κ follows the specific heat and therefore behaves as T^3.

In contrast, the behaviour of amorphous solids is completely different (Fig. 4.22(b)). No peak in κ is observed, but instead a plateau region occurs at $\simeq 10$ K, and the low-temperature region behaves approximately as T^2, rather than T^3. The cause of the plateau region (B in Fig. 4.26(a)) is not well understood; however a rapid decrease in the phonon mean free path with frequency, as shown schematically in Fig. 4.26(b), does account for this feature although its cause is not understood. Region A of κ vs. T is accounted for by a constant ℓ at high

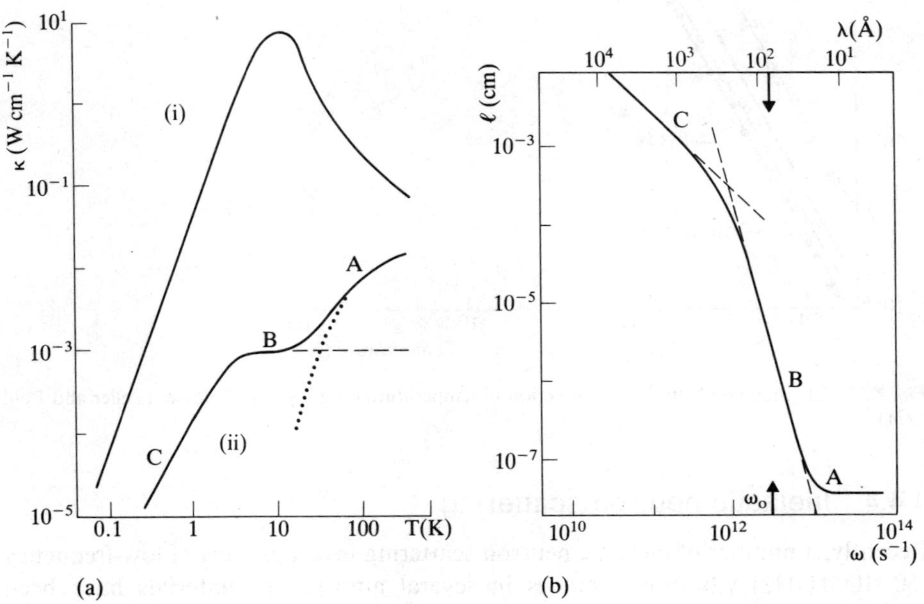

Fig. 4.26 (a) Schematic illustration of the temperature dependence of the thermal conductivity of (i) crystalline quartz and (ii) vitreous silica. Note the plateau region B in the curve for the glass. (b) Frequency dependence of the intrinsic phonon mean free path proposed to account for the κ vs. T, curve (ii), in (a). Region C is drawn with a slope -1, and region B at ω_0 is drawn with a slope -4; the whole curve is quantitatively appropriate for a borosilicate glass (A. C. Anderson, in W. A. Phillips 1981).

frequencies. The T^2 temperature dependence of κ is explained if ℓ has a T^{-1} temperature dependence (cf. [4.56]); Zeller and Pohl (1971) first suggested that this behaviour might be associated with the anomalous linear specific heat, and this is discussed in more detail in the next section.

We note finally that values of the temperature exponent of κ in region C often differ from two, values in the range 1.8–2 commonly being found. Furthermore, the thermal conductivities of amorphous materials in the T^2 region are often very similar (in contrast to the linear specific heat), as can be seen from Fig. 4.27, and may therefore be more of an 'intrinsic' property and less dependent on the impurity content.

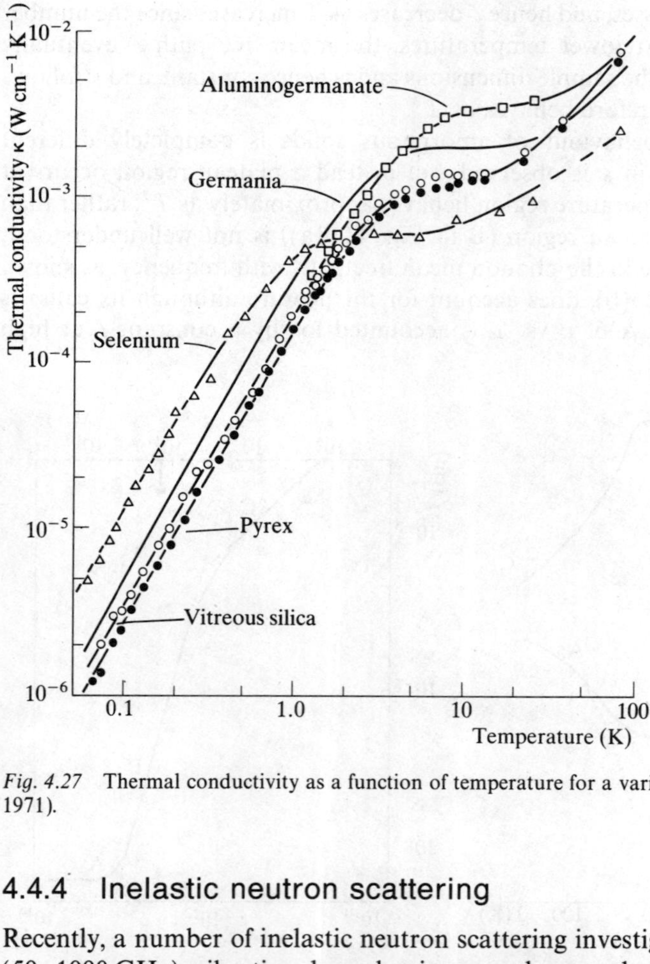

Fig. 4.27 Thermal conductivity as a function of temperature for a variety of glasses (Zeller and Pohl 1971).

4.4.4 Inelastic neutron scattering

Recently, a number of inelastic neutron scattering investigations of low-frequency (50–1000 GHz) vibrational modes in several amorphous materials have been performed; such experiments can given information on the character of these modes, and have shed much light on the origin of the low-temperature (2–50 K) behaviour discussed above. However, most investigations have concentrated on vitreous SiO_2, the archetypal inorganic glass-forming system, so here we confine the discussion to this material.

For frequencies below, say, 1 THz, five possible types of low-lying vibrational excitations could exist in amorphous solids. In addition to harmonic sound waves, which are the only such modes generally occurring in crystals at such low frequencies, additional harmonic vibrational excitations can also exist, together with anharmonic relaxations and two-level (or tunnelling) states at yet lower frequencies (Buchenau 1987); the latter two types of excitation will be discussed further in sect. 4.4.5. Furthermore, vibrational modes for frequencies down to $\simeq 1$ THz can exist as 'fractons' (Orbach 1986), or excitations on a fractal lattice (see sect. 3.6.2); this special type of vibrational excitation will be discussed more thoroughly in sect. 4.4.6. Inelastic neutron scattering in principle is able to distinguish between these various possibilities: the temperature dependence of the scattering intensity is sensitive to the degree of anharmonicity, whilst the Q-dependence provides information on the eigenvectors of the modes.

Buchenau (1985) has reformulated the expression for the inelastic neutron scattering intensity obtained by Carpenter and Pelizzari (1975b) for scattering from sound waves, extending the treatment to higher frequencies where the modes can no longer be described as plane waves, but where coherence effects are nevertheless still appreciable, i.e. where neighbouring atoms still vibrate approximately in phase with each other. The advantage of this treatment is that the density of vibrational states can be obtained without the traditional separation into coherent and incoherent contributions. The one-phonon coherent scattering differential cross-section is then given by (Buchenau 1985, Buchenau *et al.* 1986)

$$\frac{d^2\sigma^{(1)}}{d\Omega\, d\nu} = \frac{\boldsymbol{k}_f}{\boldsymbol{k}_i}\frac{3\hbar N}{2}\frac{\rho(\omega)}{\omega} n(\omega) I^{(1)}(\boldsymbol{Q}) \qquad [4.57]$$

(cf. [4.44]), where $\boldsymbol{Q} = \boldsymbol{k}_f - \boldsymbol{k}_i$ and $\boldsymbol{k}_f$ and $\boldsymbol{k}_i$ are the scattered and incident neutron wavevectors, respectively, $n(\omega)$ is the Bose factor ([4.41]), N is the number of atoms, and

$$I^{(1)}(\boldsymbol{Q}) = \left| \sum_j b_j e^{-W_j} e^{i\boldsymbol{Q}\cdot\boldsymbol{R}_j} (\boldsymbol{Q}\cdot\boldsymbol{e}_j)/M_j^{1/2} \right|^2 \qquad [4.58]$$

where b_j is the coherent scattering length, W_j the Debye–Waller factor, M_j the mass, $\boldsymbol{R}_j$ the position vector and $\boldsymbol{e}_j$ the displacement eigenvector of atom j. (The actual displacement vectors are given by $\boldsymbol{u}_j = \boldsymbol{e}_j M_j^{-1/2}$, with the normalization condition $\Sigma_j \boldsymbol{e}_j^2 = 1$.)

With sound waves, where the atomic pair correlations remain essentially unchanged by the vibrations, oscillations with Q in the scattering cross-section occur about the background $Q^2 e^{-W}$ in phase with those in the *elastic* structure factor $S_0(Q) \equiv S(Q, O) = S(Q)e^{-2W}$ (Buchenau 1987). In other words, [4.57] becomes (Buchenau *et al.* 1986)

$$\frac{d^2\sigma^{(1)}}{d\Omega\, d\omega} = \frac{\boldsymbol{k}_f}{\boldsymbol{k}_i}\frac{\hbar Q^2}{2M_{av}} n(\omega) \frac{\rho_D(\omega)}{\omega} I^{(0)}(Q) \qquad [4.59]$$

where $\rho_D(\omega)$ is the Debye-like density of states and

$$I^{(0)}(Q) = \left| \sum_j b_j e^{-W_j} e^{i\boldsymbol{Q}\cdot\boldsymbol{R}_j} \right|^2 \qquad [4.60]$$

$I^{(0)}(Q)$ is therefore proportional to $S_0(Q)$. For v-SiO_2, the first peak in $S_0(Q)$ occurs at $Q \simeq 1.6\ \text{Å}^{-1}$ (Fig. 4.28(a)), and indeed a small feature does appear in the experimental inelastic scattering at this Q-value (Buchenau *et al.* 1984) – see Fig. 4.28(b). However, it is clearly apparent from Fig. 4.28(a) that at these frequencies (0.6–0.8 THz), Debye-like sound waves contribute only a very small proportion to the observed scattering. A plot of the Q-dependent inelastic scattering intensity at these frequencies (in excess of the Q-independent multiple-scattering contribution indicated in Fig. 4.28(a)) as a function of $n(\omega)e^{-2W}$ is shown in Fig. 4.29(a); at these frequencies, the Bose factor, $n(\omega)$ ([4.41]) is proportional to temperature. The linear temperature dependence of the inelastic scattering is indicative of the fact that at $\omega \simeq 700$ GHz, the (non-sound-wave) excitations are harmonic (Buchenau *et al.* 1988a).

When the true sound-wave contribution is subtracted from the observed

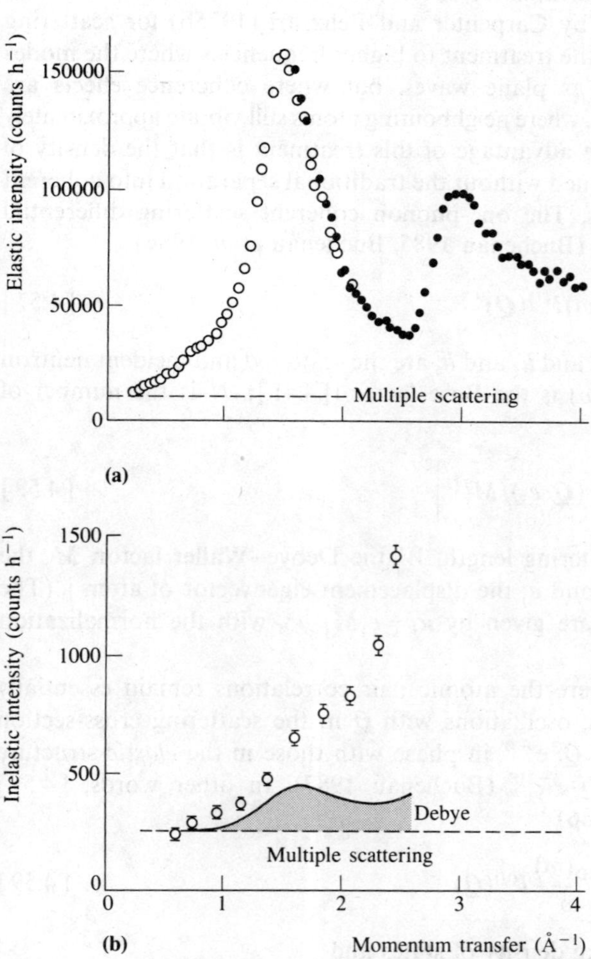

Fig. 4.28 (a) Elastic neutron scattering intensity from v-SiO_2 at 290 K.
(b) Inelastic neutron scattering from v-SiO_2 at 290 K for frequencies between 0.6 and 0.8 THz. The shaded area corresponds to the contribution to the scattering from sound waves (Buchenau *et al.* 1984).

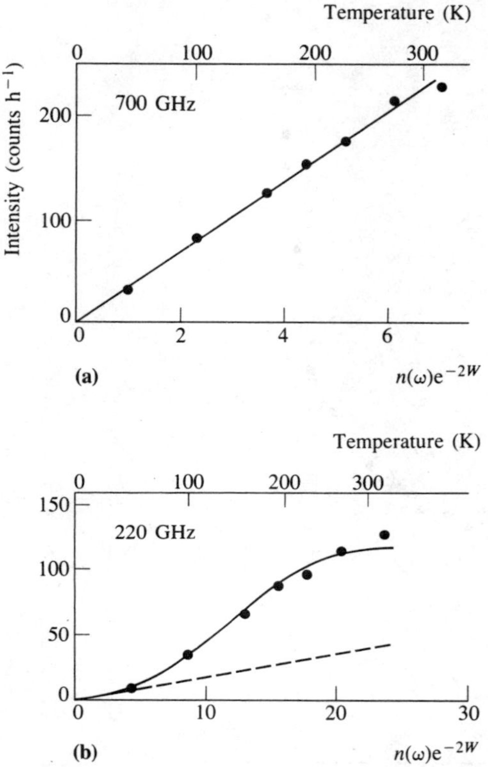

Fig. 4.29 Temperature dependence of inelastic neutron scattering intensities for v-SiO_2 for a frequency of (a) 700 GHz and (b) 220 GHz (Buchenau *et al.* 1988).

inelastic scattering, it is found that the sets of data for different frequencies all have the *same* Q-dependence, and that they can therefore be scaled so that they lie on a common curve (Fig. 4.30). The shape of this curve is markedly different from that of the elastic structure factor, $S_0(Q)$ (Fig. 4.28(a)), indicating that the eigenvectors for the modes are *not* sound-wave eigenvectors. The frequency-dependent scaling function, required to bring into coincidence on a master plot all the scattering data, is simply proportional to the *density of states* $\rho(\omega)$ of the modes responsible (Buchenau *et al.* 1986). This is shown in Fig. 4.31(a) (in which it has been assumed that only oxygen motion is involved in the excitations), and compared with the density of states inferred from specific-heat measurements on v-SiO_2; it can be seen that the two are in very close agreement, and markedly in excess of the Debye-like (sound-wave) density of states inferred from sound-velocity data:

$$\rho_D(\omega) = \frac{4\pi V}{3}\left(\frac{1}{v_L^3} + \frac{2}{v_T^3}\right)\omega^2 \qquad [4.61]$$

Conversely, the density of states for the excess modes, obtained from the inelastic neutron results, can be used to calculate the temperature dependence of the specific heat (Buchenau *et al.* 1986); this is shown in Fig. 4.31(b), from which it can be

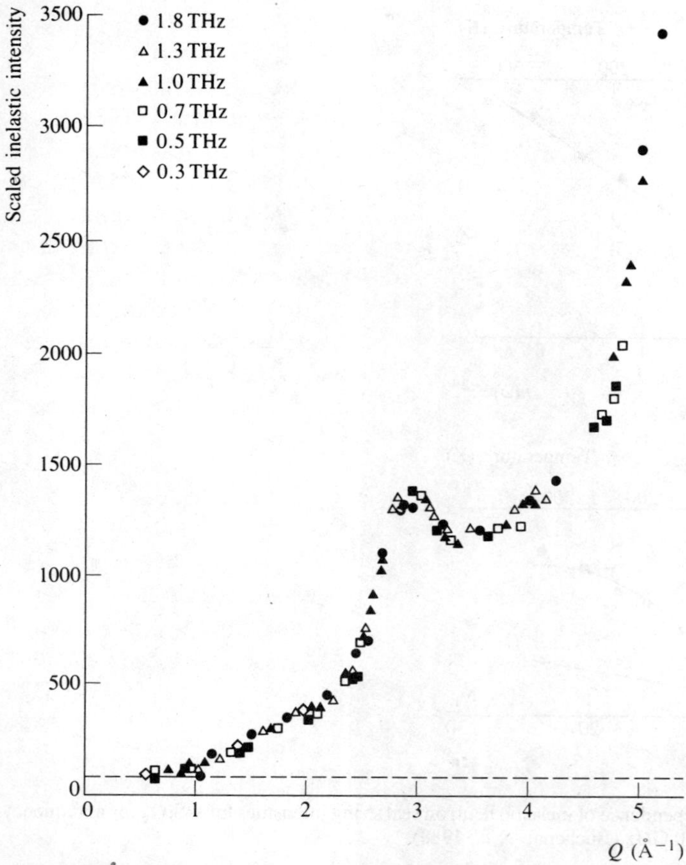

Fig. 4.30 Inelastic scattering intensities for v-SiO_2 for excess (non-sound-wave) modes at different frequencies, scaled to a common curve (Buchenau *et al.* 1986).

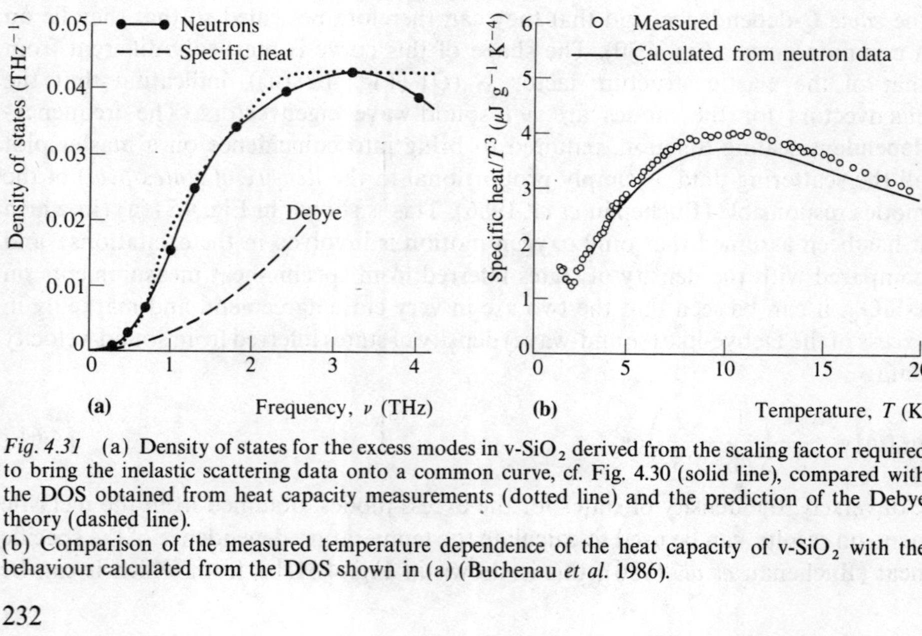

Fig. 4.31 (a) Density of states for the excess modes in v-SiO_2 derived from the scaling factor required to bring the inelastic scattering data onto a common curve, cf. Fig. 4.30 (solid line), compared with the DOS obtained from heat capacity measurements (dotted line) and the prediction of the Debye theory (dashed line).
(b) Comparison of the measured temperature dependence of the heat capacity of v-SiO_2 with the behaviour calculated from the DOS shown in (a) (Buchenau *et al.* 1986).

seen that the hump in the plot of c_v/T^3 versus T observed at $T \simeq 10$ K, mentioned in sect. 4.4.2, arises solely from these extra modes.

The question then arises of what microscopic atomic displacements give rise to these excess harmonic modes in v-SiO_2. Buchenau *et al.* (1984, 1986) have proposed that the motion responsible is the coupled libration of (at least) five corner-sharing SiO_4 tetrahedra (Fig. 4.32(a)). From [4.57] and [4.58], the scattering intensity can be calculated for such librational motion of tetrahedra; Fig. 4.32(b) shows calculated curves of $I^{(1)}(Q)/Q^2$ for motion of a single tetrahedron and two and five coupled tetrahedra, respectively; it can be seen that at least a qualitative resemblance to the experimental inelastic data is achieved for the local mode shown in Fig. 4.32(a). It remains to be seen whether more extensive calculations, using say a CRN model of SiO_2 and including perhaps more tetrahedra in the coupled librational motions, will improve the fit to experiment.

Similar low-frequency inelastic neutron scattering measurements have recently been performed on a-Ge (Buchenau *et al.* 1988b) which, because of the high average coordination number associated with the tetrahedral bonding, is expected to exhibit much smaller thermal anomalies than, say, v-SiO_2, as is observed experimentally (Graebner and Allen 1984, Van den Berg and Von Lohneysen 1985). Nevertheless, behaviour similar to that observed in vi-SiO_2 is revealed by the neutron measurements. Non-sound-wave-like modes persist down to frequencies below 1 THz, where no such modes exist in c-Ge; Buchenau *et al.* (1988b) have ascribed such modes in a-Ge to a bending motion of a Ge_3–Ge_3 cluster with the displacement vectors normal to the common Ge—Ge bond (i.e. the same

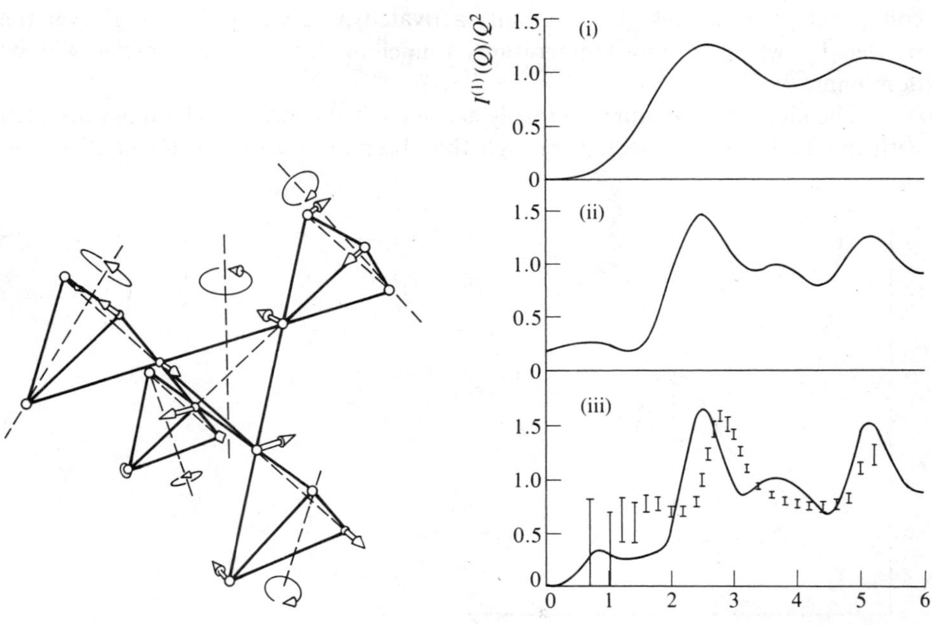

Fig. 4.32 (a) Local vibrational mode of five coupled SiO_4 tetrahedra.
(b) The inelastic neutron scattering intensity calculated for the rotation of (i) a single tetrahedron and (ii) two tetrahedra, and (iii) for the motion shown in (a). (Buchenau *et al.* 1986)

eigenvector as in the crystalline (111) zone-boundary TA mode). The density of such low-frequency non-Debye-like modes also predicts an excess specific-heat 'bump' at $T \simeq 20$ K, in agreement with experiment (Buchenau *et al.* 1988b), exactly as for v-SiO_2. However, films of a-Ge prepared by sputtering are known to contain a relatively high density of voids, and it is not clear from this single set of experiments whether the observed low-frequency modes are intrinsic to the fully connected tetrahedral network or are associated with the void surfaces.

4.4.5 Two-level systems

The low-temperature thermal properties that are observed in glasses are very different from those observed in crystals, and considerable theoretical effort has been expended in attempting to understand them. A number of theories have been proposed, the most successful of which has been that involving 'two-level systems' (TLSs), proposed independently by Anderson *et al.* (1972) and Phillips (1972).

The TLS model takes as its central hypothesis the assumption that, in a glassy system, a certain number of atoms (or groups of atoms) can occupy one of (at least) two equilibrium positions. The atoms therefore move in an anharmonic potential, such as the double-well potential shown in Fig. 4.33, which in general will be asymmetric, characterized by the 'asymmetry energy' Δ and the barrier height V_o, both of which are expected to have a distribution of values in an amorphous solid; the positional coordinate q in Fig. 4.33 is termed a 'configurational coordinate' to disguise the fact that the microscopic nature of the state is not in general known. At high temperatures, the atoms comprising the TLS will change configuration by means of a thermally activated process, by 'hopping' over the barrier V_o, whilst at low temperatures tunnelling through the barrier will be dominant.

The higher-temperature, thermally activated configurational changes associated with the TLS can be detected through the observation of *relaxational* effects, in

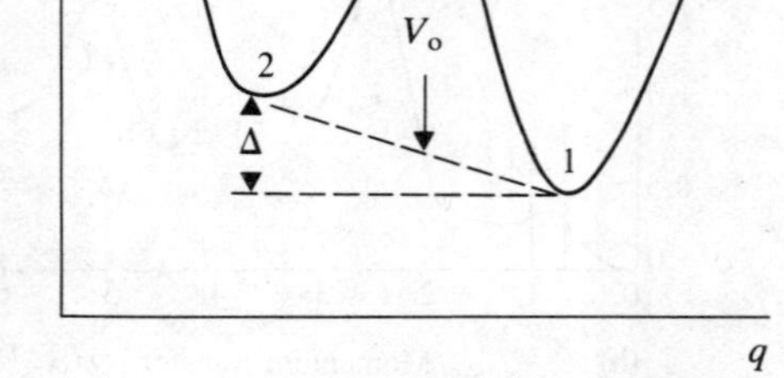

Fig. 4.33 Schematic illustration of the double-well potential characterizing a two-level system plotted as a function of configuration coordinate q. The parameters Δ (the asymmetry energy) and V_0 (the barrier height) are also indicated.

which loss peaks are manifested in mechanical or dielectric relaxation spectroscopy experiments for which the frequency corresponding to the peak maximum, ω_m, corresponds to the inverse of the relaxation time τ^{-1} characteristic of the thermally activated process at a particular temperature *T*. For v-SiO_2, ω_m is found to vary with temperature from a frequency of $\simeq 1$ kHz at about 30 K to about 10^{10} Hz at 100 K; an activation energy of about 0.04 eV can be deduced for these processes (Buchenau 1987). However, conventional relaxation spectroscopy experiments provide no information about the *nature* of the relaxational process.

Inelastic neutron scattering experiments for somewhat lower-energy (frequency) transfers than those discussed in sect. 4.4.4 can provide additional information. For instance, for v-SiO_2, the non-linear temperature dependence of the scattering intensity at 220 GHz shown in Fig. 4.29(b) indicates that, unlike at higher frequencies (Fig. 4.29(a)), the excitations being probed are strongly *anharmonic*. Furthermore, the dynamical structure factor for such low-frequency excitations is practically indistinguishable from that for the higher-frequency (non-sound-wave-like) modes (see Fig. 4.30), suggesting that the eigenvectors are the *same* in both cases (Buchenau *et al.* 1988a). This observation indicates, therefore, that the atomic motion involved in relaxational processes in v-SiO_2 is probably the same as that already discussed in sect. 4.4.4, namely a librational motion of at least five coupled tetrahedra (Fig. 4.32(a)). Buchenau *et al.* (1988a) have calculated the coherent neutron scattering cross-section expected for relaxational processes associated with the asymmetric TLS shown in Fig. 4.33, in which an atom j hops between the two potential minima at positions $\boldsymbol{R}_j + \boldsymbol{d}_j/2$ and $\boldsymbol{R}_j - \boldsymbol{d}_j/2$, giving

$$\frac{d^2\sigma}{d\Omega\, d\omega} = \frac{1}{\pi}\frac{k_f}{k_i}\int\int \frac{\tau}{1+\omega^2\tau^2}\,\mathrm{sech}^2\left(\frac{\Delta}{2k_B T}\right) I^{(1)}(Q)\, g(\Delta,\tau)\, d\Delta\, d\tau \qquad [4.62]$$

where the overall relaxation rate is given by $\tau^{-1} = \Gamma_{12} + \Gamma_{21}$ (where Γ_{ij} is the jump rate between minima i and j), and τ is given by

$$\tau = \tau_o\, \mathrm{sech}\left(\frac{\Delta}{2k_B T}\right)\exp\left(\frac{V}{k_B T}\right) \qquad [4.63]$$

where τ_o has a value typical of an inverse phonon frequency, viz. $\tau_o \simeq 10^{-13}$ s. In amorphous solids, it is expected that the TLS configuration would be different for different sites in the structure, giving rise to a distribution function $g(\Delta, \tau)$.

The *Q*-dependent contribution to the scattering cross-section for the coupled librational motion can be calculated from [4.58] and is given by (Buchenau *et al.* 1986, 1988)

$$I^{(1)}(Q) = \left|\sum_j b_j e^{-W_j} e^{i\boldsymbol{Q}\cdot\boldsymbol{R}_j} \sin(\boldsymbol{Q}\cdot\boldsymbol{d}_j/2)\right|^2 \qquad [4.64]$$

Experimentally, it is observed that this is the same for both high- and low-frequency regimes (Fig. 4.30); [4.64] is constant over the measured *Q*-range only if the jump distance in the TLS (or equivalently the extent of the librational motion) is $d \lesssim 0.8$ Å (Buchenau *et al.* 1988a). This estimate, taken together with the inelastic neutron scattering intensity, provides an estimate for the number density of active TLS centres (Buchenau *et al.* 1988a), namely $n \simeq 6 \times 10^{26}\ \mathrm{m}^{-3}$ (or about 10^{-2} states per atom). Quite why only such a small concentration of centres should exist is not clear at present, and an answer probably awaits a more detailed theoretical investigation of local relaxational and vibrational motions in large structural models which simulate well the static structural characteristics of the glass.

At still lower frequencies, i.e. for lower temperatures, *tunnelling* can occur through the potential barrier separating the two minima in a double-well potential (Fig. 4.33). In such a case, the two lowest states of the asymmetric double well (i.e. the energy splitting of the ground state introduced by the tunnelling) have energies $\pm E/2$ where:

$$E^2 = (\Delta^2 + \Delta_o^2) \qquad [4.65]$$

and where Δ_o, the coupling energy, is the energy difference between the two lowest states in the symmetric case. Δ_o is related to the parameters of the well by

$$\Delta_o = \hbar\omega_o e^{-\lambda} \qquad [4.66]$$

where

$$\lambda = (2mV_o/\hbar^2)^{1/2} d \qquad [4.67]$$

and ω_o is the frequency of oscillation in an individual well and the exponential factor in [4.66] represents the overlap between the wavefunctions for the two wells. This model can account for the low-temperature thermal anomalies since the atomic tunnelling offers a mechanism by which low excitation energies can be generated (10^{-5} eV $< E < 10^{-4}$ eV, corresponding to temperatures between 0.1 and 1 K) which therefore give rise to the anomalous specific heat. The tunnelling states at the same time act as resonant scattering centres for acoustic phonons, thereby controlling the mean free path and hence the thermal conductivity and, as we will see, the ultrasonic attenuation. The tunnelling state model will now be explored in more detail.

Although the terms 'tunnelling state' and 'two-level system' are often used interchangeably, there is in fact a distinction between the two. The tunnelling state model is the more general and assumes that in an amorphous solid there is a distribution of values of Δ_o (or equivalently V_o) as well as E and averages must be taken over both in any calculation; in practice, the distribution function for Δ_o is not known and if this averaging is neglected, the more simplified two-level model ensues in which E is the only parameter.

The free energy of a single two-level system may be written (Phillips 1972) as:

$$F(E) = -k_B T \ln[\cosh(E/2k_B T)] \qquad [4.68]$$

and since the specific heat is given by $-T(\partial^2 F/\partial T^2)$, the total specific heat is thus given by an integration over all two-level systems:

$$C = k_B \int_0^\infty n(E) \left(\frac{E}{2k_B T}\right)^2 \text{sech}^2\left(\frac{E}{2k_B T}\right) dE \qquad [4.69]$$

where $n(E)$ is the density of two-level systems per unit volume and energy interval. A similar equation holds for the case of tunnelling states, but includes also a distribution in Δ_o (or λ). If the density of states is taken to be a constant, n_o, for the range of energies of interest (i.e. 10^{-5} eV $< E < 10^{-4}$ eV), [4.69] can be readily evaluated to give:

$$C = \frac{\pi^2}{6} k_B^2 n_o T \qquad [4.70]$$

Therefore the tunnelling state (or two-level) model accounts naturally for the anomalous linear term in the specific heat with the physically plausible assumption that the density of tunnelling states is independent of energy. Specific heat data exhibiting a super-linear temperature dependence $\propto T^{1+m}$ (see Fig. 4.24) can be accommodated within the theory by relaxing the condition that $n(E)$ is independent of E, and assuming instead that it is weakly dependent on energy: $n(E) \propto E^m$, where $m = 0.1$–0.3.

However, it is not at all clear that the states giving rise to the anomalous specific heat are intrinsic in character; there is considerable evidence from sample purity variations, etc., that perhaps some of the effect may be extrinsic. The density of tunnelling states for SiO_2 determined from specific heat studies using [4.70] is $\sim 10^{21}\,\mathrm{eV^{-1}\,cm^{-3}}$ ($\sim 10^{46}\,\mathrm{J^{-1}\,m^{-3}}$). If the states are assumed to be distributed over a range of energies ~ 0.1 eV, then this corresponds to the number of states being of the order of 1/250 of the total number of SiO_2 groups, in reasonable agreement with the estimate for the number of *relaxation* centres obtained from inelastic neutron scattering, $\sim 10^{-2}$ per atom (Buchenau *et al.* 1988a), *if* the assumption is made that the tunnelling states are associated with the librational motion of coupled SiO_4 tetrahedra (Fig. 4.32(a)).

The tunnelling state model is also capable of explaining the anomalous T^2 temperature dependence of κ observed in glasses at low temperatures (see region C of Fig. 4.26(a)). The additional scattering centres which progressively limit the mean free path of the phonons at low temperatures can be identified with tunnelling states. Phonons of energy $\hbar\omega = E$, the tunnel splitting, can be scattered by a tunnelling state by a process of excitation from the ground state and subsequent spontaneous decay and emission of an incoherent phonon, i.e. 'resonance scattering'. The mean free path of phonons scattered in such a manner by a two-level system can be shown to be (see W. A. Phillips 1981):

$$\ell_{\mathrm{res}} = \frac{\rho v^3}{\pi n_o M^2 \omega} \coth\left(\frac{\hbar\omega}{2k_B T}\right) \qquad [4.71]$$

where v is the average velocity, M is the coupling constant relating E to the strain (~ 3 eV), and in this simplified model the constant density of states n_o is identical with that obtained from the linear specific heat. For thermal phonons, $\hbar\omega \simeq k_B T$, and hence $\ell_{\mathrm{res}} \propto T^{-1}$; if this is substituted into [4.56] together with $C \propto T$, the T^2 temperature dependence of κ results. If this scattering mechanism is assumed, there is no plateau in κ as a function of temperature; ℓ_{res} becomes larger without limit with increasing T at fixed ω.

There is an additional phonon scattering mechanism that is exhibited by tunnelling states. In common with any two-level system (e.g. a spin 1/2 in a magnetic field), the rate at which equilibration of the levels takes place after a change in population is given by T_1^{-1} where (W. A. Phillips 1981):

$$T_1^{-1} = \frac{M^2 E^3}{2\pi\rho\hbar^4 v^5} \coth\left(\frac{E}{2k_B T}\right) \qquad [4.72]$$

This process gives a non-resonant scattering mean free path given by (Phillips 1978) $\ell_{\mathrm{nres}} \propto T^{-3}$, and the net mean free path is $\ell^{-1} = \ell_{\mathrm{nres}}^{-1} + \ell_{\mathrm{nres}}^{-1}$. The non-resonant term limits the increase in mean free path with temperature resulting from the resonant term, and can thereby give rise to a plateau region as in Fig. 4.22(b), but its magnitude is insufficient to account entirely for the experimentally observed plateau.

As is the case for the specific heat, an energy dependence of the density of tunnelling states can be introduced into the model; if $n(E) \propto E^m$ is assumed as before, $\kappa \propto T^{2-m}$ is obtained, in agreement with many experimental data. As we have seen previously, the thermal conductivity is very similar for a variety of different materials (Fig. 4.27); furthermore, it is practically sample independent, unlike the linear specific heat, and is therefore probably more of an intrinsic phenomenon.

Further evidence which confirms the presence of two-level systems in glasses in a striking way comes from acoustic absorption experiments. Data for vitreous silica are shown in Fig. 4.34 (W. A. Phillips 1981) where the inverse mean free path of longitudinal ultrasonic waves (1 GHz) is plotted as a function of temperature. For *low* acoustic intensities, ℓ^{-1} increases as T decreases at very low temperatures. This can be understood in terms of the two phonon scattering processes discussed for thermal conductivity. The resonant inverse mean free path ([4.71]) varies as $\ell_{\mathrm{res}}^{-1} \propto \omega^2/T$ for the case of acoustic phonons for which $\hbar\omega \ll k_B T$, and this process gives rise to the rise in attenuation at very low temperatures. The increase in attenuation at higher temperatures is due to the other process for which $\ell_{\mathrm{nres}}^{-1} \propto T^3$. The startling observation is that at *high* acoustic intensities, the attenuation decreases continuously with decreasing temperature. This can be

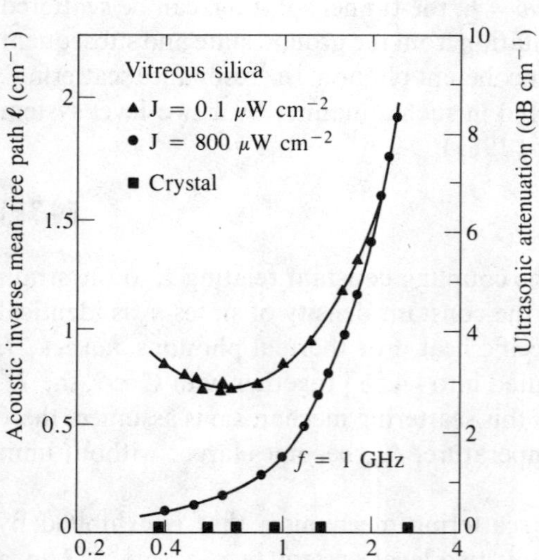

Fig. 4.34 Temperature dependence of the ultrasonic absorption in vitreous silica for longitudinal waves of 1 GHz at low temperatures. Note the decrease in absorption at the lowest temperatures for high acoustic intensities. The absorption of a quartz crystal is also shown for comparison (Hunklinger and Schickfus, in W. A. Phillips 1981).

explained by the *saturation* of the tunnelling states at high intensities when the two levels become equally populated, and hence the process leading to [4.71], namely spontaneous decay, becomes inoperative. This is incontrovertible evidence that two-level systems, whatever their origin, exist.

Although the model is phenomenologically very successful, there is no agreed explanation concerning the microscopic origin of two-level systems in general. It is observed that large, low-temperature thermal anomalies are universally found in glasses containing low coordinated atoms. Although inelastic neutron scattering cannot yet probe directly the very-low-energy excitations associated with tunnelling states, for v-SiO_2, at least, it appears that the librational motion of coupled SiO_4 tetrahedra as shown in Fig. 4.32(a), which appears to be responsible for higher-frequency excitations, could also be the origin of the tunnelling states. This cannot be a general explanation, however, since thermal anomalies are also found in metallic glasses which do not have directional bonding. Nevertheless a similar type of atom switching motion may be postulated, the atom moving from one relatively low density region of the close-packed assembly to a similar neighbouring region.

The proposition that impurities are the cause of the tunnelling states in many amorphous systems cannot be dismissed lightly. Indeed, the OH group, which is a widespread impurity in many silica samples, has a considerable effect on low temperature thermal properties, as can be seen by reference to Fig. 4.24 in which the heat capacity of an a-SiO_2 sample containing $\simeq 1000$ p.p.m. OH is some 30% higher than one containing only $\simeq 1$ p.p.m. However, there is also much evidence that a proportion of the effect is intrinsic. What also remains a mystery, besides the microscopic nature of the tunnelling states, is why the phonon coupling constant M should be approximately constant for all materials.

4.4.6 Fractons

We conclude this section on low-frequency vibrational modes with a brief discussion of 'fractons', or vibrational excitations on fractal lattices. Fractal objects are self-similar over a range of length scales and are characterized by the fractal (or Hausdorff) dimension D_H, where the mass of the object scales as $M(r) \propto r^{D_H}$ and for which $D_H \leqslant d$, where d is the normal Euclidean dimension (see sect. 3.6.2). In addition to the non-Euclidean geometric aspects of the structure of fractal objects, the *dynamics* associated with such lattices are also anomalous. Thus, the behaviour of the diffusion of an atom on a fractal structure, and equivalently of the vibrational excitations of a fractal network of bonds, are both very different from those of non-fractal homogeneous (albeit random) structures for which $D_H = d$. In particular, excitations (vibrational or diffusional) associated with fractal lattices are localized for certain fractal dimensions.

In the present context, we restrict the discussion to vibrational excitations, or fractons (Alexander and Orbach 1982, Orbach 1986). In the long-wavelength (i.e. very-low-frequency) limit, where the structure appears to be homogeneous, the vibrational excitations of even disordered materials can be described in terms of phonon (sound-wave) modes. However, at shorter length scales the structure of low-density materials in particular can sometimes be regarded as being fractal;

this self-similar regime will extend over the range of length scales $a < r < \zeta$, where a is a typical interatomic distance and ζ marks the distance beyond which the material appears to be homogeneous. For *large* length scales, the vibrational density of states will simply by Debye-like, i.e. proportional to ω^2 in three dimensions ($d = 3$) ([4.61]) or, in general,

$$\rho_D(\omega) \propto \omega^{d-1}, \qquad r > \zeta \qquad [4.73(a)]$$

whilst for a fractal lattice and at shorter length scales in the fracton regime

$$\rho_F(\omega) \propto \omega^{\tilde{d}-1}, \quad a < r < \zeta \qquad [4.73(b)]$$

where $\tilde{d}$ is the so-called 'fracton' (or spectral) dimension (Alexander and Orbach 1982), and $\tilde{d} \leqslant D_H \leqslant d$. The fracton dimension also appears in a treatment of *diffusion* on a fractal lattice, for which the diffusion coefficient becomes length-dependent, with $D(r) \propto r^{-\theta}$ (and $\langle r^2(t) \rangle \propto t^{2/(2+\theta)}$, with the exponent θ given by (Orbach 1986)

$$\tilde{d} = 2D_H/(2+\theta) \qquad [4.74]$$

The above discussion implies that there should be a cross-over from phonon-like behaviour at low frequencies to fracton-like behaviour at higher frequencies, with the cross-over frequency obeying this scaling relationship (Orbach 1986):

$$\omega_{co} \propto \zeta^{-D_H/\tilde{d}} \qquad [4.75]$$

The precise shape of the overall density of vibrational states in the region of ω_{co} has been the subject of much controversy; one possible representation is shown schematically in Fig. 4.35 (Orbach 1986). The existence of a distinct peak in the VDOS at ω_{co} is not confirmed in every case by calculations made on a variety of fractal structures, but the change in *slope* of $\rho(\omega)$ near ω_{co} does seem to be a general feature.

The significance of this picture lies in the suggestion by Alexander *et al.* (1983) that the anomalous low-temperature thermal properties which are almost universally observed in amorphous solids could be explained in terms of a common framework involving fractons. The peak in the density of states marking the phonon–fracton changeover (Fig. 4.35) could be identified with the peak in the VDOS, in for example v-SiO_2, determined from inelastic neutron scattering (Fig. 4.31(a)), and corresponding to the excess (non-Debye) bump in the specific

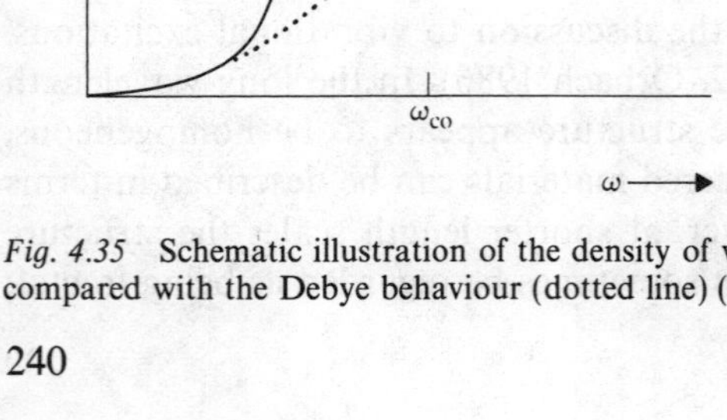

Fig. 4.35 Schematic illustration of the density of vibrational states for a fractal network (solid line) compared with the Debye behaviour (dotted line) (Orbach 1986).

heat at $\simeq 10$ K. Furthermore, the occurrence of fracton modes could also account for the temperature dependence of the thermal conductivity, in particular the plateau behaviour exhibited by most glasses (Fig. 4.27). As pointed out by Alexander *et al.* (1983), the localized nature of fracton modes provides a natural explanation for the precipitous decrease in phonon mean-free path below a certain critical wavelength, or equivalently above a critical frequency, in this case identified with ω_{co} (see Fig. 4.26(b)), where the magnitude of the mean-free path becomes of the same order as the interatomic spacing.

Although the fracton picture provides an appealing explanation for the anomalous thermal properties of glasses, a number of problems with the interpretation mean that it is probably *not* of general applicability. First, the presence of fracton modes presupposes the existence of a fractal lattice, and while sometimes a fractal network can occur as a (connectivity) subset of another structure (e.g. as a percolation pathway), more often the framework structure itself is fractal and as such should be detectable by, for example, small-angle scattering (sect. 3.6.2). For bulk glassy SiO_2 and similar materials there is no evidence that the structure is in any way fractal in character. Furthermore, peaks in the specific heat as a function of temperature are also observed in *crystalline* materials (e.g. the crystobalite polymorph of SiO_2), for which there is no suggestion that fractal/fracton effects are important. In addition, the inelastic neutron scattering measurements on v-SiO_2 by Buchenau *et al.* (1986) indicate that the frequencies where thermal anomalies are observed, namely $\simeq 0.3$ THz, excess harmonic modes coexist with propagating sound waves. This observation runs counter to the central notion in the fracton picture, namely that phonon-like and fracton-like modes cannot coexist.

The recent Brillouin light-scattering experiments by Courtens *et al.* (1987, 1988) on amorphous silica aerogels, however, appear to provide direct evidence for a phonon-fracton cross-over. Aerogels are extremely low-density products resulting from the sol–gel method of manufacture (sect. 1.3.6), in which the gel is dried 'hypercritically', i.e. under pressure and temperature conditions above the solvent critical point, thereby preserving the delicate porous microstructure of such materials. Gel densities in the range 0.1–0.4 g cm^{-3} can result, which should be compared with the density of bulk glassy SiO_2, 2.2 g cm^{-3}. Such low-density aerogels are bulk fractals; small-angle scattering measurements (sect. 3.6.2) give a fractal dimension $D_H = 2.4$ (Courtens *et al.* 1988).

Brillouin scattering on samples in the *phonon* regime results in a Lorentzian peak centred at a frequency ω from the Rayleigh (laser) line, where the value of ω observed in a given experiment depends on the scattering geometry through the scattering angle 2θ, or equivalently on the scattering vector $q = 4\pi n \sin(\theta)/\lambda$ (cf. [3.9]), where n is the refractive index of the material: a dispersion curve of ω vs. q can therefore be constructed, and from the slope of the linear portion the sound velocity can be obtained. For silica aerogels it is found that for frequencies greater than that for the phonon–fracton cross-over, $\omega > \omega_{co}$, or equivalently for scattering vectors $q < q_{co} \equiv 1/\zeta_{co}$ (where ζ_{co} is the cross-over length scale separating phonon and fracton behaviour), very different behaviour is observed (Courtens *et al.* 1987, 1988): the Brillouin peak becomes much broader and non-Lorentzian in shape, and the frequency corresponding to the peak maximum, ω_{max}, no longer scales with q, but instead saturates at a value $\omega_{max} \simeq 1.2\omega_{co}$. Figure 4.36 shows

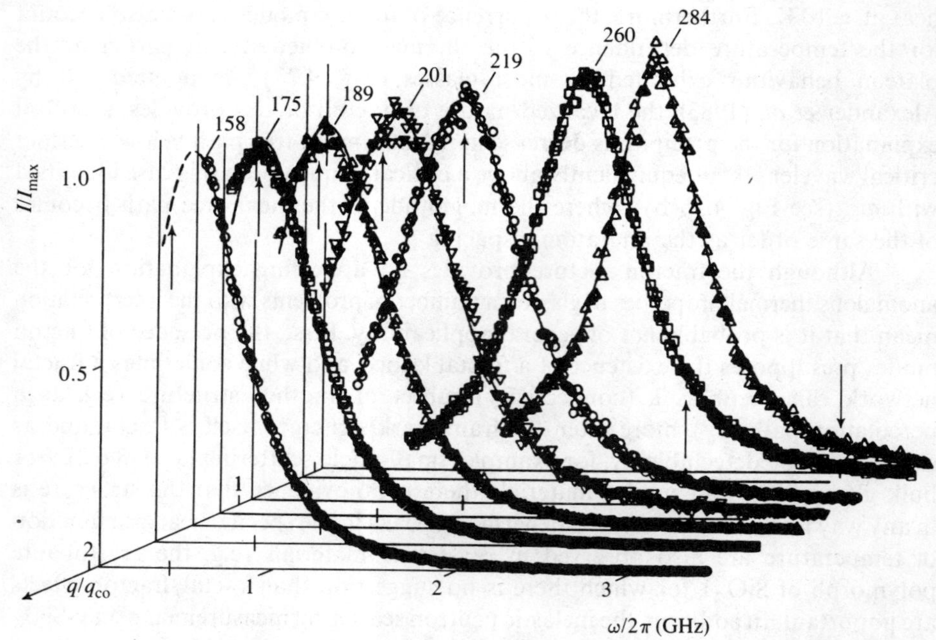

Fig. 4.36 Brillouin scattering data for silica aerogels whose densities (in kg m^{-3}) are shown in both the phonon and fracton regimes. The aerogels with densities of 201 and 219 kg m^{-3} are in the phonon–fracton cross-over region (Courtens *et al.* 1988).

this behaviour, for which the three least dense aerogels are in the fracton regime, the next two are in the cross-over region, whilst the two densest samples are in the phonon regime. The observed saturation of ω_{max} in the fracton regime would imply an accumulation of states near ω_{co} in the VDOS $\rho(w)$, seemingly in agreement with Fig. 4.35.

These data also yield an estimate for the fracton dimension, $\tilde{d}$. The sound velocity v scales with the mass density ρ as (Courtens *et al.* 1987)

$$v(\zeta)/v(a) = (\rho(\zeta)/\rho(a))^x \qquad [4.76]$$

where ζ is a correlation length for the fractal regime and a is a characteristic size below which the material is non-fractal. The exponent x in [4.76] is given, in terms of the fractal and fracton dimensions, by

$$x = (D_H - \tilde{d})/\tilde{d}(3 - D_H) \qquad [4.77]$$

Since $\rho(\zeta)/\rho(a) \propto (\zeta/a)^{D_H - 3}$, it is predicted that the phonon–fracton cross-over frequency ω_{co} scales with density as (Courtens *et al.* 1987)

$$\omega \propto \rho^y \qquad [4.78(a)]$$

where

$$y = x + 1/(3 - D_H) \qquad [4.78(b)]$$

Analysis of the data shown in Fig. 4.36 yields a value for the fracton dimension of $\tilde{d} = 1.3$, in rather close agreement with the universal value of $\tilde{d} = 4/3$ conjectured by Alexander and Orbach (1982).

II ATOMIC TRANSPORT

4.5 Ionic conduction

4.5.1 Introduction

In the first part of this chapter, we have discussed various aspects of the dynamics of atoms when they are constrained to reside at particular sites in an amorphous structure, namely vibrational excitations. In this section we lift this restriction and consider long-range translational motion of atoms (usually ions). If the diffusing atoms are charged they can carry electrical current, and in certain circumstances this ionic contribution to the electrical conductivity can greatly exceed that due to electrons, particularly in 'superionic' (or fast-ion conducting) materials, where the value of the ionic conductivity is comparable to that of molten salts or of concentrated ionic solutions, for which $\sigma \geqslant 10^{-2}\,\Omega^{-1}\,\mathrm{cm}^{-1}$. Measurement of the ionic contribution to the electrical conductivity is therefore a relatively straightforward experimental probe of the diffusional dynamics of ions in glasses, although the diffusion coefficient may also be measured directly using radio-tracer methods and indirectly using quasi-elastic neutron scattering and (pulsed-field gradient) NMR techniques. In this section, we concentrate on ionic transport since it is the most important technologically, but we will also mention atomic diffusion where appropriate.

Ionic conduction is well known in crystalline ionic solids (for reviews see Salamon 1979 and Chandra 1981), and both cations and anions can carry current (sometimes simultaneously) in certain circumstances. The mechanism giving rise to the (super)ionic conduction may be of either the 'point defect' or the 'molten sublattice' type. In the former, ionic transport arises through thermally generated defect pairs, either of the Schottky type (cation and anion vacancies, as in the alkali halides) or of the Frenkel type (interstitials and vacancies on either the cation or anion sublattices, as in the silver halides), and conduction is therefore thermally activated. In the latter case, the number of ions of a particular type is less than the number of available sites in their sublattice (i.e. the distribution of ions is therefore highly disordered, or 'liquid-like'), and consequently a large number of ions are able to conduct with a lower activation energy than is possible in defect processes. The mechanisms of transport in glassy ionic conductors are not as well understood, principally because the structure, both of the framework and also of the local environment of the ions, is in general not known.

Many types of ion can diffuse in glasses, although cation motion is the most common (see Angell 1986 and Ingram 1987 for reviews). Singly charged cations tend to diffuse the most easily, with the smallest ions, e.g. Li^+, or the most deformable, e.g. Ag^+, having the highest conductivities of all. Figure 4.37(a) shows the temperature dependence of the ionic conductivity for a number of Ag-containing 'fast-ion conducting' (FIC) glasses together with some representative crystalline

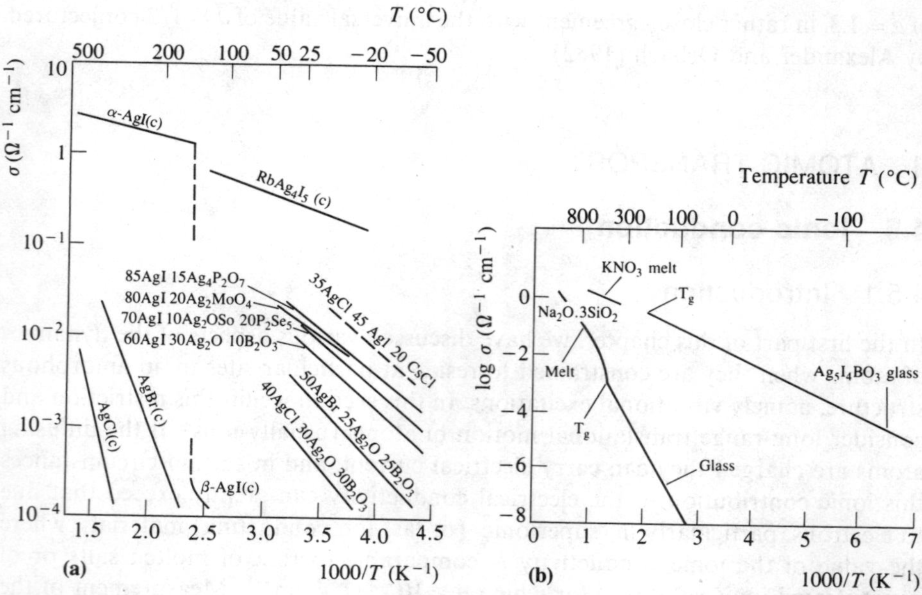

Fig. 4.37 Arrhenius plots of the d.c. conductivity in ionically conducting amorphous systems. (a) Various Ag^+ ion conducting glasses compared with certain crystals (c-) (Angell 1986). (b) Temperature dependence of the ionic conductivity of poorly conducting and superionic amorphous materials near the glass-transition temperature (Ingram 1987).

FIC materials. It can be seen that the best FIC glasses have values of ionic conductivity which are comparable to those of their crystalline superionic counterparts, although it should be noted that the highest conductivities in the glasses are achieved by the introduction of so-called 'doping salts', AgI or LiI, which by themselves in the crystalline state are superionic conductors. Glassy FIC materials have several advantages over their crystalline counterparts: the conduction in amorphous materials is *isotropic*, whereas in many crystalline superionic solids (e.g. Na-β-alumina) ionic conduction preferentially takes place along well-defined channels or planes in the crystal structure; furthermore, the composition of glasses can often be varied over rather large ranges, thereby enabling many physical properties to be controlled systematically.

The detailed temperature dependence of the ionic conductivity of two glasses, one a superionic material containing a doping salt and the other a conventional alkali silicate, is shown in Fig. 4.37(b). Two features are worthy of note. For temperatures below the glass-transition temperature T_g, the ionic conductivity of all glasses appears to obey a perfect Arrhenius law:

$$\sigma = \sigma_o \exp(-E_\sigma / k_B T) \qquad [4.79]$$

(sometimes the product σT is plotted instead of σ alone), whereas for $T > T_g$ markedly non-Arrhenius behaviour is observed, whose temperature dependence appears to obey a relation of the Vogel–Tammann–Fulcher type (cf. [2.8]). In addition, it is interesting to note that the curve for the superionic glass appears to be simply a continuation of that for the fused salt, KNO_3.

Anions as well as cations can also diffuse in certain glasses, although less work has been carried out in this area than on anionic conduction in crystalline hosts. One interesting example is the fluoride glasses, based on ZrF_4, AlF_3, YF_3, BaF_2, etc., which, in addition to having considerable promise for use in low-loss optical fibres, are also quite good F^- conductors; glasses such as $40ScF_3.20YF_3.40BaF_2$ have values of conductivity at 200°C of $\sigma > 10^{-5}\ \Omega^{-1}\ cm^{-1}$ (Ravaine *et al.* 1983).

In the next section, we discuss a number of the theories which have been proposed to account for ionic conductivity in glasses.

4.5.2 Mechanisms for ionic transport

The ionic conductivity may be written in the form

$$\sigma = nq\mu \qquad [4.80]$$

where n is the concentration of mobile ionic species carrying charge q, and μ is the mobility. A simple expression for the ionic conductivity may be obtained by assuming that the ionic motion occurs by field-enhanced, thermally activated hopping between equivalent sites in the structure. The action of the applied electric field E is to lower the energy of a site by an amount equal to qEa, where a is the intersite spacing. Therefore, the probability of ionic hopping in the direction of the field increases, and the net number of ions per volume hopping in the direction of the field, averaged over all space, is then

$$\phi = \frac{nqaE\nu_o}{6k_B T} \exp\left(\frac{-W}{k_B T}\right) \qquad [4.81]$$

where W is the mobility activation energy, i.e. the barrier to hopping (diffusion) between two sites, ν_o is the pre-exponential term in the hopping rate expression (essentially equal to the 'rattling' vibrational frequency of an ion trapped in the potential well at a site), and it has been assumed that $qaE \ll k_B T$. The current density is simply $j = qa\phi$, and the ionic conductivity is then given by $\sigma = j/E$, i.e.

$$\sigma = \frac{nq^2a^2\nu_o}{6k_B T} \exp\left(\frac{-W}{k_B T}\right) \qquad [4.82]$$

Thus, [4.82] predicts that $\ln(\sigma T)$ vs. T plots should be linear. The ionic carrier concentration may itself be thermally activated:

$$n = n_o \exp(-E_c/k_B T) \qquad [4.83]$$

and so the d.c. conductivity activation energy in this picture is given in general by the sum of two terms:

$$E_\sigma = W + E_c \qquad [4.84]$$

It should be noted that the activation energies E_c and W are in fact *enthalpies*, and the corresponding entropy terms associated with the individual processes are subsumed into the pre-exponential terms n_o and ν_o.

A number of models have been proposed to account for various aspects of ionic conduction in glasses, in particular the behaviour of the d.c. conductivity activation energy, E_σ: a review of these has been given by Ingram (1987). For simplicity, we may divide these theories into three categories:

(1) the 'strong-electrolyte' model, in which the effective carrier density n in [4.80] is assumed to be independent of ion concentration or temperature, making the mobility μ the term that varies as a function of these parameters;
(2) the 'weak-electrolyte' model, in which conversely μ is assumed to be independent of ion concentration (and only weakly temperature-dependent), whereas n depends strongly on both ion concentration and temperature;
(3) 'defect' models, which utilize concepts employed in the study of *crystalline* ionic conductors, and where the emphasis is on the particular sites in the structure participating in the diffusion process.

Although conceptually different, there are some similarities between the three approaches, as is discussed below.

We commence with a discussion of the strong-electrolyte approach, sometimes termed the Anderson–Stuart (1954) model. In this, the activation energy for the d.c. conductivity E_σ (or the diffusion coefficient) is evaluated in terms of a specific *microscopic* picture: a cation is presumed to hop from an occupied site in the vicinity of a negatively charged counterion, such as a non-bridging oxygen (NBO) site in an oxide glass, to a vacancy near another NBO site, passing through a 'gateway' formed by bridging oxygen (BO) atoms (see Fig. 4.38). Anderson and Stuart (1954) suggested that E_σ may be regarded as consisting of two terms, an electrostatic binding energy E_b required to remove a cation from an NBO site, and a strain term E_s associated with the long-range mobility ('gate-passing'). Thus, E_σ can be written as $E_\sigma = E_b + E_s$, or

$$E_\sigma = \frac{\alpha Z_+ Z_- e^2}{4\pi\varepsilon\varepsilon_o r_o} - \frac{2Z_+ Z_- e^2}{4\pi\varepsilon\varepsilon_o(l/2)} + \frac{\pi G l}{2}(r_+ - r_d)^2 \qquad [4.85]$$

where Z_+ and Z_- are the charge states of cation and NBO respectively, r_d is the 'gateway' radius between BO atoms, G is the shear modulus and ε the dielectric constant of the host, and l is a jump length given by

$$l = R - 2r_o$$

where $r_o = (r_+ + r_-)$, the sum of the cation and NBO radii respectively, and R is an intersite separation between neighbouring NBO centres. The first term in [4.85] represents the energy required to remove a cation from an NBO site and place it at infinity (the Madelung-like constant α is of the order of unity), reduced by the second term, the attractive electrostatic potential experienced by a cation midway between two NBO sites; the sum of these two terms is thus equal to the electrostatic binding energy of the cation at a site. The third term is the strain energy as calculated by McElfresh and Howitt (1986) for enlargement of a cylindrical hole (of length l) from radius r_d to r_+, and represents the strain associated with an ion moving between two sites. The energetics of the Anderson–Stuart model are illustrated in the configuration-coordinate diagram for the process (Martin and Angell 1986) shown in Fig. 4.38(a).

The hopping distance l in [4.85] would be expected to have a wide distribution of values in a glass, and yet the perfect linearity of logarithmic plots of σ_{dc} versus

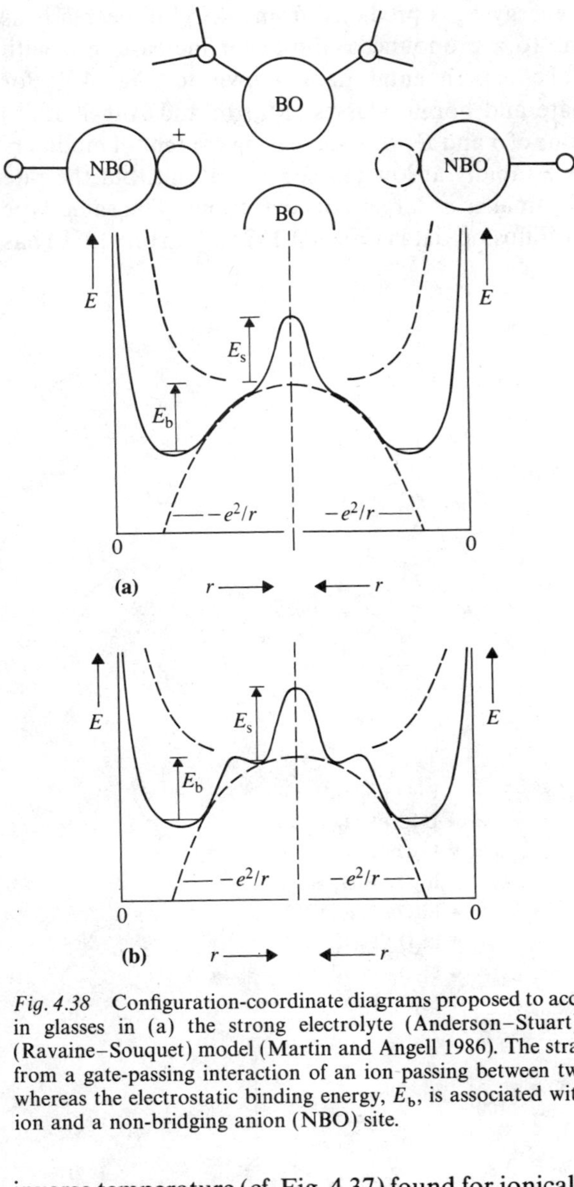

Fig. 4.38 Configuration-coordinate diagrams proposed to account for the energetics of ionic transport in glasses in (a) the strong electrolyte (Anderson–Stuart) model and (b) the weak-electrolyte (Ravaine–Souquet) model (Martin and Angell 1986). The strain-energy term, E_s, is presumed to arise from a gate-passing interaction of an ion passing between two (or more) bridging anion (BO) sites, whereas the electrostatic binding energy, E_b, is associated with the Coulomb interaction between the ion and a non-bridging anion (NBO) site.

inverse temperature (cf. Fig. 4.37) found for ionically conducting glasses (compared with the marked non-linearity observed in the extrinsic regime of crystalline ionic conductors) implies that a *single* value of l determines E_σ for a given composition. (The same is true for tracer diffusion.) Elliott (1988) has argued that, for d.c. conductivity, the correct choice for l is the critical percolation radius r_p, which in three dimensions is given by (Pramanik *et al.* 1983)

$$r_p = \left(2.7 \times \frac{3}{4\pi n}\right)^{1/3} \qquad [4.86]$$

where n is the site concentration. Thus, as the site (cation) concentration n increases, the d.c. conductivity activation energy E_σ is predicted from [4.85] to decrease, as observed experimentally, leading to a pronounced non-linear increase in σ with increasing modifier content. The experimental data shown in Fig. 4.39 for lithia-modified silicate, phosphate and borate glasses (Martin and Angell 1986) show the typical general behaviour of σ and E_σ with increasing content of modifiers or doping salts: E_σ first decreases rapidly at low ion concentrations, but the rate of decrease becomes increasingly smaller at larger concentrations. The same type of behaviour is also observed for diffusion data (Frischat 1975). Martin (1988) has

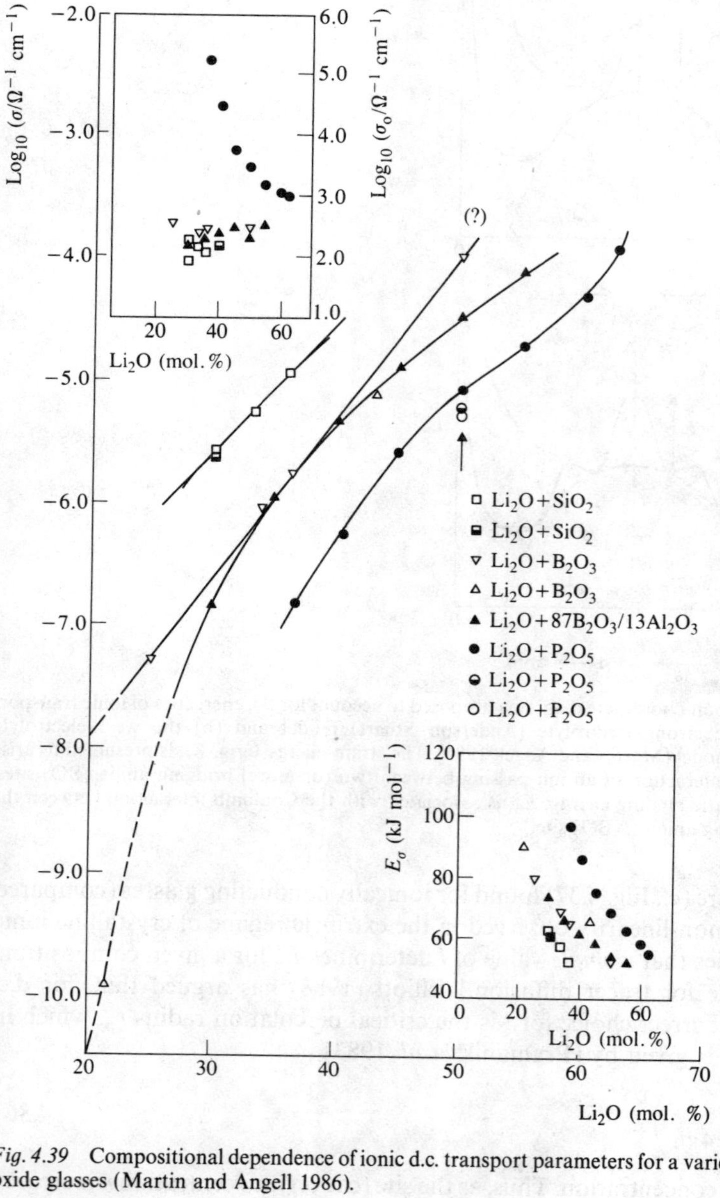

Fig. 4.39 Compositional dependence of ionic d.c. transport parameters for a variety of Li^+-conducting oxide glasses (Martin and Angell 1986).

shown that, for alkali borates at least, the main contribution to the decrease in E_σ comes from the electrostatic term in [4.85], since for alkali contents below 30 at.% the shear modulus G, and the glass-transition temperature T_g, which roughly scales with it, *increases* with increasing alkali content (as a result of the progressive alkali-induced conversion of trigonal boron sites to tetrahedral coordination), and consequently there is an approximate cancellation of the changes of G and l with n. (In other cases, e.g. the alkali silicates, the converse can be true: addition of alkali modifiers causes a progressive depolymerization of the structure, resulting in a decrease of T_g and hence, presumably, of G.)

Another interesting feature emerges from an examination of Fig. 4.39: for the same mole percentage of modifier, silicate glasses generally have the highest ionic conductivity, whilst phosphate glasses have the lowest, a result mostly of the relative values of the activation energies, E_σ, rather than of the pre-exponential factors, σ_0. This observation can also be understood on the basis of the Anderson–Stuart model by noting that, at, say, the meta-compositions Li_2SiO_3, $LiBO_2$ and $LiPO_3$, there are twice as many NBO sites per network-forming cation for Si than for B or P; as a result, the hopping length determined by the critical percolation radius r_p ([4.86]) will be the smallest, so E_σ will also be the smallest for silicates for a given modifier content. However, implicit in this treatment of the Anderson–Stuart model is the assumption that every modifier cation is mobile and contributes to the conductivity; this may not be the case, as stressed by the defect models for ionic transport.

The Anderson–Stuart model can also explain why replacement of oxygen atoms in ionic oxide glasses by sulphur to form the corresponding sulphide materials results in a marked enhancement of the ionic conductivity (Souquet 1981). For instance, glasses such as Li_2S–GeS_2–LiI and Li_2S–P_2S_5–LiI have room-temperature conductivities greater than $\simeq 10^{-3}\,\Omega^{-1}\,cm^{-1}$. The reason for this enhancement can be ascribed to the decrease in activation energy associated with an increase in dielectric constant ε in [4.85] upon incorporation of the more polarizable sulphide ions.

We turn now to a discussion of the so-called 'weak-electrolyte' model, proposed initially by Ravaine and Souquet (1977). Unlike the Anderson–Stuart model described above, the weak-electrolyte approach is based on general thermodynamic principles; microscopic mechanisms are not considered. Central to this approach is the observation that the large increase in ionic d.c. conductivity caused by the addition of modifiers or dopant salts MX (e.g. $M^+ = Li^+$; $X^- = OLi^-$ or I^-) (Fig. 4.39) are accompanied by large increases in the activity of the modifier, a_{MX}. The following equilibrium is postulated to occur in the glass:

$$MX \rightleftharpoons M^+ + X^- \tag{4.87}$$

i.e. not all modifier cations are presumed to be mobile. The equilibrium constant describing the dissociation reaction [4.87] can be written as:

$$K = \frac{a_{m^+}a_{X^-}}{a_{MX}} = \frac{\gamma_{M^+}\gamma_{X^-}[M^+][X^-]}{a_{MX}} \tag{4.88}$$

where the γ's are activity coefficients, and where $[M^+]=[X^-]$. Substitution of

$[M^+] \equiv n$ from [4.88] into [4.80] yields (Ravaine and Souquet 1977)

$$\sigma = F\mu K^{1/2} a_{MX}^{1/2} \qquad [4.89]$$

where the term $F = q/(\gamma_{M^+}\gamma_{X^-})^{1/2}$. The square-root dependence on the activity embodied in [4.89] has been verified experimentally (Ravaine and Souquet 1977) – see Fig. 4.40.

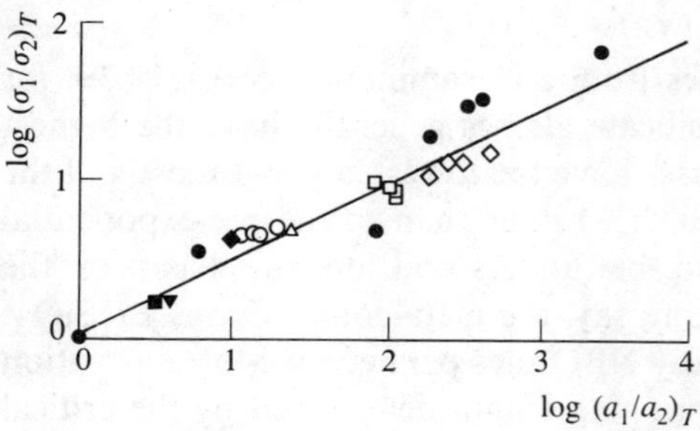

Fig. 4.40 Dependence of the ionic d.c. conductivity σ on the activity a of alkali oxide in various silicate glasses at various temperatures; the straight line has a slope of 0.5 (Ravaine and Souquet 1977).

The dissociation constant K can be rewritten in the form (Kone and Souquet 1986)

$$K = \exp(-\Delta G^\circ_{diss}/k_B T) = \exp(-\Delta H^\circ_{diss}/k_B T)\exp(\Delta S^\circ_{diss}/k_B T) \qquad [4.90]$$

and the activity of MX can also be expressed in terms of the partial free energy $\overline{\Delta G}_{MX}$:

$$a_{MX} = \exp(\overline{\Delta G}_{MX}/k_B T) = \exp(\overline{\Delta H}_{MX}/k_B T)\exp(-\overline{\Delta S}_{MX}/k_B T) \qquad [4.91]$$

Using the temperature dependence of the mobility given in [4.82], the overall activation energy of the ionic conductivity is therefore predicted to be (Kone and Souquet 1986)

$$E_\sigma = \tfrac{1}{2}(\Delta H^\circ_{diss} - \overline{\Delta H}_{MX}) \qquad [4.92]$$

In the weak-electrolyte approach, it is assumed that the dominant term in this expression is the dissociation enthalpy, ΔH_{diss}, and that the mobility is therefore practically independent of temperature, and furthermore is not a function of modifier composition.

Although at first sight the Anderson–Stuart and Ravaine–Souquet models would appear to have little in common, Martin and Angell (1986) have suggested that a configuration-coordinate diagram for the weak-electrolyte model can be constructed (Fig. 4.38(b)) which is very similar to that for the strong-electrolyte approach (Fig. 4.38(a)), with the exception of small energy barriers of unknown structural origin separating the deep trapped state and the higher-energy mobile state. In this picture, E_b is equivalent to half the dissociation energy $\Delta H^\circ_{diss}/2$, and the strain energy E_s is the origin of the mobility activation energy W. The central assumption in the weak-electrolyte model of *two* categories of modifier ion in the glass, immobile and mobile, is also a feature of some of the defect models to be discussed next, and so in principle the thermodynamic approach of the weak-electrolyte theory can be used in that context too.

Defect-based theories for the mechanism of ionic transport in glasses draw heavily on the extensive studies that have been made of ionic migration in ionic crystals, where invariably diffusion takes place by means of defect processes involving vacancies or interstitials. Interstitial ions can move either by a direct mechanism involving motion between neighbouring interstitial sites, or by an indirect 'interstitialcy' mechanism in which the arrival of an interstitial ion causes a cation to be displaced from a neighbouring lattice or another interstitial site. The simultaneous existence of mobile and immobile ionic species implicit in the weak-electrolyte approach is naturally understandable in the interstitialcy picture, where the mobile species make up a relatively small proportion of the interstitial ions.

Where the electrical conductivity arises from a diffusive mechanism, as with ionic conduction, a 'conductivity' diffusion coefficient, D_σ, can be defined via the Nernst–Einstein relation:

$$D_\sigma = k_B T\sigma/nq^2 \qquad [4.93]$$

In general, however, the tracer diffusion coefficient, D^*, will not be equal to D_σ if correlation effects are important; they are related by the so-called Haven ratio,

$$H_R = D^*/D_\sigma \qquad [4.94]$$

where $H_R \leqslant 1$. In *crystalline* materials, where the defect transport processes can be characterized with respect to a well-defined lattice, the Haven ratio can be calculated for a given type of crystalline structure (Compaan and Haven 1956, 1958), giving the following approximate relations (Jain *et al.* 1983):

$$H_R = 1 - (2/Z) \qquad [4.95]$$

for a vacancy mechanism, and

$$H_R = 2(r_t/r_q)^2[1-(1/Z)] \qquad [4.96]$$

for an interstitialcy mechanism, where r_t and r_q are the jump distances of tracer and charge respectively, and Z is the number of neighbouring sites involved. For a *direct* interstitial process, $H_R = 1$. Values of the Haven ratio differing from unity arise from correlation effects in the tracer-diffusion process, i.e. where certain types of jump are precluded following a given hop. In direct interstitial motion the atoms follow a random walk, and so the correlation factor is unity. Thus, in principle, measurement of H_R allows the diffusion mechanism to be identified.

However, for atomic diffusion in non-crystalline solids this approach is not as straightforward, since the precise value of H_R depends on the geometry of the jump process, which is not in general known for amorphous materials. In addition, the precise value of the ionic carrier density n to be used in [4.93] is also uncertain. Furthermore, it is not clear how interstitial or vacancy defects are defined in a non-crystalline host. The Haven ratio has been estimated experimentally for numerous ionically conducting glasses, particularly alkali silicates (Frischat 1975, Jain *et al.* 1983, Beier and Frischat 1985). It is found that at very low alkali ion contents $H_R \simeq 1$, whereas at higher concentrations the Haven ratio decreases until

$H_R \simeq 0.3$–0.4. Jain *et al.* (1983) have suggested a model to account for these observations. Following Charles (1961), they suppose that the cation interstitial sites consist of many equivalent positions around each NBO, with which is already associated another cation. At very low ion concentrations, the diffusive motion is mainly of direct interstitial type between interstitial positions at different NBO sites. As the cation concentration is increased, cation–cation interactions will cause the motion to become correlated, and as a consequence H_R will decrease.

4.5.3 Frequency-dependent transport

The ionic component to the d.c. conductivity is not straightforward to measure experimentally since, if electrodes are used which are blocking to the ions, the d.c. conductivity cannot be obtained by the application of a d.c. voltage because of polarization effects. Instead, the method of 'impedance spectroscopy' has to be used, in which the frequency-dependent response is measured and the d.c. value obtained as the $\omega \to 0$ limit. Essentially, an a.c. field at a circular frequency ω is applied to the sample, and both in-phase (real) and out-of-phase (imaginary) components of the current are monitored; in other words, the resistance R_s and capacitance C_s of the sample are measured. There are a number of ways of presenting the data (see Jonscher 1983 and MacDonald 1987 for details), including the complex admittance $Z^* = R_s^{-1} + i\omega C_s$, the complex impedance $Z^* = (Y^*)^{-1}$, the complex permittivity $\varepsilon^* = \varepsilon_1 + i\varepsilon_2$ and the complex modulus $M^* = M_1 + iM_2 \equiv (\varepsilon^*)^{-1}$. The d.c. conductivity can be obtained either by the $\omega \to 0$ extrapolation of complex-plane plots where real and imaginary parts of, say, Z^* or M^* are plotted against each other as an implicit function of frequency, or of the real part of the conductivity $\sigma_1(\omega)$ plotted against frequency. The complex-plane analysis has the advantage that the intrinsic dielectric response due to the sample itself should consist of a single arc (whose detailed shape will be discussed below), any spurious electrode polarization effects showing up as (parts of) other arcs.

Although impedance spectroscopy data are used mostly to obtain values for the d.c. conductivity as outlined above, they also yield frequency-dependent information in the frequency range $\simeq 10$–10^9 Hz which is interesting in its own right and can shed much light on the ionic transport mechanisms involved. Representative frequency-dependent electrical behaviour for ionically conducting glasses is shown in Fig. 4.41 for a Li_2O–P_2O_5 glass (Martin and Angell 1986). Fig. 4.41(a) shows a double-logarithmic plot of the real part of the conductivity $\sigma_1(\omega)$ against frequency at different temperatures; the data exhibit a power-law behaviour, which can be written in the form

$$\sigma_1(\omega) = \sigma(0) + A\omega^s \qquad [4.97]$$

exactly as observed for *electronic* conduction in amorphous semiconductors (sect. 6.3.3). The power-law exponent, s, is usually a decreasing function of temperature, and the d.c. conductivity $\sigma(0) = \sigma_1(\omega \to 0)$. It can be observed that the (activated) temperature dependence of the d.c. conductivity ([4.79]) is much stronger than that of the pure a.c. component. The frequency dependences of the real (M_1) and imaginary (M_2) parts of the complex electrical modulus M^* for the same sample are shown in Fig. 4.41(b) as composite plots for data taken at different

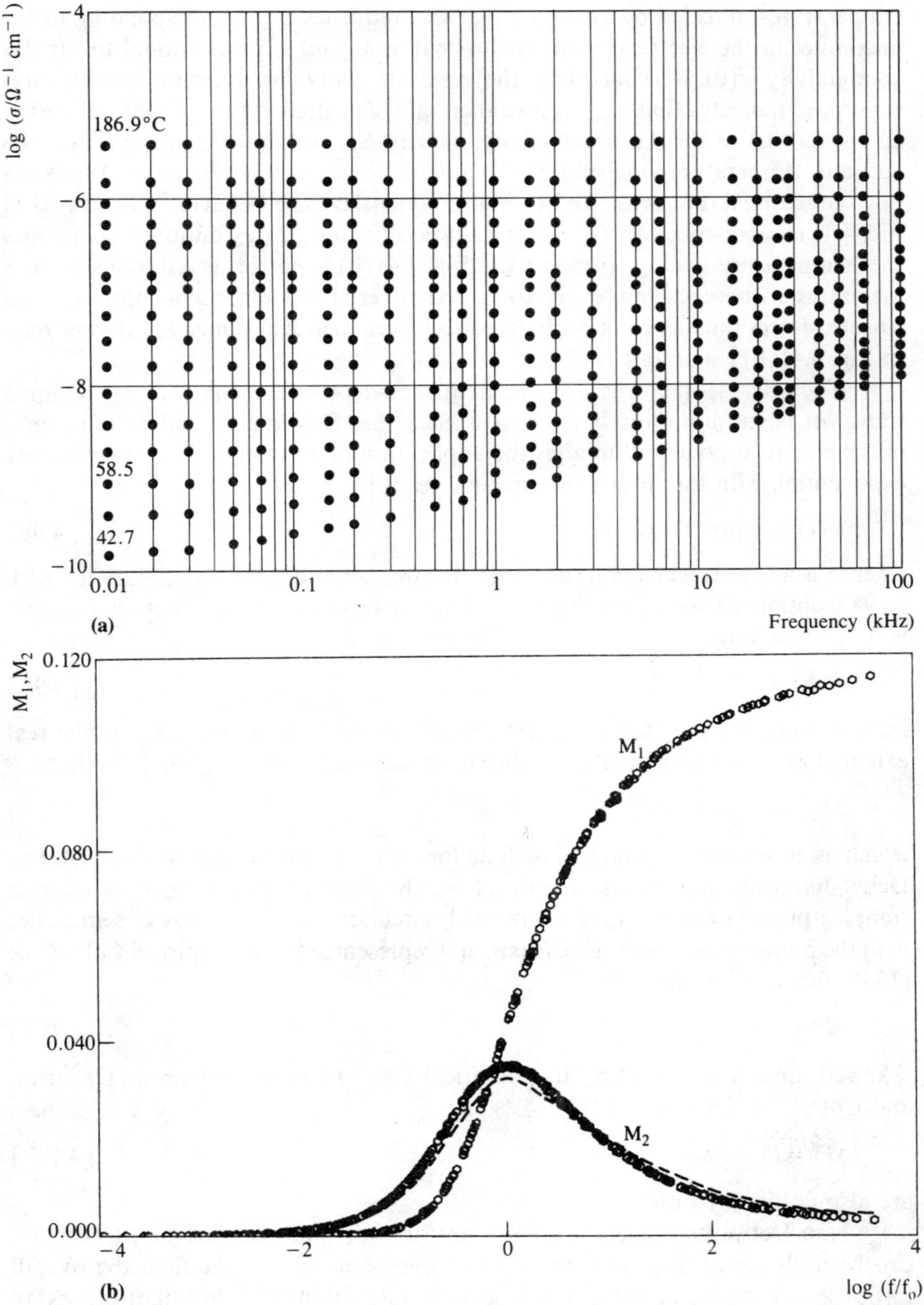

Fig. 4.41 A.C. ionic transport behaviour in Li^+-conducting phosphate glasses (Martin and Angell 1986).
(a) Frequency dependence of the a.c. conductivity of $(Li_2O)_{0.41}(P_2O_5)_{0.59}$ at various temperatures.
(b) Frequency dependence of the real (M_1) and imaginary (M_2) parts of the complex electrical modulus M^* for the same composition as in (a). The data points for different temperatures lie on these common curves if the peak positions of M_2 are scaled to the same frequency.

temperatures, rescaled by assuming that the frequency ω_{max} corresponding to the maximum in the electrical modulus loss curve $M_2(\omega)$ is proportional to the d.c. conductivity $\sigma(0)$. The fact that the resulting plots for different temperatures superpose indicates that ω_{max} is also thermally activated with the same activation energy as that of the d.c. conductivity, namely E_σ, and that the *shape* of the loss curve is temperature-independent. It should be noted that the $M_2(\omega)$ curve is not symmetrical, but is skewed to high frequencies. The behaviour illustrated in Fig. 4.41 is representative of most ionically conducting glasses, although sometimes the 'temperature scaling' embodied in Fig. 4.41(b) is not observed, the skewness increasing with increasing temperature, and at very low ion concentrations in soda silicate glasses the loss peak is observed to be symmetric and much narrower than in Fig. 4.41(b) (Simmons *et al.* 1979).

The microscopic origin of this frequency-dependent behaviour of the ionic transport is still not entirely clear. It is clear that the time-dependent relaxation of the electrical polarization after the application of an electric field is *not* simply exponential, with the relaxation function given by

$$\phi(t) = \exp(-t/\tau) \qquad [4.98]$$

where τ is a (fixed) relaxation time, since the frequency dependence associated with [4.98], obtained from a pure imaginary Laplace transform of the function $-\mathrm{d}\phi/\mathrm{d}t$, is of 'Debye' form:

$$M^* = (1 + i\omega\tau)^{-1} \qquad [4.99]$$

i.e. it is represented by a perfect semicircular arc with the centre lying on the real axis in a complex-plane plot. The dielectric loss peak is, from [4.99], given by

$$M_2 = \omega\tau/(1 + \omega^2\tau^2) \qquad [4.100]$$

which is a symmetric Lorentzian function, unlike that shown in Fig. 4.41(b). Debye-like behaviour is also ruled out for the majority of ionic glasses because complex-plane plots are very rarely perfectly circular: instead, 'depressed' semicircles with the centre lying below the real axis, and represented by the empirical Cole–Cole (1941) function, are found

$$M^* = [1 + (i\omega\tau)^{1-\alpha}]^{-1}, \quad 0 < \alpha < 1 \qquad [4.101]$$

'Skewed' arcs represented by the empirical Cole–Davidson function (Davidson and Cole 1951, Davidson 1961)

$$M^* = (1 + i\omega\tau)^{-\gamma}, \quad 0 < \gamma < 1 \qquad [4.102]$$

are also frequently observed.

Non-Debye-like dielectric behaviour (for both ionic and electronic transport) can be understood in general in terms of two scenarios. In the first, the overall response is the sum of parallel independent relaxational (Debye-like) processes, with a *distribution* of relaxation times, $n(\tau)$, producing the net non-Debye-like behaviour. In the second the response can be regarded as *serial* (Palmer *et al.* 1984), with individual relaxational processes strongly coupled so that the dynamics are severely constrained, one site needing to relax before another can do so; even if each relaxation site has the same value of τ the coupling between them ensures that the time domain is effectively stretched, thereby ensuring a non-Debye response. An empirical distribution of relaxation times $n(\tau)$ can always be found which will account for the observed experimental behaviour, but such a procedure is valid only if the particular $n(\tau)$ has a physical significance. It is much more likely that coupling between ionic motions is important because of the repulsive Coulomb interactions between diffusing cationic species, and that therefore a form of series-constrained dynamics is likely to be operative.

One rather simple microscopic model along these lines which accounts for many of the experimental observations is the so-called diffusion-controlled relaxation (DCR) model (Elliott 1988, Elliott and Owens 1989), illustrated in Fig. 4.42. It is assumed that dielectric polarization occurs principally by means of an interstitialcy-like mechanism, in which an ion diffuses to a site and causes the ion already residing at the site to move to another equilibrium position via the mutual Coulomb repulsion. The time-scale in this case is 'stretched' by the diffusion time between sites. For 3D diffusion, the decay function can be shown to be (Elliott and Owens 1989, Bordewijk 1975)

$$\phi(t) = \exp(-t/\tau')\exp[-(t/\tau'')^{1/2}] \quad [4.103]$$

where

$$\tau' = 1/(4\pi DRn) \quad [4.104(a)]$$

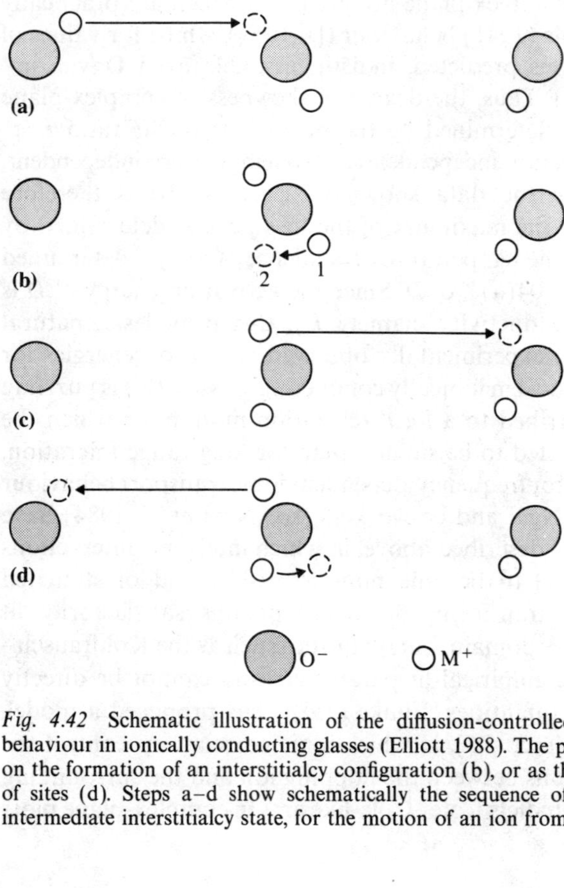

Fig. 4.42 Schematic illustration of the diffusion-controlled relaxation model for a.c. transport behaviour in ionically conducting glasses (Elliott 1988). The polarization mechanism can occur either on the formation of an interstitialcy configuration (b), or as the result of ionic hopping between pairs of sites (d). Steps a–d show schematically the sequence of possible configurations, involving an intermediate interstitialcy state, for the motion of an ion from one site to another.

and

$$\tau'' = 1/(64\pi D R^4 n^2) \qquad [4.104(b)]$$

where R is the radius of the interstitialcy site within which the relaxation occurs, and D is the ionic diffusion coefficient of the mobile ions with concentration n. Thus, the relaxation function [4.103] appears to be of the stretched exponential, Kohlrausch–Williams–Watts form [2.7], but with a time-dependent exponent β.

The frequency dependence corresponding to [4.103] is obtained by taking the Laplace transform of $d\phi/dt$, yielding (Elliott and Owens 1989)

$$M^* = \frac{1}{1+i\omega\tau'} + \frac{i\omega\tau'}{(1+i\omega\tau')^{3/2}}\left[\frac{\pi\tau'}{4\tau''}\right]^{1/2} \exp\left[\frac{\tau'}{4\tau''(1+i\omega\tau')}\right] \times \operatorname{erfc}\left[\frac{(\tau')^{1/2}}{(4\tau''(1+i\omega\tau'))^{1/2}}\right] \qquad [4.105]$$

where erfc stands for the complementary error function; the real and imaginary parts of [4.105] can be obtained numerically (Elliott and Owens 1989). Several interesting features emerge from examination of [4.105]. For values of $\tau'/\tau'' \ll 1$, pure Debye-like behaviour is recovered: from [4.104] this occurs for very small ion concentrations $n \ll (16R^3)^{-1}$, as observed experimentally for soda-doped silicate glasses (Simmons *et al.* 1979). At somewhat larger values of the ratio $\tau'/\tau'' < 1$, depressed arcs in the complex plane are predicted which are practically indistinguishable from Cole–Cole (1941) behaviour ([4.101]), whilst for values of $\tau'/\tau'' > 1$, skewed-arc behaviour is predicted, indistinguishable from Davidson–Cole (1951) behaviour ([4.102]). Thus, the degree of skewness of complex-plane arcs and loss peaks ($M_2(\omega)$) is determined by the magnitude of the ratio τ'/τ'' which, from [4.104], if n is temperature-independent, is also temperature-independent; i.e. the 'temperature-scaling' of the data shown in Fig. 4.41(b) is therefore understandable. The position of the maximum of the loss peak is determined by the condition $\omega_{max}\tau' \simeq 1$, and so the temperature dependence of ω_{max} is determined by that of τ' and hence, from [4.104(a)], of D. Since the activation energy of D is the same as that of the d.c. conductivity, namely E_σ, this provides a natural explanation for the fact that the experimentally observed activation energies for ω_{max} and $\sigma(0)$ are very often identical in ionically conducting glasses; this is puzzling if the loss-peak behaviour is ascribed to a *local* relaxation motion for which the activation energy would be expected to be smaller than for long-range migration.

A number of other theories for frequency-dependent ionic transport behaviour in glasses have been proposed. Ngai and co-workers (see Ngai *et al.* 1984) have developed a model similar to that described above, in which interionic interactions cause the response function $\phi(t)$ to become non-Debye-like and of stretched exponential form; the Fourier transforms of such functions satisfactorily fit experimental data in the frequency domain, but quantities such as the Kohlrausch–Williams–Watts exponent β are empirical-fit parameters and cannot be directly related to microscopic physical quantities. Funke (1986) has proposed a model, similar to the 'polaron' picture (sect. 5.5), in which ionic correlations control the ion jump rates through interactions between the hopping ion and the surrounding polarization cloud. This theory predicts Cole–Cole-like arcs in complex-plane plots but does not appear able to account for skewed-arc behaviour. Dyre (1986) has proposed a model based on a distribution of relaxation times, in which an ion is assumed to experience an approximately random potential energy on diffusing through the disordered structure of an amorphous solid; this model predicts an asymmetric shape for the dielectric loss peak and a power-law behaviour for the a.c. conductivity [4.97], qualitatively similar to that observed experimentally, and also a temperature independence of the shape of the loss-peak or conductivity curves, i.e. 'temperature scaling'. However, this temperature independence occurs only if a random distribution of barrier energies is chosen.

4.5.4 Mixed mobile-ion effect

It might seem reasonable to assume that, as the relative proportion of a binary mixture of, say, alkali oxide modifiers added to a glass-former is varied such that the *total* modifier content is kept constant, the physical properties should change in a slowly varying, approximately linear fashion from one extreme composition to the other. This is indeed the case for gross structural properties, such as molar volume and density; however, properties which are connected with ionic motion, such as conductivity, viscosity and dielectric loss, vary in an extremely non-linear fashion (see Fig. 4.43(a)). The ionic conductivity in such cases can vary by up to

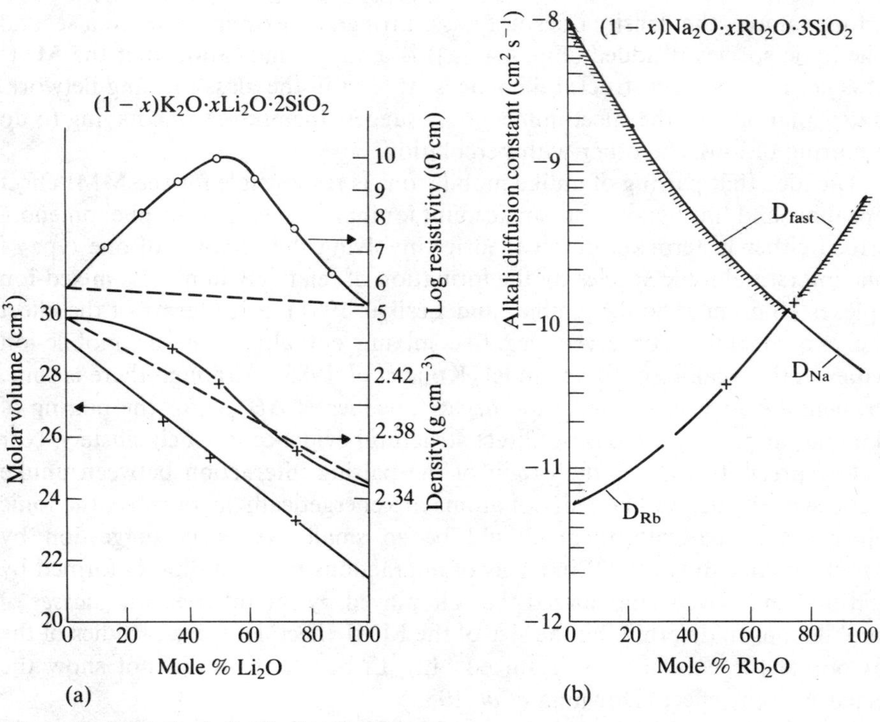

Fig. 4.43 The 'mixed-alkali' effect in silicate glasses.
(a) Electrical resistivity (150°C) compared with molar volume (20°C) and density (20°C) as a function Li:K concentration (Day 1976).
(b) Plots of the alkali diffusion constants (400°C) as a function of Na:Rb concentration. The hatched line corresponds to the dominant diffusion coefficient at a given concentration.

about five orders of magnitude, reaching a minimum near the equimolar composition if the conductivities at the extreme compositions are comparable; both the conductivity pre-exponential factor and the activation energy exhibit maxima at the same composition, and the variation in E_σ is the dominant factor. This departure from linear additivity is termed the 'mixed mobile-ion' (MMI) effect, and while it has been extensively investigated in mixed-alkali glassy systems (see Isard (1969) and Day (1976) for reviews), the MMI effect is also observed for other mixtures of mobile ions, containing e.g. H^+, Ag^+ or Tl^+, and even anion-conducting glasses show the effect (Ingram 1987). However, the phenomenon is not confined to amorphous materials: the MMI effect has also been observed in crystalline FIC materials, such as those based on β-alumina, and also in molten salts.

It is significant that the addition of a second mobile species invariably *lowers* the diffusion coefficient of the original ion (see Fig. 4.43(b)), resulting in a mobility cross-over at some intermediate composition which is not necessarily that at which the conductivity minimum occurs; in other words, the self-diffusion coefficient for an ion in excess is always greater than that of the 'impurity' ion, in marked contrast to the case of metals where interstitial impurities always, and substitutional impurities often, have the larger diffusion coefficients. It is interesting that the Haven ratio ([4.94]) passes through a *maximum* for near-equimolar compositions in mixed-alkali glasses (Jain *et al.* 1983).

The origin of the MMI effect is not properly understood. Clearly, the fact that, for example, the density does not pass through an extremum when a second mobile-ionic species is added (Fig. 4.43(a)) is a strong indication that the MMI effect is *not* due to gross structural changes, at least in the glass-forming network. Two explanations for the effect immediately suggest themselves, one having to do with pairing of ions, the other with percolation.

The idea that pairing of unlike mobile ions is responsible for the MMI effect is appealing, and this hypothesis can account for many aspects of the phenomenon, described either in terms of defect theories involving the removal of one type of mobile interstitial ionic species by the formation of relatively immobile mixed-ion complexes (Ingram 1980, Moynihan and Lesikar 1981), or in terms of the effect of the experimentally observed negative mixing enthalpies on the mobile-ion activities in the weak-electrolyte model (Kone *et al.* 1983). Although there is some experimental evidence (e.g. from the negative values of ΔH_{mix}) for the pairing of unlike ions, at present there is no direct structural evidence for such clusters. Nor is it clear precisely what is the origin of the pairing interaction between unlike ions and why, if such unlike pairs of atoms are energetically favourable, the ionic mobility associated with them should be so small. An early suggestion by Hendrickson and Bray (1972) in terms of interactions between dipoles formed by the cations and NBOs emphasized the role played by the difference in *masses* of the mobile ions in determining the size of the MMI effect. However, studies of the behaviour of the conductivity in mixed $^6Li/^7Li$ borate glass did not show the expected isotopic effect (Downing *et al.* 1982).

An alternative explanation for the MMI effect is in terms of changes in percolating conducting pathways caused by varying the relative proportions of mobile ions, as proposed by Suzuki *et al.* (1984) and Harde *et al.* (1986); the latter authors ascribe the effect to a blocking of percolative pathways by immobile mixed-mobile-ion pairs. However, although the conductivity minimum, characteristic of the MMI effect, can be reproduced by calculations on percolation networks, it is not immediately apparent how such results relate to actual transport paths in real glasses.

Problems

4.1 Deduce the dynamics of amorphous networks using a Lagrangian formulation assuming only nearest-neighbour central forces to be operative. (See M. F. Thorpe and F. L. Galeener (1980), *Phys. Rev.*, **B22**, 3078 and P. N. Sen and M. F. Thorpe (1977), *Phys. Rev.*, **B15**, 4030 for details).

Refer atomic displacements to bond directions (described by unit vectors $\hat{\boldsymbol{q}}$) and call the amplitude of the displacement q. Use the labels ℓ, ℓ' to denote *sites* and Δ, Δ' to denote *bonds*, i.e. the unit vectors are $\hat{\boldsymbol{q}}_\Delta(\ell)$ and the displacement amplitudes of the generalized coordinates $q_\Delta(\ell)$. The potential energy, for bond-stretching forces only, can be written in this coordinate system as:

$$V=\frac{\alpha}{2}\sum_{\{\ell,\ell',\Delta\}}[q_\Delta(\ell)+q_\Delta(\ell')]^2 \qquad [4.1\text{P}]$$

where each bond is only counted once.

The kinetic energy can be written in terms of conventional, orthogonal coordinates (x, y, z) as

$$T=\frac{M}{2}(\dot{x}^2+\dot{y}^2+\dot{z}^2) \qquad [4.2\text{P}]$$

(a) For a twofold coordinated atom (e.g. O), construct axes such that $\hat{\boldsymbol{y}}$ bisects the angle, θ, between two bonds $\hat{\boldsymbol{q}}_1$ and $\hat{\boldsymbol{q}}_2$, and $\hat{\boldsymbol{x}}$ is perpendicular to $\hat{\boldsymbol{y}}$ and in the same plane as $\hat{\boldsymbol{q}}_1$, $\hat{\boldsymbol{q}}_2$. Hence show that

$$T=\frac{M}{2}[a(\dot{q}_1^2+\dot{q}_2^2)+b(\dot{q}_1+\dot{q}_2)^2] \qquad [4.3\text{P}]$$

where $a=(1-\cos\theta)^{-1}$, $a+2b=(1+\cos\theta)^{-1}$, and the third generalized coordinate $q_3=z$ has been omitted because for nearest-neighbour central forces, V is independent of q_3.

Similarly show for a tetrahedrally coordinated atom (e.g. Si), that

$$T=\frac{3m}{8}(\dot{q}_1^2+\dot{q}_2^2+\dot{q}_3^2+\dot{q}_4^2) \qquad [4.4\text{P}]$$

with the constraint $q_1+q_2+q_3+q_4=0$, for a coordinate system such that $\hat{\boldsymbol{z}}$ is antiparallel to q_4 and $\hat{\boldsymbol{x}}$ is the projection of $\hat{\boldsymbol{q}}_1$ on to a plane perpendicular to $\hat{\boldsymbol{z}}$.

The Lagrangian $\mathscr{L}$ is defined by

$$\mathscr{L}([q_i,\dot{q}_i])=T([\dot{q}_i])-V([q_i]) \qquad [4.5\text{P}]$$

where [ ] denotes the whole set of $q_i, \dot{q}_i$ for all i. The equations of motion are:

$$\frac{d}{dt}\left(\frac{\partial \mathcal{L}}{\partial \dot{q}_i}\right) - \frac{\partial \mathcal{L}}{\partial q_i} = 0 \qquad [4.6P]$$

(b) Deduce the dynamical properties of a tetrahedral network by first showing that the Lagrangian may be written as

$$\mathcal{L} = \frac{3}{8}M \sum_{\ell,\Delta} [\dot{q}_\Delta(\ell)^2] - \frac{\alpha}{2} \sum_{\{\ell,\ell',\Delta\}} [q_\Delta(\ell) + q_\Delta(\ell')]^2 - \frac{\lambda}{2}\sum_{\ell}\left(\sum_{\Delta} q_\Delta(\ell)\right)^2$$

[4.7P]

where in the last term λ is a Lagrange multiplier, tending to infinity eventually, and which accounts for the constraint at each site ℓ, $\Sigma_\Delta q_\Delta(\ell) = 0$. Obtain the equation of motion from [4.6P] and look for normal modes of $q_\Delta(\ell)$ and $q_\Delta(\ell')$ having a time dependence $\exp(i\omega t)$ and hence eliminate $q_\Delta(\ell')$. By summing over all ℓ' (neighbours of ℓ), calling $Q(\ell) = \Sigma_\Delta q_\Delta(\ell)$, and rewriting $\Sigma_{\ell'} Q(\ell') = \varepsilon Q(\ell)$ show that

$$\varepsilon = \frac{1}{\lambda\alpha}[(\alpha - \tfrac{3}{2}M\omega^2)^2 + 4\lambda(\alpha - \tfrac{3}{4}M\omega^2) - \alpha^2] \qquad [4.8P]$$

Let $\lambda \to \infty$, thereby driving off to infinite frequency the unwanted modes associated with the extra degrees of freedom. Thus finally

$$M\omega^2 = \frac{\alpha}{3}(4 - \varepsilon) \qquad [4.9P]$$

The bounds to ε can be found by noting that $\Sigma_{\ell'} Q(\ell') = \varepsilon Q(\ell)$ can be written as $\mathbf{M}\boldsymbol{v} = \varepsilon \boldsymbol{v}$ where $\mathbf{M}$ is the connectivity matrix ($M_{ij} = 1$, i, j, nearest neighbours; $M_{ij} = 0$, otherwise), and is real, symmetric and positive, whereupon the Frobenius theorem can be applied, and it is then found that

$$-Z \leqslant \varepsilon \leqslant Z \qquad [4.10P]$$

where Z is the connectivity (4 for Si). Hence, the bounds to the frequencies given by [4.9P] are $0 \leqslant \omega^2 \leqslant 8\alpha/3M$.

(c) By inserting twofold coordinated atoms between the atoms to generate an AX_2 network (e.g. SiO_2) with an A—X—A bond angle θ, use the same Lagrange formalism to deduce that the frequencies of the band edges (i.e. when $\varepsilon = 0, 4$) are given by the equations:

$$\omega_1^2 = \frac{\alpha}{m}(1 + \cos\theta) \qquad \omega_2^2 = \frac{\alpha}{m}(1 - \cos\theta)$$

$$\omega_3^2 = \omega_1^2 + \frac{4\alpha}{3M} \qquad \omega_4^2 = \omega_2^2 + \frac{4\alpha}{3M} \qquad [4.11P]$$

Discuss the behaviour of the modes as $\theta \to 90°$, particularly in relation to an isolated AX_4 molecule, and find the critical bond angle in this model below and above which molecular-like and band-like modes, respectively, exist.

4.2 There is a formal analogy between the two-level system (TLS) used to describe low-temperature excitations in glasses and a spin-$\frac{1}{2}$ particle in a magnetic field, enabling much of the formalism developed for the magnetic case to be used for the two-level system model also.

The Hamiltonian for a TLS in an elastic strain field e and an electric field F can be written as:

$$\mathscr{H} = H_0 + H_1 + H_2 = \frac{1}{2}\begin{pmatrix} E & 0 \\ 0 & -E \end{pmatrix} - \frac{1}{2}\begin{pmatrix} D & 2M \\ 2M & -D \end{pmatrix} e - \frac{1}{2}\begin{pmatrix} \boldsymbol{\mu} & 2\boldsymbol{\mu}' \\ 2\boldsymbol{\mu}' & \boldsymbol{\mu} \end{pmatrix} \boldsymbol{F} \qquad [4.12\text{P}]$$

where H_0 represents the unperturbed case (E given by [4.65]), H_1 represents the perturbation due to the strain field and H_2 represents the perturbation due to the electric field. The diagonal and off-diagonal components of the deformation potential (D, M), and the permanent and induced dipole moments $(\boldsymbol{\mu}, \boldsymbol{\mu}')$ can be written in terms of TLS parameters as:

$$D = \frac{2\gamma\Delta}{E}, \qquad M = \frac{\gamma\Delta_0}{E} \qquad [4.13\text{P}]$$

$$\mu = \frac{2p\Delta}{F}, \qquad \mu' = \frac{p\Delta_0}{F} \qquad [4.14\text{P}]$$

where $\gamma = \frac{1}{2}(\partial\Delta/\partial e)$ and $p = \frac{1}{2}(\partial\Delta/\partial F)$.

(a) Show that, by analogy with the equivalent equations for magnetic resonance, the dynamics of a TLS subject to an electric strain field can be written in the form of the Bloch equations:

$$\dot{X} = -X/T_2 - (E + De)Y/\hbar \qquad [4.15(\text{a})\text{P}]$$

$$\dot{Y} = (E + De)X/\hbar - Y/T_2 - (2Me/\hbar)Z \qquad [4.15(\text{b})\text{P}]$$

$$\dot{Z} = (2Me/\hbar)Y - (Z - Z_e)/T_1 \qquad [4.15(\text{c})\text{P}]$$

X and Y represent the induced elastic (or electric) polarization, but Z can only formally be interpreted as a polarization component since the level splitting is not caused by an external field (as in the magnetic case). Z_e is the value of Z in its 'instantaneous equilibrium':

$$Z_e = -\frac{1}{2}\tanh[(E + De)/2k_B T] \qquad [4.16\text{P}]$$

(b) Discuss the dynamical solutions of [4.15P] for a TLS subject to a pulsed field of duration τ for various relative magnitudes of T_1, T_2 and τ.

(c) The contribution of N TLS (per unit volume) to the internal energy U can be written as $U = -N(2Me, 0, De)\cdot(X, Y, Z)$. The contribution to the dynamic elastic modulus is then:

$$\partial C = \frac{1}{e}\frac{\partial(\partial u)}{\partial e} = -\frac{N}{e}(2MX + DZ) \qquad [4.17\text{P}]$$

Sound dispersion and absorption are then given by:

$$\delta v(\omega) = \left(\frac{1}{2\rho v}\right) \mathrm{Re}\{\partial C(\omega)\} \qquad [4.18(a)P]$$

$$\alpha(\omega) = -\left(\frac{\omega}{\rho v^3}\right) \mathrm{Im}\{\partial C(\omega)\} \qquad [4.18(b)P]$$

where v is the velocity of sound in the absence of TLS. Using the steady state solution of X ([4.15(a)P]) which gives an absorption of resonant character, derive

$$\alpha_{\mathrm{res}}(\omega) = \frac{\pi n_{\mathrm{a}} M^2 \omega}{\rho v^3} \frac{\tanh(\hbar\omega/2k_{\mathrm{B}}T)}{\sqrt{(1+I/I_{\mathrm{C}})}} \qquad [4.19P]$$

where I is the acoustic intensity, $I_{\mathrm{c}} = \hbar^2 \rho v^3 / 2M^2 T_1 T_2$ and n_{a} is the constant density of states for the acoustically active centres. Hence account for the saturation behaviour shown in Fig. 4.34.

Bibliography

Vibrational properties

D. Beeman and R. Alben (1977), *Adv. Phys.* **26**, 339.

M. Cardona (ed.) (1975), *Light Scattering in Solids* (*Topics in Applied Physics*, vol. 8) (Springer-Verlag).

S. S. Mitra (ed.) (1976), *Physics of Structurally Disordered Solids* (Plenum).

B. T. M. Willis (ed.) (1973), *Chemical Applications of Thermal Neutron Scattering* (OUP).

Ionic transport

M. D. Ingram (1987), *Phys. Chem. Glasses* **28**, 215.

Low-temperature properties

W. A. Phillips (ed.) (1981), *Amorphous Solids: Low Temperature Properties* (*Topics in Current Physics*, vol. 24) (Springer-Verlag).

W. A. Phillips (1987), *Rep. Prog. Phys.* **50**, 1657.

5 Electrons

5.1 Introduction

It has been a great triumph of the application of quantum mechanics to solid-state physics that an understanding of why certain crystals are metals and others are insulators has been achieved. The presence of perfect periodicity greatly simplifies the mathematical treatment of the behaviour of electrons in a solid. The electron states in this case can be written as 'Bloch waves' extending throughout the crystal:

$$\psi(\boldsymbol{k}, \boldsymbol{r}) = u(\boldsymbol{k}, \boldsymbol{r}) \exp(i\boldsymbol{k}\cdot\boldsymbol{r}) \qquad [5.1]$$

where the function $u(\boldsymbol{k}, \boldsymbol{r})$ has the periodicity of the crystal lattice (in which a lattice translation vector $\boldsymbol{R}_1$ connects lattice points):

$$u(\boldsymbol{k}, \boldsymbol{r} + \boldsymbol{R}_1) = u(\boldsymbol{k}, \boldsymbol{r}) \qquad [5.2]$$

and this modulates the term $\exp(i\boldsymbol{k}\cdot\boldsymbol{r})$ representing a plane wave. The allowed wavevectors $\boldsymbol{k}$ of the electrons are intimately related to the symmetry of the underlying crystal lattice since a reciprocal lattice (related to the unit cell parameters) can be established in reciprocal or $\boldsymbol{k}$-space.

The allowed energies of the electrons can thus be represented by means of a 'band structure' in $\boldsymbol{k}$-space. A free electron has an energy $E(k) = \hbar^2 k^2/2m$, but this parabolic dependence is distorted considerably if the electron experiences a scattering potential. Of course for a crystalline solid, these potentials arise from the periodic array of atom centres, and the electron waves can Bragg reflect from the lattice planes. This results in energy gaps opening up at values of k corresponding to certain values of reciprocal lattice vector ($2\pi/a$ for a linear array of atoms of separation a), i.e. at the edge of the 'Brillouin zone'. The occurrence of energy ranges for which there are *no* allowed electron states can be thought of as being due to destructive interference of the Bragg reflected electron waves; in the 'nearly-free electron' model, the size of the energy gap is determined by the Fourier component of the potential corresponding to the Bragg condition (see e.g. Madelung 1978). The difference between metals and insulators then simply amounts to whether there are sufficient electrons available to fill all the states in a Brillouin zone; if the band is only partly filled the solid is a metal, whereas if the band is completely filled, there is a gap between occupied and unoccupied electron states, and the material is an insulator at $T = 0$.

Let us now consider the case of amorphous materials. For these, as we saw in Chapter 3, there is no periodicity and hence there can be no reciprocal or ($\boldsymbol{k}$-) space, and the electrons cannot be represented as Bloch states ([5.1]). Since, as we have seen, the occurrence of band gaps could be viewed as arising from Bragg reflection of electron waves and hence is a direct consequence of periodicity, the question therefore arises: Should band gaps occur in amorphous materials too? The fact that, for example, window-glass (silica) is transparent to visible light is direct experimental proof that a band gap $\gtrsim 2$ eV must exist for this material (in fact the gap is nearer 10 eV). Conventional (crystalline) solid-state physics is incapable of accounting for this behaviour, and we shall see that concepts more akin to those of chemistry can resolve the dilemma.

We have seen that the absence of periodicity in an amorphous solid dictates that there can be no reciprocal space. In this case, electron states cannot be represented by a band structure in the form $E(\boldsymbol{k})$. This problem is entirely analogous to that encountered for phonon dispersion curves $\omega(\boldsymbol{q})$ in Chapter 4. The quantity that *is* equally valid as a description of electron states for both crystalline and amorphous solids is, as in the phonon case, the density of states. This can be written in the form:

$$g(E)=\frac{1}{V}\sum_{n}\delta(E-E_n) \qquad [5.3]$$

where $g(E)$ is the density of states per unit volume per unit energy interval, and V is the volume of the system. The quantity more often used is the integrated density of states:

$$N(E)=\int_{-\infty}^{E} g(E)\,\mathrm{d}E \qquad [5.4]$$

We can now address the problem of whether a gap can exist in the density of states of an amorphous covalently bonded material. Weaire and Thorpe (1971) first showed that if short-range interactions between electrons are dominant, then it is the *short-range order* which mainly determines the electronic density of states. In particular they showed using a simple model Hamiltonian for the electron interactions that a gap *is* expected for an ideal tetrahedrally coordinated amorphous solid, providing that the interactions are of a certain magnitude. The amorphous structure is taken to be that in which each atom is in a perfect tetrahedral environment, with presumably a wide distribution of dihedral angles necessary to generate a random network. (Whether in practice a CRN could be constructed without bond-angle distortions remains unclear, but this assumption simplifies the treatment.)

The alternative approach to the nearly-free electron approximation, namely the tight-binding LCAO approach, is adopted, in which the basis functions are localized at each atomic site rather than being extended plane (Bloch) waves. The two interactions considered in the model are an intrasite 'banding' interaction V_1, responsible for the width of the bands, and an intersite 'bonding' interaction V_2, responsible for the separation of bonding and antibonding bands; they are shown schematically in Fig. 5.1, where the basis functions are taken to be sp^3 hybridized orbits localized at each site. The Hamiltonian is thus written as (Weaire and Thorpe 1971):

$$\mathscr{H}=\sum_{\alpha ij} V_1|\alpha i\rangle\langle\alpha j| + \sum_{\alpha\beta i} V_2|\alpha i\rangle\langle\beta i| \qquad [5.5]$$

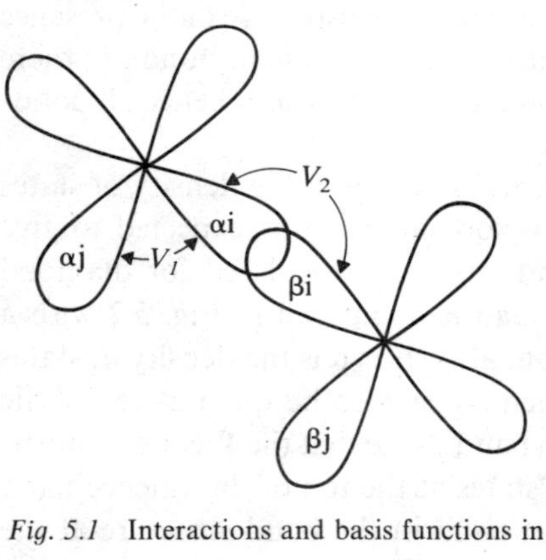

Fig. 5.1 Interactions and basis functions in the Weaire–Thorpe Hamiltonian.

All site orbitals are assumed to be orthogonal, and V_1 and V_2 are assumed to be the same for all atoms of the network. By consideration of an isolated atom, it can be shown that V_1 must be negative (so that the s-states are lower in energy than the p-states), and V_2 must also be negative (so that bonding orbitals are lower in energy than antibonding orbitals). Instead of using the states $|\alpha i\rangle$ as a basis, we can equally use bonding (B) and antibonding (A) orbitals associated with pairs of neighbouring atoms:

$$|B,\alpha i\rangle = \frac{1}{\sqrt{2}}(|\alpha i\rangle + |\beta i\rangle)$$
$$|A,\alpha i\rangle = \frac{1}{\sqrt{2}}(|\alpha i\rangle - |\beta i\rangle) \qquad [5.6]$$

If the assumption is made that the valence band is solely constructed from *bonding* orbitals, i.e. $|\psi\rangle = \Sigma_{\text{bonds}} b_{\alpha i}|B,\alpha i\rangle$, then the expectation value of the energy for this state is:

$$E = \frac{\langle\psi|H|\psi\rangle}{\langle\psi|\psi\rangle} = \frac{\Sigma b^*_{\alpha' i'} b_{\alpha i}\langle B,\alpha' i'|H|B,\alpha i\rangle}{\langle\psi|\psi\rangle} \qquad [5.7]$$

where the sum is over all pairs of bonds. Contributions of V_2 are obtained from the terms for which $\alpha i = \alpha' i'$, and $\frac{1}{2}V_1$ from terms for which $\alpha = \alpha'$, $i \neq i'$, giving:

$$E = V_2 + \tfrac{1}{2}V_1\left[\sum_{\alpha}\sum_{i\neq i'} b^*_{\alpha i'} b_{\alpha i}\right] \bigg/ \tfrac{1}{2}\sum_{\alpha}\sum_{i}|b_{\alpha i}|^2 \qquad [5.8]$$

It can be shown (see problem 5.1) that the valence band limits must lie between $(V_2 - V_1)$ (when each $\Sigma_i b_{\alpha i} = 0$) and $(V_2 + 3V_1)$ (when each $b_{\alpha i} = b_{\alpha i'}$). Similarly, by assuming the conduction band is constructed solely from *antibonding* orbitals, the band limits are found to be $(-V_2 + 3V_1)$ and $(-V_2 + V_1)$.

Thus, if $|V_2| > 2|V_1|$, there is no overlap between the bands, i.e. a true band gap of magnitude

$$E_g \geqslant 2|V_2 - 2V_1| \qquad [5.9]$$

must exist. Note that this model is concerned only with the short-range structure and says nothing at all about longer range structure, i.e. the results obtained hold for amorphous *and* crystalline tetrahedral systems alike, irrespective of the presence of periodicity. This is in accord with the chemist's view of covalent bonding; there are four states per atom in the valence (and conduction) band, and hence all bonds are satisfied in the valence band.

Although providing an 'existence theorem' for a gap in the density of states of an amorphous semiconductor, the model is too simple to be expected to give quantitative results. As an example, the band structure calculated for diamond cubic Ge using the Weaire–Thorpe Hamiltonian is compared in Fig. 5.2 with a more sophisticated pseudopotential calculation; also shown is the density of states obtained from such band structures. Agreement is seen to be qualitative for the valence band, but very poor for the conduction band. Note that the Weaire–Thorpe model gives a delta function in the density of states at the top of the valence band (resulting, in the crystalline case, from the flat band in the band structure at the

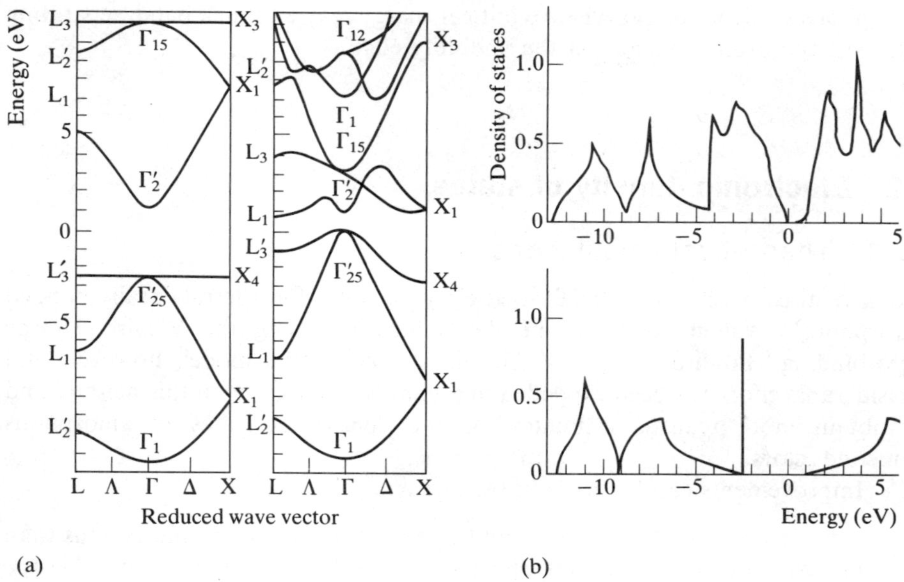

Fig. 5.2 (a) Band structure of crystalline Ge (diamond cubic structure) calculated using the Weaire–Thorpe Hamiltonian and pseudopotential theory.
(b) (top) The density of states for these band structures (the zero of energy marks the Fermi level). (bottom) The density of states calculated using the Weaire–Thorpe Hamiltonian (Weaire *et al.* 1972).

same position), which contains pure p-like bonding states. This region is relatively insensitive to the detailed structure (although the presence of like-atom bonds in an alloy does affect it); in contrast, as we shall see later, the deeper-lying states in the valence band density of states (mainly s-like for the deepest band and mixed s- and p-like for the intermediate band) *are* very sensitive to structural variations. In order to improve on these calculations for amorphous solids, a more realistic Hamiltonian is required, which remedies at least two deficiencies, namely the lack of the inclusion of longer range interactions and the absence of variations in the interactions arising from fluctuations in the structure, e.g. bond-angle distortions (see Yonezawa and Cohen 1981).

The object of this chapter is to discuss the electronic properties of amorphous materials, both amorphous semiconductors and amorphous metals. In addition, we will consider almost exclusively 'ideal' amorphous structures, in which there are no structural defects such as dangling bonds or voids, etc., only structural fluctuations such as bond-angle variations or like-atom bonds consistent with a CRN; defect-controlled electronic properties will be dealt with in Chapter 6.

In this chapter, several universal features exhibited by amorphous materials will be explored. The first is the fact that randomness causes the electron wavevector, $\boldsymbol{k}$, to be no longer a good quantum number (hence making band structures meaningless); the *density of states*, however, does remain a valid concept and in fact is determined primarily by the local order in an amorphous solid, as we have already seen in this section. The second universal feature is that disorder can, under certain circumstances, cause *localization* of the electronic wavefunction for electron energies through much of the band of states. This is particularly important for amorphous semiconductors where a critical energy may exist in a band, separating localized and extended states at the 'mobility edge'.

5.2 Electronic density of states

5.2.1 Theoretical calculations

We have already seen that the allowed electron states for a tetrahedrally bonded amorphous covalent solid can be determined by using the Weaire–Thorpe tight-binding Hamiltonian [5.5]. This is a very simple model, however, and considerable effort has been expended in attempts to improve on this method and to obtain more realistic estimates for the density of states of amorphous semiconductors.

Improvements can be made in two areas:

(a) A more realistic Hamiltonian can be used, involving more interactions than just the two inter- and intrasite terms considered by Weaire and Thorpe (Fig. 5.1), in particular including interactions involving more distant neighbours (see e.g. Bullett and Kelly 1979).

(b) *Topological* disorder can be introduced quantitatively by considering a more realistic structural model than that used by Weaire and Thorpe, which assumed perfect nearest-neighbour tetrahedral order, but did not address the problem of longer range structure (since only *short-range* interactions were included). Although this independence of structure is a feature of the Weaire–Thorpe model, giving rise to an energy gap whatever the structure (if $|V_2| > 2|V_1|$), it is intuitively obvious that details of the structure, such as ring statistics for a covalent solid, would be expected to influence the detailed shape of the density of states. Thus the atomic coordinates from a structural model which fits well experimental scattering data, such as a continuous random network for an amorphous semiconductor or a dense random packed mode for an amorphous metal, are a better starting point. In this manner, *quantitative* disorder in the form of variations of the value of interactions (e.g. V_1 and V_2) are naturally included as a result of the presence of bond-angle and dihedral-angle variations for a CRN, or packing variations for a DRP model.

The use of large structural models in density of states calculations, however, necessitates the employment of sophisticated numerical techniques in order to cope with the diagonalization of the large matrices involved; a general review of these techniques can be found in Kramer and Weaire (1979). Essentially the same methods can be employed as are used in the *vibrational* density of states calculations that we discussed in section 4.3. The Lanzcos method at the heart of the negative eigenvalue method (sect. 4.2.3) and the equation-of-motion method (sect. 4.2.4) for example, have both been used in this regard. More widely used, however, are two equally successful techniques which calculate the *local* density of states of a cluster ([4.29]), rather than effect the diagonalization of large matrices corresponding to the whole cluster. These are the 'recursion' method and the 'cluster-Bethe lattice' method.

The recursion method has already been reviewed in section 4.2.5; an extensive discussion of its application to electronic density of states calculations for amorphous solids has been given by Bullett and Kelly (1979) and Kelly (1980). The recursion scheme to calculate the local density of states commences with a choice of $|u_o\rangle$, the orbital of interest (e.g. an sp^3 hybrid for a tetrahedral solid), together with an appropriate Hamiltonian. If the starting vector is constructed from an equal contribution from each orbital in the system, but with a random phase factor $\exp(i\delta)$ with $0 \leqslant \delta_i \leqslant 2\pi$, a good approximation to the total density of states is obtained since the starting orbital picks up an equal contribution to the spectrum from each distinct energy. The local density of states is related to the diagonal Green function matrix element $n(E, m) = -1/\pi \, \mathrm{Im}\langle\phi_m|(E + i\varepsilon - H)^{-1}|\phi_m\rangle$ (cf. [4.29]) where $|\phi_m\rangle = |u_o\rangle$ in the recursion scheme. The average over a unit cell of a *crystalline* local density of states is proportional to the total density of states; for an amorphous solid, on the other hand, an average should be performed over as many atomic sites as possible since each $n(E)$ reflects its own particular environment, although 20 sites seem to be sufficient in practice. Boundary effects are minimized by choosing sites near the centre of the cluster since, although interactions with distant sites are included in the continued fraction, their contribution becomes negligibly small if the cluster is large enough (say, a few hundred atoms). Thus, there is no need for structural models with periodic boundary conditions (e.g. the Henderson model) as required by those methods more commonly used for crystalline systems (e.g. using pseudopotentials).

As an example of the recursion method, we show in Fig. 5.3 calculations of the valence-band densities of states for silicon in various structural forms. The two crystalline forms are the diamond lattice and the ST-12 lattice (so named because it has a simple tetragonal unit cell containing 12 atoms), which differ in two respects; the diamond lattice is formed from sixfold rings with each atom having the same tetrahedral bond angle, whereas the higher density ST-12 structure has a variety of ring sizes, even and odd (the smallest being five-membered) with a spectrum of bond angles $\simeq \pm 25\%$ about the tetrahedral value, $109°28'$. The model amorphous structures employed are the Polk–Boudreaux (even–odd ring) model and the Connell–Temkin (even ring) model. The first point to note is that the local densities of states of all these structural forms are quantitatively similar, supporting Weaire and Thorpe's contention that it is the short-range order, i.e. the tetrahedral coordination, which determines the gross features. However, there are distinct differences, despite the fact that the same model Hamiltonian was used in each case, which must arise therefore from topological differences between the various structural forms. In particular, the amorphous forms are differentiated from the diamond cubic form by a filling-in of the dip between s-states (at X_1) and a distinct skewing of the p-state distributions at the top of the valence band towards the gap. These trends are shown schematically in Fig. 5.4, and are also observed experimentally in photoemission studies of Ge and Si (see later), lending credence to the validity of the theoretical calculations and suggesting that odd-membered rings must be present in the amorphous phase. Note, however, from Fig. 5.4(b) that these trends are different for the case of heteropolar systems (e.g.

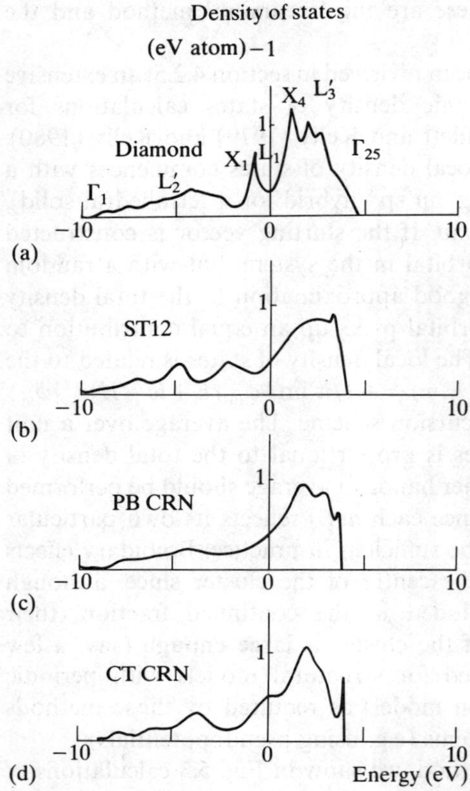

Fig. 5.3 Densities of states for the valence band calculated using the recursion method for various crystalline modifications and amorphous models of Si (Kelly 1980): (a) diamond cubic; (b) ST12 crystalline modification; (c) Polk–Boudreaux (1973) CRN; (d) Connell–Temkin (1974) even-ring CRN. The zero of energy is the self-energy of an isolated bond.

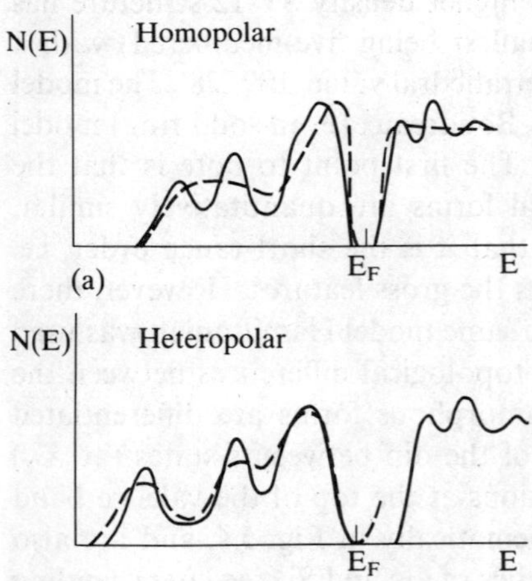

Fig. 5.4 Schematic density of states for tetrahedrally bonded semiconductors, (a) homopolar and (b) heteropolar, in both the amorphous (dashed line) and crystalline (solid line) phases (after Joannopoulos and Cohen 1976).

GaAs), principally because of the difference in ionicity of the anion and cation and the consequent avoidance of like-atom bonds and hence odd-membered rings.

Another method commonly used to obtain the local density of states is the 'cluster-Bethe-lattice' method – for a review of this approach see Joannopoulos and Cohen (1976) and Joannopoulos (1979). In this, a small symmetrical cluster (containing a few tens of atoms) has Bethe lattices (or Cayley trees) attached to the surface dangling bonds; the Bethe lattice is characterized by having the same connectivity as the host network, but contains *no* closed rings of atoms (Fig. 4.7). The method has the following advantages:

(a) The density of states of a Bethe lattice can be calculated exactly for a variety of model Hamiltonians, and furthermore generally yields a smooth and a featureless spectrum.
(b) The local density of states of the atom at the centre of the cluster in a given environment (say in a fivefold ring) can be calculated analytically, without attendant problems of boundary effects, because of the attached Bethe lattices which simulate the rest of the network.

The local density of states is calculated in a manner similar in spirit to that of the recursion method, namely via a continued fraction, although the details differ. The Green function involving the model Hamiltonian $\mathscr{H}$ may be written as a Dyson equation:

$$\mathbf{G} = (E - \mathscr{H})^{-1} = 1/E + \mathscr{H}\mathbf{G}/E \qquad [5.10]$$

so that:

$$E\langle i|\mathbf{G}|j\rangle = \delta_{ij} + \sum_{n} \langle i|\mathscr{H}|n\rangle\langle n|\mathbf{G}|j\rangle \qquad [5.11]$$

where n runs over the sites which interact with $|i\rangle$. If we consider a Bethe lattice of coordination z, with a single orbital $|i\rangle$ per site which interacts only with its nearest neighbours with strength h, then generally:

$$E\langle n|\mathbf{G}|0\rangle = h\langle n-1|\mathbf{G}|0\rangle + (z-1)h\langle n+1|\mathbf{G}|0\rangle \qquad [5.12]$$

Equation [5.12] is solved successively to give a continued fraction expansion of $\langle 0|\mathbf{G}|0\rangle$, which of course gives the local density of states (cf. [4.29]). Simple tight-binding Hamiltonians are used in this method, but complicated Hamiltonians describing additional interactions can be dealt with, although less easily than in the recursion method.

An interesting recent development by Hickey and Morgan (1986) has been to use the equation-of-motion method (see sect. 4.2.4) to obtain for a-Si not just the electronic density of states $g(E)$, but also the *spectral function* $\rho(k, E)$, i.e. the probability that an electron with energy E has a momentum $\hbar k$, where

$$g(E) = \frac{1}{2\pi^2}\int k^2 \rho(k, E)\,\mathrm{d}k \qquad [5.13]$$

by using pseudopotentials and performing the calculations in *reciprocal space*. It was assumed for simplicity that the pseudopotentials at each site are identical and equal to that for crystalline Si. Disorder was therefore introduced only in topological form through the use of CRN models for a-Si, namely those constructed by Wooten *et al.* (1985) containing periodic boundary conditions. The density of states so obtained is shown in Fig. 5.5(a), where it can be seen that a well-defined gap in $g(E)$ separating valence and conduction bands appears to be predicted for a-Si, in accordance with its observed semiconducting properties. (The noisy appearance of the calculated DOS for a-Si results from the small size of the model used.) It can also be seen that there is a close correspondence with $g(E)$ calculated

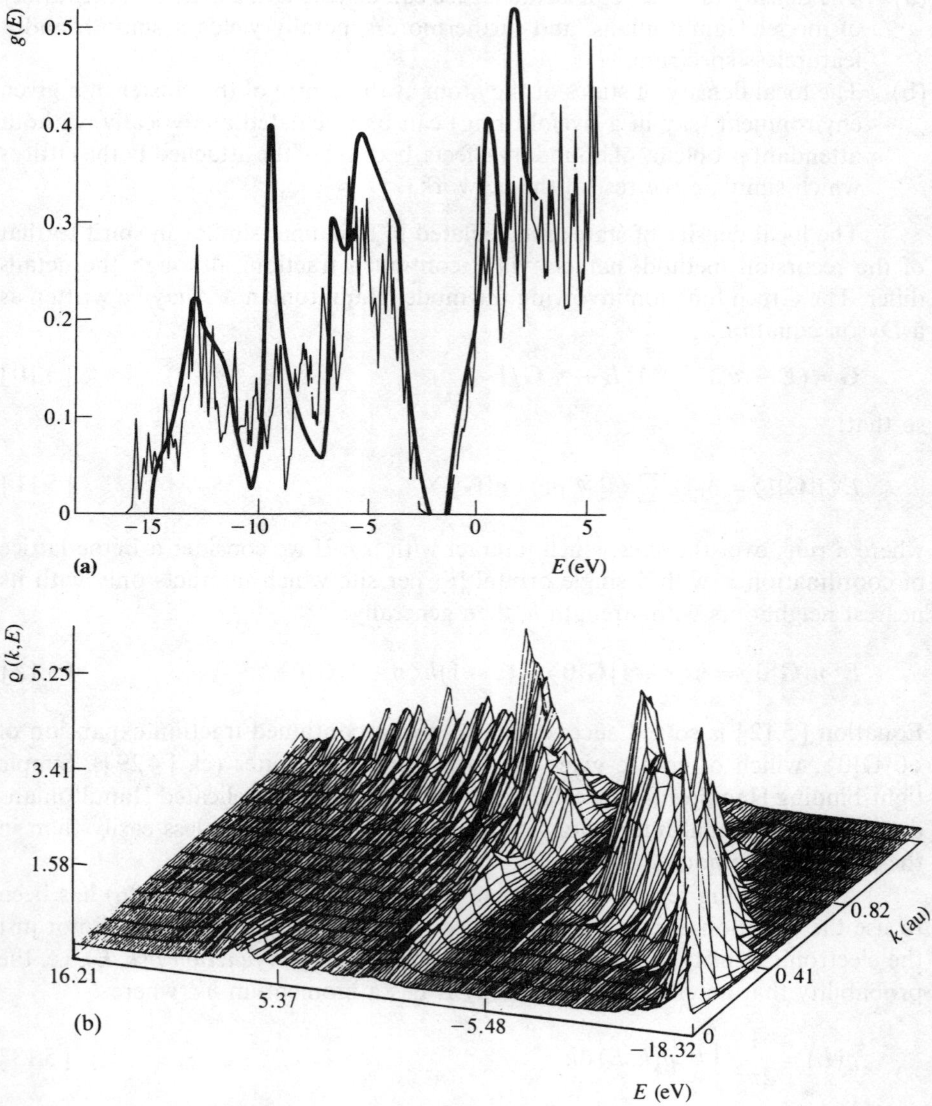

Fig. 5.5 (a) Calculated electronic density of states for a-Si, compared with that for c-Si. (b) Spectral function $\rho(k, E)$ for a-Si (Hickey and Morgan 1986).

for crystalline Si using the same pseudopotential, with the exception of the sharp peak about 6 eV below the top of the valence band for c-Si.

The calculated spectral function $\rho(k, E)$ for a-Si is shown in Fig. 5.5(b); interestingly it has an essentially free-electron-like parabolic behaviour, but appreciably broadened with a large spectral width. The existence of free-electron-like states in amorphous semiconductors runs counter to the assumption of tight-binding states employed in the local picture used by Weaire and Thorpe (1971) to account for the origin of the band gap in amorphous semiconductors (sect. 5.1). Hickey and Morgan (1986) have instead ascribed the existence of band gaps to the behaviour of the spectral function $\rho(k, E)$. This can be written as (Hickey and Morgan 1986)

$$\rho(k, E) = \frac{\Gamma_{\text{I}}}{\pi[(E - k^2 - \Gamma_{\text{R}})^2 + \Gamma_{\text{I}}^2]} \tag{5.14}$$

(cf. [4.29]), where Γ_{R} and Γ_{I} are the real and imaginary parts of the self-energy, respectively. A gap can arise, i.e. $g(E) \to 0$, by Γ_{R} becomining infinite, but it is more likely that instead Γ_{I} vanishes whilst the term $(E - k^2 - \Gamma_{\text{R}})$ remains non-zero in the gap region. The precise behaviour of $\rho(k, E)$ is difficult to ascertain from the computed results for a-Si because of the noise problems alluded to above. However, it does seem that, in the vicinity of the gap, the amplitude of the spectral function decreases but the width of $\rho(k, E)$ appears not to broaden, seemingly supporting the hypothesis that it is the behaviour of Γ_{I} that controls the gap behaviour. The results of this interesting approach need to be corroborated by further studies on larger models and different systems.

One of the principal uses of theoretical densities of states calculations is to act as a test of structural models for amorphous materials. If the electronic Hamiltonian (or dynamical matrix) is realistic (as may be ascertained by a comparison of calculated crystalline band structures (or dispersion relations) and densities of states with experiment), calculations using various amorphous structural models provide a sensitive probe of the topology, e.g. ring statistics, of such models. In fact, the sensitivity of the electronic or vibrational states to the ring statistics can be significantly higher than, say, is the RDF.

5.2.2 Experimental determination

In this section we will be concerned with those experimental techniques which probe the gross features of the electronic structure and consequently give a measure of the density of states. Two such techniques are photoemission and UV and X-ray absorption. We will not discuss those experiments which probe the band-edge region for semiconductors, e.g. optical absorption at lower energies, since these will be dealt with later in this chapter.

The technique most widely used to explore the valence and conduction band densities of states is photoemission; as its name implies, this measures the energy distribution of photo-emitted electrons as a function of incident photon energies. The interested reader is referred to the two books edited by Cardona and Ley (1978) for details. The commonest photon sources are the UV line spectra of rare-gas discharge lamps, or X-ray anodes; these give rise, respectively, to two conventional branches of photoemission, namely 'ultra-violet photoemission spectroscopy' (UPS) and 'X-ray photoemission spectroscopy' (XPS). The UV photon energy range is 10–50 eV, whereas the K_α X-ray lines commonly employed are 1486.6 eV (Al) and 1253.6 eV (Mg), although this gap can be bridged by the use of synchrotron radiation, rendering the difference between the two techniques less distinct. The photo-emitted electrons can be energy analysed either by a retarding grid analyser or a dispersive electrostatic analyser. The experiments have to be conducted in ultra-high vacuum to minimize surface contamination, since the escape depth of the photo-electrons is very small ($\simeq 5$ Å for UPS and $\simeq 50$ Å for XPS).

The photoemission process can be understood in terms of the 'three-step model', namely: (1) optical excitation of an electron; (2) its transport through the solid (including the possibility of inelastic scattering by other electrons; and (3) escape through the sample surface into the vacuum, although this approach drastically approximates the many-body processes that take place. The 'energy distribution curves' (EDCs) of the photo-emitted electrons are given by:

$$I(E, \hbar\omega) = P(E, \hbar\omega) T(E) D(E) \qquad [5.15]$$

where $P(E, \hbar\omega)$ represents the distribution of photo-electrons of energy E excited by a photon of energy $\hbar\omega$, $T(E)$ is a transmission function (weakly and smoothly varying with E), and $D(E)$ is an escape function which is also a smooth function of E. Thus it is only $P(E, \hbar\omega)$ which contributes structure to the EDCs, and this may be written, when $\boldsymbol{k}$-conservation is not important, as is the case for amorphous materials, as:

$$P(E, \hbar\omega) \propto N_c(E) N_v(E - \hbar\omega) M^2(E, E - \hbar\omega) \qquad [5.16]$$

where N_c and N_v are the conduction and valence band densities of states, respectively, and M is a matrix element which may be taken to be constant for limited photon energy ranges. Thus we see that the photoemission EDCs are determined by a *joint* density of states involving a convolution of occupied and unoccupied electron states.

The origin of $P(E, \hbar\omega)$ is illustrated schematically in Fig. 5.6, where the features in the EDCs can be traced back to structure in the densities of states of the bands. The valence band density of states has been raised by $\hbar\omega$ with respect to the conduction band to account for the term $N_v(E - \hbar\omega)$ and this is then convoluted with $N_c(E)$ and $T(E)$ to give $P(E, \hbar\omega)$ (assuming $D(E)$ and $M(E, E - \hbar\omega)$ to be constant). It can be seen from Fig. 5.6 that features in the EDCs obtained from UPS which *do* shift in position when $\hbar\omega$ is varied are to be associated with peaks in the *valence* band density of states, whereas conversely those features which do *not* shift in position are associated with maxima in the *conduction band* density of states, although conduction band states below the vacuum level of the semiconductor are inaccessible to the technique and can only be rendered accessible by means of a layer of caesium evaporated on to the surface to lower the work function. In this way the valence band and conduction band densities of states can be extracted independently from the EDCs. An illustration of this is given in Fig. 5.7(a) for the case of amorphous and crystalline Ge (Spicer 1974), where it is seen that the EDCs for both materials alter dramatically (in

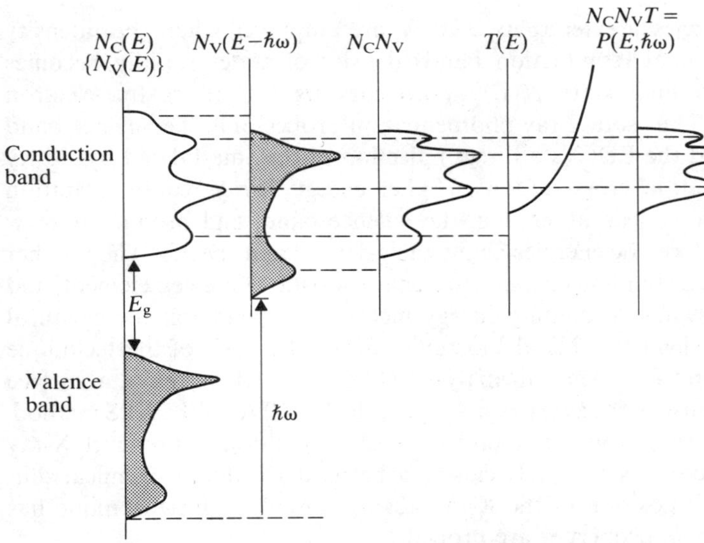

Fig. 5.6 Schematic illustration of the origin of the features which are observed in photoemission EDCs, $P(E, \hbar\omega)$ (after Mott and Davis 1979).

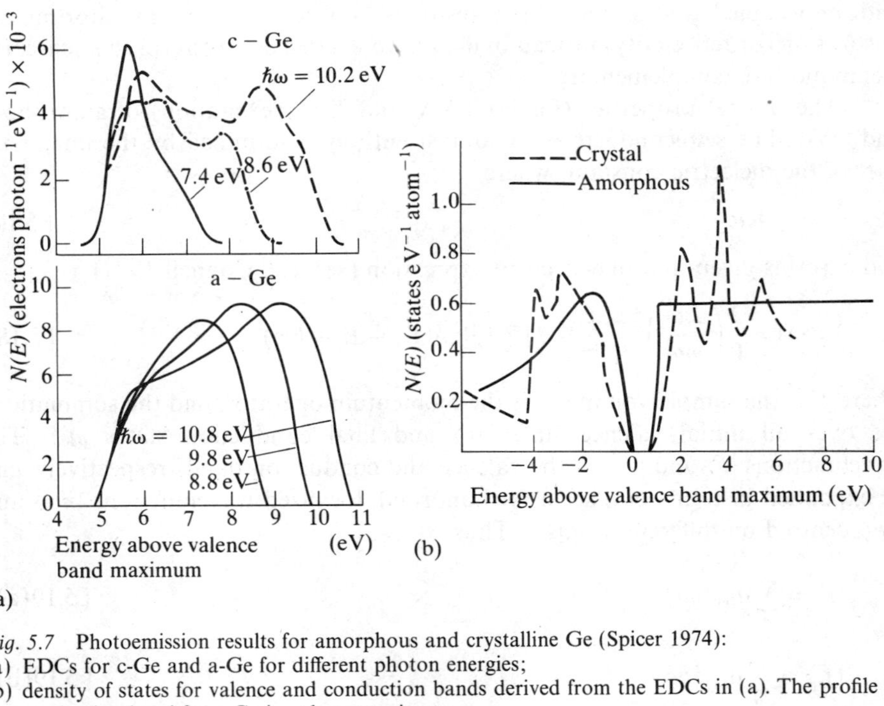

Fig. 5.7 Photoemission results for amorphous and crystalline Ge (Spicer 1974):
(a) EDCs for c-Ge and a-Ge for different photon energies;
(b) density of states for valence and conduction bands derived from the EDCs in (a). The profile for the conduction band for a-Ge is only approximate.

different ways) upon varying the UV photon energy. The densities of states deduced from these spectra are shown in Fig. 5.7(b) using [5.16] for the amorphous case (although not for c-Ge, where $\boldsymbol{k}$ is conserved and a different equation must be employed).

If photon energies greater than $\simeq 20$ eV are employed, then the intensity modulation by the final (conduction band) density of states rapidly becomes unimportant as the final state $N(E)$ approaches its featureless free-electron dependence $\propto E^{1/2}$. Thus soft X-ray photoemission probes only the valence band density of states, and the EDC is a direct reflection of this, modulated by slowly varying photo-ionization cross-sections. Higher energy X-rays cause excitation from deep-lying core states, rather than the valence band, and produce narrow peaks in the EDCs at kinetic energies below the valence band spectra. The number of core levels, and their binding energies, are characteristic of a given element, and moreover the exact value of binding energy measured depends on the chemical environment of the element. This phenomenon forms the basis of the technique 'electron spectroscopy for chemical analysis' (ESCA) which is really a surface probe, however, because of the short sampling depth (10–60 Å) of the XPS method. It should be noted in passing that under certain conditions, core-level X-ray absorption spectroscopy (sect. 3.2.3) can also be used to obtain chemical-shift information from the position of the X-ray absorption edge; this technique has the advantage that *bulk* properties are probed.

Ultra-violet and X-ray absorption spectroscopies are techniques analogous to photoemission in that photons cause electronic transitions between occupied and unoccupied states, but the transitions are detected by monitoring the absorption (or reflectivity) instead of the photo-excited electrons; in this sense, the techniques are complementary.

The optical properties (for both UV and X-ray excitation) of amorphous and crystalline semiconductors are almost entirely determined by the imaginary part of the dielectric constant, where:

$$\varepsilon = \varepsilon_1 + i\varepsilon_2 \qquad [5.17]$$

and $\varepsilon_2(\omega)$ is given in a one-electron expression (see, e.g. Connell 1979) as:

$$\varepsilon_2(\omega) = \frac{2}{V}\left(\frac{2\pi e}{m\omega}\right)^2 \sum_i \sum_f |\langle f|P|i\rangle|^2 \delta(E_f - E_i - \hbar\omega) \qquad [5.18]$$

where V is the sample volume, P is the momentum operator, and the summations are over all initial valence states $|i\rangle$ and final conduction states $|f\rangle$. The wavefunctions $|i\rangle$ and $|f\rangle$ of the valence and conduction bands respectively, can be expanded in terms of a set of orthonormal, localized wavefunctions $|nv\rangle$ and $|nc\rangle$ centred on different atoms n. Thus

$$|i\rangle = \sum_n a_{inv}|nv\rangle \qquad [5.19(a)]$$

$$|f\rangle = \sum_n a_{inf}|nc\rangle \qquad [5.19(b)]$$

For a crystal, the a's are plane waves and $|nv\rangle$ and $|nc\rangle$ are Wannier functions. For an amorphous solid however, writing $a = Ae^{i\delta}$, the phases δ vary randomly from site to site if the uncertainty in $\boldsymbol{k}$ is of the order of the wavevector itself, $\delta k \simeq k$ (for strong scattering); the amplitude A may also vary randomly from atom to atom, although being of the same magnitude for extended states. This 'random-phase approximation' (RPA) is an embodiment of the rule introduced by Ioffe and Regel that $k\ell \gtrsim 1$, i.e. when the wavefunction loses phase memory from atom to atom, the mean free path ℓ cannot be less than the interatomic spacing. The RPA is not in general a realistic model since it does not take account of the considerable degree of short-range order exhibited by amorphous semiconductors (Ch. 3), although it is a useful starting point.

The momentum matrix element, averaged over an ensemble of random systems, can be evaluated for transitions between extended (delocalized) states:

$$\langle |\langle f|P|i\rangle|^2 \rangle_{\text{ensemble}} = \frac{a^3}{V}\left\langle \sum_{n'} |\langle n'c|P|nv\rangle|^2 \right\rangle_{\text{ensemble}} \qquad [5.20]$$

where a is the interatomic spacing. When $\langle nc|P|nv\rangle = P_{\text{cv}} \neq 0$, this term will dominate the sum, and so:

$$\langle |\langle f|P|i\rangle|^2 \rangle_{\text{ensemble}} = \frac{a^3}{V} P_{\text{cv}}^2 \equiv |P_{\text{am}}(\omega)|^2 \qquad [5.21]$$

where $P_{\text{cv}} \sim \hbar/a$.

Substituting this in [5.18] for $\varepsilon_2(\omega)$ gives for the amorphous case:

$$\varepsilon_2^{\text{am}} = \frac{2}{V}\left(\frac{2\pi e}{m\omega}\right)^2 |P_{\text{am}}(\omega)|^2 \sum_i \sum_f \delta(E_f - E_i - \hbar\omega) \qquad [5.22(a)]$$

$$= 2\left(\frac{2\pi e}{m\omega}\right)^2 a^3 P_{\text{cv}}^2 \int_0^{\hbar\omega} \mathrm{d}E\, g_{\text{v}}(-E) g_{\text{c}}(\hbar\omega - E) \qquad [5.22(b)]$$

where we have introduced the density of states as defined by [5.3], and the zero energy is taken at the top of the valence band edge.

We see from [5.22] that $\varepsilon_2(\omega)$ is a function of a *joint* density of valence and conduction band states, much as was the case for UPS, [5.16]. This is also true for the optical absorption α, since this is related to $\varepsilon_2(\omega)$ by the following relation:

$$\varepsilon_2(\omega) = \frac{n_0 c \alpha(\omega)}{\omega} \qquad [5.23]$$

where n_0 is the refractive index and c the speed of light. In practice, the absorption is often obtained by a Kramers-Kronig transformation of reflectivity data.

It is interesting to compare the equation for $\varepsilon_2(\omega)$ for the amorphous case ([5.22]) with that for the crystalline case. This is obtained from [5.18] by requiring that optical transitions conserve momentum (or $\boldsymbol{k}$) (i.e. are 'vertical'), which leads to:

$$\varepsilon_2^{\text{cr}}(\omega) = \frac{2}{V}\left(\frac{2\pi e}{m\omega}\right)^2 |P_{\text{cr}}(\omega)|^2 \sum_i \sum_f \delta(E_f(\boldsymbol{k}) - E_i(\boldsymbol{k}) - \hbar\omega) \qquad [5.24(a)]$$

$$= 2\left(\frac{2\pi e}{m\omega}\right)^2 |P_{\text{cr}}(\omega)|^2 \frac{1}{(2\pi)^3} \int_{\text{B}} \mathrm{d}^3\boldsymbol{k}\, \delta(E_f(\boldsymbol{k}) - E_i(\boldsymbol{k}) - \hbar\omega) \qquad [5.24(b)]$$

where $|P_{\text{cr}}(\omega)|^2$ is independent of $\boldsymbol{k}$, and the integral is taken over the Brillouin zone. Thus, *crystalline* spectra are expected to show sharp features (Van Hove singularities) when $\hbar\omega$ is in the vicinity of critical points in the *joint* density of states where $\nabla[E_i(\boldsymbol{k}) - E_f(\boldsymbol{k})] = 0$, as well as when the joint density of states has structure, whereas *amorphous* solids should exhibit features in the spectra *only* at energies where the joint density of states has structure. This difference in ε_2 spectra is strikingly shown in Fig. 5.8(a) for amorphous and crystalline Ge (Spicer and Donovan 1970). Although the spectra are qualitatively similar, that for amorphous Ge is featureless with no trace of the sharp peak observed at 4.5 eV in crystalline Ge. This peak is known to be enhanced due to an Umklapp enhancement of the crystalline matrix element superimposed on an otherwise monotonically decreasing

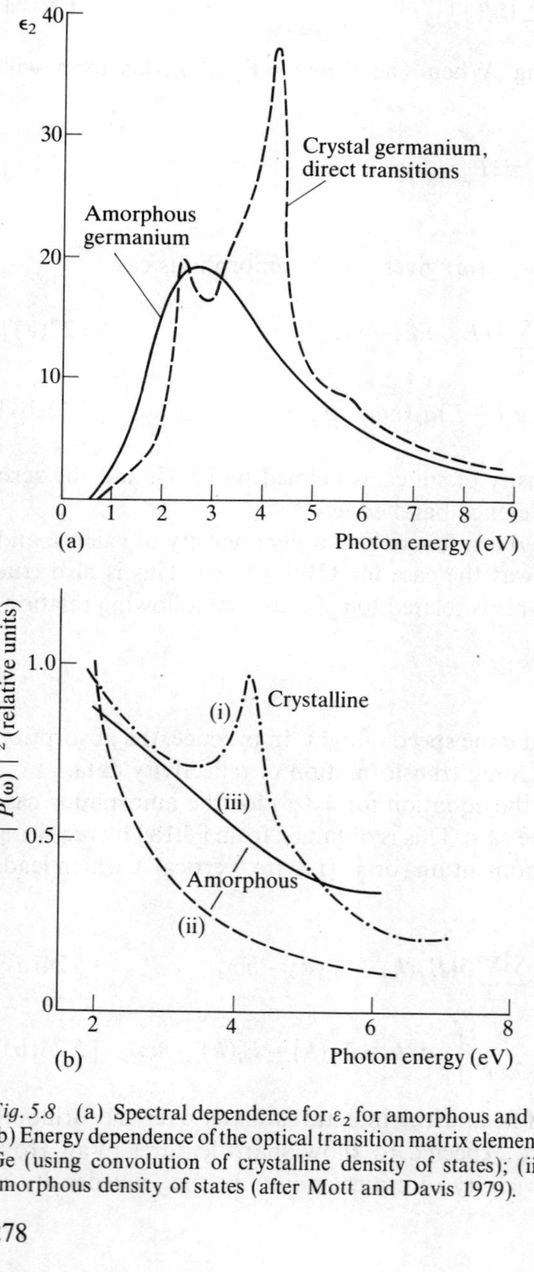

Fig. 5.8 (a) Spectral dependence for ε_2 for amorphous and crystalline Ge (Spicer and Donovan 1970). (b) Energy dependence of the optical transition matrix elements for Ge: (i) crystalline Ge; (ii) amorphous Ge (using convolution of crystalline density of states); (iii) amorphous Ge (using convolution of amorphous density of states (after Mott and Davis 1979).

ω-dependence (see Fig. 5.8(b)) often taken as $|P_{am}(\omega)|^2 \propto \omega^{-2}$. This dependence is predicted by the simple isotropic 'Penn model' of a semiconductor, in which it is assumed that just two bands exist, and the Penn gap $\hbar\omega_g$ represents an *average* gap between valence and conduction bands in a real material, and is a measure of the cohesive energy. The Brillouin zone (or Jones zone for a heteropolar system) is assumed to be spherical, with a magnitude of the wavevector, k_F, at the Fermi level. The matrix element for this model is then:

$$|P_{Penn}(\omega)|^2 = \hbar k_F \left(\frac{\omega_g}{\omega}\right)^2 \qquad [5.25]$$

The density of states can be determined from ε_2 spectra using an estimate for the matrix element, either a suitably averaged crystalline value or from the Penn model ([5.25]). The *valence band* density of states can be obtained by using [5.22(b)], and by making the approximation that the conduction band density of states can be reasonably well represented as a step-function $g_c(E_c)$ at an energy E_c above the valence band edge (see Fig. 5.7(b) for an indication of the validity of this), the following expression for the valence band density of states $g_v(E)$ in terms of ε_2 is thus obtained:

$$g_v(E_c - \hbar\omega) \propto \frac{d}{d\omega}\left[\frac{\omega^2 \varepsilon_2(\omega)}{|P_{am}(\omega)|^2}\right] \qquad [5.26]$$

The valence band density of states for a-Ge obtained from ε_2 data using [5.26] is shown in Fig. 5.9(a), compared with results obtained from XPS. The agreement between the two methods is excellent, the only differences arising from the neglect of the width of the tail to the conduction band edge in the derivation of [5.26].

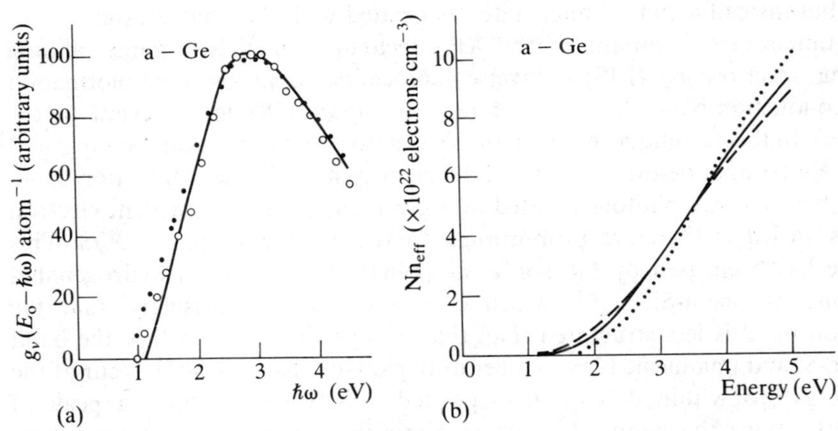

Fig. 5.9 (a) Valence band density of states of a-Ge measured from the conduction band edge. The points refer to data obtained from optical measurements (as in Fig. 5.8(a)) on sputtered material deposited on to substrates at 25°C (solid circles) and 350°C (open circles) (W. Paul, G. A. N. Connell and R. J. Temkin (1973) *Adv. Phys.* **22**, 529, together with data obtained by photoemission on evaporated material (L. Ley, S. Kowalczyk, R. Pollak and D. A. Shirley (1972) *Phys. Rev. Lett.* **29**, 1088).
(b) The effective number of electrons involved in transitions up to energy $\hbar\omega$ in a-Ge (dashed curve $T_d = 25$°C, solid curve $T_d = 350$°C; dotted curve crystalline Ge) (G. A. N. Connell, R. J. Temkin and W. Paul (1973) *Adv. Phys.* **22**, 643).

The effective number, n_{eff}, of free electrons per atom contributing to the optical absorption up to an energy $\hbar\omega_0$ is given by the following plasma sum rule:

$$n_{\mathrm{eff}} = \frac{m}{2\pi^2 N e^2} \int_0^{\omega_0} \omega \varepsilon_2(\omega)\, \mathrm{d}\omega \qquad [5.27]$$

where N is the atomic density. n_{eff} is shown in Fig. 5.9(b) for the same samples of a-Ge as were measured to give Fig. 5.9(a), together with data for c-Ge for comparison. It can be deduced that n_{eff} reaches 1.85 electrons per atom at 4.5 eV, a value in accord with our theoretical understanding of the make-up of the upper part of the valence band (all that is shown in Fig. 5.9(a)), namely that it is a broadened δ-function derived from p-states and therefore contains 2 electrons per atom (i.e. the s^2p^2 electronic configuration characteristic of group IV elements).

Thus far, we have concentrated on optical transitions from valence states to conduction states, produced by UV excitation. If photons of higher energy, e.g. X-rays, are used, the initial states become the narrow core levels situated at energies well below the valence band, and the technique becomes X-ray absorption spectroscopy (sect. 3.2.3). The advent of tunable X-ray synchrotron radiation sources has meant that this form of spectroscopy has become much more feasible. The *conduction band* density of states is explored by varying the X-ray photon energy and exciting from a given core level. This technique suffers, however, from two disadvantages. The first is that often the density of states near the edge appears to be enhanced as compared with results obtained using UPS; this effect has been ascribed to core exciton formation. The second disadvantage is more serious in that it casts some doubt on the ability of the technique to give a true representation of the conduction band density of states. The spectra of some alloys, e.g. a-As_2Se_3 (Bordas and West 1976) are identical to the sum of the individual spectra of the constituent elements, suggesting that the conduction band of a-As_2Se_3 is *not* being probed, but instead a sum of final states associated with the excited atoms.

A time-reversed variant of the XPS technique, namely 'bremsstrahlung isochromat spectroscopy' (BIS), or inverse photoemission, can also give information on the conduction-band density of states (see Smith 1988 for a review of the technique). In this technique, monoenergetic electrons impinge upon a sample and generate X-rays after de-excitation and filling empty states in the conduction band. The number of X-ray photons emitted at a given energy as the incident electron energy is varied is therefore proportional to the conduction-band DOS. This technique has been used by Jackson *et al.* (1984) in the study of hydrogenated amorphous silicon, a-Si:H, for which it is found, not surprisingly, that the conduction band is less structured than that of crystalline Si, but that the band edges of c-Si and unannealed low-temperature glow-discharge a-Si:H occur at the same energy (to within 0.2 eV). As-deposited a-Si:H films exhibit a peak of 4 eV width above the band edge, which diminishes appreciably on annealing. Jackson *et al.* ascribed this peak to the presence of Si-H antibonding states, probably associated with polyhydride $(SiH_2)_n$ species, which are the most susceptible to annealing resulting in effusion to H_2.

We conclude this survey of experimental techniques which probe one or other of the bands of amorphous materials by summarizing the capabilities of each method in Table 5.1.

Table 5.1 Experimental techniques for density of states determination

Technique	*Band probed*
Ultra-violet photoemission spectroscopy (UPS)	Valence and/or conduction
X-ray photoemission spectroscopy (XPS)	Conduction
Ultra-violet absorption spectroscopy	Valence and conduction
X-ray absorption spectroscopy	Conduction (?)

5.3 Localization

5.3.1 Introduction

In the last section we considered the theoretical and experimental determination of the overall shape of the electronic density of states in amorphous solids, primarily semiconductors, yet paid no attention to the *nature* of such electron states, i.e. whether they are spatially extended throughout a sample or localized in the vicinity of a given atom. In this section, we show that the presence of disorder can greatly influence the nature of the electron wavefunctions. As before, we will consider only the effect of topological (or other) disorder in 'ideal' random structures and leave a discussion of the electron states associated with structural defects until Chapter 6.

5.3.2 Effects of disorder

The presence of disorder is manifested in the electron states in a variety of ways. We have seen that sharp features in the crystalline density of states – the Van Hove singularities – become smeared out if there is no long-range order. Furthermore, fluctuations in short-range order, such as bond-angle distortions, lead to 'tailing' of states into the gap at the band edges (Fig. 5.10(a)); this may be seen from a consideration of the Weaire–Thorpe Hamiltonian ([5.5]) in which

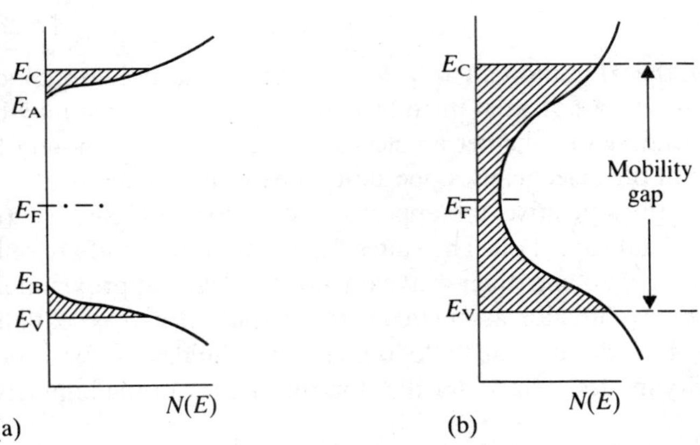

Fig. 5.10 Models for the density of states of an amorphous semiconductor.
(a) A true gap in the density of states with some band tailing.
(b) The Cohen–Fritzsche–Ovshinsky (1969) model of overlapping valence and conduction band tails.

variations of V_1 cause a broadening of the bands and consequent tailing of states into the gap. Band tails are even more pronounced if chemical disorder, such as the presence of 'wrong bonds', is present, as may be seen by reference to Fig. 5.4(b). The states at the top of the valence band in heteropolar systems such as a-GaAs derive primarily from cation-cation p-like bonding states which are extremely sensitive to the presence of cation like-bonds (Joannopoulos and Cohen 1976). Another cause of band tailing can be the occurrence of short-range disorder in *non-bonded* atom distances. This is particularly important for chalcogen atoms in amorphous semiconductors, which have one non-bonding p-like orbital (forming the top of the valence band), in addition to two predominantly p-like bonding orbitals per atom. If two such atoms are forced close to each other in the random network the interaction between the non-bonding lone-pair orbitals causes the energy of such states to be raised, again producing tailing of states from the valence band edge into the gap. In the extreme case these effects give rise to the Cohen–Fritzsche–Ovshinsky (1969) (CFO) model for the density of states in the gap region of an amorphous multicomponent (chalcogenide) semiconductor (Fig. 5.10(b)). In this, the band tailing is so pronounced that the tails arising from the valence and conduction bands actually overlap in mid-gap; this model will be discussed in more detail in Chapter 6, since defect states can also give rise to appreciable densities of states in the gap.

Another important consequence of disorder is that the electrons may become 'localized', i.e. spatially confined to the vicinity of predominantly a single atomic site. This occurrence is more probable the greater the degree of disorder in the potentials experienced by the electrons, and so is more likely in the band tails, since these in general are associated with the most distorted sites. Criteria which dictate whether electron states should be localized or non-localized (i.e. extended) will be discussed later.

It should be stressed at the outset that localized states *per se* are not unique to the amorphous phase, but can also occur in crystalline materials, albeit generally containing a degree of disorder. For instance, if a single defect is introduced into an otherwise perfectly periodic lattice the Schrödinger equation becomes

$$[\mathcal{H}_0 + U(r)]\psi = E\psi \qquad [5.28]$$

where the solutions of the Hamiltonian $\mathcal{H}_0$ for the perfect lattice are Bloch functions, and $U(\boldsymbol{r})$ is the extra potential introduced by the defect, which may be positive or negative depending on whether an electron is repelled by or bound to the defect. The presence of the defect causes one state to be split off from the band under consideration: if $U(\boldsymbol{r})$ is positive, the uppermost state splits off, and if $U(\boldsymbol{r})$ is negative the lowermost state splits off. The states deep within the band are only marginally displaced in energy and their wavefunctions remain approximately Bloch-like. However, the wavefunction associated with the split-off state is spatially localized in the vicinity of the defect. (An analogous situation holds also for lattice vibrations where impurity modes are also localized in the vicinity of the impurity-atom defect.)

In contrast, for systems having a considerable degree of topological or energetic disorder, e.g. amorphous semiconductors, *all* the states in a band may be localized if the disorder is sufficiently great, even though no defects may be present. Although this statement must be qualified to specify the dimension of the system, the type of disorder and even the length scale of the sample under consideration, nevertheless it is true under certain conditions and is a very surprising observation for those versed in the physics of crystalline solids. However, in most circumstances for three-dimensional systems, the criteria for localization are not met for states throughout an entire band, and only states near the band edges are localized; hence critical energies (labelled E_v and E_c in Fig. 5.10) separate localized from extended states.

Criteria for localization must be established before any discussion concerning the amount of disorder necessary to induce it can be pursued. A variety have been suggested. The first such criterion was put forward by Anderson (1958) in his now classic paper on electron localization in random lattices, namely the absence of diffusion at $T=0$: if an electron is placed on an atomic site (or equivalently in a volume V such that an energy E is at the centre of the band) at time $t=0$, then the state (at energy E) is deemed to be localized if as $t \to \infty$ the electron has not diffused away and has a finite probability $\propto e^{-2\alpha r}$ of remaining at distance r within the volume; if there is a finite chance of diffusion at $T=0$ then the state is delocalized or extended. The quantity α^{-1} is known as the localization length. The exponentially decreasing nature of the localized wavefunction, $\psi \sim e^{-\alpha r}$, has suggested another criterion, namely that the energy levels of localized states should be insensitive to boundary conditions (see e.g. Thouless 1979). The *size* of the sample is thus seen to be crucial; if the localization length is much greater than the sample size, $\alpha^{-1} \ll L$, it becomes very difficult to determine whether the state is localized or extended. Another measure of the degree of localization that has been used is the 'participation ratio' P (or its inverse) that has already been mentioned in connection with vibrational modes ([4.24]), which is roughly speaking the number of sites over which a localized function has significant amplitude. Thus a localized state is characterized by the participation number $P \propto [\Sigma_i|\psi_i|^2]^2/|\Sigma_i|\psi_i|^4]$ tending to zero, whereas an extended state has P finite. A final criterion for localization may be chosen to be the behaviour of the d.c. electrical conductivity as $T \to 0$; essentially this is related to the electron diffusion. States of energy E are deemed to be localized if the ensemble average of the d.c. conductivity $\langle \sigma_E \rangle$ is zero at $T=0$, whereas extended (metallic-like) states have a finite conductivity at $T=0$. (In other words, conduction between localized states can only take place by means of thermal assisted 'hopping' as we shall see later.) The various criteria are summarized for convenience in Table 5.2.

We can now turn to the problem of how to obtain a numerical estimate for the degree of disorder necessary to localize, say, all states in a band, having at our

Table 5.2 Criteria for distinguishing localized and extended electron states (after Kramer and Weaire 1979)

Name		*Localized regime*	*Extended regime*
Inverse localization length	$\begin{smallmatrix}\alpha\\ \beta\end{smallmatrix}\vert\Psi\vert \sim r^{-\beta}e^{-\alpha r}$	finite	0
Prefactor index		finite	0
Participation ratio	$P \propto [\Sigma_i \vert\Psi_i\vert^4]^{-1}$	0	finite
Conductivity	$\langle \sigma_E \rangle$	0	finite

disposal means of distinguishing between localized and extended states. However, in order to do this, we must address the question of dimensionality, since this profoundly influences the behaviour.

5.3.3 One dimension

We consider firstly one-dimensional systems for which the results are the most clear-cut. Put simply, any disorder, however small, is sufficient to localize *all* states, as first shown by Mott and Twose (1960); extended states are only obtained if all the potentials in a 1-D chain are of equal magnitude and spacing (the Kronig–Penney model). This result can be proved rigorously for the case of one dimension; a review is given by Ishii (1973).

Much work has been devoted to the theoretical consideration of thin metal wires, which Thouless (1977) first showed should exhibit *non-metallic* conduction, i.e. localized behaviour, if the *resistance* of the wire is greater than several tens of kilohms. This prediction, which has been borne out by experiment, is sufficiently startling and interesting that we shall examine it further here, for, although not concerning amorphous materials *per se*, it does shed some useful light on the localization process. The essence of the argument lies in the localization criterion that electron energy levels are sensitive to boundary conditions, since it can be shown that this sensitivity is closely related to the resistance of the system at that energy.

Consider a metal wire of length L and cross-sectional area A, which has sufficient disorder (impurities, etc.) that the bulk material has a finite conductivity σ at $T=0$. Now an electron wave-packet diffusing along the length of the wire is insensitive to the boundary for the time t it takes to travel the length of the wire, and so, by the uncertainty principle, the energy levels can be shifted only by an amount ΔE where:

$$\Delta E = \frac{\hbar}{t} = \frac{\hbar D}{L^2} \qquad [5.29]$$

where D is the diffusion constant for the electron. This can be related to the conductivity, and the density of states $\mathrm{d}n/\mathrm{d}E$ (per unit volume per unit energy interval) by the Einstein relation

$$\mu = \frac{eD}{n}\frac{\mathrm{d}n}{\mathrm{d}E} \qquad [5.30]$$

and $\sigma = ne\mu$ ([4.80]), i.e.

$$\sigma = e^2 D \frac{\mathrm{d}n}{\mathrm{d}E} \qquad [5.31]$$

Combining [5.29] and [5.30] yields:

$$\Delta E = \frac{\hbar}{e^2}\frac{\sigma}{L^2}\frac{\mathrm{d}E}{\mathrm{d}n} = \frac{\hbar}{e^2 R}\Delta W \qquad [5.32]$$

where R is the resistance of the wire and $\Delta W = \mathrm{d}E/\mathrm{d}N$ (where N is the total number of states) is the average spacing between energy levels. The significance of the ratio $\Delta E/\Delta W$ is that it can be regarded as a measure of the extent of coupling between two wires, each of length L, joined together to form a new wire of twice the length. If $\Delta E/\Delta W \gg 1$, the coupling is strong and the electron states in the joined wire are made up of those of the individual wires; the new wire is similar to each original wire, except that now ΔW has half and t four times the former value (cf. [5.29]). Thus, [5.32] shows that in this case the resistance of the joined wire is simply twice that of the original wires, i.e. the resistance scales normally with size. However, in the other limit, $\Delta E/\Delta W \ll 1$, the coupling is very weak, and consequently the states in one wire do not interact strongly with those in the other wire. In this case, the states can be regarded as localized, and consequently the resistance should increase much faster with increasing length than the normal linear dependence. The condition necessary for this to occur is:

$$R_c \gg \hbar/e^2$$

or in other words the maximum resistance for metallic behaviour of a wire should be $\simeq 5\ \mathrm{k\Omega}$, and the localization length for such a system is such that it corresponds to a resistance of this amount. Later work, which considered that the time between inelastic collisions (rather than the diffusion time) limits the metallic behaviour, raised the critical resistance to $36.5\ \mathrm{k\Omega}$ (Thouless 1980). Such effects should only be observable at low temperatures, since phonons will induce transitions between localized states at higher temperatures.

Anderson *et al.* (1980) further proposed that in one dimension the resistance should scale as the inverse localization length, α, according to:

$$R(L) = \frac{\pi\hbar}{e^2}\{\exp(\alpha L) - 1\} \qquad [5.33]$$

This scaling parameter can be rewritten in a form proportional to $\ln(1 + R)$, and for small resistances the scaling reduces to R (i.e. normal additivity of resistances), but for large resistances, it reduces to $\ln R \propto \alpha L$, reflecting exponential localization. The length L is the length of the wire at $T = 0$, or the distance between inelastic collisions at finite temperatures. If the electronic motion is diffusive this gives:

$$L_i \simeq \sqrt{(D\tau_i)} \qquad [5.34]$$

where τ_i is the inelastic scattering time. The localization length, α^{-1}, is related to the maximum metallic resistance:

$$R_c = \rho_0 \frac{\alpha^{-1}}{A} \qquad [5.35]$$

where ρ_0 is the bulk resistivity and A is the cross-sectional area, and is also related to the (large-scale) resistance R_0 of a wire of length L

$$\alpha^{-1} = \frac{\pi\hbar}{e^2}\frac{L}{R_0} \qquad [5.36]$$

Inserting these expressions into [5.33], expanded to the quadratic term, gives:

$$R \simeq R_0\left(1+\frac{\rho_0\sqrt{(D\tau_i)}}{2R_c A}\right) \qquad [5.37]$$

i.e. the extra resistance associated with localization should be inversely proportional to A. (NB This expression for the extra resistance differs by a factor of 4 from that given by Thouless 1980.) Wires of several hundred ångströms diameter and a few micrometres in length made from Au–Pd alloys have been studied, and typical results of the resistance as a function of A^{-1} are shown in Fig. 5.11(a) for different temperatures. The temperature dependence of the extra resistance is seen to be proportional to $T^{-1/2}$ (Fig. 5.11(b)); this behaviour can be understood if the scattering time is shorter and increases less rapidly with temperature than is expected for phonon processes, and it has been suggested that 'two-level systems' are responsible, for which $\tau_i \propto T^{-1}$.

5.3.4 Two and three dimensions

The one-dimensional case is exceptional in that any amount of disorder, no matter how small, will strongly localize the electron states. It was originally thought that the behaviour in two and three dimensions was qualitatively different, namely that a certain critical degree of disorder is necessary to induce localization in an entire band, otherwise the electron states are delocalized. It is now known, originating with the scaling arguments of Abrahams *et al.* (1979) (see later), that this is *not* the case: it is only in 3D that a true localization (Anderson) transition takes place; in 2D, the states are always localized (as in 1D), but there is a cross-over in behaviour between strongly localized and weakly localized regimes.

The question of localization in random lattices was first addressed by Anderson (1958), who considered the following simple tight-binding Hamiltonian:

$$\mathscr{H} = \sum_i \varepsilon_i |i\rangle\langle i| + V\sum_{i,j} |i\rangle\langle j| \qquad [5.38]$$

The basis states $|i\rangle$ are located on the sites of a *periodic* lattice and are coupled by constant nearest-neighbour (off-diagonal) interactions V. The disorder in this model occurs in the form of random site energies ε_i (or diagonal elements) termed 'diagonal disorder'. The ε_i are usually taken to be uniformly distributed over a certain energy width W, i.e.

$$\begin{aligned} p(\varepsilon) &= 1/W \qquad |\varepsilon| < W/2 \\ &= 0 \qquad\;\; |\varepsilon| > W/2 \end{aligned} \qquad [5.39]$$

and the strength of the disorder is then conveniently characterized by the ratio W/V. This model is illustrated schematically in Fig. 5.12. It can be seen that the problem is essentially a competition between the two terms in [5.38]; the first in isolation produces a localized (single) state, whereas the second produces an extended (Bloch) state, and the value of W/V is obviously crucial in deciding which of these two effects is dominant. Thus, for a given amount of disorder, all the states in the upper and lower tails of a band having energies greater than or less than critical energies, E_c and E_c', will be localized, whereas states between

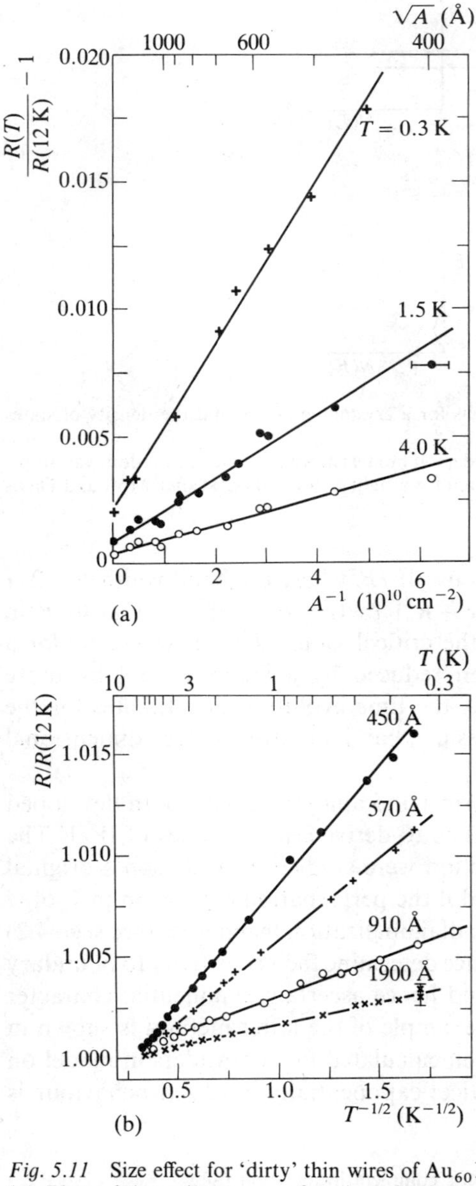

Fig. 5.11 Size effect for 'dirty' thin wires of $Au_{60}Pd_{40}$ (Giordano 1980).
(a) Plots of the resistance as a function of $1/A$ at several temperatures.
(b) Temperature dependence of the resistivity of wires of varying diameters.

these energies nearer the centre of the band will be delocalized (extended). As the degree of disorder increases, the energies E_c and E'_c move towards each other until, at a critical amount of disorder, *all* states in the band are Anderson localized. This value of W/V is the quantitative criterion by which the results of calculations using the various methods detailed later are compared. The localization criterion may be rewritten in the general form W/zV so that localization on lattices having different values of coordination number, z, may be compared directly. An alternative

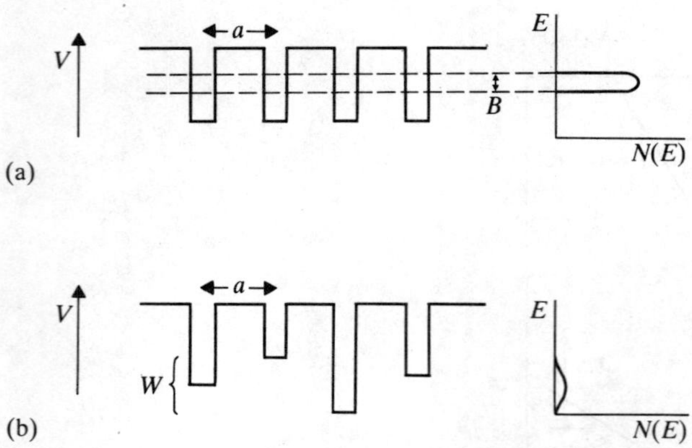

Fig. 5.12 (a) Representation of the potential wells for a crystalline lattice and the density of states expected for a tight-binding model.
(b) Representation of the potential wells in the Anderson model (in which there are random variations in the site energies) and the density of states expected for a tight-binding model (after Mott and Davis 1979).

formulation (Mott and Davis 1979) is to use W/B, where the band width $B = 2zJ$ and J is the overlap integral, $J = \int \psi^*(\boldsymbol{r} - \boldsymbol{R}_n)\mathcal{H}\psi(\boldsymbol{r} - \boldsymbol{R}_{n+1})\mathrm{d}^3x$, equal to V in this case. Anderson's first estimate for the critical value of W/V was $\simeq 62$ for a simple cubic lattice; this has since been reduced by a factor of $\simeq 4$ by more sophisticated calculations. Table 5.3 lists the time evolution of estimates for the critical values of W/V to localize all states in a band for various three-dimensional lattices.

A variety of techniques, both numerical and analytical, have been developed to study the problem of localization, and so to derive critical values of W/V. The analytical properties of the Green's function were studied in Anderson's original work, states being deemed to be localized if the perturbation expansion in V of G converged, G being therefore real. Matrix diagonalization techniques (see sect. 4.2) can be used to obtain eigenvalues (and hence determine their sensitivity to boundary conditions) or to obtain eigenvectors (and hence ascertain exponential character or calculate the participation ratio). An example of the latter method is shown in Fig. 5.13 where the eigenvectors have been calculated for the Anderson model on a two-dimensional 100×100 square lattice; exponentially localized behaviour is

Table 5.3 Critical strengths of disorder (W/V) for complete localization (after Kramer and Weaire 1979)

	Anderson (1958)	*Ziman (1969)*	*Herbert and Jones (1971)*	*Licciardello and Economou (1975)*
Square lattice	28	22	—	7.2
Diamond cubic lattice	32	22	—	8.2
Simple cubic lattice	62	32.4	40, 24, 21	14.5

References: P. W. Anderson (1958), *Phys. Rev.* **109**, 1492; J. M. Ziman (1969), *J. Phys.* **C2**, 1230; D. C. Herbert and R. Jones (1971), *J. Phys.* **C4**, 1145; D. C. Licciardello and E. N. Economou (1975), *Phys. Rev.* **B11**, 3697.

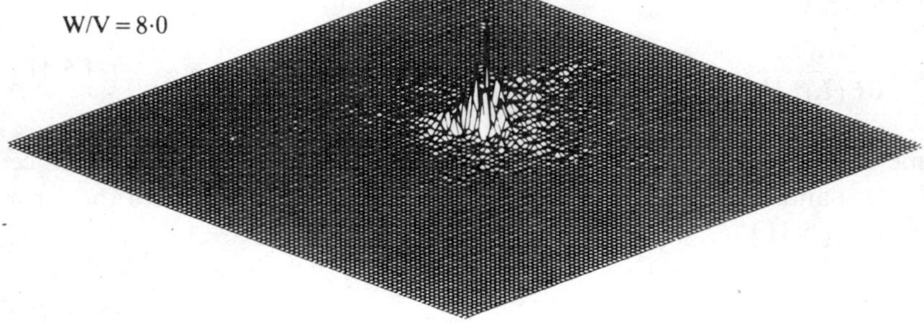

Fig. 5.13 Wavefunction amplitude for the Anderson model on a square lattice, showing a localized state with $W/V = 8.0$ (Yoshino and Okazaki 1977).

clearly seen for values of W/V in excess of 8.0 (Yoshino and Okazaki 1977). Although the work by Yoshino and Okazaki (1977) appeared to show a transition to extended-state behaviour for somewhat smaller values of W/V than for that corresponding to Fig. 5.13, subsequent investigations of the Anderson model by MacKinnon and Kramer (1983) on much larger systems containing $\simeq 10^9$ sites led to the conclusion that only with such large systems can the true (weakly) localized character of states in 2D be identified.

The Anderson model is highly simplistic in considering only *diagonal* disorder, i.e. a random variation in the site energies ε_i. A more realistic model Hamiltonian for amorphous systems would be one which incorporates *off-diagonal* disorder, i.e. random variations in the inter-site interaction V (now becoming V_{ij}) caused by a random spatial distribution of centres. This form of disorder, however, is found to be much less effective in inducing Anderson localization; indeed there are some doubts whether all states in a band can be localized in this manner, although band-tail states certainly become localized by this mechanism.

The localization transition can obviously be regarded as a phase transition. The critical behaviour on the localized side of the transition in 3D can be expressed in terms of the divergence of a scaling length, in this case the localization length α^{-1}. This may be written in the form

$$\alpha \propto (E - E_c)^{\nu} \qquad [5.40]$$

where E_c is the critical value of W/zB. The critical exponent has been found from the numerical results obtained for very large systems by MacKinnon and Kramer (1983) to be $\nu = 1.5$, although other workers (e.g. Vollhardt and Wölfle 1982) find $\nu = 1$.

Scaling theories for the localization transition developed by Abrahams *et al.* (1979) give results for three dimensions which are in marked contrast to those for one and two dimensions. The approach is based on the ideas of Thouless that have already been discussed in sect. 5.3.3 on 1-D systems, and on the renormalization-group scheme developed for phase transitions. Small blocks of material of size L^d (where d is the dimensionality and $L > \lambda$, the mean free path) are coupled together to form larger blocks, and the strength of this coupling, g, is taken to be the ratio of the fluctuation in energy levels ΔE (resulting from say a change in boundary conditions) to the mean spacing of the levels dE/dN, i.e.

$$g(L) = \frac{\Delta E(L)}{dE(L)/dN} \qquad [5.41]$$

This parameter g is a dimensionless conductance (related to the real conductance G by $g = \hbar G/e^2$) and is related to the conductivity σ by an expression of the form already given in [5.31]

$$g(L) = \frac{\hbar}{e^2}\sigma L^{d-2} \qquad [5.42]$$

The fundamental assumption made is that the coupling strength of the *larger* blocks, i.e. the scaled system, should only depend on the *original* g parameter. Put another way, in continuous terms:

$$\frac{d(\ln g(L))}{d(\ln L)} = \beta(g(L)) \qquad [5.43]$$

Thus, the scaling function $\beta(g)$ is all that is required to ascertain how the coupling depends on L, and hence whether states are localized or not. The behaviour of β with g can be ascertained from a consideration of its asymptotic behaviour as $g \to 0$ and ∞. For large coupling, i.e. large g, macroscopic transport theory is valid (via extended states), so that $G(L) = \sigma L^{d-2}$ and hence from [5.42]

$$\lim_{g \to \infty} \beta(g) = d - 2$$

In fact it can be shown that this is the zeroth order term in a perturbation series, which to first order is:

$$\lim_{g \to \infty} \beta(g) = (d - 2 - a/g + \ldots) \qquad [5.44]$$

In 2-D, $a = \pi^{-2}$ and in 3-D, $a = 2^{1/2} 4\pi^2$ (see Mott and Kaveh 1981). In the limit of small coupling, i.e. small g, the states are exponentially localized. Thus, if $g(L) \sim e^{-\alpha L}$,

$$\lim_{g \to 0} \beta(g) \sim \ln g \qquad [5.45]$$

Hence the limiting behaviour is in accord with Fig. 5.14, which shows calculated curves of $\beta(g)$ for 1-D, 2-D and 3-D, the results of numerical computations for the Anderson Hamiltonian by MacKinnon and Kramer (1983) on very large systems. It can be seen that only in *three* dimensions does the function $\beta(g)$ have a zero value for which, from [5.43], $g(L)$ = constant, i.e. the conductivity vanishes as $L \to \infty$, or in other words the electron states become localized. On the localized side of the transition ($\beta < 0$), the conductance depends exponentially on the length scale (Abrahams *et al.* 1979):

$$g \simeq g_c \exp(-\alpha L) \qquad [5.46]$$

where α^{-1} is the localization length ([5.40]) whose critical exponent is ν. If it is assumed that β varies linearly at the localization transition $g = g_c$, i.e. that $\beta = s \ln(g/g_c)$, the critical exponents can be obtained from the slope at g_c of the curve for 3-D in Fig. 5.14 since $\nu = 1/s$: the results of MacKinnon and Kramer (1983) indicate that $\nu = 1.5$, whereas Abrahams *et al.* (1979) argue that the slope s should increase near g_c, making $\nu \leqslant 1$.

In contrast to the behaviour in 3-D, all the results from scaling theories indicate that in the case of 1-D and 2-D, $\beta(g)$ is *always* negative, i.e. the electron states are always localized for arbitrarily small amounts of disorder. It can be seen from Fig. 5.14 that in 2-D β *almost* becomes zero; however, instead of a transition to extended-state behaviour at large g, the states become *weakly* localized. In this regime, the conductance has a *logarithmic* dependence on length scale; from [5.43] and [5.44],

$$g = g_0 - A \ln L \qquad [5.47]$$

instead of the exponential dependence ([5.46]) characteristic of the strongly localized regime. These logarithmic size corrections in the weak localization regime will be discussed more fully in sect. 5.4.1. It should be stressed, however, that the discussion so far has neglected electron–electron interactions, which can also lead to logarithmic corrections (Efros and Pollak 1985) – see sect. 5.4.1.

It should be noted that mechanisms other than the one derived from the Anderson model [5.38] can also lead to electron localization. Thus, electrons diffusing on a *fracal* lattice will be localized if the fracton dimension $\tilde{d} < 2$ (Rammal and Toulouse 1983); this should be as true for diffusing electrons as for the

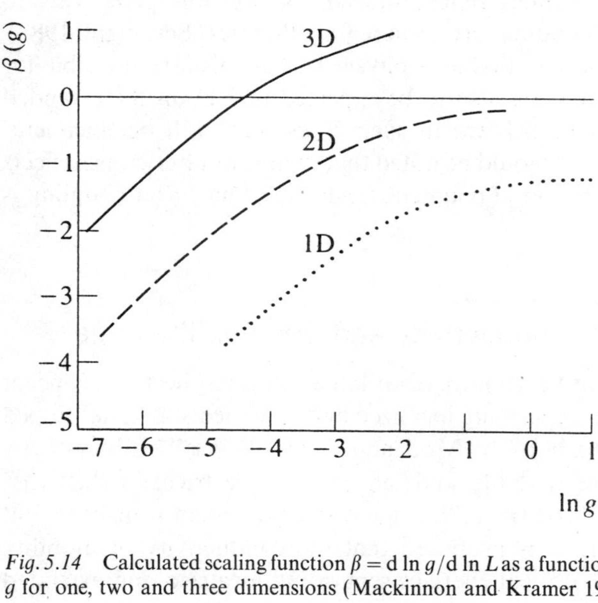

Fig. 5.14 Calculated scaling function $\beta = \mathrm{d}\ln g/\mathrm{d}\ln L$ as a function of the logarithm of the conductance g for one, two and three dimensions (Mackinnon and Kramer 1983).

vibrational and mass diffusion behaviour discussed in sect. 4.4.6. An example of a fractal structure for which such ideas should apply is the critical percolation cluster in, say, a random resistor network (see sect. 5.3.6).

Another mechanism involves quantum-interference effects or so-called '$2k_F$ scattering'. This phenomenon is perhaps most easily understood by considering the diffusive motion in 2-D of an electron, where randomly distributed impurities cause it to be scattered elastically (thereby conserving energy but changing the direction of momentum or $\boldsymbol{k}$). One can consider a closed path for the scattering events by which an electron may return to the origin at 0 (Fig. 5.15(a)). *Classically*, the probability of returning to the origin is given by $p_c = 1/4\pi Dt$ (where D is the diffusion constant) and is made up of equal contributions from each of the paths followed ($0 \to 1 \to 2 \to \cdots \to 0$ or $0 \to 1' \to 2' \to \cdots \to 0$). However, if the electron is treated as a *quantum* particle, two partial waves can be considered to propagate around the path shown in Fig. 5.15(a), and if the system is time-invariant, the two partial waves arrive back at the origin in phase and with the same amplitude. The amplitudes at the origin therefore add *coherently*, yielding a resultant probability for return to the origin of $p_q = 1/2\pi Dt$, or *twice* that for classical diffusion (Bergmann 1983). At any other point except the origin, the partial waves are incoherent and therefore only their intensities add, giving a net probability of arrival little different from that expected classically. Thus, the probability distribution of a diffusing electron starting from the origin $\boldsymbol{r} = 0$ at time $t = 0$ being found at position $\boldsymbol{r}$ a time t later is as shown in Fig. 5.15(b), where it can be seen that quantum diffusion results in an enhanced probability that the electron will remain at the origin, i.e. be localized. Similar arguments can be employed for k-space (Bergmann 1983), in which an electron with initial wavevector $\boldsymbol{k}_F$ at the Fermi surface becomes preferentially backscattered, i.e. with wavevector $-\boldsymbol{k}_F$, after a series of intermediate scattering events through a coherent superposition of the scattered wave amplitudes; this process is referred to as $2k_F$ scattering (Morgan and Hickey 1985). Quantum-interference-induced localization gives rise to logarithmic corrections to the conductivity, as in scaling theories (Bergmann 1983), and in 2-D it can therefore be regarded as a physical mechanism responsible for weak localization. Similar effects are also to be expected in 3-D on the extended side but near the localization transition; in such cases there will be 'incipient' localization behaviour. Finally, it should be noted that quantum-interference effects can be destroyed by the application of magnetic fields or if spin–orbit coupling is strong (Bergmann 1983).

5.3.5 Minimum metallic conductivity and the mobility edge

A central theme of the theory of electrons in disordered solids has been the concept of the so-called 'mobility edge' separating localized and extended states, developed by Mott and expounded in the book by Mott and Davis (1979). This hypothesis asserts that at $T = 0$ the electron mobility, and hence the d.c. electrical conductivity, should be finite on the delocalized (metallic) side of the Anderson transition, but *zero* for localized states. Thus, it is proposed that the conductivity or mobility changes *discontinuously* at the critical energies separating localized and extended

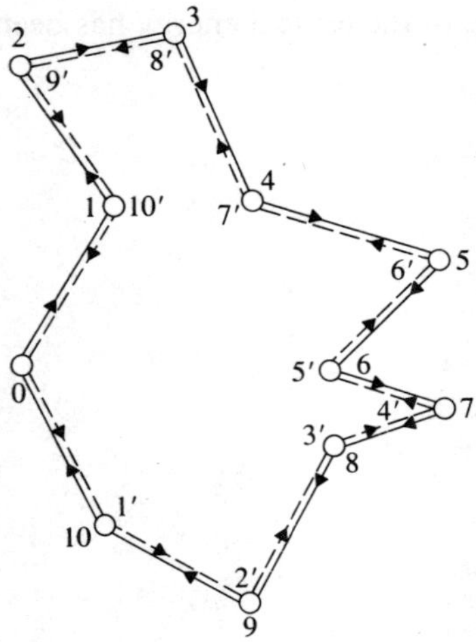

(a)

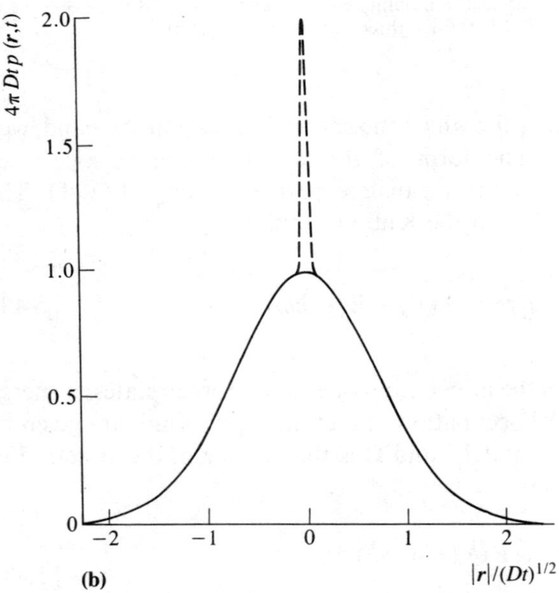

(b)

Fig. 5.15 Illustration of the quantum-interference effect (Bergmann 1983).
(a) Diffusion paths of a conduction electron in opposite senses around a closed loop of sites.
(b) Probability distribution of a diffusing electron starting at $r = 0$ at time $t = 0$. In quantum diffusion (dashed peak), the probability for return to the origin is twice that for classical diffusion (solid curve).

states (E_v and E_c in Fig. 5.10), even though the density of states is continuous there. This behaviour is shown schematically in Fig. 5.16, together with the power-law behaviour that is suggested by scaling theories of localization

(Abrahams *et al.* 1979). Note that if the Fermi level can be moved in some way through E_c or E_v, a metal–insulator transition occurs; the situation when E_F lies among a set of localized states on the localized side of the critical energy has been termed a 'Fermi glass'.

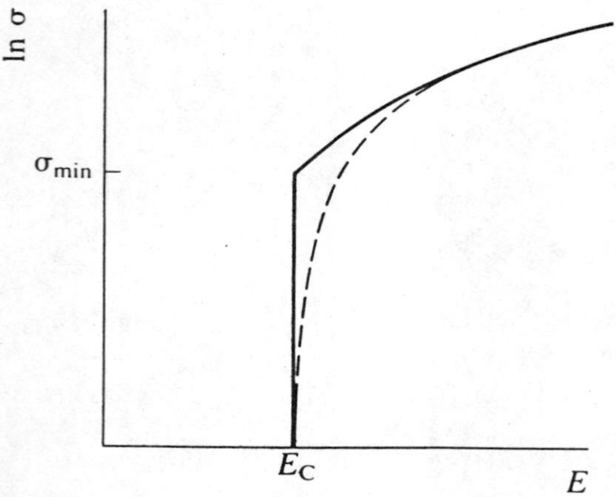

Fig. 5.16 Behaviour of the d.c. conductivity near a mobility edge as suggested by Mott (———) and scaling theories (– – –) (Abrahams *et al.* 1979) for a three-dimensional system.

Consider first extended states for which the energy lies deep in the band, well away from the mobility edge. The form of the wavefunctions is as shown schematically in Fig. 5.17(a), and the mean free path ℓ is long ($k\ell \gg 1$). The conductivity σ_E can be evaluated from the Kubo formula:

$$\sigma(\omega) = \frac{e^2\pi}{m^2\omega\Omega}\sum_{\alpha}\sum_{\beta\neq\alpha}|\hat{p}_{\alpha\beta}|^2(f_\beta - f_\alpha)\delta(E_\alpha - E_\beta - \hbar\omega) \tag{5.48}$$

where $\hat{p}_{\alpha\beta}$ is the matrix element of the momentum operator between states of energy E_α and E_β, whose probabilities of occupation in thermal equilibrium are given by the Fermi–Dirac distributions f_α and f_β, and Ω is the volume of the system. For low temperatures, this may be written as:

$$\sigma(\omega) = \frac{\pi e^2\hbar\Omega}{2m^2}\int_{E_F-\hbar\omega}^{E_F}\frac{|\hat{p}|^2_{av}N(E)N(E+\hbar\omega)\,dE}{\hbar\omega} \tag{5.49}$$

In the random-phase approximation, in which the wavefunctions are written as a sum of plane waves, with random amplitudes distributed in a gaussian fashion, $|\alpha\rangle = \Sigma a_k^\alpha|k\rangle$ the conductivity can be evaluated to be (see problem 5.3 and Thouless 1979):

$$\sigma(\omega) = \frac{k_F^3 e^2\tau}{3\pi^2 m}\frac{1}{(1+\omega^2\tau^2)} \tag{5.50}$$

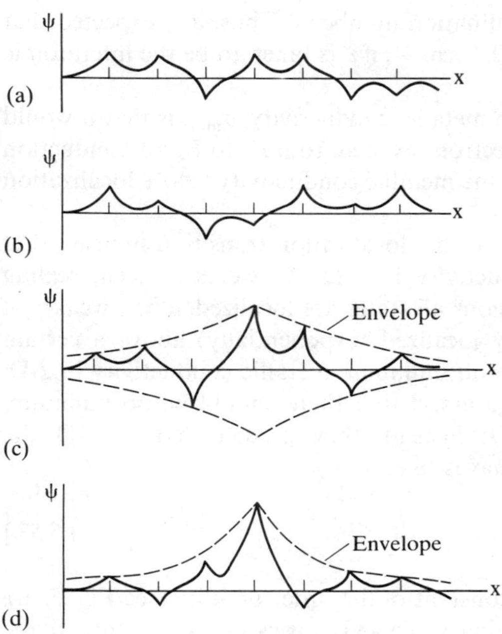

Fig. 5.17 Schematic illustration of the form of the electron wavefunction in the Anderson model (after Mott and Davis 1979).
(a) when $\ell \simeq a$;
(b) when states are just non-localized ($E \gtrsim E_c$);
(c) when states are just localized ($E \lesssim E_c$);
(d) strong localization.

This is of the Drude form, and the relaxation time τ is given by $\tau = (m\ell/\hbar k_F)$. For zero frequency, and a spherical Fermi surface of wavevector $k_F = (3\pi n)^{1/3}$ and area S_F, this reduces to the formula that can be obtained also by the Boltzmann formulation, i.e.:

$$\sigma_B = \frac{e^2 k_F^2 \ell}{3\pi^2 \hbar} = \frac{e^2 \ell S_F}{12\pi^3 \hbar} \qquad [5.51]$$

As the amount of disorder is increased, the mean free path decreases until it reaches the Ioffe–Regel minimum, such that $k\ell \simeq 1$, where $\ell \simeq a$, the interatomic spacing (or L_i, the inelastic diffusion length, whichever is the shorter). The conductivity at this point is given approximately, therefore, by substituting these limits into [5.51], i.e. $\sigma_B \simeq e^2/3\hbar L_i$. Any further increase in disorder only serves to broaden the bands, thereby lowering $N(E)$ from its free electron value by a factor Γ, rather than decreasing ℓ further. Thus, the 'minimum metallic conductivity', σ_{min}, is the minimum conductivity at the mobility edge on the delocalized side before all states become localized and the conductivity $\langle \sigma_E \rangle$ drops precipitously to zero (Mott and Davis 1979, Mott 1981); for three dimensions it can be written as:

$$\sigma_{min}^{3\text{-D}} \simeq \left(\frac{e^2}{3\hbar\ell}\right)\Gamma^2 \qquad [5.52]$$

where $\Gamma \simeq \frac{1}{3}$ but depends on the coordination number z. Thus, it is expected that $\sigma_{min}^{3\text{-D}}$ should lie in the range 300–600 Ω^{-1} cm^{-1}, if ℓ is taken to be the interatomic spacing, $\simeq 3$ Å.

The significance of the minimum metallic conductivity, σ_{min}, is that it would be the pre-exponential factor σ_0 for electrons excited from E_F to E_c for conduction in non-metals, or the limiting value of the metallic conductivity before localization sets in for the case of metals.

However, recent scaling theories of the localization transition indicate that there is no minimum metallic conductivity in 3-D. As we have seen, scaling arguments predict that in two dimensions all states are localized, albeit weakly, if any disorder is present, and strongly localized (exponentially) above a certain degree of disorder. Thus, there can be no minimum metallic conductivity in 2-D. However, Abrahams *et al.* (1979) also assert that there should be no minimum metallic conductivity, either, in 3-D; instead, they predict that in 3-D the conductivity near a mobility edge behaves like:

$$\sigma_E \simeq \sigma_0 \left[\frac{(E - E_c)}{B} \right]^{\nu} \qquad [5.53]$$

where B is the bandwidth and ν is a constant of the order of unity (see Fig. 5.16). This expression is obtained by assuming β ([5.43]) varies as $\beta \simeq 1/\nu \ln(g/g_c) \simeq (g - g_c)/\nu g_c$, where g_c is the value of g where $\beta = 0$; integration of [5.43] then gives [5.53].

Figure 5.18 qualitatively shows this behaviour in the system crystalline Si:P (in which the P donors produce an impurity band having a random array of

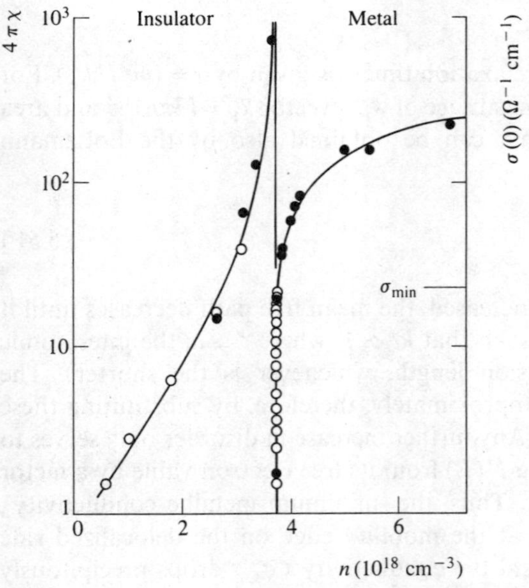

Fig. 5.18 D.C. conductivity $\sigma(0)$ and the donor dielectric susceptibility $4\pi\chi$, extrapolated to $T = 0$ K, respectively, above and below the metal–insulator transition in crystalline Si:P as a function of the donor concentration. The data points represented by open circles for $\sigma(0)$ were obtained from stress measurements (Hess *et al.* 1982).

potentials much as in the Anderson model). Note the several experimental points for which the conductivity is *less* than σ_{min}. (It should be noted that σ_{min} for impurity bands in crystalline semiconductors, as in Fig. 5.18, is considerably smaller, by a factor of ten or so, than for amorphous semiconductors, since in the former the value of ℓ in [5.52] corresponds to the separation between impurities, whilst in the latter case it is of the order of the interatomic spacing.) The metal–insulator transition characteristic of the localization transition is signified by the divergence of the dielectric susceptibility $4\pi\chi$ as the transition is approached from the insulating side (Fig. 5.18). However, the presence of compositional or other fluctuations might also lead to the absence of σ_{min} and a rounding of the transition at E_c due to the existence of percolation channels, which will be discussed in the next section.

The behaviour near the localization transition in 3-D is still not properly understood. Whilst the concept of a well-defined minimum metallic conductivity, corresponding to a discontinuous change in conductivity at the transition, is *not* supported by the scaling theories of localization, there are nevertheless several instances where the conductivity pre-exponential factors do have values corresponding to [5.52] (Mott 1987a). The reasons for this widespread occurrence are not clear since σ_{min} now no longer has the universal validity it was once believed to possess (see e.g. Mott and Davis 1979). Of course, the concept of a metal–insulator transition at the localization transition in 3-D is not in doubt, and neither is the fact that at the transition the electronic mobility or conductivity can change very rapidly, if not discontinuously (see Fig. 5.18), and that therefore the idea of a mobility edge at E_c or E_v still remains approximately valid, particularly at low temperatures. Another unresolved problem associated with the localization transition lies with the values of the critical exponents, for, say, the conductivity, which do not appear to agree with the predictions of scaling theory.

5.3.6 Percolation

We have seen in Chapter 3 that structural and compositional fluctuations may exist in an amorphous solid. If the length scale of these fluctuations is of the order of hundreds of ångströms, then tunnelling of electrons between such regions is essentially precluded, and a *classical* approach may be adopted. The long-range potential fluctuations we are considering are shown schematically in Fig. 5.19 (Fritzsche 1973); note that these represent only *symmetric* fluctuations due to electrostatic charges, and the antisymmetric potential fluctuations due to density variations which shift the valence and conduction band states in different directions, together with short-range fluctuations, are omitted for clarity. Electrons of a given energy E are only able to reside in those regions where the potential energy is lower than E. At low energies, the allowed regions would form isolated pockets in space, and as E is increased these regions would grow and merge until, at a critical energy, E_c, an allowed channel would be formed throughout the volume of the sample and the electron could therefore move anywhere. This phenomenon is termed 'percolation', and the occurrence of a percolation threshold at E_c can, in this model, be identified with the mobility edge or the localization transition separating localized and extended behaviour. The allowed channel is termed the 'infinite percolation cluster'.

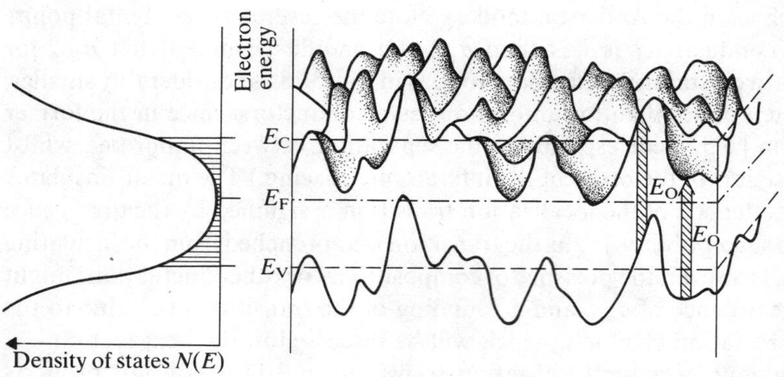

Fig. 5.19 Schematic illustration of the symmetric part of long wavelength potential fluctuations. Short-range fluctuations, as well as antisymmetric fluctuations which shift valence and conduction band states in opposite directions, are omitted for clarity. E_0 corresponds to an average optical gap, E_c and E_v are percolation thresholds or mobility edges. The density of states corresponding to this picture is shown on the left-hand side (Fritzsche 1973).

Percolation is a subject in its own right (see Stauffer 1985 for a general review) and has applications in very many areas of science. Within the rather restricted confines of the subject-matter of this book, percolation aspects are important in phenomena as disparate as the glass transition and the free-volume theory for it, the sol–gel transition, electronic hopping conduction in amorphous semiconductors, and the Anderson localization transition; fractal dimensionality is also intimately involved with such considerations. In this book, we will not give an exhaustive treatment of percolation theory, but merely give some important results, and indicate those cases where a percolation treatment can be useful. Zallen (1983) gives a good account of the applications of percolation theory to problems associated with disordered systems, and the interested reader is referred there for more detail.

There are essentially two types of percolation problem: *site* percolation, where if a site has a probability p of being 'occupied', then the percolation probability function $P(p)$ is the probability that a given occupied site belongs to the infinite percolation cluster; conversely, in *bond* percolation all sites in a lattice are supposed 'occupied', and if a bond connecting two such sites has a probability p of being 'open' then $P(p)$ is the probability that such a favourable bond forms part of the infinite percolation cluster linked by such bonds. The percolation probability $P(p)$ is identically zero until $p = p_c$, the critical percolation threshold, whereupon it rises steeply until eventually $P = p = 1$. The critical percolation concentration, p_c, obviously varies, depending on whether site or bond percolation is involved, and also p_c is different for different dimensionalities and for different types of lattice in a given dimension. Table 5.4 lists values of p_c for several types of lattice in both 2-D and 3-D, and it can be seen that in general $p_c(\text{bond}) < p_c(\text{site})$ for a given lattice of particular dimensionality, p_c decreases as the connectivity or local coordination of the lattice increases, and for a given connectivity p_c decreases as the dimensionality increases. (It is interesting to note in passing, however, that for a Cayley tree (or Bethe lattice) $p_c(\text{bond}) = p_c(\text{site})$.)

Table 5.4 Critical percolation concentrations p_c for a variety of lattices

Dimensionality	*Lattice*	p_c *(bond)*	p_c *(site)*	*Coordination*
2	Triangular	0.3473	0.5000	6
2	Square	0.5000	0.593	4
3	f.c.c.	0.119	0.198	12
3	b.c.c.	0.179	0.245	8
3	s.c.	0.247	0.311	6
3	Diamond	0.388	0.428	4

Close to the percolation threshold, quantities related to percolation can be expressed as a function of the departure from the threshold, $\xi = p - p_c$, and for $\xi \ll 1$, such quantities can be written in power-law form (as in phase-transition theory). Thus, the percolation probability P, the conductivity σ and the mean cluster size s can be expressed as

$$P \propto (p - p_c)^{\beta} \qquad [5.54]$$

$$\sigma \propto (p - p_c)^{t} \qquad [5.55]$$

$$s \propto (p_c - p)^{-\gamma} \qquad [5.56]$$

where P and σ are non-zero for $p > p_c$ and s is non-zero for $p < p_c$. The critical exponents β, t and γ are given in Table 5.5. It is interesting to note that the critical percolation exponents, unlike the critical concentrations p_c, are *independent* of the type of lattice, and depend only on the dimensionality.

The reason why the conductivity behaviour does not follow that of the percolation probability just beyond threshold (i.e. $t > \beta$) is that, although the fraction of allowed channel volume does increase sharply at $p = p_c$, the first percolation channels will be very convoluted in shape and hence will contribute little to conduction; thus $\sigma(p)$ increases much less rapidly than $P(p)$ for $p \geqslant p_c$. Hence percolation alone cannot be responsible for the metal–insulator transition associated with localization, as shown in Fig. 5.18. The critical exponents for the conductivity $\sigma \propto (E - E_c)^t$ and for the wavefunction localization length, v [5.40] should be the same (sect. 5.3.4); numerical calculations by MacKinnon and Kramer (1983) for Anderson localization have indicated that $v = t = 1.5$ in 3-D, reasonably close to the value of $t = 2.0$ characteristic of classical percolation (Table 5.5). However, the experimental data for localization in the impurity band of Si:P (Fig. 5.18) indicate that the conductivity exponent is near $t = 0.5$ (Hess *et al.* 1982).

Table 5.5 Critical exponents for percolation-related quantities (Stauffer 1985)

		Dimension	
Quantity	*Exponent*	d = 2	d = 3
P	β	0.14	0.40
σ	t	1.3	2.0
s	γ	2.4	1.8

Thus, the experimentally observed localization transition is considerably sharper than predicted from scaling or percolation theories; this marked discrepancy may be due to the neglect of electron–electron interactions.

5.4 Transport properties

We turn our attention now to the electrical transport properties of *ideal* amorphous semiconductors and glassy metals. We regard materials as 'ideal' if they are free of structural defects, such as dangling bonds, which give rise to electron states lying deep in the energy gap of semiconductors, and hence states which *do* exist in the gap arise only from band tails. Defect-related transport properties will be discussed in Chapter 6. Transport properties such as d.c. electrical conductivity, thermopower and Hall effect will be discussed in the light of the models for electron states in amorphous materials that we have developed in the preceding sections. Extensive reviews of the subjects may be found in Mott and Davis (1979), Nagels (1979) and Pollak (1987) for amorphous semiconductors and Rao (1983) for glassy metals.

5.4.1 D.C. electrical conductivity

We commence this section with a discussion of the effects of quantum interference and electron–electron interactions on the metallic conductivity in the weak-localization regime, and then consider the conductivity of amorphous semiconductors. In the normal metallic regime, the d.c. conductivity is given by the Drude formula, as obtained for example from the Boltzmann transport equation,

$$\sigma_B = ne^2\tau/m \qquad [5.57]$$

which can be rewritten in the form given by [5.51]. However, even in the highly conducting regime, quantum corrections apply to [5.57]. Scaling theories of localization (Abrahams *et al.* 1979) and quantum-interference considerations (Bergmann 1983) lead to the following modified forms for the conductivity:

$$\sigma_3 = \sigma_B\left[1 - \frac{3}{(k_F\ell)^2}\left(1 - \frac{\ell}{L}\right)\right] \qquad [5.58]$$

for three dimensions and

$$\sigma_2 = \sigma_B - \frac{e^2}{\hbar\pi^2}\ln\left(\frac{L}{\ell}\right) \qquad [5.59]$$

for two dimensions, where ℓ is the elastic electron mean free path, and L is either the size of the specimen or the inelastic diffusion length $L_i = (\ell\ell_i/2)^{1/2}$, where ℓ_i is the inelastic mean free path [5.34]) or the cyclotron radius $(c\hbar/eH)^{1/2}$ in a magnetic field H, whichever is the smaller. The length corrections for 1-D transport have already been given in [5.37]. Abrahams *et al.* (1979) and Shapiro and Abrahams (1981) have derived these equations using scaling arguments (cf. [5.44]), whilst Bergmann (1983) obtained the 2-D result from quantum-interference considerations (sect. 5.3.4). Since inelastic scattering events destroy the phase coherence required to give localization effects, L_i and τ_i ([5.34]) act as upper limits beyond which localization does not occur. The inelastic scattering time is temperature-dependent, and if phonon relaxation is dominant then (Anderson *et al.* 1979)

$$\tau_i(T) \propto T^{-p} \quad [5.60]$$

where $p = 2$, 3 or 4, depending on the temperature. The temperature dependence of the conductivity in 2-D in the weak-localization regime is then (Anderson *et al.* 1979)

$$\sigma_2^{WL}(T) = \sigma_B(T_0) + \frac{pe^2}{2\pi^2\hbar}\ln\left(\frac{T}{T_0}\right) \quad [5.61]$$

where T_0 is some reference temperature.

However, scaling theories and equivalently quantum-interference effects are non-interaction theories; electron–electron interactions can also produce corrections to the conductivity (Altshuler and Aronov 1985) which are logarithmic for 2-D:

$$\sigma_2^{EE}(T) = \sigma_B(T_0) + \frac{Ae^2}{4\pi^2\hbar}\ln\left(\frac{T}{T_0}\right) \quad [5.62]$$

where A is a constant of order unity. It can be seen, therefore, that a logarithmic temperature dependence is predicted for *both* weak-localization and electron-interaction effects. In 3-D, electron–electron interactions give rise to the correction term (Altshuler and Aronov 1985)

$$\sigma_3^{EE}(T) = \sigma_B(T_0) + \frac{Be^2}{4\pi^2\hbar}\left(\frac{k_B T}{\hbar D}\right)^{1/2} \quad [5.63]$$

where B is another constant and D is the electron diffusion coefficient. The $T^{1/2}$ temperature dependence in [5.63] arises from a minimum in the electronic density of states at E_F, the so-called Coulomb gap, caused by the electron–electron interactions. Electron–electron effects are expected to be important at very low temperatures.

Experimental evidence for the $T^{1/2}$ temperature dependence embodied in [5.63] and exhibited by amorphous metals and disordered crystalline metallic alloys is shown in Fig. 5.20(a) (Cochrane and Strom-Olsen 1984). Howson and Greig (1984) have found similar behaviour in other amorphous metals, but they also observe a cross-over to a linear temperature dependence at higher temperatures (Fig. 5.20(b)). This latter behaviour has been ascribed by Howson (1984) to a scaling correction term resulting from incipient localization effects, for which [5.58] can be rewritten approximately as (Shapiro and Abrahams 1981)

$$\sigma_3 = \sigma_B + \frac{e^2}{\pi^2\hbar L_i} \quad [5.64]$$

At low temperatures (but higher than those where electron–electron effects are important), the inelastic scattering is controlled by electron–electron interactions for which $L_i \propto T^{-1}$ (Mott 1987a); this accounts for the observed linear temperature

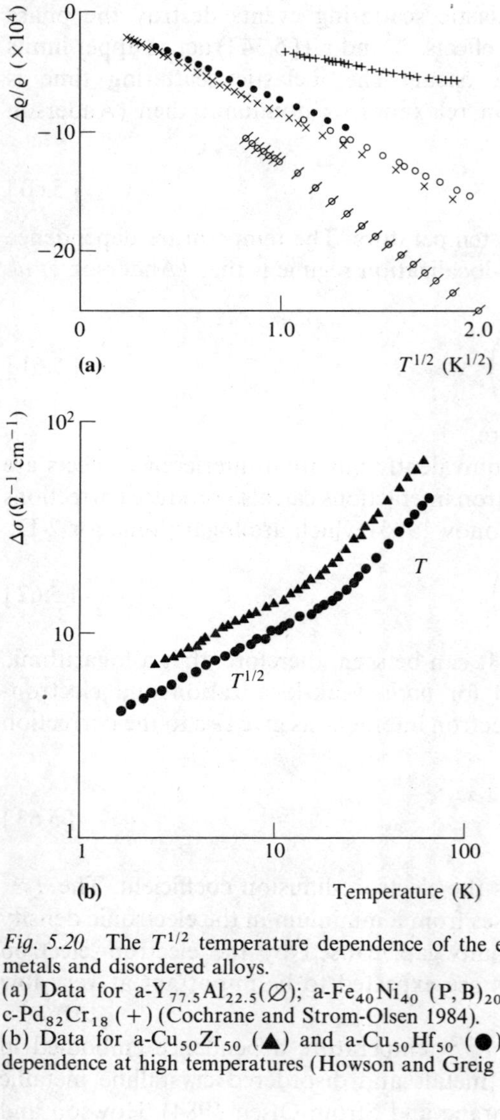

Fig. 5.20 The $T^{1/2}$ temperature dependence of the electrical resistivity ρ exhibited by amorphous metals and disordered alloys.
(a) Data for a-$Y_{77.5}Al_{22.5}$(∅); a-$Fe_{40}Ni_{40}$ (P, B)$_{20}$ (×); a-$Cu_{60}Zr_{40}$ (⊙); a-$Fe_{79}Cr_1B_{20}$ (●); c-$Pd_{82}Cr_{18}$ (+) (Cochrane and Strom-Olsen 1984).
(b) Data for a-$Cu_{50}Zr_{50}$ (▲) and a-$Cu_{50}Hf_{50}$ (●) showing cross-over to a linear temperature dependence at high temperatures (Howson and Greig 1984).

dependence of σ (Fig. 5.20(b)). At still higher temperatures phonon collisions become dominant for which $L_i \propto T^{-1/2}$ (Mott 1987a), and therefore $\sigma \propto T^{1/2}$ again.

So, for amorphous (or disordered crystalline) metals where the resistivity is quite large and where the electron mean free path is correspondingly short (comparable to the interatomic spacing), the 'temperature coefficient of the resistivity' (TCR), defined as

$$\alpha_\rho = \frac{1}{\rho}\frac{\partial \rho}{\partial T} \qquad [5.65]$$

is predicted to be *negative* if the conductivity $\sigma \propto T^{1/2}$ (cf. [5.63] and [5.64]). Thus, this explanation in terms of incipient-localization (quantum-interference) and electron–electron effects can account naturally for the almost universal observation of negative TCRs at low temperatures (Cochrane and Strom-Olsen 1984).

At room temperature, negative TCRs can also be found for amorphous metals, but now there appears to be a correlation, first discovered by Mooij (1973), that the *sign* of the TCR appears to be correlated with the *magnitude* of the resistivity, negative values of TCR being the rule for metals with resistivities greater than $\simeq 150\ \mu\Omega$ cm (Fig. 5.21).

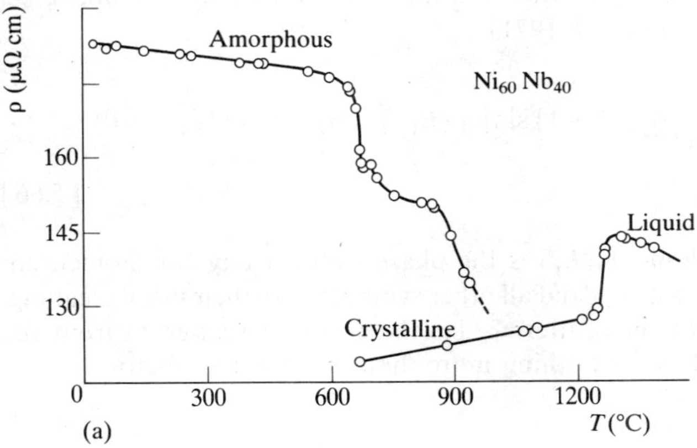

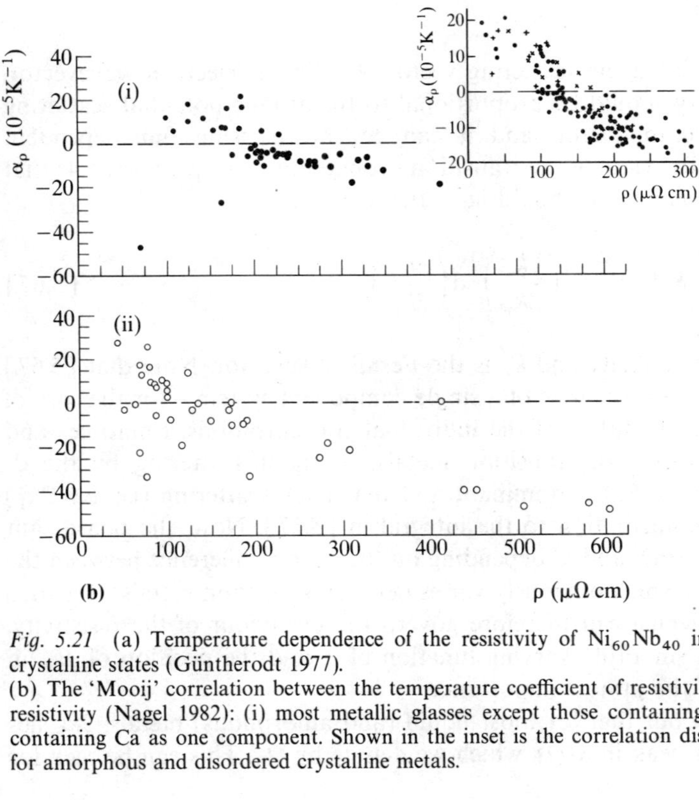

Fig. 5.21 (a) Temperature dependence of the resistivity of $Ni_{60}Nb_{40}$ in amorphous, liquid and crystalline states (Güntherodt 1977).
(b) The 'Mooij' correlation between the temperature coefficient of resistivity α_ρ and the value of the resistivity (Nagel 1982): (i) most metallic glasses except those containing Ca; (ii) metallic glasses containing Ca as one component. Shown in the inset is the correlation discovered by Mooij (1973) for amorphous and disordered crystalline metals.

This behaviour can also be accounted for by a theory first developed by Ziman (1961) to describe the resistivity of simple-metal *liquids*. The theory is based on a *diffraction* model, in which the scattering of electrons, viewed as plane waves characterized by wavevector k, which gives rise to the conductivity can be conveniently viewed as arising from two mechanisms: (1) scattering from the atomic potentials, and (2) coherent scattering, i.e. diffraction, from assemblies of atoms. The potential contributing to the first type of scattering is the pseudopotential $V(k)$ for the case of simple metals (Ziman 1961), but for transition metals in which the scattering is due to the d-electrons, the potential can be replaced by an expression in terms of scattering phase shifts for non-overlapping potentials, i.e. the so-called 't-matrix' (Evans *et al.* 1971):

$$t(k)=\frac{2\pi\hbar^3}{m_e(2m_eE_F)^{1/2}\Omega}\sum_l(21+1)\sin[\eta_l(E_F)]\exp[i\eta_l(E_F)]P_l(\cos\theta) \qquad [5.66]$$

where Ω is the atomic volume, $\eta_l(E_F)$ is the phase shift for angular momentum quantum number l evaluated at E_F, and all other symbols have their usual meaning. The second contribution to the scattering, i.e. the coherent scattering from the atoms in the condensed phase, is nothing more than the structure factor:

$$S(\boldsymbol{k})=\frac{1}{N}\left|\sum_i\exp(i\boldsymbol{k}\cdot\boldsymbol{r}_i)\right|^2$$

where we have associated the scattering vector Q, with the electron wavevector, k. Thus, the resistivity should be proportional to the atomic potential scattering factor and the structure factor, and it can be shown using time-dependent perturbation theory to evaluate the transition probabilities (see problem 5.4) that the resistivity of a *liquid* metal should have the form:

$$\rho_{liq}=\frac{12\pi\Omega}{e^2\hbar v_f^2}\int_0^1 S(k)|t(k)|^2\left(\frac{k}{2k_F}\right)^3 d\left(\frac{k}{2k_F}\right) \qquad [5.67]$$

where v_f is the Fermi velocity and k_F is the Fermi wavevector. Note that [5.67] is derived strictly only for the case of a single component system – for the case of alloys, account has to be taken of the individual concentrations, t-matrices and partial structure factors. For transition metals, resonant scattering by the d-electrons means that $\eta_2(E_F)$ is dominant and that back-scattering (i.e. $k=2k_F$) gives the dominant contribution to the integral in [5.67]. Now, the phase shift varies smoothly between 0 and π, depending on the energy difference between the d-band centre and E_F which obviously varies between transition metals. It is thus the structure factor which can therefore govern the behaviour of the resistivity, since it is a far from smoothly varying function of k, and the position of $2k_F$ in relation to $S(k)$ is of the utmost importance.

In fact, it turns out that for most liquid (and amorphous) metals, $2k_F$ lies very close to the first peak in $S(Q)$, which we denote by Q_p. This can be seen for example by calculating the Fermi wavevector using the relation

$$k_F = \left(\frac{3\pi^2 Z}{\Omega}\right)^{1/3} \qquad [5.68]$$

where Z is the (effective) valence of the metal, and for the case of Ni, for which $Z = 1$ may be taken, a value for $2k_F = 2.8$ Å^{-1} is found. This is to be compared with the position of the first peak in $S(Q)$ for liquid Ni, i.e. $Q_P = 3.1$ Å^{-1} (Waseda 1980). In fact, since the structure in $S(Q)$, and hence Q_P, is relatively insensitive to the type of atom involved, monovalent metals have $2k_F$ generally lying to the left of Q_P while divalent and higher valency metals have Q_P lying progressively to the right of Q_P. Thus, by alloying a metalloid ($Z = 4$) with, say, a transition metal ($Z = 1$), $2k_F$ can be caused to move through the first peak in $S(Q)$ (Fig. 5.22(a)).

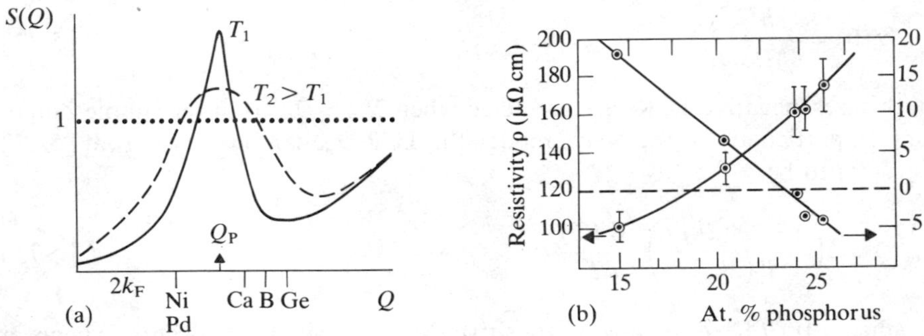

Fig. 5.22 (a) Schematic illustration of the relation of the position of $2k_F$ for different valence elements with the position (Q_p) of the first peak of the structure factor. Also shown is the broadening (dashed curve) of $S(Q)$ at higher temperatures.
(b) Example of a change of sign of the TCR, α_ρ, as the metalloid content is changed in the alloy a-$Ni_{1-x}P_x$ (Cote and Meisel 1981).

The most strongly temperature-dependent quantity in the expression for the resistivity ([5.67]) is $S(Q)$, for which the Debye–Waller factor causes peaks to decrease in height and broaden in width. Thus, for those materials for which $2k_F = Q_P$, the TCR ([5.65]) is predicted to be *negative* because of the dominant influence of the Debye–Waller factor in determining $\rho(T)$. This is Ziman's explanation for the negative TCRs observed in liquid metals.

It is an obvious extention to apply the theory to account for the similar effects observed in amorphous metals too (see e.g. Cote and Meisel 1981). However, this necessitates several modifications to the simple Ziman diffraction theory, principally because generally the temperature is less than the Debye temperature θ_D, which means that the structure factor $S^\rho(Q)$ appearing in [5.67] can no longer be identified simply with the X-ray structure factor $S^x(Q)$, as was done for the case of liquids. Nagel (1977) suggested the relation

$$S^\rho(Q) \simeq 1 + [S_E(Q) - 1]e^{-2W(T)} \simeq 1 + [S_0(Q) - 1]e^{-2[W(T) - W(0)]} \qquad [5.69]$$

where S_E and S_0 are the X-ray structure factors at $T = 0$ K, being the equilibrium function and that including zero-point motion, respectively. The Debye–Waller factor $W(T)$ in the Debye approximation is well known (see e.g. Ziman 1972)

$$W(T)=\frac{3\hbar^2k^2T^2}{2Mk_B\theta_D^3}\int_0^{\theta_D/T}\left(\frac{1}{e^z-1}+\frac{1}{2}\right)Z\,dZ \qquad [5.70]$$

where M is the atomic mass, and $Z=\hbar\omega/k_BT$. The asymptotic behaviour of $W(T)$ is:

$$W(T)\to W(0)+\frac{2\pi^2}{3}W(0)\left(\frac{T}{\theta_D}\right)^2 \qquad T\ll\theta_D \qquad [5.71(a)]$$

$$W(T)\to W(0)+4W(0)\left(\frac{T}{\theta_D}\right) \qquad T\gg\theta_D \qquad [5.71(b)]$$

where

$$W(0)=\frac{3\hbar^2k^2}{8Mk_B\theta_D} \qquad [5.72]$$

In both cases, negative TCRs are predicted when $2k_F\simeq Q_P$, as in the simple Ziman theory. In particular, for the Nagel model, the TCR, α_ρ, can be shown from [5.67] and [5.69] to be

$$\alpha_\rho\simeq 2\left(\frac{1-S^\rho(2k_F)}{S^\rho(2k_F)}\right)\frac{\partial W(T)}{\partial T} \qquad [5.73]$$

and since $\partial W(T)/\partial T$ is always positive, a negative TCR occurs whenever $S^\rho(2k_F)>1$.

A good example of a change in sign of α_ρ upon altering the composition of an alloy is shown in Fig. 5.22(b) for the case of a-$Ni_{1-x}P_x$, where the progressive addition of the metalloid causes $2k_F$ to increase and hence move through the first peak in $S(Q)$, whose position is relatively unaffected by the presence of the metalloid atoms. Note that the magnitude of α_ρ in amorphous metals is generally of the order $10^{-4}\,K^{-1}$ irrespective of the sign.

In addition, these modified diffraction theories predict that (cf. [5.71(b)]) for amorphous metals the resistivity should vary as T at high temperatures (whether α_ρ is negative or positive), which is observed experimentally (Fig. 5.20(b)). Thus the temperature dependence of ρ for amorphous metals is considerably weaker than that for crystalline metals, where at low temperatures $\rho\propto T^5$. However, as the resistivity rises, the mean free path decreases, until for resistivities greater than $\simeq 100\,\mu\Omega\,cm$ it becomes comparable to the interatomic spacing. Under these conditions of extremely strong scattering, many of the assumptions underlying the weak-scattering Ziman theory become invalid.

We turn now to a discussion of the behaviour of the d.c. conductivity in amorphous semiconductors, i.e. those materials where the Fermi level lies in the gap between valence and conduction bands. We consider here only 'ideal' amorphous semiconductors, namely those which do not contain defects which give rise to defect states in the gap, and defer until Chapter 6 a discussion of the transport properties of Fermi glasses, where E_F lies in a band of (defect-related) localized states.

The total conductivity at a finite temperature can be written as:

$$\sigma = -\int \sigma_E \frac{\partial f(E)}{\partial E} \, dE \qquad [5.74]$$

where σ_E is the conductivity for states with energy E, and where $f(E) = (1 + \exp(E - E_F)/k_B T)^{-1}$, the Fermi–Dirac distribution function, and hence $\partial f(E)/\partial E = -f(E)[1 - f(E)]/k_B T$. Conduction involving localized states can only occur by phonon-assisted hopping, and this will be left for consideration until later. Thus, the only current-carrying states in this model are the extended states beyond E_v and E_c in the valence and conduction bands, respectively. Equation 5.74 can be rewritten in the form

$$\sigma = e \int N(E)\mu(E) f(E)[1 - f(E)] \, dE \qquad [5.75]$$

where $\mu(E)$ and $N(E)$ are the mobility and density of states at an energy E. If the mobility edges at E_v and E_c are sharp, and if conduction takes place far enough away from E_F that $f(E)[1 - f(E)] \simeq \exp(-(E - E_F)/k_B T)$ [5.75] may be integrated to give (for electrons):

$$\sigma_{ext} = eN(E_c)\mu_e k_B T \exp[-(E_c - E_F)/k_B T] \qquad [5.76]$$

where μ has been taken to be zero in the mobility gap, and to be some average value μ_e in the band.

A proper theoretical understanding of the conductivity pre-exponential factor σ_0 is still lacking, however, now that Mott's idea of a minimum metallic conductivity σ_{min} (for which $\sigma_0 = \sigma_{min}$), associated with a discontinuous change in conductivity on passing through the localization transition, is no longer supported by scaling ideas of localization. Within the spirit of the model assumed above, where conduction is presumed to take place in extended states beyond a reasonably well-defined mobility edge, an estimate for μ_e in [5.76] can be obtained by assuming that the electron (or hole) transport is diffusive when the mean free path is rather short. In this case, the mobility is given by the Nernst–Einstein relation (cf. [4.80] and [4.93])

$$\mu_e = \frac{eD}{k_B T} \qquad [5.77]$$

where the diffusion constant is given by $D = (1/6)\nu_{el} a^2$, and ν_{el} is an electronic frequency ($\nu_{el} \simeq \hbar/m_e a^2 \simeq 10^{15}\,s^{-1}$). Hence (see e.g. Mott and Davis 1979):

$$\mu_e^D \simeq \frac{e\hbar}{6m_e k_B T} \qquad [5.78]$$

which gives a value $\mu_e \simeq 7.5\,cm^2\,V^{-1}\,s^{-1}$ at room temperature. (For comparison, the electron mobility in, say, crystalline GaAs is $\mu_e \simeq 9 \times 10^3\,cm^2\,V\,s^{-1}$.) The frequency ν_{el} may also be written in the form $\sim JF/\hbar$, where J is an electronic transfer integral and F is a factor accounting for the fact that increasing disorder decreases the carrier diffusion (Emin 1976). In the random phase model, the role of F is played by the factor $[a^3(2zJ)N(E_c)]$, so that the mobility can be written:

$$\mu_e^{RPA} = \frac{2\pi e a^5 z J^2 N(E_c)}{3\hbar k_B T} \qquad [5.79]$$

Mott (1987b) has also suggested, from scaling considerations, that the pre-exponential factor should have the form (cf. [5.64])

$$\sigma_0 = Ce^2/\hbar L_i \qquad [5.80]$$

where L_i is the inelastic diffusion length [5.34], and where C is a constant of the order of $\simeq 0.3$. Values for σ_0 then lie in the range 100–200 Ω^{-1} cm^{-1}.

In fact, the experimentally observed prefactor contains other terms in addition to the theoretical terms arising from details of the transport process and the nature of the localized states, [5.76] and [5.80] (Stuke 1987). These extra terms arise from the fact that the various energies associated with the band gap shift with temperature, and two effects can be distinguished (Overhof and Beyer 1983). The first is the so-called 'statistical shift' of the Fermi level (Spear and Le Comber 1976). In an amorphous semiconductor with localized states tailing into the gap at the band-edges and defect states in the gap, the resultant DOS distribution need not be symmetrical with respect to the band centre. At $T = 0$ K, therefore, the Fermi level will in general lie not at mid-gap but at a position either above or below mid-gap, dictated by the shape of the DOS involved, particularly in the mid-gap region where the DOS is defect-related. (Intrinsic a-Si:H, for example, is n-type, indicating that E_F lies in the upper half of the gap.) However, as the temperature increases, electrons and holes are thermally excited to states at the band edges, and charge neutrality demands that there is a statistical shift of E_F towards mid-gap since the DOS at the band edges is more symmetrical with respect to the mid-gap position than is the defect-related DOS. In a-Si:H, therefore, this shift will be in the downward direction, from conduction band to mid-gap; the temperature dependence of E_F can therefore be approximated by a linear dependence,

$$E_F(T) = E_F(0) - \gamma_F T \qquad [5.81]$$

The other effect involves a temperature shift of the band edges or, if all states are assumed to shift rigidly with the bands, a shift of E_c and E_v. This effect has two origins (Overhof and Beyer 1983): the first is the thermal expansion of the structure, which produces a change in one-electron potentials; the other is electron–phonon coupling via Debye–Waller and self-energy corrections. Consequently, the temperature shift of the conduction band edge (mobility edge) can be written as

$$E_c(T) = E_c(0) - \gamma_c T \qquad [5.82]$$

at temperatures above, say, the Debye temperature θ_D; for $T < \theta_D$, E_c is practically temperature-independent. From optical measurements as a function of temperature, the temperature coefficient of the *gap* is $\simeq 4$–8×10^{-4} eV K^{-1}; if E_F lies near mid-gap, the temperature coefficient of the band edge with respect to E_F, namely γ_c, would therefore be about half this value.

The inclusion of these temperature-dependent effects has important ramifications for the interpretation of measured values of both the activation energy

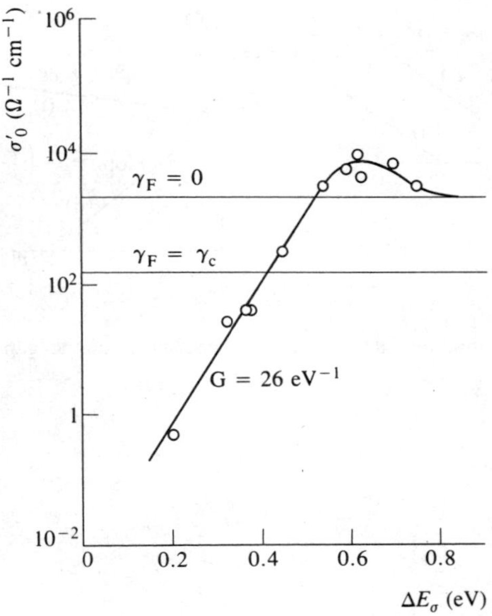

Fig. 5.23 Relation between the d.c. conductivity pre-exponential factor σ'_0 and the activation energy ΔE_σ for a-Si:H (Stuke 1987). The linear behaviour exhibited for values of ΔE_σ less than 0.6 eV is an example of the Meyer–Neldel rule.

and the pre-exponential factor of the d.c. conductivity. The activation energy is thus, from [5.76],

$$(E_c - E_F)_T = (E_c - E_F)_0 - \gamma T \qquad [5.83]$$

where $\gamma = \gamma_c - \gamma_F$, and all shifts are measured with respect to mid-gap. The *observed* d.c. electrical conductivity activation energy is therefore simply $\Delta E_\sigma = (E_c - E_F)_0$, which is the intercept of the linearly temperature-dependent term $(E_c - E_F)_T$ extrapolated to $T = 0$ K (see Fig. 5.24(a)); in itself, therefore, this quantity has *no* physical significance. The measured pre-exponential factor is changed as well, since it will contain terms involving the temperature coefficients:

$$\sigma'_0 = \sigma_0 \exp(\gamma/k) \qquad [5.84]$$

The statistical shift of the Fermi level has been shown to be primarily responsible for an interesting correlation observed between the measured values of the conductivity pre-exponential factor σ'_0 and the activation energy ΔE_σ for a-Si:H (Overhof and Beyer 1983): the data seem to obey the relation

$$\sigma'_0 = \sigma_0 \exp(G\,\Delta E_\sigma) \qquad [5.85]$$

where the slope parameter generally has the value $G \simeq 25\,\text{eV}^{-1}$ (Fig. 5.23). This behaviour has been termed the Meyer–Neldel rule after similar behaviour observed in crystalline semiconductors (Meyer and Neldel 1937). (In fact, an exponential relation between the pre-factor and the activation energy of the form of [5.85] is rather widespread, and is observed in time-dependent transport behaviour, as well as in steady-state properties such as σ_{dc} (Jackson 1988).) It is plausible that the magnitude of the statistical shift of the Fermi level, $\gamma_F = \partial E_F/\partial T$ (cf. [5.81]) should

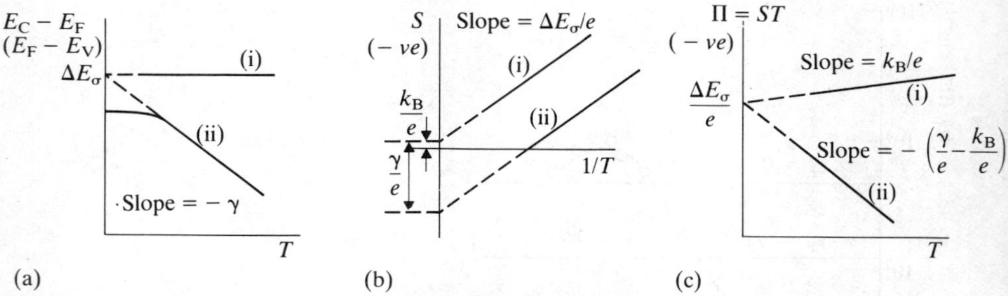

Fig. 5.24 Schematic illustration of methods of obtaining the temperature coefficient of the gap (curves (ii)); curves (i) arise if $\gamma = 0$.
(a) Electrical conductivity, $E_c - E_F = \Delta E_\sigma - \gamma T$

(b) Seebeck coefficient $S = \frac{-k_B}{|e|}\left(\frac{\Delta E_\sigma}{k_B T} - \frac{\gamma}{k_B} + 1\right)$

(c) $\Pi = -\frac{\Delta E_\sigma}{|e|} + \left(\frac{\gamma}{|e|} - \frac{k_B}{|e|}\right)T$

(after Mott and Davis 1979).

scale inversely with the size of the band gap, since the probability of electron excitation into the tail states increases as the gap decreases; if such an effect is dominant the relationship given by [5.85] is obtained. Kikuchi (1988) has further stressed the need for a deep minimum in the gap-state DOS for the observation of the Meyer–Neldel rule.

Meyer–Neldel correlations between σ_0' and ΔE_σ can occur in a-Si:H for a variety of reasons: for example, a collection of differently prepared samples can contain different concentrations of dopant impurities, thereby shifting E_F, or of defects, thereby changing the gap-state DOS; alternatively, the effect may be observed in a single sample in which the (dangling-bond) defect density can be changed in some way, e.g. by irradiation, either by light (the Staebler–Wronski (1980) effect) or by electrons. Although the observed values of σ_0' can vary by three orders of magnitude (Fig. 5.23), all the data can be fitted with a 'true' pre-factor with a value $\sigma_0 \sim 150$–$200\,\Omega^{-1}\,\mathrm{cm}^{-1}$.

So far we have developed the theory for d.c. electronic transport in the framework of a sharp mobility edge at E_c. One might ask whether a constant activation energy for conduction would still be observed if, instead, the mobility edge were more gradual, for example at a percolation threshold or at a scaling localization transition. However, it can be shown that if the mobility increases as, say, $(E - E_c)^m$, integration of [5.75] introduces a term $(k_B T)^m$ into the pre-factor making it temperature-dependent, and leaving a constant activation energy as before. Döhler (1979) has considered such cases, ascribing the continuous increase in μ_e to the influence of electron–phonon interactions in causing activated hopping processes below E_c, and also in reducing the mobility in extended states above E_c.

Thomas and co-workers have also considered electron–phonon interactions in the single-phonon approximation, and argued that 'phonon-induced delocalization' occurs at finite temperatures (Müller and Thomas 1984, Fenz *et al.* 1985, Dersch and Thomas 1987); it should be remembered that, for example, scaling theories of localization (sect. 5.3.4) are strictly valid only at $T=0$ K. A state can be regarded as being localized if a charge-density fluctuation associated with it does not decay in time (cf. the analogous argument for density–density correlations for the glass transition, sect. 2.3.4). At finite temperatures, 'delocalization' of the previously localized electron states occurs because the electron–phonon interaction allows an extra channel by which such charge-density fluctuations may decay (by a hopping conduction process). Thus, the differential conductivity σ_E (cf. [5.74]) at energies E below the localization transition point at E_c is due to incoherent hopping processes and decreases with energy below E_c more slowly than is predicted by scaling theories at $T=0$ K, namely as $\sigma \propto (E-E_c)^t$. Above E_c, the dominant conduction processes are determined by coherent (tunnelling) electronic transport (Müller and Thomas 1984); in such cases, a near-constant d.c. conductivity activation energy results (Dersch and Thomas 1987).

One may also ask about the effect on the d.c. conductivity if fluctuations in potential (caused, perhaps, by a random distribution of charged centres) are present, as in Fig. 5.19. It is assumed that such fluctuations are sufficiently long-range to cause only a random local perturbation in the energies of the bands, and hence in the local mobility edge E_c, without inducing further localization (Overhof and Beyer 1981). Overhof and Beyer (1983) showed that the d.c. conductivity activation energy would still be approximately temperature independent, but would contain an extra term involving Δ, the width of such charge-induced fluctuations in E_c:

$$\Delta E_c = (E_c - E_F) + 0.2k_B T - \Delta/4 \qquad [5.86]$$

Finally, we include in Table 5.6 a selection of values for the d.c. conductivity activation energy, ΔE_σ, for those materials in which extended state conduction is observed, together with the optical gap and other parameters.

5.4.2 Thermopower

Thermoelectric power (or thermopower for short) is another transport property, closely related to σ, which gives a great deal of information about transport mechanisms in amorphous semiconductors. The thermopower or Seebeck coefficient, denoted by S, is the constant of proportionality between the voltage (in the absence of a current) and the temperature gradient which causes it. The Peltier coefficient, Π, is the constant of proportionality between the heat flux transported by the electrons and the current density, and S and Π are simply related via:

$$S = \Pi/T \qquad [5.87]$$

The thermopower may be written as (Mott and Davis 1979):

$$S = -\frac{k_B}{|e|\sigma}\int \sigma_E \frac{(E-E_F)}{k_B T}\frac{\partial f(E)}{\partial E}\,dE \qquad [5.88]$$

where $|e|$ is the value of the electronic charge and where σ_E may be written as $\sigma_E = eN(E)\mu(E)k_B T$, as before; substituting this and [5.75] for σ gives, in the

Table 5.6 Electrical and optical parameters for some amorphous semiconductors

Material	*Preparation technique*	ΔE_σ (eV)	σ(300 K) (Ω^{-1} cm^{-1})	E_o(300 K) (eV)	*Comments*
Se	LQ	1.10(1)	$\simeq 10^{-16}$(1)	2.05(2)	—
As	VT	0.81(3)	4×10^{-9}(3)	1.4(3)	Bulk (H-transported)
P	S	1.03(4)	$\simeq 10^{-13}$(4)	1.72(4)	Annealed
Si	E	†	2×10^{-3}(5)	1.08(5)	Oxygen-free
Si:H	GD	0.6(6)	2.5×10^{-9}(6)	1.64(7)	$T_d = 600$ K. Intrinsic
Ge	E	†	$\simeq 10^{-5}$(8)	1.00(9)	Annealed at 670 K
Ge:H	GD	0.43(10)	$\simeq 10^{-4}$(10)	0.92(11)	$T_d = 500$ K. Intrinsic
As_2S_3	LQ	1.14(12)	$\simeq 10^{-17}$(12)	2.32(13)	—
As_2Se_3	LQ	0.91(12)	$\simeq 10^{-12}$(12)	1.76(13)	—
As_2Te_3	LQ	0.42(12)	$\simeq 10^{-4}$(12)	0.83(14)	—

Key: † = $T^{-1/4}$ conductivity; VT = vapour transport; S = sputtered; E = evaporated; GD = glow discharge; LQ = liquid quench.

References: (1) J. L. Hartke (1962), *Phys. Rev.* **125**, 1177; (2) E. A. Davis (1970), *J. Non-Cryst. Sol.* **4**, 107; (3) G. N. Greaves, S. R. Elliott and E. A. Davis (1979), *Adv. Phys.* **28**, 49; (4) P. Extance and E. A. Davis (to be published); (5) S. K. Bahl and S. M. Bhagat (1975), *J. Non-Cryst. Sol.* **17**, 409; (6) W. E. Spear and P. G. Le Comber (1976), *Phil. Mag.* **33**, 935; (7) R. J. Loveland, W. E. Spear and A. Al-Sharbaty (1973/4), *J. Non-Cryst. Sol.* **13**, 55; (8) H. Mell (1974), *Proc. 5th Int. Conf. on Amorphous and Liquid Semiconductors*, J. Stuke and W. Brenig (eds) (Taylor and Francis) p. 203; (9) M. L. Theye (1974), *Proc. 5th Int. Conf. on Amorphous and Liquid Semiconductors*, J. Stuke and W. Brenig (eds) (Taylor and Francis) p. 479; (10) D. I. Jones, W. E. Spear and P. G. Le Comber (1976), *J. Non-Cryst. Sol.* **20**, 259; (11) D. I. Jones – private communication; (12) C. H. Seager and R. K. Quinn, *J. Non-Cryst. Sol.* (1975) **17**, 386; (13) N. F. Mott and E. A. Davis (1979), *Electronic Processes in Non-Crystalline Materials* (OUP); (14) K. Weiser and M. H. Brodsky (1970), *Phys. Rev.* **B1**, 791.

non-degenerate limit when the current path is far from E_F:

$$S = -\frac{k_B}{|e|}\frac{\int \mu(E)N(E)[(E-E_F)/k_BT]\exp[-(E-E_F)/k_BT]\,dE}{\int \mu(E)N(E)\exp[-(E-E_F)/k_BT]\,dE} \qquad [5.89]$$

For the case of band conduction of electrons in extended states at and beyond E_c, assuming that $N(E)$ and $\mu(E)$ are *constant*, [5.89] can be readily integrated to give:

$$S_{ext} = -\frac{k_B}{|e|}\left(\frac{E_c - E_F}{k_BT} + A\right) \qquad [5.90]$$

where Ak_BT is the average energy of the transported electrons with respect to E_c; for amorphous semiconductors, and conduction in extended states, $A \simeq 1$. Note that if $(E_c - E_F)$ is temperature dependent ([5.83]), then:

$$S_{ext} = -\frac{k_B}{|e|}\left(\frac{\Delta E_\sigma}{k_BT} - \frac{\gamma}{k_B} + 1\right) \qquad [5.91]$$

and it is immediately apparent that the slopes of the curves $\ln\sigma$ vs. $1/T$ and S vs. $1/T$ should be the *same* if conduction is in *extended* states. If both electrons and holes contribute to the conductivity, the thermopower consists of the weighted sum:

$$S = \frac{S_e\sigma_e + S_h\sigma_h}{\sigma_e + \sigma_h} \qquad [5.92]$$

A most important point to note is that the sign of the thermopower is reversed if holes carry the current, and thus the thermopower is a reliable guide to the nature of the dominant charge type, whether n- or p-type. For instance, most chalcogenide glasses exhibit a positive thermopower, indicating p-type conduction. Note that, unlike the case for crystalline materials, the Hall effect *cannot* be used in this way, as we will see in the next section.

Measurements of the thermopower can also be used to obtain a value for γ, the temperature coefficient of $(E_c - E_F)$, from the *intercept* of S vs. $1/T$ or the slope of Π vs. T (Fig. 5.24).

In sect. 5.4.2 it was pointed out that true values of energy-gap or pre-exponential parameters could not be obtained from Arrhenius plots of the d.c. conductivity owing to the temperature-dependent shifts of E_F and E_c ([5.81, [5.82]). Döhler (1979) and Overhof and Beyer (1981) have shown that a particular combination of conductivity and thermopower values obviates this problem. Specifically, the value of the function

$$Q(T) = \ln \sigma(T) + eS(T)/k_B \qquad [5.93]$$

is independent of the Fermi-level position, and hence of γ_F, as long as unipolar conduction is dominant (cf. [5.83] and [5.91]). For a-Si:H, for example, Overhof and Beyer (1984) find that $Q(T)$ is essentially linear:

$$Q(T) = Q_0 - E_Q/k_B T \qquad [5.94]$$

where the constant term appears to have a quasi-universal value, $Q_0 = \ln \sigma_0 + A \simeq 11$, whilst values for the activation energy lie in the range $0.05 \leqslant E_Q \leqslant 0.2$ eV. Overhof and Beyer (1983) have ascribed this behaviour to the effect of long-range potential fluctuations modulating the mobility edges (see sect. 5.4.1), with $E_Q \simeq 1.25\Delta$ (where Δ is the width of the energy-fluctuation distribution). Fenz *et al.* (1985) have shown that the same behaviour can result from a shift in the conduction path below E_c due to phonon-induced delocalization (see also sect. 5.4.1). The observation of a non-zero value for E_Q implies that the characteristic energy associated with the thermopower is not exactly the same as the activation energy for the conductivity; this results from the use of [5.89] when the conduction path can fall below E_c.

5.4.3 Hall effect

Measurements of the Hall voltage are a natural complement to conductivity measurements in transport studies of *crystalline* semiconductors, since, if unipolar conduction occurs, the Hall coefficient provides a reliable estimate of the carrier concentration n and the carrier type:

$$R_H = \frac{E_y}{HJ_x} = \frac{1}{ne}\frac{\mu_H}{\mu_e} \qquad [5.95]$$

where μ_H and μ_e are the Hall and conductivity mobilities, and R_H is negative for electrons and positive for holes. Equation 5.95 is valid only if the mean free path is sufficiently long for Boltzmann transport theory to apply; it is *not* appropriate for amorphous semiconductors. It is found, moreover, that the sign of R_H is always *opposite* to that of the thermopower in all the amorphous semiconductors examined so far except for certain compositions in the alloy system, $CdGe_xAs_2$, in which both n-type S and R_H are observed. This 'sign anomaly' is especially dramatic in a-Si:H where it is found that on changing the dopant from B (p-type) to P (n-type), the sign of the Hall effect changes from negative to positive in opposition to the change in sign of the thermopower (Le Comber *et al.* 1977). The sign of R_H for P-doped a-Si:H reverts however to its 'proper' negative value on recrystallization (Reilly and Spear 1978). This behaviour, completely at variance with that observed in crystals, is a major challence to any theory of the Hall effect. A review of the subject has been given by Friedman (1987).

Consider first the case of carriers excited to a mobility edge where, although the states are extended, the scattering is very strong and the mean free path is very short. It is a central feature of all calculations that transitions between *two* sites do *not* contribute to the Hall voltage; instead a minimum of *three* sites is necessary, for which the carrier is given a choice of which way to proceed in the direction perpendicular to both the applied electric and magnetic fields, $\boldsymbol{E}$ and $\boldsymbol{H}$, either moving directly between initial and final states, or alternatively via occupation of one or more other sites as virtual states. The three sites contributing to the Hall effect may be three of the atoms (or their bonds) forming a ring in the structure, or alternatively for group IV or V elements, they may be three of the bonds emanating from a single given atom.

Using the random-phase model, Friedman (1971) first showed that in this regime the Hall mobility should be

$$\mu_H \simeq 2\pi\left(\frac{ea^2}{\hbar}\right)\left(\frac{B_0}{B}\right)\left(\frac{\eta\bar{z}}{z^2}\right) \qquad [5.96]$$

Here B_0 is the bandwidth in the absence of disorder ($=2zJ$), where z is the coordination number and J the overlap integral, B is the bandwidth with disorder ($\simeq(a^3N(E_c)^{-1}$), $\bar{z}$ is the average number of closed three-site paths about an arbitrarily chosen site, and η is the projection of the area of a three-site path in the direction perpendicular to the field ($\simeq 1/3$ for 3-D). Putting all these factors together gives:

$$\mu_H \simeq \frac{4\pi}{3}\frac{ea^5}{\hbar}JN(E_c)\left(\frac{\bar{z}}{z}\right) \qquad [5.97]$$

It has been assumed that the phase correlation of wavefunctions extends to nearest neighbours, but that there is no correlation between the phase differences of distinct pairs of sites (Friedman 1980), which is a somewhat less stringent condition than that used in the original random phase model, for which the phases of individual site wavefunctions are uncorrelated and hence J would vanish on averaging over the random phases. Note that [5.97] predicts that μ_H should be *temperature independent* (in contrast to $\mu_e \propto 1/T$, [5.77–5.79]), and if it is assumed that $\bar{z} \simeq z$ and $a = 3$ Å, $\mu_H \simeq 10^{-1}$ cm^2 V^{-1} s^{-1} (cf. $\mu_e \simeq 10$ cm^2 V^{-1} s^{-1}). Comparison with the random phase model prediction for μ_e([5.79]), indicates that:

$$\frac{\mu_H}{\mu_e} \simeq \frac{2}{z}\frac{k_B T}{J} \qquad [5.98]$$

where generally $J \gg k_B T$.

The Hall mobility is commonly found to decrease with decreasing temperature from a temperature-independent value at high temperatures. In the 'two-channel' model, this observation is reconciled with the random phase model results by assuming that, at high temperatures, carriers are excited to the mobility edge, [5.97] should be obeyed, and μ_H should be temperature independent. It was tacitly assumed for a long time that the Hall mobility for carriers hopping among localized states (in the band tails) would be negligibly small, and hence the decrease in μ_H at low temperatures is simply a reflection of the decreasing number of carriers excited to the mobility edge. The total Hall mobility, like the thermopower ([5.92]) can be written as a sum of various contributions, weighted by the conductivity:

$$\mu_H^{Tot} = \frac{\mu_H^{ext}\sigma_{ext} + \mu_H^{hop}\sigma_{hop}}{\sigma_{ext} + \sigma_{hop}} \simeq \frac{\mu_H^{ext}}{1 + (\sigma_{hop}/\sigma_{ext})} \qquad [5.99]$$

if $\mu_H^{hop} \ll \mu_H^{ext}$.

There has been much recent work on the size of the Hall effect expected for carriers hopping among disorder-localized states. The consensus of opinion seems to be that μ_H^{hop} is very small, but not negligible ($\simeq 10^{-4}$ cm^2 V^{-1} s^{-1} at room temperature), although various authors differ as to the precise temperature dependence to be expected. Various attempts have been made to observe the Hall effect for hopping conduction of electrons between localized states, so far without success. Klein (1985), for example, detected no Hall effect for conduction in the impurity band of p-type crystalline Ge; the upper limit of sensitivity of the equipment used was about an order of magnitude below the theoretical estimate for the Hall mobility calculated for hopping conduction. However, a Hall effect has recently been found for a glassy *ionic* conductor, the $AgPO_3$–AgI system (Clement *et al.* 1988), although the observed mobiliy is very low, $\mu_H \simeq 10^{-4}$ cm^2 V^{-1} s^{-1}. Since the mobile Ag^+ ions can be treated as classical rather than quantum particles, the quantum three-site interference responsible for the Hall effect in electronic conduction cannot be operating in ionic conduction (where presumably there is no phase memory or correlation between sites). There is no theory available for the ionic Hall effect, and the situation is complicated by the fact that the conduction mechanism in (super)ionic glassy systems is not understood (see sect. 4.5.2); however, the observation of an ionic Hall effect may be indicative that a contribution from hopping (albeit of ions) can exist.

We turn our attention now to the intriguing problem of the sign anomaly. It was originally believed (Friedman 1971, Holstein 1973) that the sign of the Hall effect should be *negative* for *both* electrons and holes for a three-site process. The ratio of the signs of R_H for electrons and holes moving in the same band was shown to be equal to $(-1)^{n+1}$, where n is the number of transfer integrals involved in the closed loop process; for three-site interference $n = 3$ and hence $(-1)^{n+1} = +1$, and both holes and electrons should have the same sign of R_H, namely negative. This is obviously contrary to experimental observation, in which several materials exhibit a positive R_H.

Emin (1977) showed that the absolute sign of R_H is given by:

$$\mathrm{sgn}(R_H) = \mathrm{sgn}\left[\varepsilon^{n+1} \prod_{i=1}^{n} J_{i,i+1}\right] \qquad [5.100]$$

where $\varepsilon = (q/|q|)$ and is -1 if the carrier is an electron moving in conduction band states or $\varepsilon = +1$ if the carrier is a hole in valence band states, and $J_{i,i+1}$ denotes the transfer integral connecting orbitals i and $i+1$. $J_{i,i+1}$ is *negative* for interactions between (s-like) bonding orbitals (since the energy entering into the transfer integral, the local atomic potential, is negative), but *positive* for interactions between (p-like or sp^3-like) antibonding orbitals. This approach is powerful in that the sign of R_H depends on the local geometry, as well as on the nature and relative orientation of the local orbitals between which the carrier moves. For atoms comprising a threefold ring, connected by s- or p-like (or hybrid) σ bonding orbitals, the sign of all the orbitals is the same, say positive, and hence J is negative, and if holes are the carriers ($\varepsilon = +1$), [5.100] gives a *negative* R_H (i.e. n-type), although the thermopower would be positive. For the case of electron transport between conduction (antibonding) states in a threefold ring, J is positive, $\varepsilon = -1$, and R_H is predicted to be *positive*, i.e. p-type. Chalcogenide glasses almost invariably exhibit p-type thermopower, but an n-type Hall coefficient. This might appear to an example of the first case, but the top of the valence band in these materials is composed of non-bonding, lone-pair p-like orbitals, rather than bonding orbitals. This apparent contradiction can be understood if the atoms do form a threefold ring formed by p-like σ-bonding orbitals, but with the non-bonding p-orbitals forming π-bonds in addition; in this case, the first situation is recovered. Alternatively, the three-site interaction may take place between three directed orbitals emanating from the *same* atom, for which J is negative, and if holes are the carriers ($\varepsilon = +1$), a negative R_H is also predicted, although the Hall mobility for this case would be small. A negative R_H would also arise if lone-pair p-orbitals on different atoms were arranged such that they are directed towards a common point or aligned roughly parallel.

So far, we have developed a theory for the Hall effect, and other d.c. transport properties such as conductivity and thermopower in amorphous semiconductors, in terms of extended electron states in the conduction or valence bands. Current scaling ideas of electron localization no longer recognize a minimum metallic conductivity and abrupt mobility edge as originally proposed by Mott, even at $T = 0$ K, and particularly at higher temperatures where phonon-induced delocalization sets in. This means that the 'two-channel' model (see e.g. Nagels 1979), where two independent conduction processes operate in parallel – namely (diffusive) conduction in extended states beyond the mobility edge and hopping conduction in the localized (tail) states below such a demarcation energy – no longer has any validity. Instead, in the phonon-delocalization picture of Thomas and co-workers, a continuous evolution in the nature of the states involved in the conduction process takes place with changing temperature as the conduction path moves in energy relative to the bands (Döhler 1979, Fenz *et al.* 1985). A completely different picture is that in which the coupling between the electrons and the lattice is supposed to be so strong that charge carriers can be self-trapped by the lattice distortion that they cause locally; the motion of the resulting 'polarons' is consequently very different from that of free carriers. Polaronic effects are discussed in sect. 5.5.

5.4.4 Frequency-dependent conductivity

Two types of frequency-dependent conductivity behaviour may be observed in amorphous materials. When conduction takes place in extended states, in either metallic or semiconducting materials, at very high frequencies ($\omega \geqslant 10^{12}\,\mathrm{s}^{-1}$), Drude-like behaviour is expected:

$$\sigma^* = \frac{\sigma(0)}{1 + i\omega\tau} \qquad [5.101]$$

where $\sigma(0) = ne^2\tau/m_e$, n being the electron density. This is the only frequency-dependent conductivity behaviour expected to occur in homogeneous materials containing no localized states (e.g. arising from defects). However, at lower frequencies ($10 < \omega < 10^9\,\mathrm{s}^{-1}$), another type of a.c. conductivity behaviour is universally observed in amorphous semiconductors (and ionically conducting glasses, see sect. 4.5.3), namely a power-law dependence ([4.97]). Although for amorphous semiconductors this behaviour is often ascribed to the hopping motion of electrons between localized (defect) states (sect. 6.3.4) (or of ions between non-bridging anion sites in ionic glasses, sect. 4.5.2), such a power-law frequency dependence can also arise from diffusive motion of charge carriers, but in *inhomogeneous* materials.

The study of the conduction behaviour of inhomogeneous media has a very long history going back to Maxwell. The a.c. response of an inhomogeneous material composed of a volume fraction c of medium 1 and $1 - c$ of medium 2 in the form of spherical aggregates was investigated in terms of effective-medium theory by Springett (1973), who obtained the following expression for the complex dielectric constant of the effective medium ε_m^* in terms of those of the constituents:

$$c\frac{\varepsilon_1^* - \varepsilon_m^*}{\varepsilon_1^* + 2\varepsilon_m^*} + (1-c)\frac{\varepsilon_2^* - \varepsilon_m^*}{\varepsilon_2^* + 2\varepsilon_m^*} = 0 \qquad [5.102]$$

where

$$\varepsilon_j^* = \varepsilon_j - \frac{i\sigma_j}{\varepsilon_0\omega} \qquad [5.103]$$

ε_j and σ_j being the real parts of the dielectric constant and conductivity of component j, respectively. This model predicts approximately power-law-like behaviour $\sigma(\omega) = A\omega^s$, where $s \leqslant 1$, over a reasonably wide frequency range between dispersionless regions where $\omega \to 0$ and $\omega \to \infty$ which correspond to the individual conductivities of media 1 and 2.

In the present context of a discussion of conduction in band states in amorphous semiconductors, the inhomogeneities may be either in the form of *compositional* fluctuations (e.g. in chalcogenide alloys), or in the form of *potential* fluctuations arising from randomly distributed charges (e.g. dopant-related) in an otherwise homogeneous material (see Fig. 5.19). Pistoulet *et al.* (1984) and Long (1989) have both considered the latter possibility, and have found that a power-law behaviour of $\sigma(\omega)$ can be obtained in this way. Long (1989) has assumed that if the potential fluctuations obey an exponential distribution

$$P(E - E_0) = \frac{C_0}{k_B T_0} \exp\left(-\frac{E - E_0}{k_B T_0} \right)$$

with an energy width $k_B T_0$ characteristic of the distribution, and where $\Delta E = E - E_0$, the conductivity of a particular region j can be written as

$$\sigma_j = \sigma_b \exp(\Delta E / k_B T)$$

where σ_b is the average conductivity of the background material; the overall conductivity is then evaluated using [5.102] and [5.103]. It was found that a power-law behaviour resulted for $\sigma_m(\omega)$ of the effective medium, with the exponent given by

$$s = 1 - (T/T_0) \qquad [5.104]$$

This particular form of temperature dependence of the exponent s has been found to occur for a-Si:H (Shimakawa *et al.* 1987), so it may arise from potential fluctuations in conduction (or valence) band edges, as indicated above. However, it is not clear why the distribution of such fluctuations should be of the exponential form required to give [5.104]; a gaussian distribution, for example, does not give rise to the same type of behaviour.

5.5 Small polarons

A charge carrier in a given state will always distort its surroundings, and the distortion will be more pronounced the more the carrier is localized. The distortion will tend to lower the energy of the carrier, and in extreme cases will lead to the carrier being self-trapped by the distortion its presence causes, and a 'polaron' is formed. The name comes from the early conjecture that self-trapping of an excess carrier might occur in a crystalline ionic (or polar) lattice; this circumstance is shown schematically in Fig. 5.25(a), and we refer to it as a 'dielectric polaron'. Polaron formation in its widest sense of course is not restricted solely to polar materials, but can occur also in covalent materials, in which the distortion is confined to neighbouring atoms which form some sort of bond when the carrier is trapped. Examples of this are V_k centres in alkali halide crystals, in which a hole trapped at a Cl^- ion causes a neighbouring Cl^- to be attracted (Fig. 5.25(b)); a similar circumstance is for holes trapped in solid rare gases, and dangling bonds in chalcogenide glasses (see sect. 6.2.2) – Mott and Davis (1979) refer to these as 'molecular polarons'. The spatial extent of the distortion may be considerable, or it may be localized in the vicinity of the carrier, giving rise to polarons termed, not surprisingly, large and small, respectively. In both cases, the polaron has a larger effective mass than the free carrier since the induced distortion is carried with it as the carrier moves through the lattice. Emin has argued that the electronic transport properties of many non-crystalline semiconductors can be understood on the basis of the motion of small polarons, without the need to introduce a mobility edge (for reviews see Emin 1973, 1976, and Austin and Mott 1969). Thus, in the expression relating the conductivity to the charge-carrier density n and the

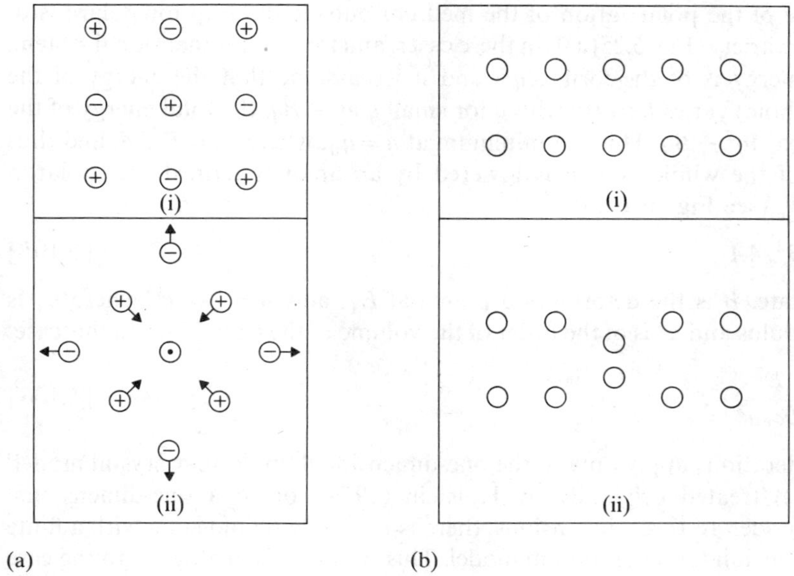

Fig. 5.25 Schematic illustration of polaron formation.
(a) 'Dielectric' polaron, in which an excess electron added to the ionic lattice shown in (i) causes the ionic readjustment shown in (ii);
(b) 'Molecular' polaron, in which an excess hole added to an array of rare gas atoms (i) causes the atomic rearrangement shown in (ii).

mobility μ [4.80], namely $\sigma = ne\mu$, two complementary scenarios are possible for electronic transport in amorphous semiconductors. In the conventional band picture discussed in sect. 5.4.1, conduction in extended states is characterized by a relatively *high* mobility (which is weakly temperature-dependent) coupled with a *low* concentration of charge carriers (which is strongly temperature-dependent because of the thermal activation of electrons across the gap). In the polaron picture, though, the converse is the case: the carrier concentration can be *high* (and temperature-independent) since in principle every site can form a polaron, but the mobility is *low* (and thermally activated) because a moving polaron must carry its distortion cloud with it (see later). (Ionic conduction in glasses can be formally regarded in similar terms: the band picture for electronic conduction (high μ, low n) is equivalent to the weak-electrolyte model, and conversely the polaron picture (high n, low μ) is equivalent to the strong-electrolyte model (see sect. 4.5.2).)

In the *adiabatic* regime, in which, because the electronic overlap (transfer) integral, J, is large, the vibrational motion of the atoms is so slow compared with that of an electron that the electron adjusts to the instantaneous position of the atoms (i.e. the Born–Oppenheimer approximation is obeyed), we may treat the system in the following manner. The energy of the system may be written in terms of a 'configurational coordinate', q, which may be taken to be the *displacement* of an atom in a molecular polaron of the V_k type (Fig. 5.25(b)), or the *dilatation* within the volume of a tap in band-tail states driven by the deformation potential (in which the potential energy of a carrier is proportional to the volume strain), or a measure of the polarization of the medium outside the trap for polarons of the dielectric variety (Fig. 5.25(a)). In these cases, and for a one-dimensional system, the lattice energy is of the form Aq^2, and if we assume that the energy of the electron (or hole) varies *linearly* with q for small q as $-Bq$, the total energy of the system is then $Aq^2 - Bq$. This is a minimum at $q = q_0$, where $q_0 = B/2A$, and thus the energy of the whole system is lowered by an amount termed the 'polaron energy', $-W_p$ (see Fig. 5.26(c)):

$$W_p = B^2/4A \qquad [5.105]$$

For a tail state, B is the deformation potential E_1, and $A = \frac{1}{2}\kappa_T a^3$, where κ_T is the bulk modulus and a^3 is of the order of the volume of the trap. Thus, in this case:

$$W_p = \frac{E_1^2}{2\kappa_T a^3} \qquad [5.106]$$

These considerations apply only to the one-dimensional 'molecular crystal model' (Fig. 5.25(b)) treated originally by Holstein (1959), or to a one-dimensional continuum model. In three dimensions, there is no stable bound state with a finite radius of distortion for a continuum model. This situation is analogous to the case of a quantum (square) well which only has a bound state if the depth and width are greater than certain critical values. Exactly the same is true here: a stable polaron (with $W_p < 0$) is formed only if q is greater than a critical value q_c (see Fig. 5.26(a)). However, if q_c is large, no stable polaron can form (Fig. 5.26(b)). On the other hand, it is believed (see e.g. Mott and Davis 1979) that for a strongly localized *defect*, $q_c = 0$, and energy will always be gained by distortion and a self-trapped carrier will be stable.

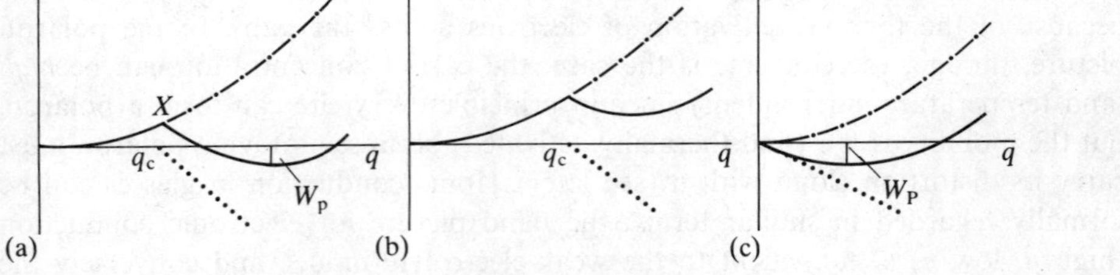

Fig. 5.26 The energy of a charge carrier in a deformable three-dimensional lattice. Solid curves represent the total energy as a function of the configurational coordinate q made up of a sum of two terms, electronic energy (dotted curve) and elastic energy (dash-dot curve).
Formation of a V_k-type polaron, (a) stable and (b) unstable. (c) Same quantities for a carrier trapped at a defect (after Mott and Davis 1979).

We now consider how a small polaron may prove through a lattice, and to do so we focus first on a polaron in the molecular crystal model moving through a periodic lattice, for which there are two distinct modes of transport. The first arises when the carrier and its accompanying lattice distortion move between sites *without* emission or absorption of phonons; such a process is called 'diagonal'.

The eigenstates of such a deformable lattice can be expressed by Bloch-type wavefunctions, as for a rigid lattice, but the bandwidth is $12Je^{-S}$ for a simple cubic lattice in the tight-binding scheme, where J is the electronic overlap integral and e^{-S} is the overlap integral between the two wavefunctions which describe the vibrational state of the system when the carrier resides on one or other of two neighbouring sites. This is reduced from the rigid-lattice bandwidth, $12J$, by the factor e^{-S} which can be considerable ($\simeq 10^{-4}$–10^{-5}), and so the small-polaron bandwidth is very narrow (or alternatively the effective mass is high). The small polarons in this limit move as (heavy) free carriers, suffering occasional scattering by atomic vibrations. However, if the vibrational energies (or the random fluctuations in the site energies for a disordered solid) are comparable to the small-polaron bandwidth, 'non-diagonal' processes become dominant, in which phonons are exchanged when the carrier moves to a neighbouring site; such a process is termed 'hopping'. Thus, in crystals, small-polaron band motion will only occur at low temperatures ($k_B T \lesssim 10^{-4}$ eV), and it is debatable whether in amorphous semiconductors it should ever occur since the fluctuations in site energies in the band tails almost certainly must be much greater than $\simeq 10^{-4}$ eV.

For a hopping event to occur, there must be an atomic distortion which momentarily brings the electronic energy level of an occupied site into coincidence with a neighbouring unoccupied site (Fig. 5.27), and this appreciable thermal distortion energy is the activation energy for the hopping process. In the simplest case, when the energy difference between electrons on neighbouring sites is zero ($W_D = 0$) and if the centres are not so close that the distortion regions overlap, then it can be shown (see problem 5.2) that the hopping energy is half the polaron energy:

$$W_H = \tfrac{1}{2} W_p \qquad [5.107]$$

The mobility of hopping polarons μ_p may be obtained from the Einstein relation and the jump probability P. The latter consists of the product of two terms, the first of which, P_1, is the probability of a coincidence event,

$$P_1 = \nu_0 \exp(-W_H / k_B T) \qquad [5.108]$$

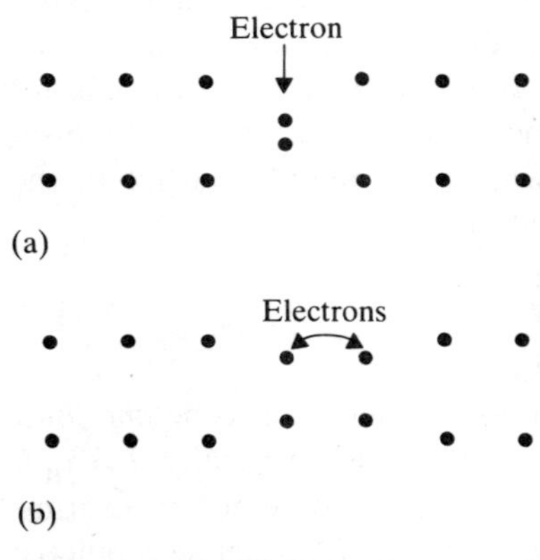

Fig. 5.27 Schematic illustration of a molecular polaron (a) and the excited state (b), i.e. the momentary coincidence of two polaron wells, which must be formed for an electron hop to take place.

where ν_0 is a typical phonon frequency. The second, P_2, is the probability of charge transfer (Holstein 1959). In the *non-adiabatic* regime (in which during each fluctuation the chance of an electron transfer is small) is

$$P_2 = \frac{1}{\hbar \nu_0}\left(\frac{\pi}{W_H k_B T}\right)^{1/2} J^2 \qquad [5.109]$$

where J, the electronic overlap integral, can be written as $J = J_0 \exp(-\alpha r)$, whereas in the *adiabatic* regime (in which during the time that an activated state persists, the electron makes several transitions back and forth), the probability of charge transfer is large, i.e. $P_2 = 1$. If we assume non-adiabatic motion, the mobility is finally:

$$\mu_p = \frac{ea^2}{k_B T \hbar}\left(\frac{\pi}{W_H k_B T}\right)^{1/2} J^2 \exp(-W_H/k_B T) \qquad [5.110]$$

The thermally activated nature of the carrier mobility is an essential feature of small-polaron theory (cf. the predicted $1/T$ behaviour of free carriers excited beyond the mobility edge in the absence of polaron formation, [5.78, 5.79]). In the high-temperature regime in which hopping transport is dominant, it is expected that the d.c. conductivity should vary as:

$$\sigma = \sigma_0 \exp\left[-\frac{(E + W_H)}{k_B T}\right] \qquad [5.111]$$

where E is the activation energy required to produce an equilibrium number of carriers, and the pre-exponential factor is given by:

$$\sigma_0 = \frac{Ne^2 a^2 \nu P}{k_B T} \qquad [5.112]$$

where the polaron hopping activation energy from the term P_1 ([5.108]) has been made explicit in [5.111]. Note that although the mobility in the small-polaron picture is very small, the carrier density, $N \simeq 10^{22}\ \text{cm}^{-3}$, is much higher than in non-polaronic theories to account for the experimentally observed pre-exponential factors.

The Hall effect for small-polaron transport has also been calculated. Band motion will result in the usual formula for the Hall coefficient R_H being obeyed ([5.95]) if $k_B T$ is small compared with the polaron bandwidth. For the case of small-polaron hopping transport, the three (or more) site interference mechanism must be employed, and for three-site interference on a triangular lattice in the non-adiabatic regime (Friedman and Holstein 1963):

$$\mu_H = \frac{ea^2}{\hbar} J \left[\frac{\pi}{12 k_B T W_H}\right]^{1/2} \exp(-W_H/3k_B T) \qquad [5.113]$$

Note that the activation energy for the Hall mobility is predicted to be *one-third* that of the conductivity mobility μ_p (cf. [5.110]), and since generally $J \gg k_B T$, therefore $\mu_H > \mu_P$. The site geometry, however, is crucial to the exact temperature dependence that is predicted, since $\mu_H \propto T^{-3/2} \exp(-W_H/3k_B T)$ for cubic geometry (see Emin 1973). The decrease in Hall mobility with temperature at low temperatures observed in many amorphous semiconductors has been interpreted by Emin in terms of small-polaron transport (hole-like for the case of chalcogenide glasses). Consequently, μ_H is thermally activated according to [5.113], rather than decreasing with temperature as a result of the carrier freeze-out effect associated with the two-channel model, in which band transport takes place in extended states beyond a mobility edge, and by phonon-assisted hopping in localized tail states below such an energy and for which the Hall mobility is assumed to be negligible compared with that for band conduction ([5.99]). The various considerations pertinent to the sign of R_H discussed in section 5.3.3 are also applicable to small polarons (and indeed were first formulated for that case).

Another feature of electron transport which small-polaron theory is capable of explaining, without recourse to the concept of mobility edges, is the difference between the activation energy for conduction and that for thermopower often observed in amorphous semiconductors. In the mobility edge model, this is accounted for by assuming transport occurs by hopping in the band-tail states. No such assumption is required in the small-polaron model, since even in the hopping regime, the polaron hopping energy W_H does *not* enter into the expression for the thermopower (see Emin 1973, 1976), which is found to be of the form (cf. [5.111]):

$$S = -\frac{k_B}{|e|}\left(\frac{E}{k_B T} + A\right) \qquad [5.114]$$

This equation is valid if the initial and final states interact equally with lattice vibrations (i.e. there is *no* vibrational energy transferred with a hop); the 'heat of transport' term, A, depends on the detailed density of states involved in the hopping, tending to zero if the bandwidth is much less than $k_B T$, but is generally expected to be of the order 1–10.

It is still uncertain how important are small-polaron effects in electron transport in amorphous semiconductors. However, there are some glasses in which small-polaron formation indisputably occurs; these contain transition metal ions in two charge states, e.g. V^{4+}/V^{5+}, Fe^{2+}/Fe^{3+}, Cu^{+}/Cu^{2+}, in glassy systems such as V_2O_5–P_2O_5 and FeO–P_2O_5. Conduction is by the motion of electrons between, say, V^{4+} and V^{5+} ions which, since lattice distortion is important, form small polarons.

Mott has suggested that for this process the d.c. electrical conductivity should be given by (Austin and Mott, 1969, Mott and Davis 1979):

$$\sigma = c(1-c)\frac{e^2 \nu_{el}}{R k_B T}\exp[-2\alpha R - W/k_B T] \qquad [5.115]$$

where c is the concentration of one type of ion, R is some average interionic separation and α^{-1} is the radius of the localized wavefunction. The activation energy is given by:

$$W = W_H + \tfrac{1}{2}W_D \qquad [5.116]$$

for the case when there is a disorder energy W_D between sites. Mixed vanadate–phospate glasses appear to obey [5.115] reasonably well, and exhibit an activation energy for conduction which decreases with decreasing temperature (see Mott and Davis 1979). This behaviour can be interpreted either according to conventional small-polaron theory which dictates that the activation energy should decrease to zero (band conduction) at low temperatures, or else in terms of 'variable-range hopping' of the small polarons (see sect. 6.3.3) which also predicts a continuously decreasing activation energy with decreasing temperature (Greaves 1973).

The thermopower of vanadate–phosphate glasses obeys the temperature-independent Heikes–Ure (1961) formula (Vomvas and Roilos 1984):

$$S = \frac{k_B}{e} \ln\left(\frac{c}{1-c}\right) \qquad [5.117]$$

This equation is only valid if $W_D \ll k_B T$, i.e. if all the sites have essentially the same energy.

The behaviour of the Hall effect for glassy $(V_2O_5)_{0.8}(P_2O_5)_{0.2}$, however, does not appear to obey very well the predictions of the small-polaron theory (Vomvas and Roilos 1984). The sign of the Hall coefficient was found to be negative, i.e. the same as that of the thermopower, but the activation energy of μ_H (~ 0.2 eV) was found to be about half the (high-temperature) value of the conductivity activation energy (~ 0.44 eV), whereas the predicted ratio is only one-third (see [5.113]). The curvature in the Arrhenius plot of σ_{dc} suggests the possibility of a variable-range hopping mechanism for the small-polaron transport, in which the temperature dependence of the conductivity is given by [6.52]:

$$\ln \sigma_{dc} \propto -(T_0/T)^{1/4}$$

Grünewald *et al.* (1981) have shown that, under the same conditions, the temperature dependence of the Hall mobility should be of the form

$$\ln \mu_H \propto -\tfrac{3}{8}(T_0/T)^{1/4} \qquad [5.118]$$

i.e. the slopes of the respective $T^{-1/4}$ plots should have a ratio of $\simeq 0.37$. Vomvas and Roilos (1984) found for the V_2O_5–P_2O_5 glass that the data for both σ_{dc} and μ_H do indeed fall on straight lines when plotted logarithmically versus $T^{-1/4}$, but that the ratio of the slopes was 0.55, i.e. greater than predicted theoretically (Fig. 5.28). Thus, the Hall effect arising from small-polaron transport is not entirely understood.

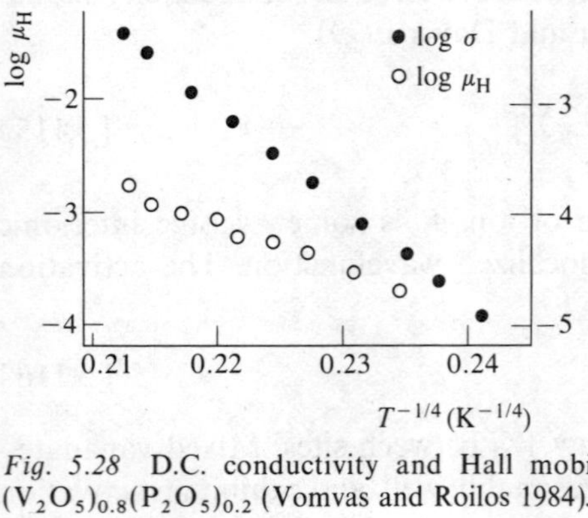

Fig. 5.28 D.C. conductivity and Hall mobility plotted logarithmically versus $T^{-1/4}$ for v-$(V_2O_5)_{0.8}(P_2O_5)_{0.2}$ (Vomvas and Roilos 1984).

5.6 Optical properties

We conclude this chapter with an account of the optical properties of amorphous materials. In a sense, we have touched on this subject several times already in this book, albeit for different frequency regimes. Thus, IR absorption, at one end of the frequency spectrum, was discussed in the context of lattice vibrations (Ch. 4), and at the other end of the scale, UV and X-ray absorption were considered in the section dealing with experimental probes of the gross features of the electronic structure (sect. 5.2.2). We shall therefore not discuss these in detail again, but will rather concentrate first on intraband optical processes and interband processes that occur for photons having energies comparable to that of the band gap of amorphous semiconductors. For most amorphous semiconductors, this means photons in the visible region of the spectrum since band gaps often lie in the range 1–3 eV, although obviously for insulators the energies are considerably larger (e.g. $\simeq 10$ eV for SiO_2). The form of the energy dependence of the optical absorption typically observed in amorphous semiconductors is shown schematically in Fig. 5.29. We will divide this part of the section into two, dealing respectively with optical transitions at energies above and below the mobility gap. The reason for this division is that different microscopic mechanisms are responsible in each case, as will become apparent. The low-energy (Urbach-edge) region will be discussed in sect. 6.3.1 in connection with various other gap-state spectroscopies. The last part of this section deals with intraionic luminescence and absorption.

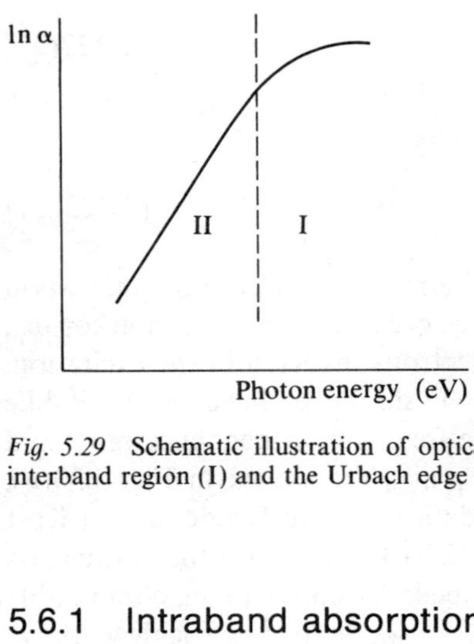

Fig. 5.29 Schematic illustration of optical absorption in amorphous semiconductors showing the interband region (I) and the Urbach edge region (II).

5.6.1 Intraband absorption

For photon energies which are sufficiently small for interband transitions between filled and empty states not to be excited, intraband excitations (plasma oscillations) can occur within a filled band of delocalized states in the conduction band of metals, or in principle in the valence band of semiconductors. Then, if the states can be treated as being nearly-free-electron-like with $k\ell \gg 1$ (ℓ being the electron mean-free path), the Drude equation ([5.101]) can be obtained by considering the Kubo–Greenwood formula ([5.48]) to be the frequency-dependent generalization of the Boltzmann expression for the conductivity ([5.57]). The real and imaginary parts of the complex conductivity $\sigma^* = \sigma_1 + i\sigma_2$ and dielectric constant $\varepsilon^* = \varepsilon_1 + i\varepsilon_2$ are related via

$$\varepsilon_2 = \sigma_1/\omega\varepsilon_0 \qquad [5.119(a)]$$

and

$$\varepsilon_1 = 1 + (\sigma_2/\omega\varepsilon_0) \qquad [5.119(b)]$$

and, furthermore, the dielectric constant is related to the optical constants (refractive index) by

$$\varepsilon^* = (n + ik)^2 \qquad [5.120]$$

i.e. $\varepsilon_1 = n^2 - k^2$ and $\varepsilon_2 = 2nk$. Thus, measurements of reflectivity and absorption coefficient can be used to obtain values for n and k (or equivalently ε_1 and ε_2, etc.) – see e.g. Ward (1988) for details. The real and imaginary parts of a particular function are not independent, however, but are related via Kramers–Kronig transforms. Thus, for the dielectric constant, for example,

$$\varepsilon_1(\omega) = 1 + \frac{2}{\pi}\int_0^\infty \frac{\Omega\varepsilon_1(\Omega)}{\Omega^2 - \omega^2}\,d\Omega \qquad [5.121]$$

and an equivalent expression relates ε_2 to ε_1. From [5.101] and [5.119], the real and imaginary parts of the dielectric constant in the Drude model are given by

$$\varepsilon_1(\omega) = 1 - \frac{\omega_p^2\tau^2}{1 + \omega^2\tau^2} \qquad [5.122(a)]$$

and

$$\varepsilon_2(\omega) = \frac{\omega_p^2\tau}{\omega(1 + \omega^2\tau^2)} \qquad [5.122(b)]$$

where the plasma frequency $\omega_p = (ne^2/\varepsilon_0 m_e)^{1/2}$ and τ is the electronic relaxation (scattering) time; an effective mass $m^* \neq m_e$ may occur in the expression for ω_p, reflecting the effective number of conduction electrons involved in the excitation.

From [5.122(b)], a plot of $1/\omega\varepsilon_2(\omega)$ vs. ω^2 should be linear if Drude-like behaviour is present. Experimental data for the frequency dependence of ε_1 and ε_2 for the transition-metal–metalloid glass $Pd_{81}Si_{19}$ are shown in Fig. 5.30(a), and it can be seen from Fig. 5.30(b) that the data obey the Drude law, at least for low energies up to 0.65 eV. From a fit to [5.122(b)], values for the parameters $\hbar\omega_p = 7.3$ eV and $\tau = 8.8 \times 10^{-16}$ s can be obtained; for comparison, comparable values for crystalline Pd are $\hbar\omega_p = 6.5$ eV and $\tau = 4.1 \times 10^{-15}$ s (Schlegel *et al.* 1979). The departures from Drude behaviour for $\hbar\omega > 0.7$ eV in Fig. 5.30 correspond to the onset of interband transitions. Similar behaviour has been found for the glassy transition-metal–metal alloy $Cu_{50}Ti_{50}$ (Lapka *et al.* 1985).

The Drude model breaks down, however, when $k\ell$ is no longer large, i.e. when the electron scattering is strong, or in other words when τ is very small. Thus, it is unlikely that the model is appropriate for intraband excitations within the bands of amorphous semiconductors, particularly for electron energies close

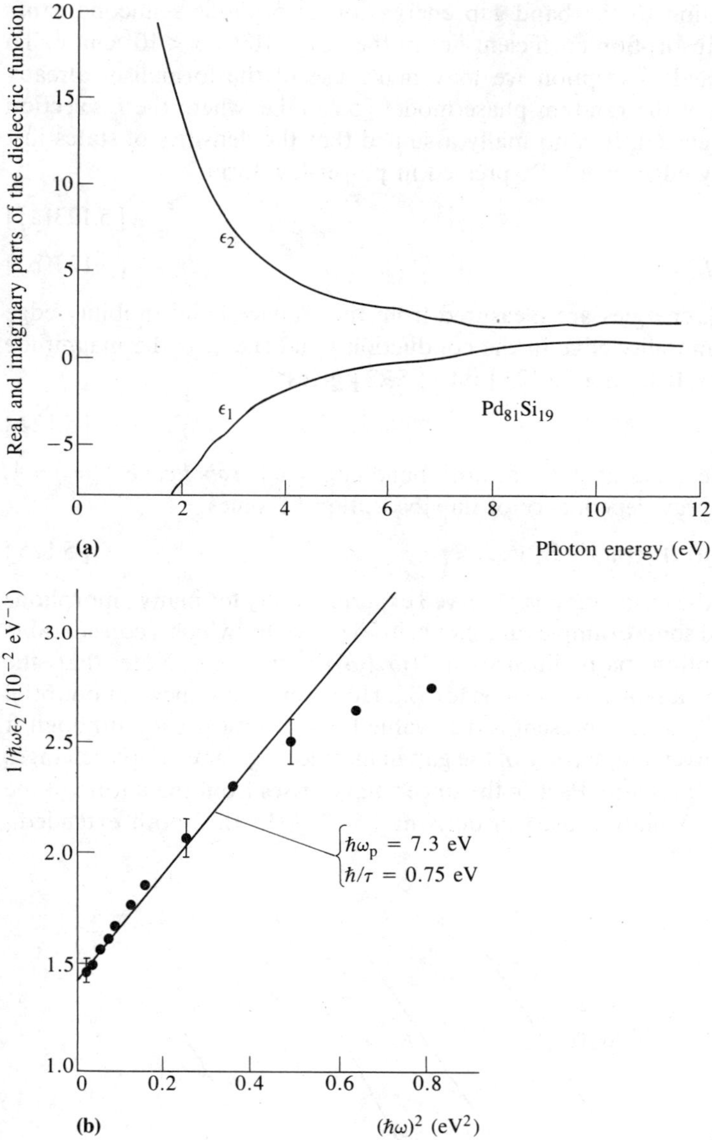

Fig. 5.30 Optical properties of the metallic glass $Pd_{81}Si_{19}$ (Schlegel *et al.* 1979): (a) real and imaginary parts of the dielectric function $\varepsilon(\omega)$ as a function of photon energy; (b) Drude plot, $1/\hbar\omega\varepsilon_2$ versus $\hbar^2\omega^2$, for the data shown in (a).

to a localization edge, and is perhaps of marginal applicability to amorphous metals where the electron scattering relaxation times have values $\tau \leqslant 10^{-15}$ s.

5.6.2 Interband absorption

Interband absorption in general has been dealt with in section 5.2.2; we are concerned here with the low-energy portion of that absorption occurring at photon energies corresponding to the band-gap energy for amorphous semiconductors and for which the absorption coefficient lies in the range $10^4 < \alpha < 10^6\ \text{cm}^{-1}$. To calculate the expected absorption we may make use of the formalism already developed in terms of the random phase model [5.22], i.e. where the *k*-selection rule is completely relaxed. It is normally assumed that the densities of states just beyond the mobility edges can be expressed in power-law form:

$$g_v(-E) \propto E^p \qquad [5.123(a)]$$

$$g_c(E) \propto (E - E_0)^q \qquad [5.123(b)]$$

where, as in [5.22], energies are measured from the valence band mobility edge and E_0 marks the mobility edge in the conduction band (i.e. it is the magnitude of the mobility gap). Inserting [5.123] into [5.22] gives:

$$\omega^2 \varepsilon_2(\omega) \propto (\hbar\omega - E_0)^{p+q+1} \qquad [5.124]$$

If the form of the densities of states at both band edges is *parabolic*, i.e. $p = q = \frac{1}{2}$, then the photon energy dependence of the absorption becomes:

$$\omega\alpha(\omega) \propto \omega^2 \varepsilon_2(\omega) \propto (\hbar\omega - E_0)^2 \qquad [5.125]$$

This form for the absorption is observed experimentally for many amorphous semiconductors, and some examples are shown in Fig. 5.31, for which a considerable part of the absorption plots linearly as $(\omega\alpha(\omega))^{1/2}$ vs. $\hbar\omega$. Note that the extrapolation of such a plot gives a value for E_0. However, it is somewhat doubtful whether in reality E_0 does represent a true value for the optical gap, although it is often used as a convenient marker of the gap in monitoring, for example, changes with temperature or pressure. Part of the uncertainty arises from the nature of the random phase approximation used in deriving [5.125]. In this, both extended–

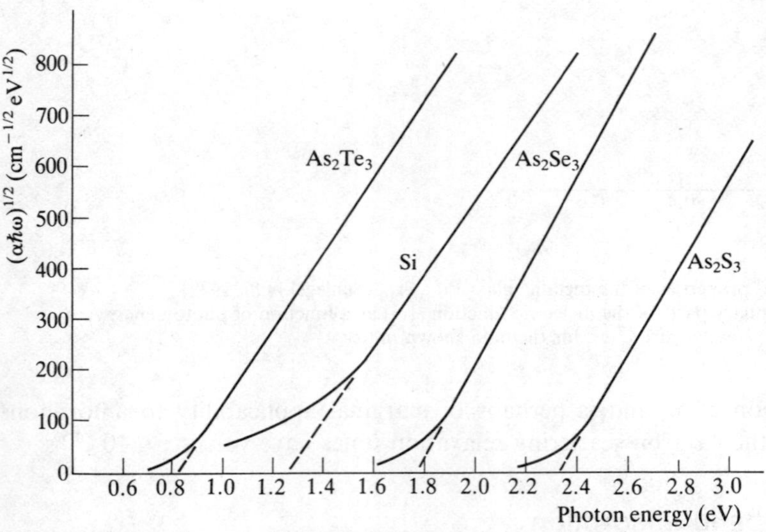

Fig. 5.31 Examples of optical absorption edges of amorphous materials whose spectral dependence is given by $\alpha\hbar\omega = C(\hbar\omega - E_o)^2$ showing the extrapolation to give values for the optical gap E_o (after Mott and Davis 1979).

extended transitions *and* localized–extended transitions are possible, with the same matrix element, and hence transitions between, say, states at the valence band mobility edge and the localized conduction band tail states having a density of states $g_c \propto (E - E_0)$ will also give a quadratic energy dependence of $\omega\alpha(\omega)$ as in [5.125], only here E_0 represents the energy separation between E_v and E_A (Fig. 5.10). However, the magnitude of the absorption due to the latter process is likely to be considerably smaller than that for extended–extended transitions due to the low density of states in the tail state region.

Dersch *et al.* (1987) have calculated the optical spectrum expected for a representative disordered structure, namely a Bethe lattice, assuming a tight-binding electronic Hamiltonian with diagonal site disorder, as in [5.38]; this is the same form as originally used by Abe and Toyozawa (1981) in their investigation of the optical properties of amorphous semiconductors using the coherent-potential approximation (CPA) approach. Both Dersch *et al.* (1987) and Abe and Toyozawa (1981) found that the random-phase approximation breaks down for photon energies comparable to the mobility gap E_0 (the separation between mobility edges in conduction and valence bands); in other words, the optical absorption is no longer determined principally by the joint density of states, as in [5.124], but matrix-element effects are significant for such interband transitions. In fact Dersch *et al.* (1987) found that the matrix element $M(E)$ is rather strongly peaked as a function of energy in the region of E_0, whereas at lower energies, in the region of tail-state absorption, $M(E) \simeq 1$, i.e. the absorption spectrum should reflect the joint density of states more accurately.

The quadratic energy dependence of the absorption embodied in [5.125] is not universally observed: a-Se exhibits a linear dependence, and multicomponent chalcogenide alloys (e.g. Ge–As–Te–Si) which exhibit electronic swtiching behaviour show a cubic energy dependence. Still other materials, e.g. a-Ge, exhibit absorption edges which cannot be fitted to any simple power law, a reflection, presumably, of the fact that the densities of states cannot be described by a simple power law either. In this case, it is not possible to obtain a value representative of the optical gap from plots such as in Fig. 5.31; instead, an alternative option is to take the energy at which the absorption coefficient reaches some value (usually 10^4 cm^{-1}), and the optical gap is then conventionally denoted E_{04}.

The interband transitions discussed above would, by themselves, lead to rather sharp optical absorption edges marking the onset of optical transitions. In practice, the absorption edge is found experimentally to decrease with energy very much less rapidly than predicted by, say, [5.125]. This ‘tailing’ into the gap – the Urbach-edge region – arises from somewhat different mechanisms, and will be discussed more fully in the context of gap-state spectroscopies in Chapter 6.

5.6.3 Intraionic absorption and luminescence

In this section we deal briefly with the intraionic optical spectra resulting from transitions which occur between the electron levels of transition metals or rare earths dissolved purposely or inadvertently in glassy matrices. The optical absorption properties of such ions, particularly transition metals, are immensely important since they are the colouring agents in oxide glasses, and have been employed in artistic and functional applications of 'glass' ever since it was first manufactured. The scientific interest in such absorption processes stems not so much from a desire to understand these processes (which are in fact well understood from studies in the crystalline state), but rather in their use as a probe for the local structure, since the absorption spectra are affected by the local configuration in subtle ways.

The interest in the intraionic luminescent properties of such ions comes about because of the considerable potential that rare earth (e.d. Nd) doped glasses have as high-power lasing materials for possible use in thermonuclear fusion programmes. In this case, the scientific effort has been directed into understanding the luminescent process itself, in order to optimize the lasing action.

We discuss first the absorption spectra of transition metal (TM) and rare earth (RE) doped glasses. A comprehensive survey of the early experimental data on this subject is contained in the classic monograph by Weyl (1959) and more recent developments are given by Bamford (1977) and Wong and Angell (1976). If we consider firstly the 3d-transition series for simplicity, the 3d levels which are degenerate in the atomic state become split in the condensed state by the crystal fields exerted by neighbouring, negatively charged anions (or 'ligands'). Whether the energy of a given d-orbital is raised or lowered with respect to the free-atom level depends on the configuration of the ligands around the transition metal ion; generally either octahedral or tetrahedral coordination by the ligands is preferred, although distortions from the symmetrical configurations are not uncommon in glassy systems. If the transition metal ion has an incomplete d-shell, transitions between the crystal-field split d-levels are possible and are responsible for the intraionic absorption we wish to discuss. However, all d–d transitions are formally forbidden by the so-called Laporte rule, which is simply the quantum-mechanical selection rule, $\Delta l = \pm 1$ for electronic transitions. However, the transitions are weakly allowed if some mixing of the 3d-transition metal orbitals with metal 4p- or ligand p-orbitals can take place. This can arise in several ways: orbital mixing cannot occur in ideal centrosymmetric (e.g. octahedral) complexes under static conditions, but thermal vibrational disorder removes the centre of symmetry as does the topological disorder inherent in a glass; furthermore, tetrahedral ligand coordination allows much more d–p mixing, with the result that typically the absorption intensity due to ions in a tetrahedral configuration is much more intense (by a factor of $\simeq 10\text{–}10^2$) than that due to the same ions in an octahedral environment. Thus, this effect alone can give some structural information if the absorption spectra of an ion in the *same charge state* in different network-forming matrices are compared. (N.b. This behaviour is exactly equivalent to that observed in the pre-edge structure in X-ray absorption spectroscopy (sect. 3.2.3).)

The particular charge state of a transition metal ion, as well as the nature and coordination of the surrounding ligand atoms, are obviously factors which determine the energy of the resulting absorption band. In order to understand these factors more fully, it is necessary to describe briefly the ligand-field theory of the d-level splitting in various transition metal complexes; extensive treatments of this topic are given in standard inorganic chemistry texts (e.g. Cotton and Wilkinson 1980) and so only a brief outline will be given here. The simplest version of ligand field theory (crystal field theory) treats the ligand ions as point charges placed at the centroid positions of the ions, considers only electrostatic interactions and neglects any covalent interactions between metal atom and ligand. This simple theory gives a clearer physical picture of the nature of the d-level splittings than is afforded by the more quantitatively accurate molecular orbital theory. Consider firstly a d^1 configuration (i.e. Ti^{3+}) in an *octahedral* environment. The d_{z^2} and $d_{x^2-y^2}$ orbitals which project along the axes will interact strongly with the two negative ligand ions at the polar positions, and hence the coulomb repulsion for these two (equivalent) states will be high; the three remaining orbitals, d_{xy}, d_{yz} and d_{zx}, are radially directed between the x, y and z axes and therefore do not interact appreciably with either the two polar or the four equatorial ligand ions. Hence the coulomb repulsion is less than for the other two d-orbitals, and the d_{xy}, etc., orbitals are said to be 'stabilized'. The splitting between the two sets of levels for the octahedral case is termed Δ_0 and is illustrated in Fig. 5.32, where it can be seen that the (d_{z^2}, $d_{x^2-y^2}$) orbitals (given the group symbols e_g) are pushed up in energy more than the (d_{xy}, d_{yz}, d_{zx}) orbitals (given the symbol t_{2g}) are pushed down; this preservation of the 'centre of gravity' of the levels occurs whenever the interactions are purely electrostatic in origin. It should be noted at this stage that a d^9 configuration ion (e.g. Cu^{2+}) can be treated in the same manner as the one-electron case, except that now there is a single *hole* in an otherwise filled d-band; this has the effect of *reversing* the orbital ordering shown in Fig. 5.32. If we now consider a d^1 ion in a *tetrahedral* environment, the ligand ions now lie at four of the corners of a cube with its principal axes lying parallel to the coordinate axis system. Now it is the (d_{xy}, d_{yz}, d_{zx}) orbitals which suffer more coulomb repulsion than the (d_{z^2}, $d_{x^2-y^2}$) set and which are therefore stabilized, in contrast to the case of octahedral symmetry (see Fig. 5.32); note too that the energy splittings themselves are inverted with respect to the octahedral case and that the overall tetrahedral splitting Δ_t is smaller than Δ_0 i.e.:

$$\Delta_t = \tfrac{4}{9}\Delta_0 \qquad [5.126]$$

(The g subscript – *gerade* – is omitted for the tetrahedral configuration because it refers to orbitals which are centro-symmetric, which they are not in the tetrahedral case.)

For the case of more than one d-electron (or hole), the situation is more complicated since now the coupling of the electronic orbital, angular and spin momenta must be considered. In the Russell–Saunders (L–S) approximation (in which coupling between L and S is neglected), the states of the many d-electron

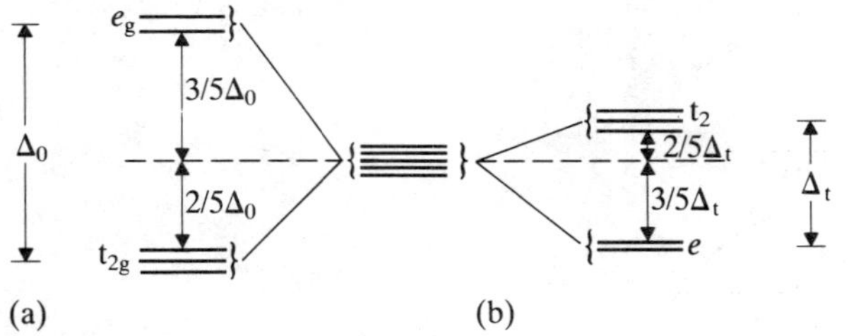

Fig. 5.32 Energy level diagram showing the splitting of a set of d-orbitals by (a) octahedral and (b) tetrahedral crystal fields (Cotton and Wilkinson 1980).

free ion, labelled S, P, D, F and G, in turn can be split by ligand fields, and each of the crystal field states has the same spin multiplicity as the free ion states from which they derive. An example of the type of splitting expected is illustrated in Fig. 5.33 for the case of a d^2 configuration. Although the number of possible optical transitions for such a level diagram appears to be very large, in fact the additional selection rule which only allows transitions between states having the same spin multiplicity means that in this case the $^3T_1(F) \rightarrow {}^3T_2(F)$, $^3T_1(F) \rightarrow {}^3T_1(P)$ and $^3T_1(F) \rightarrow {}^3A_2(F)$ transitions will be spin allowed. For the case of a d^n ion in a *tetrahedral* environment the same level inversion occurs as for the d^1 case; hence the $^3A_2(F)$ state becomes the ground state of a d^2 ion in a tetrahedral crystal field (or of a d^8 ion in an octahedral site). In the simple crystal field model, the overall stabilization energy E_s of a d^n ion in, say, an octahedral field can simply be calculated by adding a stabilization energy of $\frac{2}{5}\Delta_0$ for each electron in a t_{2g} state and subtracting a destabilization energy of $\frac{3}{5}\Delta_0$ for each electron in an e_g state (cf. Fig. 5.32). Note that for the d^5 configuration (i.e. Mn^{2+}) if it is in the 'high-spin' state, i.e. with an electron in each of the t_{2g} and e_g levels and therefore with none paired, $E_s = 0$ and no transitions at all are allowed since they are all spin forbidden. Thus very little absorption is expected from this ion in an octahedral site, as indeed is observed experimentally (Fig. 5.34(a)); Mn^{2+} in a (distorted) tetrahedral configuration is much more absorptive because of the enhanced d–p orbital mixing now involved. Thus Mn^{2+} is an excellent ion for probing the local site geometry in various glassy matrices, even though its overall absorptivity is very low. In this

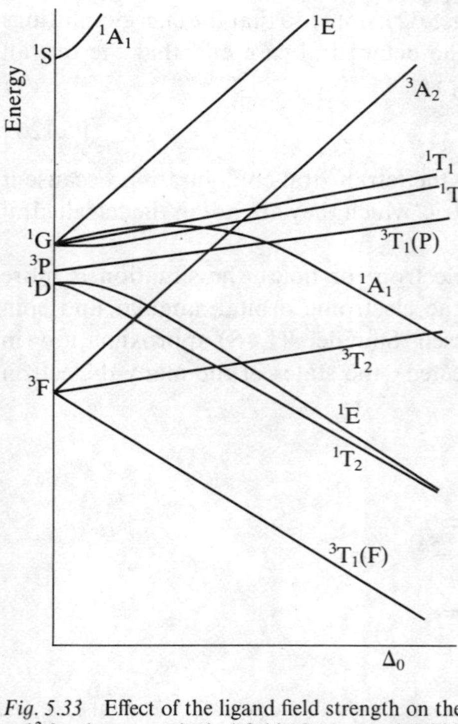

Fig. 5.33 Effect of the ligand field strength on the energies of the Russell–Saunders free ion states for a d^2-ion in an octahedral field (Cotton and Wilkinson 1980).

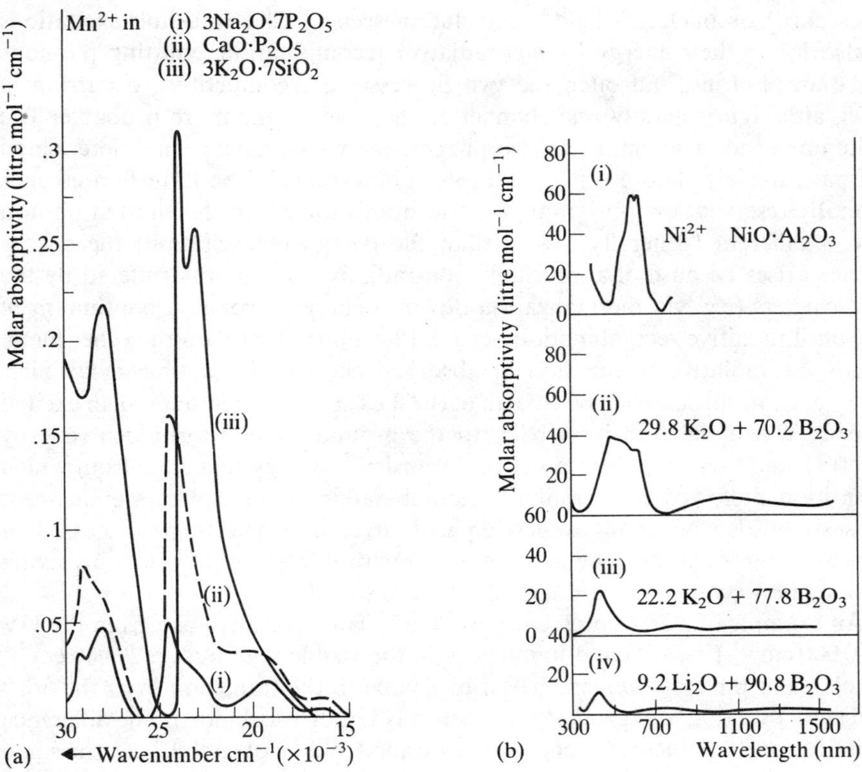

Fig. 5.34 Intraionic absorption for transition metal ions in glassy matrices.
(a) Mn^{2+} in phosphate and silicate glasses. Octahedral sites favoured in (i) and tetrahedral in (ii).
(b) Ni^{2+} in borate glasses and crystalline $NiO{\cdot}Al_2O_3$. (i) Regular tetrahedral site; (ii) tetragonally distorted tetrahedral site; (iii) tetragonally distorted octahedral site; (iv) regular octahedral site (Wong and Angell 1976).

regard, Ni^{2+} and Co^{2+} are even better candidates since their absorptivities are typically three orders of magnitude greater than Mn^{2+} and they have characteristically different absorption bands when occupying either a tetrahedrally or an octahedrally coordinated site. However, because the d^8 configuration (i.e. Ni^{2+}) is overwhelmingly stabilized in the octahedral configuration ($E_s = \frac{6}{5}\Delta_0$), it is not often found in tetrahedral sites, a problem not shared by Co^{2+} (d^7), although it in turn suffers from the disadvantage that the absorption bands for octahedral and tetrahedral configurations tend to overlap in energy. This site-selective characteristic of Ni^{2+} is illustrated in Fig. 5.34(b), where tetrahedral symmetry is imposed on the Ni^{2+} by the crystal lattice in crystalline $NiO.Al_2O_3$, and the absorption band observed is due to the ${}^3T_1(F) \rightarrow {}^3T_1(P)$ transition (the splitting is due to spin–orbit coupling effects), whereas in the glasses, the Ni^{2+} tends to favour an octahedral site and a much weaker absorption band appears at a higher energy (due to ${}^3A_2(F) \rightarrow {}^3T_1(P)$ transitions).

We turn now to luminescent processes. Luminescence is a general term describing the radiative recombination of carriers (electrons and holes) excited by a variety of methods: light (photoluminescence), high-energy electrons (cathodoluminescence), or an electric field (electroluminescence). Electron–hole excitations may also lower their energy by *non*-radiative recombination, emitting phonons rather than photons, and often the two processes are competitive, occurring in parallel, although generally one channel or the other is the more probable. The term luminescence also embraces two phenomena which have been differentiated in the past, namely 'fluorescence' and 'phosphorescence'. The term fluorescence is generally reserved for those radiative transitions for which the emitted photon energy is different (generally lower) than the energy of excitation; the energy difference arises because of appreciable non-radiative energy loss suffered by the excited carriers (e.g. via thermalization down in energy through a continuum of states) until radiative recombination occurs. Phosphorescence, on the other hand, refers to the radiative re-emission of absorbed energy with a time delay after (optical) excitation; such behaviour can occur if a carrier is promoted to an excited state from which the transition back to the ground state is forbidden (e.g. by symmetry), and therefore there must be a transfer of energy to a state from which the transition is allowed before radiative recombination can take place. Luminescent processes can also occur in connection with structural defects, as discussed in section 6.3.1, although here we are concerned with intraionic processes involving transitions between crystal-field split d or f levels.

An example of a 3d fluorescent ion is Cu^+. This typically absorbs in the UV at $\simeq 40\,000\ cm^{-1}$ ($\simeq 5$ eV) and luminesces in the visible region at $\simeq 20\,000\ cm^{-1}$ ($\simeq 2.5$ eV), and the emission is believed to be due to the transition from the $3d^{10}$ excited state to the $3d^9 4s^1$ ground state, which is Laporte forbidden. The difference in excitation and luminescence energies is ascribed to a 'Stokes shift', i.e. the same two levels are involved in both absorption and emission processes but because of a strong coupling between the d-electrons of the ion and the surrounding lattice (through the ligands), there is a change in structural configuration when the ion is electronically excited and hence the minima in the vibronic potential wells for each of the two levels do not coincide as a function of configuration coordinate; as a consequence, the excitation energy is greater than the luminescent energy (see Fig. 6.30).

However, lanthanides (rare earths) are perhaps more important than transition metals as luminescent species because of the potential that complex silicate glasses doped with rare earths have as high-power laser materials. In the lanthanides, the intraionic electronic transitions of interest are f–f transitions, since the 4f levels are also split by the local crystal fields experienced by an ion. However, unlike the d-electrons in transition metals, the 4f-orbitals in the lanthanides have appreciable electron density close to the nucleus and so the outer 5s and 5p electrons shield them to a certain extent from surrounding ligand fields; thus the f-level splittings are much less dependent on the environment (such as coordination and type of ligand) than is the case for 3d-splittings in transition metal ions. For this reason the site symmetries of lanthanide ions are less readily obtained from absorption spectra. Luminescence can also result from f–f transitions, and it is these which make lasing action in 4f-ion doped glasses (and crystals) possible. For example, in Nd^{3+}-doped (K, Ba silicate) glass, the lasing luminescent transition occurs at 1.06 μm and arises from the ${}^4F_{3/2} \to {}^4I_{11/2}$ transition of the Nd^{3+} ion.

However, the utility of intraionic luminescent (and absorption) transitions as local structural probes for glasses is hampered to a considerable extent by the fact that, because the optically active ions can occupy a large variety of sites in the amorphous state, the resulting luminescent (or absorption) bands are significantly inhomogeneously broadened. Nevertheless, intraionic optical transitions can be used as the basis of a *site-selective* spectroscopy using the technique of 'fluorescent line-narrowing' (FLN) – for a recent review, see e.g. Imbusch (1987). In this, only a sub-set of ions within the inhomogeneous absorption profile is probed using a particular excitation energy from a tunable narrow-band laser; by varying the laser wavelength, the whole manifold of sites may be investigated. The resulting FLN signals are significantly narrower, with a width determined by the intrinsic *homogeneous* broadening involved in the transition and the bandwidth of the laser (see Fig. 5.35(a)). The process shown in case (i) of Fig. 5.35(b) represents a *resonant* FLN process, in which the energy of the luminescent (fluorescent) transition is exactly the same as that of the exciting light. Non-resonant FLN spectra can arise if the radiative transition involves a third set of energy levels (see cases (ii) and (iii) in Fig. 5.35(b)), although the widths of these FLN lines can be appreciably broader than in resonant FLN spectra.

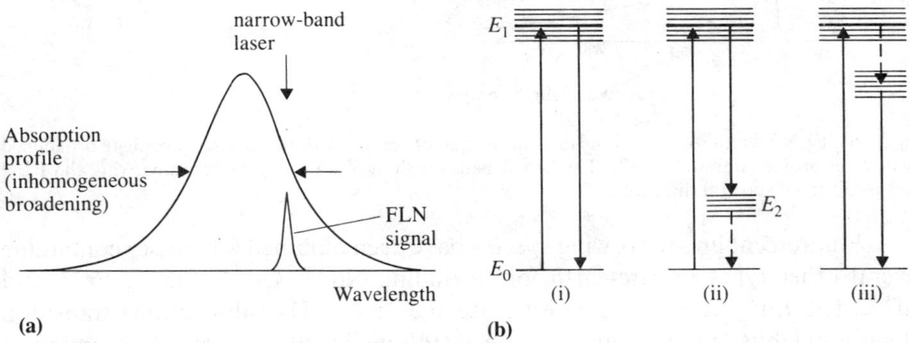

Fig. 5.35 Schematic illustration of fluorescence line narrowing (FLN) spectroscopy.
(a) Narrow-band laser excitation of a sub-set of ions within an inhomogeneously broadened absorption profile results in a narrow FLN luminescent emission band.
(b) Resonant (i) and non-resonant (ii, iii) FLN transitions.

Examples of FLN spectra are shown in Fig. 5.36, where the sharp FLN lines resulting from resonant excitation of Yb^{3+} in a silicate glass by two different laser frequencies lying within the broad inhomogeneous absorption profile are illustrated. The Yb^{3+} ion is particularly interesting to study by FLN since it has a very simple energy level structure consisting of just two levels, $^2F_{5/2}$ and $^2F_{7/2}$. Although the resonant FLN lines are clearly distinguishable in the spectra shown in Fig. 5.36, a very broad but much less intense component can also be seen in each FLN spectrum. It arises, in this case, from 'spectral diffusion': the sub-set of resonantly excited ions may decay in one of two ways: either *radiatively*, thereby producing the resonant FLN line, or *non-radiatively*, whereby the excited ions ('donors') may transfer their excitation energy to nearby ions of the same (or different) type; these 'acceptor' ions will not necessarily be found at the same type of sites as the donors, and consequently the overall acceptor-based luminescence profile will have a shape similar to that of the inhomogeneously broadened (absorption) band.

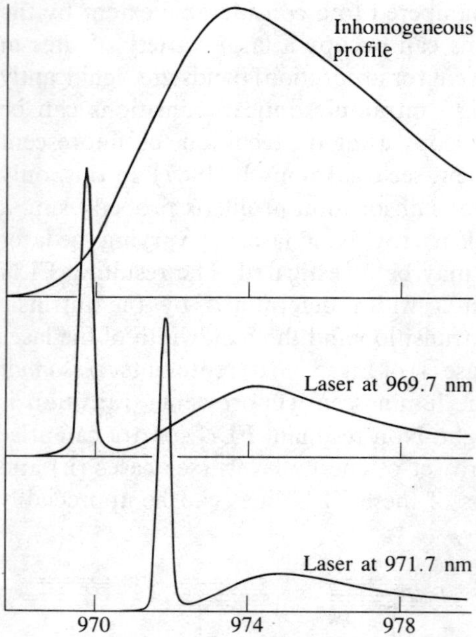

Fig. 5.36 FLN spectra for Yb^{3+} ions in a silicate glass together with the inhomogeneously broadened absorption profile (Imbusch 1987). The broad bands at longer wavelengths accompanying the FLN lines result from spectral diffusion.

Fluorescent line-narrowing spectra have been obtained for glasses containing several other types of rare-earth ion, including Nd^{3+}, Gd^{3+}, Sm^{3+}, Pr^{3+} and Eu^{3+}. The Eu^{3+} ion is of interest since the $^7F_0 \rightarrow {}^5D_0$ (absorption) transition which it exhibits is unique among rare-earth ions in that the two states involved have only a single Stark component; excitation of the $^7F_0 \rightarrow {}^5D_0$ transition enables the site-dependent Stark splitting of the resultant $^5D_0 \rightarrow {}^7F_{1,2}$ non-resonant FLN spectra to be studied, the positions of which allow the local structural environments of the Eu^{3+} ions to be obtained via consideration of the relevant crystal-field parameters (Brecher and Riseberg 1976).

There have been fewer FLN studies of glasses containing transition-metal ions, principally because of the greater inhomogeneous broadening characteristic of such systems. The most-studied ion is Cr^{3+}; examples of FLN spectra for Cr^{3+}-containing glasses are shown in Fig. 5.37. Other ions studied by this technique include Bi^{3+} (Boulon *et al.* 1980), which is interesting because the electronic transitions involved are not within the same shell, as they are for rare-earth and transition-metal ions, but between $6s^2$ and 6s6p configurations ($^1S_0 \leftrightarrow {}^3P_1$); the transitions are therefore much more sensitive to local variations in crystal-field interactions and hence to the local structure.

In the limit when instrumental and laser bandwidths are very small, the measured FLN linewidth will be determined by the intrinsic homogeneous linewidth of the electronic transition involved. Such homogeneous linewidths have been measured for a number of rare-earth-containing glasses as a function of

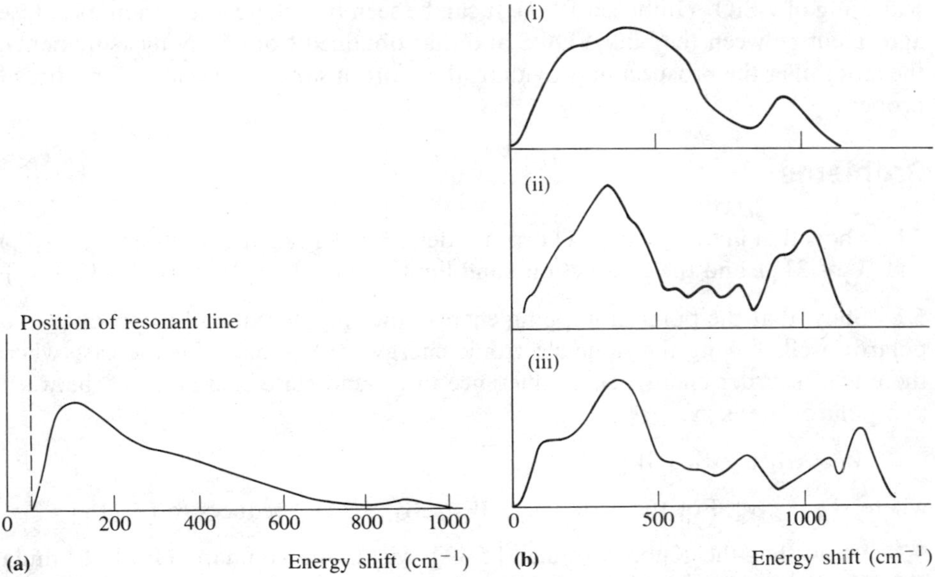

Fig. 5.37 Site-selective vibrational spectroscopy using FLN (Imbusch 1987).
(a) The vibrational side-band accompanying the narrow resonant FLN line for Cr^{3+} in a glass silicate matrix.
(b) The density of vibrational states (i) deduced from the side-band shown in (a), compared with (ii) the HV Raman spectrum of the host silicate glass and (iii) the VDOS obtained from inelastic neutron scattering of v-SiO_2 (Galeener *et al.* 1983).

temperature. They are considerably broader for glassy matrices than for crystalline hosts; moreover, the width varies as T^2 over a rather large temperature range ($\simeq$10–100 K), as first found by Selzer *et al.* (1976) for a Eu^{3+}-silicate glass. Although a two-phonon Raman relaxation mechanism, as operates in crystals, could account for this behaviour at high temperatures, it cannot be operative at the lowest temperatures. It appears, therefore, that additional relaxation centres are present in the glassy state; it has been suggested that such centres might be the two-level systems (or tunnelling modes) which are manifested in low-temperature thermal properties (see sect. 4.4.5).

Finally in this section, we discuss another origin of broad satellite bands accompanying a narrow FLN line, in addition to that involving spectral diffusion discussed above. The additional broad band shown in Fig. 5.37(a) for a Cr^{3+}-containing glass is a *vibrational* side-band to the main FLN line and results from a coupling of the optically active ion to the lattice and its vibrations. If the strength of this coupling is taken to be determined by the *amplitude* of the vibrational displacements, it can be shown that the side-band intensity is related to the vibrational density of states (Imbusch 1987):

$$I(\omega) = C|\langle V_1 \rangle|^2 \rho(\omega)/\omega \qquad [5.127]$$

where V_1 is the first-order coupling constant and C is a constant. The density of vibrational states obtained using [5.127] from the FLN side-band spectrum of Cr^{3+} in a silicate glass shown in Fig. 5.37(a) is shown in Fig. 5.37(b), together with the HV Raman spectrum and the VDOS obtained from neutron inelastic scattering of v-SiO_2 (Imbusch 1987). It can be seen that there is a reasonably close agreement between the bulk VDOS and that obtained from FLN measurements; the latter offer the prospect of providing, therefore, a *site-selective* local vibrational probe.

Problems

5.1 Show that in the Weaire–Thorpe model, the valence band limits are $(V_2 - V_1)$ and $(V_2 + 3V_1)$, and the conduction band limits are $(-V_2 + 3V_1)$ and $(-V_2 + V_1)$.

5.2 Show that the polaron hopping energy (the energy required to produce two polaron wells having the same electronic energy, cf. Fig. 5.36) for the case when there is a disorder energy, i.e. a difference in ground state energies, W_D, between two polaron wells, is:

$$W = W_H + \tfrac{1}{2}W_D + W_D^2/16W_H \qquad [5.1P]$$

where $W_H = \frac{1}{2}W_p$. For the case where $W_H \gg W_D$, [5.1P] reduces to [5.116].

5.3 Show that the Kubo formula ([5.49]) yields the standard Drude formula ([5.52]) for the frequency-dependent conductivity of an electron gas, given that the electron wavefunctions are given as

$$|\alpha\rangle = \sum_k a_k^\alpha |\boldsymbol{k}\rangle \qquad [5.2P]$$

where the a_k^α are independent random gaussian variables whose variance is given by:

$$\langle |a_k^\alpha|^2 \rangle = \left(\frac{\pi}{lk^2\Omega}\right)\frac{1}{[(k-k_\alpha)^2 + l^2/4]} \qquad [5.3P]$$

Bibliography

M. H. Brodsky (ed.) (1979), *Amorphous Semiconductors* (*Topics in Applied Physics*, vol. 36) (Springer-Verlag).

N. E. Cusack (1988), *Physics of Structurally Disordered Matter* (Adam Hilger).

J. C. Garland and D. B. Tanner (1978), *Electrical Transport and Optical Properties of Inhomogeneous Media* (AIP).

S. S. Mitra (ed.) (1976), *Physics of Structurally Disordered Solids* (Plenum).

N. F. Mott (1987), *Conduction in Non-Crystalline Materials* (Clarendon Press).

N. F. Mott and E. A. Davis (1979), *Electronic Processes in Non-Crystalline Materials* (OUP).

J. Pankove (ed.) (1984), *Semiconductors and Semimetals*, vol. 21 (Academic).

M. Pollak (ed.) (1987), *Noncrystalline Semiconductors* (CRC Press).

J. Wong and C. A. Angell (1976), *Glass: Structure by Spectroscopy* (Dekker).

F. Yonezawa (ed.) (1981), *Fundamental Physics of Amorphous Semiconductors* (*Solid-State Sciences*, vol. 25) (Springer-Verlag).

R. Zallen (1983), *The Physics of Amorphous Solids* (Wiley).

J. M. Ziman (1979), *Models of Disorder* (CUP).

6 Defects

6.1 Introduction

Thus far in this book, we have dealt in the main with 'ideal' amorphous solids, considering their structure and their electronic and vibrational properties. However, for real amorphous materials the introduction of structural defects or imperfections is often an unavoidable consequence of the drastic methods of preparation needed in producing an amorphous solid to avoid the initiation of crystallization. In addition, structural defects can be considered to be present in thermal equilibrium in certain liquids or even solids and these will be frozen-in on rapid quenching.

The notion of a defect in the structure of an amorphous solid might at first sight seem somewhat strange. Surely, the reader may ask, if the structure of the solid is random, is it not rather a contradiction in terms to talk about a defect in an amorphous matrix? Or phrased in a different way, might not the structure of a non-crystalline solid be thought of as being *completely* defective, as in the so-called 'dislocation model' of amorphous structures (in which a crystal is considered to contain so many intersecting dislocations that an aperiodic structure results (Ninomiya 1977))? These points of view are not, however, particularly helpful; instead we assert that the concept of a defect in an amorphous material *is* a valid one, although of course any given defect can only be defined with reference to some non-defective state. For crystalline materials, this is obviously the perfect single crystal lattice. For non-crystalline solids, we may take 'ideal' amorphous structures represented for example by a continuous random network for a covalent material or a dense random packing of spheres for a metal, as discussed extensively in Chapter 3.

The importance of defects, and the reason why we devote a chapter to them, lies in the fact that many properties of amorphous materials can be defect controlled, as indeed is also the case for crystalline solids. Among these number magnetic properties, opto-electronic behaviour, vibrational properties and mechanical characteristics. In many cases we shall see that the behaviour resulting from the presence of defects can completely dominate that due to the intrinsic material. Furthermore, for certain materials, e.g. the chalcogenide glasses, the 'ideal' amorphous state appears to be impossible to achieve experimentally since structural defects are present even in thermal equilibrium in the melt and are consequently frozen-in on vitrification.

The first part of this chapter is therefore devoted to a description of the different types of defect that are believed to occur in amorphous solids, with examples of their occurrence in a variety of materials. Some of these kinds of defect are analogous to those found in crystals and are familiar from conventional materials science (see e.g. Henderson 1972, Stoneham 1975); others are unique to the amorphous phase. The last part of the chapter deals with a discussion of the various properties that are exhibited by such defects, stressing in particular differences in behaviour from that expected for intrinsic (defect-free) materials. Comprehensive reviews of structural defects and their influence on physical properties have been given by Davis (1979), Mott (1980) and Robertson (1982).

6.2 Types of defect

6.2.1 Coordination defects (positive U)

In covalently bonded materials with pronounced directional bonding and consequently a well-defined local geometry, one type of point defect (i.e. that in which the deviation from the ideal structure is localized to a volume containing one or two atoms) concerns the atomic *coordination.* Such coordination defects are conceptually perhaps the most straightforward to understand. They can be either of two types: involving under-coordination or over-coordination with respect to the normal coordination of the ideal structure. Under-coordinated defects in covalent solids are simply dangling (or broken or unsatisfied) bonds (see Fig. 6.1); such a concept has no validity for materials containing non-directional bonds, such as metals, ionic salts or rare-gas solids. Over-coordinated defects are perhaps less common. They can arise when an additional covalent bond is formed datively between a normally coordinated atom and another (utilizing non-bonding

Fig. 6.1 (a) Isolated dangling bond in a tetrahedrally coordinated continuous random network simulating the structure of a-Si or a-Ge.

lone-pair electrons); the formation of such defects provides the mechanism for producing a negative effective correlation energy for dangling bonds in chalcogenide glasses (see Fig. 6.7(a)). A similar type of over-coordinated site has been postulated to exist in tetrahedrally coordinated materials, e.g. a-Si; this is the fivefold-coordinated site (Pantelides 1986), which can be regarded either as a Si self-interstitial configuration or, equivalently, as a close interaction between a dangling bond and a normally bonded, fourfold-coordinated site (see Fig. 6.2(a)). In another type of over-coordinated defect site in a-Si:H which has been proposed, the hydrogen interacts interstitially with long (i.e. weak) Si–Si bonds (Hishikawa *et al.* (1987) – see Fig. 6.2(b)).

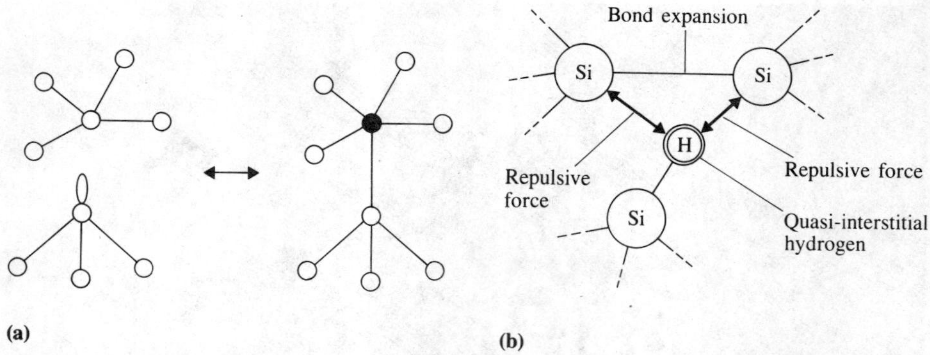

Fig. 6.2 Overcoordinated defect configurations in a-Si:H.
(a) Fivefold-coordinated site involving 'floating bonds', resulting from Si self-interstitials (Pantelides 1986).
(b) Five fold coordinated site resulting from hydrogen in monohydride configurations acting interstitially (Hishikawa *et al.* 1987).

There is an important difference between coordination defects in amorphous and crystalline solids: *isolated* defects can occur in the non-crystalline state, but not in ordered structures. An example of an isolated under-coordinated defect in an amorphous matrix is shown in Fig. 6.1, and a representation of an isolated over-coordinated defect is shown in Fig. 6.2(a). The corresponding defects in crystalline solids can only be created several at a time. For example, in the diamond-cubic structure of crystalline Si, if an atom is removed to create a vacancy, *four* dangling bonds are formed which point into the vacancy (see sect. 6.2.2); this is the least number that can result from the introduction of vacancies. Equally, line defects such as dislocations in crystalline covalent materials also result in the introduction of many dangling bonds along the length of the defect. Similar considerations apply to over-coordinated defects: an interstitial atom in a tetrahedral configuration in c-Si produces *four* fivefold-coordinated Si atoms surrounding the central interstitial atom.

A simple dangling bond normally contains *one* electron and is electrically neutral; however it is amphoteric (being able to donate or accept an electron), and under certain circumstances the electronic occupancy can change, varying concomitantly the charge of the centre. A convenient nomenclature for labelling isolated coordination defects in amorphous covalent materials is in terms of the type of atom that is the site of the defect, together with its coordination number and charge state as subscript and superscript respectively. Thus, for tetrahedrally coordinated materials such as Si or Ge, the different charge states of (under-coordinated) dangling bonds are denoted by T_3°, T_3^+ and T_3^- (normally bonded atoms are denoted by T_4°); the over-coordinated 'floating-bond' defect postulated by Pantelides (1986) can be represented as T_5°. In simple chalcogenide glasses, e.g. a-Se, it is believed that the coordination of 'dangling-bond' defects changes with the electron occupancy (see later), resulting in the configurations C_3^+, C_1° and C_1^- (with the normally bonded site represented as C_2°). Such a simple labelling scheme is really appropriate only for monatomic systems; for materials containing more than one type of element, the kind of atoms which are nearest neighbours of the defect site also need to be specified.

The energy levels for electron states associated with an isolated dangling bond may be discussed in terms of a simple molecular orbital picture. If we consider the case of a tetrahedral semiconducting material, e.g. a-Si, having an atomic electronic configuration s^2p^2, the atomic levels hybridize to form four sp^3 molecular hybrids, each of which may admit a bonding or antibonding orbital; solid-state interactions then broaden the molecular levels into bands separated by a band gap (Fig. 6.3(a)). A dangling bond, or non-bonding orbital, containing a single electron will therefore have an energy level lying at the zero energy for the sp^3 hybrids, viz. near the *middle of the gap* (if atomic relaxations are neglected). Thus, structural defects such as dangling bonds are expected to introduce electron states deep into the gap (which is otherwise empty in the ideal case except for band tailing – see Fig. 5.10(a)); the precise position of the energy levels, however, will depend crucially on factors such as structural relaxation around the defect or the electronic character of the states at the top of the valence band and the bottom of the conduction band from which the eigenfunction of the defect state derives. Note the analogy between the behaviour of electron states and vibrational modes associated with local defects already mentioned in section 4.2.3; like the vibrational excitations in that case, we expect the electron states in the mobility gap arising from dangling bond defects to be highly localized.

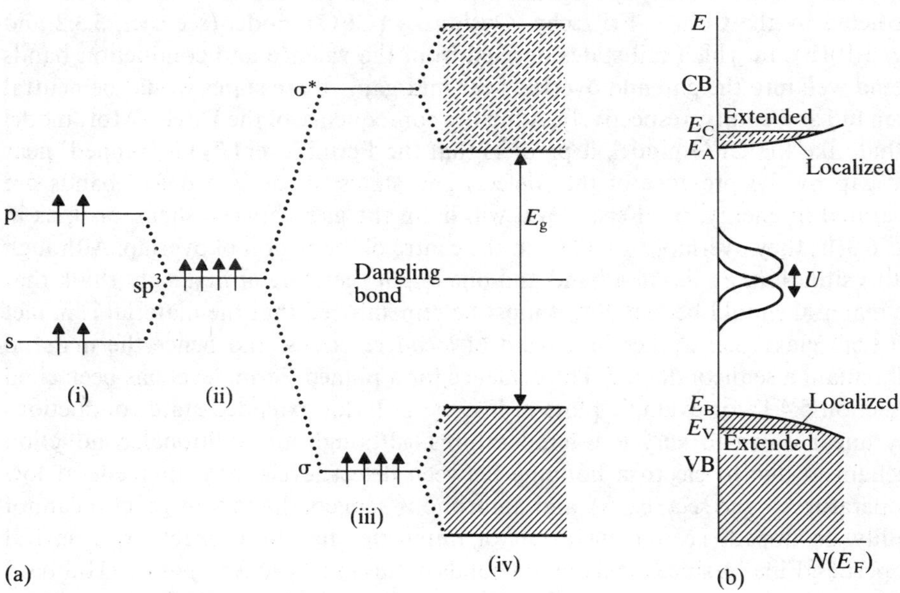

Fig. 6.3 (a) Schematic representation of the origin of valence and conduction band states for a tetrahedrally bonded semiconductor (i) atomic s- and p-states; (ii) sp^3-hybrid states; (iii) bonding (σ) and antibonding (σ^*) states; (iv) broadening of σ- and σ^*-states into valence band (VB) and conduction band (CB).
(b) Density of states for such a band scheme, showing the localized band-tail states. A dangling-bond band is also near mid-gap, together with the band for double occupancy higher by U in energy (for a positive-U system).

The density of states in the gap for an amorphous semiconductor containing isolated dangling bond defects might be as shown schematically in Fig. 6.3(b). The dangling bond level is broadened by disorder (e.g. by bond-angle fluctuations, etc.) into a band; the lower mid-gap band in Fig. 6.3(b) corresponds to the neutral dangling bond containing a single electron (spin) and is 'donor-like' (i.e neutral when occupied), and the upper band corresponds to a different charge state of the *same* defect, namely when an extra electron is placed in it, and is 'acceptor-like' (i.e. neutral when empty). Under normal circumstances, the net energy cost for this addition is positive; it costs an extra energy U (the Hubbard or correlation energy):

$$U = \langle e^2/4\pi\varepsilon_0 \varepsilon r_{12} \rangle \qquad [6.1]$$

where r_{12} represents the appropriate separation for two electrons at the same site, and a configurational average has been taken in evaluating U.

The density of states depicted in Fig. 6.2(b) was first proposed by Davis and Mott (see Mott and Davis 1979) and is based on the assumption that mobility edges exist in the valence and conduction bands, marking the demarcation between strongly delocalized (extended) states. This model predicts that the material would be expected to be optically transparent for photons of energy less than that corresponding to the mobility gap, although theoretical considerations by Dersch *et al.* (1987) indicate that optical transitions involving the localized tail states, which are 'phonon-delocalized' at elevated temperatures, can give rise to appreciable absorption tails in the gap region. This latter circumstance is similar to that predicted by the Cohen–Fritzsche–Ovshinsky (CFO) model (see sect. 5.3.2 and Fig. 5.10(b)) in which tail states derived from the valence and conduction bands extend well into the gap and overlap near mid-gap – these states would be neutral when full and empty, respectively. A further consequence of the Davis–Mott model is that, like the CFO model, it predicts that the Fermi level (E_F) is 'pinned' near mid-gap by the presence of the (defect) gap states; if the two defect bands are separated in energy, the Fermi level will lie in the gap between them, or if, as in Fig. 6.3(b) they overlap, E_F will lie in the centre of the region of overlap. Although in this situation, E_F lies in a band and one might therefore at first sight think that the material should be metallic, it must be remembered that the material is in fact a 'Fermi glass', i.e. E_F lies in a band of *localized* states, and hence the material will remain a semiconductor. The evidence for a pinned Fermi level has been cited in section 5.4.1; for example plots of $\ln \sigma$ vs. $1/T$ (for extended-state conduction) stay linear down to very low temperatures (although an additional conduction mechanism due to electron hopping between defect levels may intercede at low temperatures – see sect. 6.3.3), and amorphous semiconductors in general cannot readily be doped, i.e. the inclusion of impurities has little effect on electrical properties. Finally, since the dangling bonds in this model suffer a *positive* Hubbard U, most of them will be neutral, possessing a single electron spin. Therefore these centres will be paramagnetic, exhibiting electron spin resonance signals and Pauli paramagnetism (sect. 6.3.1).

The materials which come closest to realizing the Davis–Mott model for states in the gap are probably evaporated or sputtered a-Si or Ge films. These amorphous thin films exhibit a large ESR signal ($\sim 10^{19}$–10^{20} spins cm^{-3}) in their as-prepared state, and a concomitant high density of states in the gap, as measured by techniques such as the 'field effect' (for an illustration of its application and use see sect. 6.3.1). No structure in the density of states, such as that shown schematically in Fig. 6.3(b), has come to light using this or other techniques for *un*hydrogenated a-Si or Ge; indeed the density of states appears to be so high that little variation in the field effect can be detected, although in the energy region which can be explored, the density of states appears to be uniform. The very high densities of gap states in these materials naturally explain the observation that they cannot be doped: any impurity which is introduced will donate or accept an electron to or from the defect gap states, but the density of states is so high and the number of impurity atoms so relatively small, that the Fermi level is not shifted significantly and the material is not doped.

This is *not* the case for *hydrogenated* a-Si:H alloys, prepared for example by the glow-discharge decomposition of silane, which have such a low residual density of gap states that the substitutional introduction of P, As or B can easily dope the materials n- or p-type, changing the electrical conductivity by many orders of magnitude (see Fig. 6.4) (Spear and Le Comber 1976). Presumably, the presence of the hydrogen acts in the main part to satisfy any dangling bonds, thereby removing the defect states in the gap, replacing them with Si—H bonding states deep in the valence band, and rendering the material dopable. This simple picture is not sufficient to account completely for the role of the hydrogen, however, since it is found that some 10 times more H ($\simeq 1$–10 at. %) is introduced by the glow-discharge process into electronic grade a-Si:H than is needed to satisfy all the dangling bonds evinced by ESR in evaporated a-Si; indeed, NMR experiments (Boyce and Stutzmann 1985) and calorimetric experiments (Lohneysen *et al.* 1984, Graebner *et al.* 1985) have demonstrated that about 5% of the hydrogen in a-Si:H occurs in the form of molecular H_2 trapped in microvoids in the matrix of amorphous silicon. Moreover the introduction of H progressively erodes the top

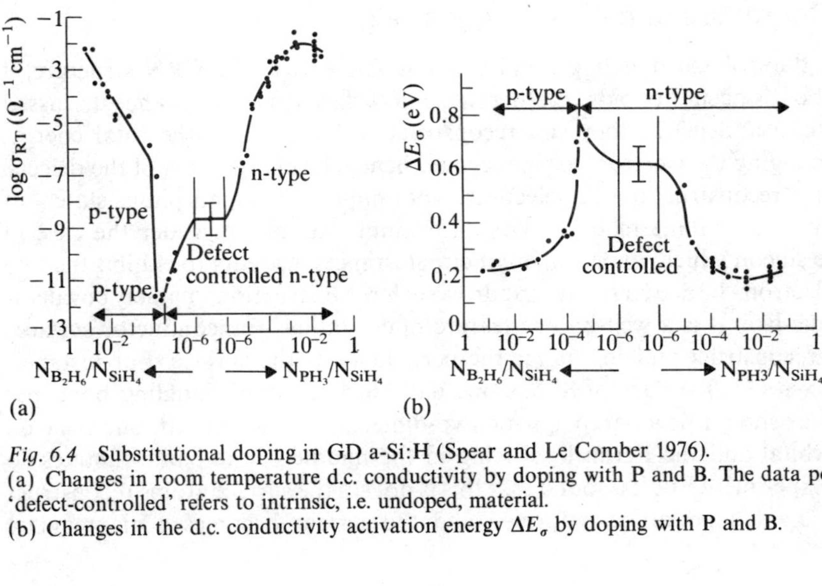

(a) (b)

Fig. 6.4 Substitutional doping in GD a-Si:H (Spear and Le Comber 1976).
(a) Changes in room temperature d.c. conductivity by doping with P and B. The data point labelled 'defect-controlled' refers to intrinsic, i.e. undoped, material.
(b) Changes in the d.c. conductivity activation energy ΔE_σ by doping with P and B.

of the valence band as well as removing states from the gap, as shown by photoemission studies and the finding that the optical gap increases on hydrogenation by $\simeq 0.4$ eV for $\simeq 15$ at.% H (Moustakas 1979). The role of hydrogen in the deposition process and the subsequent electrical activity of this potentially useful electronic material is still a subject of some considerable controversy and active study. Although there is incontrovertible experimental evidence that amorphous hydrogenated silicon can be doped (Fig. 6.4), the doping mechanism appears to be much more complicated than that for substitutional doping in, say, crystalline Si; it appears that coordination defects play an essential role in mediating the doping process in the amorphous phase (see sect. 6.2.6).

There is a considerable body of experimental evidence that the defects in a-Si:H exhibit a *positive* correlation energy, with $U \simeq 0.4$ eV (see sect. 6.3.1), and such defect states have been ascribed to simple dangling bonds, T_3°. However, *ab initio* total-energy calculations for the T_3° centre, modelled as a monovacancy in the diamond-cubic crystalline Si structure with three of the resulting dangling bonds saturated by H atoms, have indicated that the Si dangling bond exhibits a *negative* effective correlation energy, $U_{\text{eff}} \simeq -0.12$ eV (see sect. 6.2.2), in other words that the T_3° centre is energetically unstable with respect to disproportionation into the diamagnetic charged centres (T_3^+ and T_3^-) (Bar-Yam and Joannopoulos 1986). This inconsistency has led Pantelides (1986) to propose that, instead, over-coordinated T_5° configurations are responsible for the defect-related electronic behaviour observed in a-Si:H. The T_5° centre is characterized by an additional 'floating bond', composed of a linear combination of five sp^3 hybrids pointing towards the central T_5° site (see Fig. 6.2(a)); in its neutrally charged state, T_5° is expected to be paramagnetic (and ESR-active), with a *positive* $U \simeq 0.3$ eV (Pantelides 1986). However, it is not clear that the ESR signature expected from a T_5° centre is in accord with that observed experimentally, and consequently the microscopic origin of the electronically active defects in a-Si:H, remains somewhat unclear.

6.2.2 Coordination defects (negative *U*)

Although the isolated dangling bond is a possible entity in a CRN structure, if dangling bonds occur at voids or other internal surfaces (a circumstance discussed in sects. 6.2.2 and 6.3.2), they can reconstruct so as to lower the total energy, thereby changing the electron occupancy and hence the charge state of the defects. Such atomic reconstructions at defects are not unique to the amorphous state, but also occur in crystalline materials. As an example we may consider the case of crystalline silicon which, under different circumstances, appears to exhibit the two forms of electron–lattice coupling that drive such reconstructions, namely covalent-like or ionic-like. It is a well-known feature of crystalline semiconductor surfaces that they reconstruct and so change the periodicity at the surface. For instance, the *unrelaxed* (111) surface of Si has one half-filled (neutral) dangling bond per site and thus should be a paramagnetic two-dimensional metal (with one electron in each orbital and the Fermi level lying in the middle of the band of states so formed). Experimentally, it is found that this is not so; the surface atoms reconstruct so that a 2×1 superlattice is formed and the surface becomes a diamagnetic semiconductor. One model for this reconstruction is based on an ionic-like electron–phonon coupling mechanism; alternate rows of atoms are raised or lowered, changing their hybridizations towards s^2p^3 or sp^2 respectively, and consequently also changing the electron occupancies of the dangling bonds (Fig. 6.5(a)) (Applebaum and Hamann 1976). The raised atoms (s^2p^3) acquire a net negative charge, and we may refer to the configuration to which they tend as T_3^-; correspondingly, the lowered surface atoms (sp^2) tend towards a T_3^+ configuration. The ionic character of the electron–lattice coupling leading to this sort of reconstruction is thus clearly seen. Recent angle-resolved photoemission results have cast doubt on the ionic buckling model, and another form of reconstruction involving π-bonding (Fig. 6.5(a)) has been proposed (Pandey 1981). An illustration of the other type of electron–lattice coupling, namely covalent-like, is afforded by the divacancy in crystalline Si. The removal of two neighbouring atoms from the lattice results in six dangling bonds pointing into the resulting divacancy. These dangling bonds can then form weak 'bent' bonds between neighbouring hybrid orbitals and even weaker bonds between more distant orbitals, if the electron occupation of the orbitals is such as to allow the bond formation (Fig. 6.5(b)).

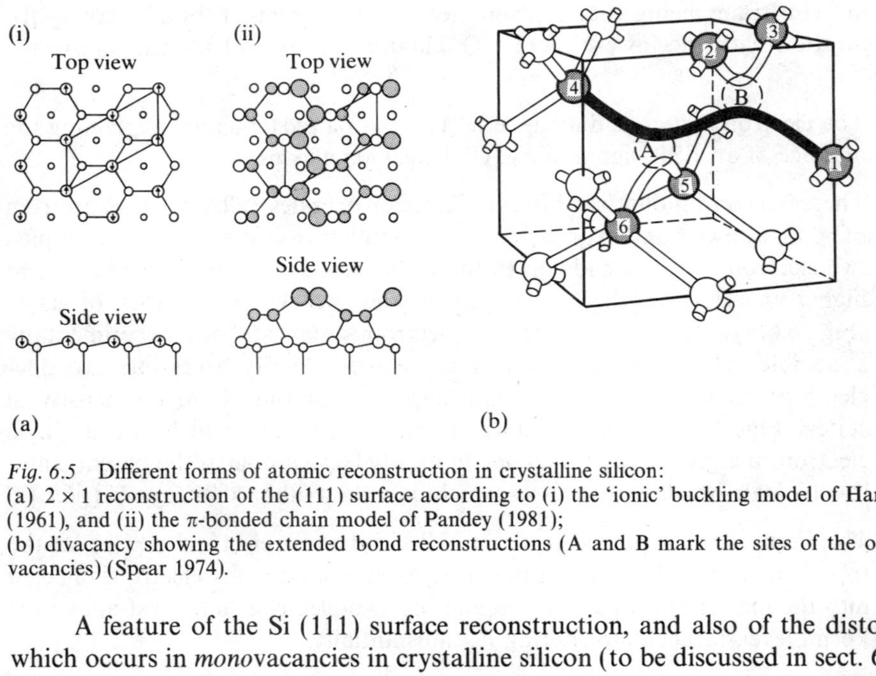

Fig. 6.5 Different forms of atomic reconstruction in crystalline silicon:
(a) 2 × 1 reconstruction of the (111) surface according to (i) the 'ionic' buckling model of Haneman (1961), and (ii) the π-bonded chain model of Pandey (1981);
(b) divacancy showing the extended bond reconstructions (A and B mark the sites of the original vacancies) (Spear 1974).

A feature of the Si (111) surface reconstruction, and also of the distortion which occurs in *mono*vacancies in crystalline silicon (to be discussed in sect. 6.2.3), is that the defects exhibit a so-called 'negative effective Hubbard or correlation energy', U_{eff} (Anderson 1975). Spin pairing of electrons occurs at the defect centres, and the net process is energetically exothermic because of the stronger (ionic or covalent) electron–phonon coupling which is a feature of such systems, despite the endothermic cost in energy to place two electrons on the same site ([6.1]).

The concept of electron pairing mediated by a negative U_{eff} mechanism was first proposed by Anderson (1975) to apply to certain covalent amorphous materials, particularly chalcogenide glasses. Certain experimental techniques (e.g.

photoluminescence, sect. 6.3.2 and a.c. conductivity, sect. 6.3.5) demonstrated for these materials that there exist states in the gap due to defects, as in the case of a-Ge or a-Si, but no ESR signal can be detected, in marked contrast to the large signals observed in the tetrahedrally bonded amorphous semiconductors. This apparently contradictory behaviour is understandable if the electron–phonon interaction is sufficiently strong in these materials to produce a negative U_{eff} and hence render spin pairing energetically favourable.

Anderson's theory is in fact couched in the most general terms and we will discuss this first before turning our attention to the applications that have been made of the theory to specific point-defect configurations. The model considers a random lattice of sites connected by a *spectrum* of 'bonds', each populated by a pair of up- and down-spin electrons, and which span the range of electron pair configurations from conventional covalent bonds, through 'weak' bonds, to non-bonding lone pairs; the electron pairing mechanism is a strong electron–phonon (polaron) coupling leading to a negative U_{eff}. Note that the concept of a well-defined *structure* of an amorphous semiconductor does not apply in this model, nor is there a distinction between defects and the ideal structure of the bulk material; large variations in bond lengths, bond angles and coordination number are allowed, the only constraint being that a given electron pair forms a 'bond' having the configuration with the lowest energy. Qualitative features of the model are the following:

(a) The electron system is diamagnetic (i.e. spin paired), thereby explaining the absence of an ESR signal in e.g. chalcogenide glasses.

(b) There is a continuous distribution of electron energies, reflecting the spectrum of bonds allowed, and in particular in the simplest case, the density of occupied *two*-electron states extends up to the Fermi level, with unoccupied states at higher energies; i.e. there is *no* gap in the two-electron density of states (Fig. 6.6(a)). However, these two-electron states are not experimentally accessible. The only states which are experimentally accessible are one-electron states for which there is a gap of magnitude U in the density of states (Figs 6.6(b) and (c)) since thermal excitations which add a single electron to a given state (and concomitantly leave a hole in the parent state) involve pair-breaking processes which cost the Hubbard energy U ([6.1]).

(c) The Fermi level is pinned in the middle of the one-electron gap, i.e. at the top of the occupied two-electron density of states, since any electrons injected into the material will pair up, populating two-electron states just above the Fermi level, rather than shifting E_F substantially.

Note the similarities between the density of states in the Anderson model (Fig. 6.6(a)) and the Cohen–Fritzsche–Ovshinsky (CFO) model (Fig. 5.10(b)); both have electron states which extend throughout the gap region, thereby producing a 'pseudo-gap' rather than a true gap. The difference between the two is that whereas in the CFO model these are *one*-electron states, in the Anderson model they are *two*-electron states although there *is* a true gap in the one-electron density of states in the latter case.

Let us now consider the Anderson model in more detail. The Hamiltonian

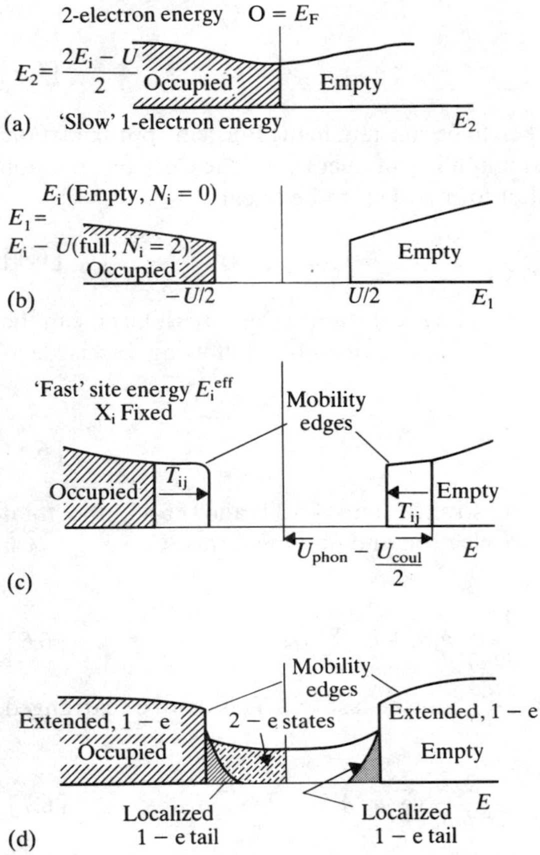

Fig. 6.6 Schematic densities of states for an amorphous semiconductor suffering a negative effective Hubbard energy (Anderson P. W. (1976) *J. de Phys.* **C4**, 339). (N.b. $U = U_{phon} - U_{coul}$.)
(a) Two-electron density of states. Note that the occupied levels extend to the Fermi level.
(b) One-electron density of states (for 'slow' states, i.e. excluding hopping) showing the gap of magnitude U between occupied and unoccupied states.
(c) The effect on the one-electron density of states of including hopping, i.e. 'fast', states.
(d) Overall density of states with the inclusion of randomness, i.e. band tails.

can be written as a sum of three parts, an electronic term, a contribution pertaining to the lattice (or phonons), and a term describing the electron–phonon interaction. The electronic term can be written as:

$$\mathscr{H}_e = \sum_{i,\sigma} E_i \hat{n}_{i\sigma} + U \sum_i \hat{n}_{i\sigma}\hat{n}_{i-\sigma} + \sum_{i,j,\sigma} T_{ij} c^{\dagger}_{i\sigma} c_{j\sigma} \qquad [6.2]$$

where the summations are over the sites i (or j), and over the spin states of the electrons σ. The parameter E_i is the energy of an electron localized on site i, U is the Hubbard electron repulsion energy ([6.1]) which occurs for double occupancy at a site, T_{ij} represents the hopping interaction with neighbouring sites which is neglected henceforth, and $\hat{n}_{i\sigma}$ $(= c^{\dagger}_{i\sigma} c_{i\sigma})$ is the occupancy of site i by an electron with spin σ, where $c^{\dagger}_{i\sigma}(c_{i\sigma})$ is the creation (annihilation) operator. The lattice or phonon term can be written as:

$$\mathscr{H}_1 = \hbar\omega \sum_i b_i^\dagger b_i \tag{6.3}$$

where ω is the phonon frequency, taken to be constant in the Einstein approximation, and $b_i^\dagger$, b_i are phonon creation and annihilation operators. The electron–phonon coupling term can be, in the simplest form, taken to be linear:

$$\mathscr{H}_{e-1} = \lambda \sum_{i,\sigma} \hat{n}_{i\sigma}(b^\dagger + b) \tag{6.4}$$

where λ is the electron–phonon coupling constant. This cross term can be eliminated by introducing a displaced phonon operator, following Licciardello *et al.* (1981):

$$d_i = b_i + \frac{\lambda \hat{n}_i}{\hbar\omega} \tag{6.5}$$

where $\hat{n}_i = (\hat{n}_{i\sigma} + \hat{n}_{i-\sigma})$. If this is substituted into [6.3] and [6.4], the total Hamiltonian reduces to a sum of pure electron and phonon terms $\mathscr{H} = \tilde{\mathscr{H}}_e + \tilde{\mathscr{H}}_1$, i.e.:

$$\mathscr{H} = \sum_{i,\sigma} E_i \hat{n}_{i\sigma} + U \sum_i \hat{n}_{i\sigma}\hat{n}_{i-\sigma} - \frac{\lambda^2}{\hbar\omega} \sum_i \hat{n}_i \hat{n}_i + \hbar\omega \sum_i d_i^\dagger d_i \tag{6.6}$$

where the first three terms represent $\tilde{\mathscr{H}}_e$ and the last $\tilde{\mathscr{H}}_1$. This can be rearranged to give:

$$\mathscr{H} = \sum_{i,\sigma} \left(E_i - \frac{\lambda^2}{\hbar\omega} \hat{n}_{i\sigma} \right) \hat{n}_{i\sigma} + \left(U - \frac{2\lambda^2}{\hbar\omega} \right) \sum_i \hat{n}_{i\sigma}\hat{n}_{i-\sigma} + \hbar\omega \sum_i d_i^\dagger d_i \tag{6.7}$$

Hence the *effective* Hubbard correlation energy for two spin-paired electrons at the same site is given by:

$$U_{\text{eff}} = U - \frac{2\lambda^2}{\hbar\omega} \tag{6.8}$$

which is negative if $(2\lambda^2/\hbar\omega) > U$, i.e. for strong electron–phonon coupling.

In general, the equilibrium eigenfunction can be written as $|n_i, v_i\rangle$, where n_i and v_i are the number of electrons and phonons respectively, localized at site i. In the low frequency limit, $v_i = 0$, and the energy eigenvalues of the Hamiltonian ([6.6]) for n electrons occupying a site are then given by:

$$E_n = nE + U\delta_{2,n} - \frac{\lambda^2 n^2}{\hbar\omega} \tag{6.9}$$

where δ is the Kronecker delta. Alternatively, we may write the energy eigenvalues of the Hamiltonian, expressed as [6.7], explicitly:

$$\begin{aligned} n &= 0 \qquad E_0 = 0 \\ n &= 1 \qquad E_1 = E_{\text{eff}} \\ n &= 2 \qquad E_2 = 2E_{\text{eff}} - U_{\text{eff}} \end{aligned} \tag{6.10}$$

where we have written

$$E_{\text{eff}} = E - \frac{\lambda^2}{\hbar\omega} \qquad [6.11]$$

Thus, we see from [6.10] that in those materials which have a negative U_{eff}, it is energetically favourable for electrons to pair up rather than remain unpaired, such that at certain sites there are two electrons and at others none, i.e. $E_2 + E_0 < 2E_1$.

The electron–lattice coupling may arise from two distinct mechanisms, as we mentioned earlier. The electron pairing may take the form of a covalent bond between two dangling bond orbitals, in which case an antisymmetric phonon mode couples with the electrons by modulating the two-centre covalent interaction which Harrison (1980) refers to as V_2; in this case we may write $\mathscr{H}_{e-1} = \Sigma_{i,\sigma} V_2^i(q) n_{i\sigma}$ (equivalent to [6.4]), where q is a 'configurational coordinate', e.g. a bond displacement. Alternatively, a lattice mode can modulate the energy of an orbital, thereby causing some orbitals to be doubly occupied (and negatively charged) and others unoccupied (and positively charged). In this ionic mechanism, the orbital hybridization depends on q via the bond angle, and $\mathscr{H}_{e-1} = \Sigma_{i,\sigma} V_3^i(q) n_{i\sigma}$ where V_3 is the ionic energy (Harrison 1980). Ngai *et al.* (1978) have further suggested that tunnelling states (sect. 4.5.4), called by them 'local rearrangement modes', may also couple to electrons in a similar manner to phonons and like them mediate a negative U_{eff} interaction.

The optical and thermal electronic excitations associated with the Anderson model are complicated, and the interested reader is referred to the articles by Anderson (1979) and Licciardello *et al.* (1981). It should be pointed out, however, that the Anderson model remains essentially a conceptual theory, and it has not been used to account in detail for experimental results for amorphous systems, e.g. chalcogenide glasses. Furthermore, it is not universally accepted that the generality of the Anderson model in predicting a very broad spectrum of 'bonds' in a negative U_{eff} material is a realistic description of *solid* amorphous semiconductors, although it may be a reasonable approximation for *liquid* semiconducting systems. Instead, it is generally contended that a solid amorphous structure with well-defined bond lengths and reasonably well-defined bond angles is a better description; disorder in this case is mostly concentrated at point (coordination) defects, e.g. dangling bonds, and it is at these that electrons can suffer a negative U_{eff} via the Anderson polaronic mechanism. This conservation of bond length is probable in a solid because the bond energy is a highly non-linear function of bond length, and it is only in liquids, where there is considerable thermal energy available, that appreciable bond-length distortions are likely to be energetically favourable.

Recently, Bar-Yam *et al.* (1986) have proposed a model in which the electronic structure and the atomic structure of amorphous semiconductors are strongly inter-dependent, being determined thermodynamically (see sect. 6.2.6), i.e. band-edge (tail) states and defect states lying deeper in the energy gap result from thermal excitations of the ideal amorphous structure. The idea that the atomic structure, and consequently the electronic structure too, can be regarded as a broad spectrum of structural configurations, ranging from a perfect fully connected structure at one extreme to coordination defects at the other, is obviously common to the approaches of both Anderson (1975) and Bar-Yam *et al.* (1986), even though the details of the interactions involved differ.

The importance of discrete dangling bond defects in amorphous semiconductors at which electrons experience a negative effective correlation energy was first stressed by Street and Mott (1975) and Mott *et al.* (1975a) and later by Kastner *et al.* (1976). These models were originally applied to the case of chalcogenide glasses, but various other authors have extended the idea of spin-paired electrons at dangling bond sites to other amorphous semiconductor systems as well. These bonding models have the virtues of being conceptually simple and of having been compared in detail with a variety of experimental data.

The essential features of the 'charged dangling bond' model may be understood by consideration of the monatomic system amorphous Se, illustrated in Fig. 6.7. The structure of amorphous Se, which is twofold coordinated, is believed

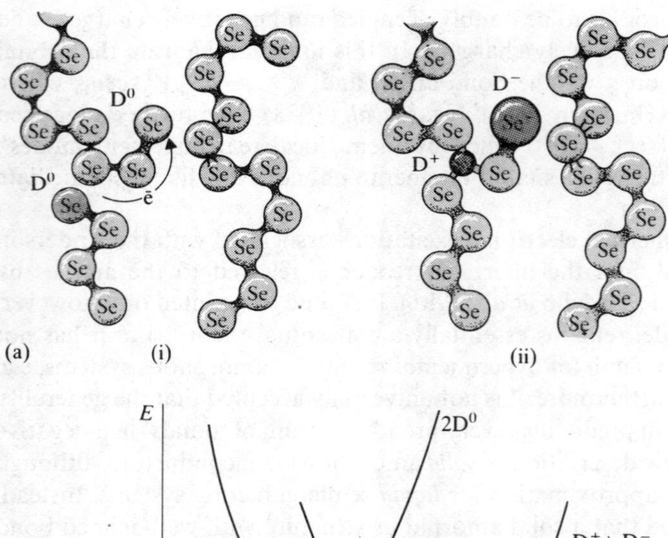

Fig. 6.7 Formation of charged defects (valence alternation pairs) in chalcogenide glasses.
(a) Illustration of the formation of threefold coordinated $D^+(C_3^+)$ and singly coordinated $D^-(C_1^-)$ defect centres by the exchange of an electron between two $D^\circ(C_1^0)$ centres (after D. Adler (1977) *Scientific American*, **236**, 47).
(b) Configuration-coordinate diagram for the formation of a D^+–D^- pair. Exchange of an electron between two D° centres to give a D^+–D^- pair at the *same* configuration costs the Hubbard energy U. The D^+–D^- centres subsequently relax to a different configuration and the overall energy is lowered by the effective correlation energy U_{eff}.

to consist mainly of chains (with perhaps a few rings). Any chain end will be the site of a dangling bond orbital, which in the simplest case will contain an unpaired electron and be electrically neutral relative to the bulk; this dangling bond defect can be written as D^0, or equivalently C_1^0. Mott *et al.* (1975) postulated that, following Anderson (1975), electrons residing at D^0 centres should experience a negative U_{eff} and electron pairing should be energetically favourable as a result of atomic rearrangements. Electron-lattice coupling is strong in amorphous chalcogenides because of the low atomic coordination resulting in a high degree of network flexibility, but more importantly there exist non-bonding, lone-pair $p\pi$ orbitals at the chalcogen atoms which may be utilized in forming additional bonds, and which are energetically accessible since they are high-lying, forming the top of the valence band (Kastner 1972). The transfer of an electron from one D^0 (C_1^0) centre to another produces one site which has the original dangling bond orbital containing two spin-paired electrons and which is consequently negatively charged (D^- or C_1^-), and the other which has an empty orbital which is then free to form a dative bond with the lone pair of a fully connected neighbouring atom; the defect now becomes *threefold* coordinated and positively charged (D^+ or C_3^+) (see Fig. 6.7(a)). Note that in the latter case the site of the defect moves from the originally dangling bond to the atom providing lone-pair electrons for the dative bond which correspondingly becomes over-coordinated; this feature has important ramifications for systems containing more than one type of atom (see later). The repulsive Hubbard energy U involved in placing an extra electron on the same site to form a D^- centre is postulated to be outweighed by the energy gained in forming the extra bond at the D^+ site, rendering the reaction

$$2D^0 \rightarrow D^+ + D^- \qquad [6.12]$$

exothermic with an overall negative effective correlation energy U_{eff} (Fig. 6.7(b)). This process can be illustrated in terms of a configurational coordinate diagram (Fig. 6.7(b)). The appropriate configurational coordinate, q, in this case may be taken to be the sum of the distances between two D^0 centres and their respective nearest neighbour, but non-directly bonded, atoms, $q_1 + q_2$. Transfer of an electron between D^0 centres *without* a change in configuration costs the Hubbard energy U, but the lowest energy state for the two resulting oppositely charged centres lies at a *different* configurational coordinate because of the atomic relaxation (bonding) that takes place at the D^+ centre; the energy for this equilibrium configuration is lower by U_{eff} than that for the two D^0 centres.

There are in fact two possible configurations for the neutral defect D^0 in chalcogenide glasses, one being under-coordinated (C_1°) and the other over-coordinated (C_3°). Kastner *et al.* (1976) proposed that the symmetric C_3° centre had the lowest total energy, whereas Mott *et al.* (1975a) believed that the actual configuration lay between the two, with one elongated bond. However, realistic total-energy calculations by Vanderbilt and Joannopoulos (1980) have subsequently shown that the C_1° centre is the lowest-energy configuration, principally because of π-bonding interactions between the dangling-bond orbital on the C_1° and a lone pair on the neighbouring C_2° atom.

Although the simple chemical-bond considerations of Kastner *et al.* (1976) and Mott *et al.* (1975a) indicate that dangling-bond defects in chalcogenide glasses should exhibit negative-U behaviour, the total-energy calculations by Vanderbilt and Joannopoulos (1983) have indicated that for the representative chalcogenide glass a-Se, the correlation energy appears in theory to be *positive*, with $U_{\text{eff}} \simeq 0.45$ eV. As expected, there is very little relaxation associated with the transformation $C_1^\circ \rightarrow C_1^-$, and although the process $C_1^\circ \rightarrow C_3^+$ involves a large (negative) energy of relaxation ($\simeq -0.5$ eV), the intra-site Coulomb repulsion energy ([6.1]) is much larger ($U_{\text{Coul}} \simeq 0.9$ eV), making the overall process endothermic.

Although theoretical considerations indicate that a-Se may be a positive-U system, a negative U_{eff} is not ruled out for compound chalcogenide glasses (e.g. a-As_2Se_3) since in many ways a-Se is anomalous. For example, it has a much smaller dielectric constant (by about a factor of two) than the compound materials which would tend to make the positive Coulomb contribution U_{Coul} much bigger in Se for other materials; furthermore, the defect wavefunction for C_1 defects in a-Se is much more localized than in other materials, which would tend to make U_{Coul} even larger (see [6.1]). Experimentally, a-Se may be anomalous too: its electronic behaviour is strongly affected by the presence of incidental impurities (e.g. O or Cl) and the observed negative-U_{eff} behaviour (e.g. absence of ESR signals) may not actually be intrinsic.

The total-energy calculations by Vanderbilt and Joannopoulos (1983a) indicate that the formation energy of a relaxed neutral dangling-bond defect C_1^0 in a-Se is only 0.8 eV (due to π-bonding interactions). In compound chalcogenide glasses, U_{eff} might be expected to be negative, since the formation of a valence-alternation pair (VAP), C_3^+ and C_1^-, from two normally bonded C_2^0 atoms via the reaction

$$2C_2^0 \rightarrow C_3^+ + C_1^-$$

involves essentially just a bond-flip, conserving the total number of bonds, and therefore the VAP creation energy E_{VAP} might also be expected to be low (Kastner *et al.* 1976). Consequently, a relatively high density of coordination defects (charged in the case of compound materials) should be present in thermal equilibrium in chalcogenide melts, which will be frozen-in on quenching to form a glass, with a concentration depending on the temperature at which equilibrium was last established, namely the glass transition (sect. 6.2.6). Thus, the concentration of *randomly* distributed VAP defect centres may be given by

$$n_{\text{R}} \simeq n_0 \exp(-E_{\text{VAP}}/2k_{\text{B}}T_{\text{g}}) \qquad [6.13]$$

where n_0 is the total (chalcogen) atomic density. Since the VAP centres C_3^+ and C_1^- are equally and oppositely charged, there is a tendency for them to pair up in the melt under the action of their mutual Coulomb interaction, thereby lowering the creation energy even further; the concentration n_{I} of such 'intimate valence-alternation pairs' – IVAPs (Kastner *et al.* 1976) would therefore be higher than n_{R}.

Experimental evidence for the existence of thermally generated defects in As_2Se_3 has come from transient photoconductivity experiments (see sect. 6.3.1) in the glassy state below T_{g} and in the liquid state above T_{g} (Thio *et al.* 1984). If neutral defects (C_1^0) are the dominant recombination centres, measurement of the recombination time from transient photoconductivity measurements yields an estimate for the concentration n of such centres. Thio *et al.* (1984) found that n

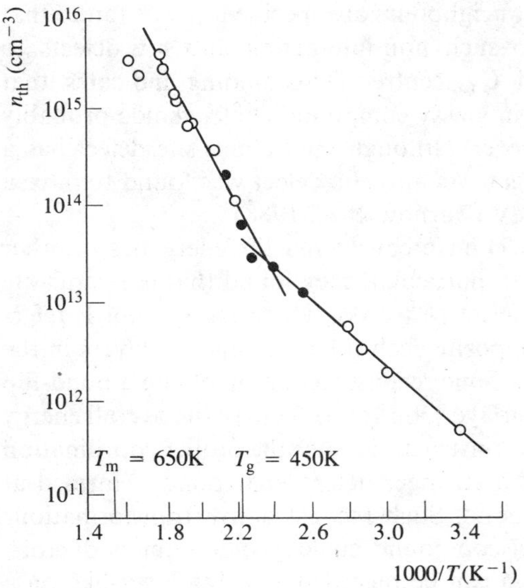

Fig. 6.8 Temperature dependence of the density of thermally-generated recombination centres for As_2Se_3 in both glassy and liquid states obtained from transient photoconductivity measurements (Thio *et al.* 1984).

was thermally activated (Fig. 6.8), with an activation energy of 0.35 eV and a pre-exponential factor of $\simeq 4 \times 10^{17}$ cm^{-3} for $T < T_g$, whereas for $T > T_g$ the activation energy found was 0.8 eV and the pre-factor $\simeq 6 \times 10^{22}$ cm^{-3}. Above T_g defects are created thermally from normally bonded atoms with a number density of $\simeq 10^{22}$ cm^{-3}; the activation energy involved in the creation of C_1^0 recombination centres from C_3^+ or C_1^- centres will therefore be $(|U_{eff}| + E_{VAP})/2$, which is identified with the value of 0.8 eV found experimentally. However, for temperatures below T_g, the total defect concentration is frozen-in (at the value it had at $T \simeq T_g$), but the conversion of C_1^0 centres from C_3^+ or C_1^- centres continues to be thermally activated, with activation energy $|U_{eff}|/2$. Thus, the results of Thio *et al.* (1984) provide an estimate for the energies involved in a-As_2Se_3, namely $E_{VAP} \simeq 0.9$ eV and $U_{eff} \simeq -0.7$ eV.

Consideration of defects in compound chalcogenide materials is considerably more complicated than for elemental materials such as a-Se since in principle both cation-based (e.g. As, Ge) and anion-based (e.g. S, Se) coordination defects can exist; the situation is further complicated by the different types of nearest neighbours each defect can have. Moreover, like-atom bonds themselves also act as defects in heteroatomic systems and can give rise to states in the gap. Vanderbilt and Joannopoulos (1981) have discussed the wide variety of such defects which can occur in glassy As_2Se_3. For some time, this complexity precluded any realistic calculations of the total energies, and hence estimates for U_{eff}, of defects in such compound chalcogenides. However, recently Tarnow *et al.* (1988) have shown theoretically that for *crystalline* As_2Se_3, the Se anti-site defect has a *negative* effective correlation energy, $U_{eff} \simeq -0.3$ eV. The neutral Se anti-site defect arises when a Se atom is substituted for an As atom; its configuration is therefore denoted by $C_3^0(3Se)$, where the three Se nearest neighbours are specified. It was found that it is energetically favourable for two such non-interacting anti-site defects to reconstruct, producing $C_3^+(3Se)$ and C_1^- centres. This finding indicates that chalcogen-based coordination defects in glassy compound chalcogenide probably do indeed have a negative U_{eff}. However, although the Se anti-site defect has a negative correlation energy, the conjugate As anti-site defect was found to have a *positive* correlation energy, $U_{eff} \simeq 0.4$ eV (Tarnow *et al.* 1988).

Subsequently Tarnow *et al.* (1989) have considered the energetics of other types of defect reactions in c-As_2Se_3. In particular, they found that a Se anti-site defect ($C_3^0(3Se)$) and an As anti-site defect ($P_2^0(2As)$, where the symbol P refers to the pnictogen atom, As), situated opposite each other in adjacent layers in the crystal structure, can undergo a facile atomic reconstruction involving a bond-flip and the consequent formation of an interlayer bond (Fig. 6.9(a)); the overall energy is lowered as a result of this bond-flip reconstruction, since the proper coordination of the atoms involved is restored and a stronger heteropolar bond is formed at the expense of a weaker homopolar bond. Such reconstructive transformations could also occur in the glassy state of compound chalcogenide semiconductors, since it is well established that they too can be regarded as being layer-like on a local scale. The particularly interesting feature of the bond reconstruction shown in Fig. 6.9(a) is that it apparently occurs with essentially no energy barrier (see Fig. 6.9(b); Tarnow *et al.* (1989) have suggested that such configurations could be candidates for the two-level (tunnelling) states which dominate the low-temperature thermal properties in chalcogenide (and other) glasses (sect. 4.4).

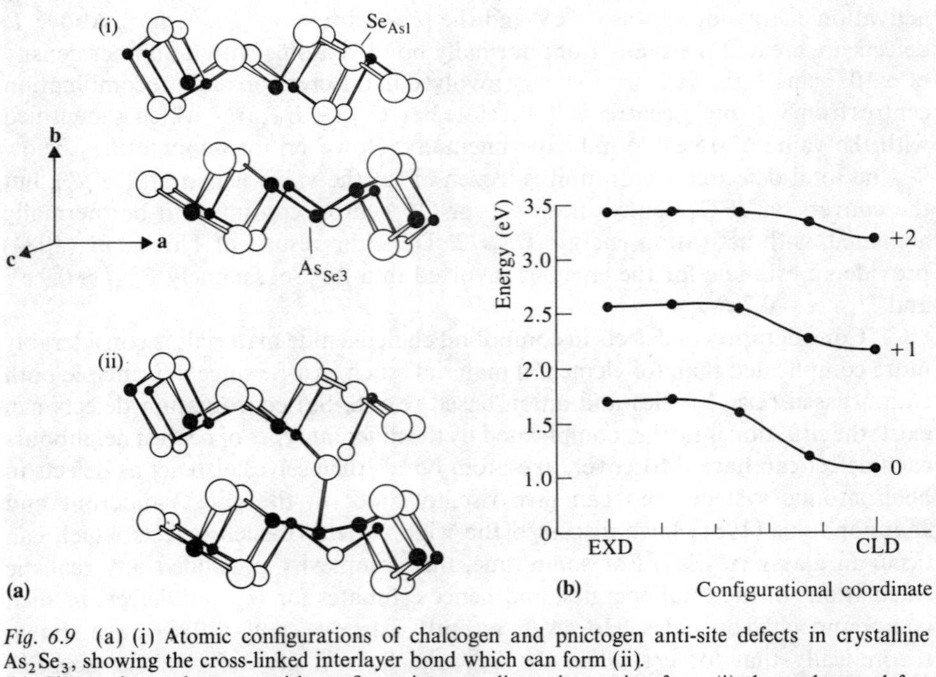

Fig. 6.9 (a) (i) Atomic configurations of chalcogen and pnictogen anti-site defects in crystalline As_2Se_3, showing the cross-linked interlayer bond which can form (ii).
(b) Change in total energy with configuration coordinate in passing from (i) the exchange defect (EXD) to (ii) the cross-linked defect (CLD). Note the absence of an activation barrier (Tarnow *et al.* 1988).

The existence of VAP defects in a-SiO_2 has been suggested by Greaves (1978, 1979) and Lucovsky (1979a,b, 1980), since the oxygen atoms are also chalcogens with a lone-pair valence band maximum. Both models agree that the negatively charged centre is a non-bridging oxygen (C_1^-), but disagree over the nature of the positively charged centre centre, Greaves suggesting T_3^+ and Lucovsky $C_3^+(T_3)$. We will discuss these valence alternation centres in more detail in connection with the radiation-induced, so-called E′ cation centre observed in both vitreous and crystalline silica, together with a rival vacancy model, in section 6.2.3. (The letter E stands for electron centre and the single prime represents the occupancy.) Silica, in both amorphous and crystalline forms, exhibits a plethora of radiation-induced defects (see e.g. Wong and Angell 1976, Griscom 1978, 1985). In addition to the E′ centre in which an electron is trapped at a Si atom, radiation-induced paramagnetic centres are also observed in a-SiO_2 at which a hole is trapped at a charged non-bridging oxygen (C_1^-). These oxygen hole centres may be of two types: in 'wet' silicas containing a high proportion of —OH impurities, the defect is believed to be a $C_1^0(T)$ in the vicinity of a neutral hydroxyl network terminator $>$Si—OH; in high purity 'dry' silicas with a very low OH content, the defect is believed to be a dangling bond on a peroxyl bridge, i.e. $C_1^0(C)$ (Stapelbroek and Griscom 1978, Robertson 1982). In addition to the intrinsic defects mentioned above, there are also many radiation-induced defects associated with impurities, such as Al or B; since these impurity-associated defects give rise to optical absorption bands in the visible, they are often termed 'colour centres' (see Wong and Angell 1976 and Griscom 1978).

We conclude this introductory section on the charged dangling bond model with a discussion concerning the ability of these centres to pin the Fermi level in the gap. We have alluded to this property of negative U_{eff} states previously when discussing Anderson's polaronic mechanism, but the concept is easier to understand for the case of dangling bond defects which give rise to discrete levels (or narrow bands) in the gap of an amorphous semiconductor. Whether U_{eff} is positive or negative affects dramatically the movement of the Fermi level when electrons are injected into the system either by the application of an external electric field (as in the field-effect experiment) or by the addition of dopant atoms. Adler and Yoffa (1976) approached this problem by considering a single defect capable of being unoccupied, singly or doubly occupied by electrons, and used the grand partition function to calculate the Fermi level position as a function of the number of electrons per defect, n $(0<n<2)$:

$$E_F = E_{eff} - k_B T \ln\left\{\frac{2|n-1|}{1-|n-1|}\right\} \qquad 0<n<1 \qquad [6.14(a)]$$

$$E_F = E_{eff} + U_{eff} + k_B T \ln\left\{\frac{2|n-1|}{1-|n-1|}\right\} \qquad 1<n<2 \qquad [6.14(b)]$$

for the case where $U_{eff} > 0$, and

$$E_F = E_{eff} - \frac{|U_{eff}|}{2} - \frac{k_B T}{2} \ln\left\{\frac{2}{n} - 1\right\} \qquad [6.15]$$

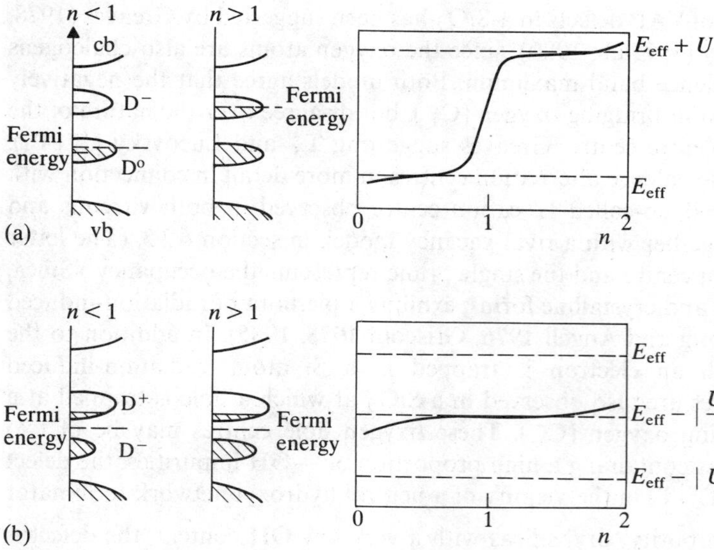

Fig. 6.10 Density of states and Fermi level as a function of electron occupation, *n*, for (a) $+U_{eff}$ defects; (b) $-U_{eff}$ defects (Adler and Yoffa 1976).

for the case where $U_{eff} < 0$. For *positive* U_{eff} materials there is a level at E_{eff} for singly occupied states and lying above it at a separation U_{eff} a level corresponding to doubly occupied centres (see Fig. 6.10(a)). As the electron concentration, *n*, increases from zero at $T = 0$ K, the Fermi level rises slowly in the lower singly occupied level. Note that the density of upper levels is equal to the density of the *filled* lower levels (a state cannot be doubly occupied for a positive U_{eff} material until it is first singly occupied). As soon as $n = 1$, the Fermi level jumps discontinuously across the gap of magnitude U_{eff} to the upper level, and subsequently rises gradually through this level until $n = 2$ and all states are doubly occupied. This behaviour can be seen by reference to [6.14] and is illustrated in Fig. 6.10(a). For *negative* U_{eff} defects, however, the levels corresponding to double occupancy lie *below* those for single occupancy, and the density of lower levels is equal to the number of electrons introduced and thus this band is always full (Fig. 6.10(b)). The injection of electrons into such a material therefore results in immediate double occupancy of levels without an intermediate $n = 1$ stage. Thus, E_F remains pinned midway between the two levels at an energy $E_{eff} - U_{eff}/2$ for all values of *n*, except when *n* is very near 0 or 2, as given by [6.15] and shown in Fig. 6.10(b). Thus in a field-effect experiment, for example, the material *behaves* as if there were a high density of states deep in the gap, pinning the Fermi level there, whereas in reality there are *no* one-electron states at the energy (Frye and Adler 1981) – see Fig. 6.10(b). For a further discussion of the properties of negative U_{eff} charged dangling bond states see section 6.3.1, in particular Fig. 6.32.

6.2.3 Atomic vacancies

Atomic vacancies are a widespread form of point defect in covalent, ionic and metallic crystalline materials. They are responsible for many of the electrical and optical properties of such materials, in addition to frequently controlling mass transport phenomena, such as diffusion, and hence many mechanical properties, e.g. creep (see e.g. B. Henderson 1972). Vacancies can be thermally generated, but are also created by radiation damage processes, and they may occur either singly as monovacancies or as clusters, divacancies, etc. We discuss them here because after the single dangling bond, they are the most simple form of structural defect likely to be found in amorphous materials, and also they necessarily contain in the case of covalent materials a number of dangling bonds, which may then reconstruct through the Anderson mechanism.

Studies on one of the most intensively studied vacancies in crystalline semiconductors, namely the monovacancy in Si, have indicated that this is a negative U_{eff} system in the Anderson (1975) sense. The monovacancy produced by electron irradiation at low temperatures has four sp^3 hybrid dangling bond orbitals pointing into the vacancy, and depending on the position of the Fermi level in the gap (determined by p- or n-type dopants) the vacancy may possess a variety of electron occupancies and hence charge states, the existence of which may be confirmed using ESR (for paramagnetic configurations) and diffusion (for diamagnetic configurations). In this way, the existence of V^{2-}, V^- V^0, V^+ and V^{2+} charge states of the vacancy has been inferred.

Consider first the series of charge states V^{2+}, V^+ and V^0. It has been proposed (Baraff *et al.* 1980a,b) that the ESR-active V^+ charge state should always be metastable and as the Fermi level is moved through the gap by p-type doping, either the V^{2+} or the V^0 centre would be the more stable, the V^{2+} centre being more stable for E_F very close to the valence band edge (i.e. p^+-doping), and the V^0 centre in turn being more stable if E_F lies deeper than $\simeq 0.1$ eV from the valence band edge. The reason for this negative U_{eff} behaviour, as we shall see in a moment, is the Jahn–Teller splitting of certain degenerate electron energy levels accompanying a lattice distortion. Note that this system is of 'pure' Anderson type: either charge state V^{2+} *or* V^0 is stable with respect to V^+ because of lattice distortion and consequent spin pairing, rather than the reaction $2V^+ \rightarrow V^0 + V^{2+}$ being exothermic, creating *pairs* of differently charged entities as in the charged dangling bond model.

The undistorted, 'ideal' monovacancy has four dangling bands centrally directed into the vacancy, and has T_d symmetry; this is the configuration adopted by the V^{2+} centre, where the two electrons occupy in a spin-paired fashion the lowest singly degenerate symmetric A_1 state which lies near the top of the valence band, leaving the triply degenerate T_2 state lying in the gap, unoccupied (see Fig. 6.11(a)). The T_2 level is susceptible to a Jahn–Teller distortion upon its occupation by any electrons, which in this case will be tetragonal with D_{2d} symmetry, and the level splits into a lower T_x state and a degenerate T_y and T_z state. The third electron in a V^+ centre will go into the T_x state, and the fourth in a V^0 centre will also go into the T_x level, further increasing the distortion and pairing the spins (Fig. 6.11(b)). Thus, the vacancy can be regarded as having distorted to give rise to a rebonding of the dangling bond orbitals. The Jahn–Teller distortion is such that as electrons are progressively introduced into the system by movement of the Fermi level through the gap, the V^{2+} centre transforms directly into the V^0 centre, without the production of the intermediate V^+ centre. Direct evidence for this behaviour has come from ESR studies (Watkins and Troxell

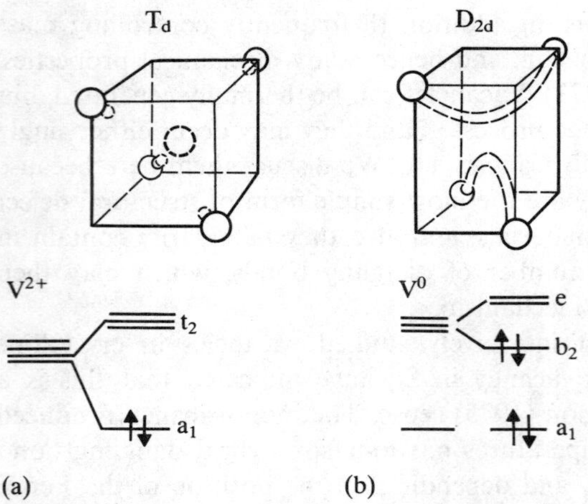

Fig. 6.11 Atomic configurations and level occupancies of a monovacancy in crystalline Si (Robertson 1982):
(a) undistorted V^{2+} centre (T_d symmetry);
(b) tetragonally distorted (D_{2d}) V^0 centre.

1980), and it has also been suggested that the V^0, V^+ and V^{2-} series of charge states is similarly subject to a negative U_{eff}, accompanying a trigonal, rather than tetragonal, distortion (Baraff *et al.* 1980b). It is interesting to speculate whether such Jahn–Teller driven reconstructions might also occur in a-Si and other similar materials.

Vacancy-related defects have been much studied in vitreous SiO_2, and the principal such defect is the so-called E′ centre. In its paramagnetic state, the ESR spectrum of the E′ centre is characterized by a hyperfine doublet, the magnitude of the splitting of which indicates that the unpaired spin is localized on an sp^3 dangling hybrid, interacting only with a single silicon (^{29}Si, $I = \frac{1}{2}$) nucleus (Griscom 1985); it is therefore a T_3^0 centre (see Fig. 6.12). It has been suggested that electronic transitions involving the E′ centre may also be responsible for the mid-gap optical absorption band at 5.8 eV observed in vitreous SiO_2 (see e.g. Greaves 1978). Although the nature of the paramagnetic state of the E′ centre is well understood, this is not the case for its precursor state, which must be diamagnetic and therefore ESR-inactive. E′ centres can be rendered ESR-active (i.e. paramagnetic) only through the interaction of radiation of various types with the silica.

Three types of radiation-induced defect-creation mechanism can be envisaged: displacement (knock-on) damage processes, radiolysis and electronic rearrangement. In the first, the incident particle of radiation (e.g. γ-ray photons or neutrons) must be sufficiently energetic to displace an oxygen or a silicon atom by breaking two or four Si–O bonds, respectively, whereas in the third process (X-ray or UV) radiation causes an electronic excitation, the electrons or holes resulting from which can either be trapped at pre-existing diamagnetic defects, converting them into paramagnetic centres, or can cause (weak) bonds to break, thereby creating a pair of paramagnetic neutral dangling bonds. The radiolysis process is somewhere between these two extremes, in that local atomic displacements can occur, facilitated by electronic rearrangements.

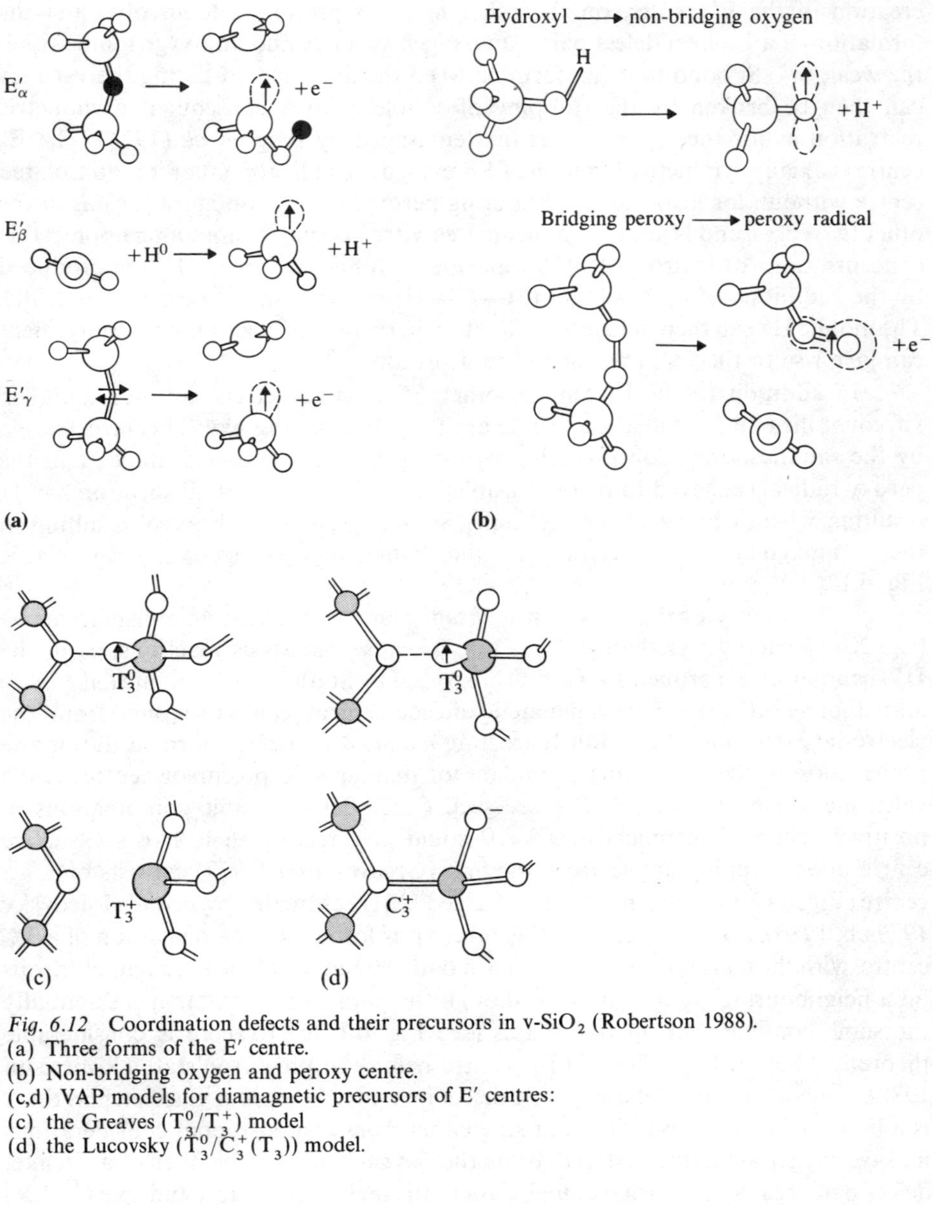

Fig. 6.12 Coordination defects and their precursors in v-SiO_2 (Robertson 1988).
(a) Three forms of the E′ centre.
(b) Non-bridging oxygen and peroxy centre.
(c,d) VAP models for diamagnetic precursors of E′ centres:
(c) the Greaves (T_3^0/T_3^+) model
(d) the Lucovsky ($T_3^0/C_3^+(T_3)$) model.

Griscom (1985) has found that there are two subtle differences in the ESR signatures of the E′ centres in different types of glassy silica, and also when different types of radiation are used; these have been labelled E'_α, E'_β and E'_γ (see Fig. 6.12(a)). The defect process thought to result in the E'_α centre is a radiolysis reaction, involving X-ray absorption, in which an oxygen vacancy is created by hole-trapping, inducing the migration of the oxygen to form a peroxyl group attached to the Si atom which has become the paramagnetic E′ centre; the other Si atom bordering the vacancy (a T_3^+ site) undergoes a relaxation towards an almost *planar* sp^2 configuration (as also postulated for the c-Si (111) surface). The γ-ray-induced creation of the E'_γ centre, on the other hand, is presumed to involve first the formation of a Frenkel defect pair – an oxygen vacancy and an oxygen interstitial; the weak Si—Si bond that can form between the Si atoms bordering the vacancy can then be broken by the trapping of a hole, with a consequent asymmetric relaxation as for the E'_α centre, as first envisaged by Fiegl *et al.* (1974). The E'_β centre is axially symmetric (from the ESR evidence), indicating that it is an isolated centre without, for instance, neighbouring peroxyl or T_3^+ configurations as in the other two cases, and is found in X-irradiated vitreous silicas containing appreciable concentrations of hydroxyl (OH) impurities. Atomic hydrogen (H^0) is produced by the radiolysis of these silanol (Si—O—H) groups (and is detected by ESR). The mobile H^0 can then diffuse to a T_3^+ site, whereupon an electronic rearrangement can give rise to the E'_β (T_3^0) centre and a proton.

In addition to the E′ centres, other ESR-active defects are also found in vitreous silica after irradiation. These are the non-bridging oxygen centre, caused by the radiolysis reaction involving hydroxyl groups mentioned above, and the peroxy radical (believed to be responsible for the 7.6 eV optical absorption band), resulting when a bridging peroxyl unit (Si—O—O—Si) is broken, resulting in the paramagnetic peroxy centre and the diamagnetic (relaxed) T_3^+ centre (see Fig. 6.12(b)).

If the energy of the radiation is *insufficient* to induce atomic displacements (e.g. X-ray energy less than $\simeq 1$ keV) or to cause radiolysis reactions, as in the UV-irradiation experiments on v-SiO_2 reported by Stathis (1988), Imai *et al.* (1988) and Tsai *et al.* (1988), paramagnetic defect centres can arise only from the electronic excitation of previously existing *diamagnetic* defects formed during the preparation of the glass. One candidate for diamagnetic precursor centres is the valence-alternation pair (VAP – see sect. 6.2.2). Two possible combinations of positively charged members of a VAP could give rise to an E′-like ESR-active centre upon trapping an electron: the first (Greaves 1978, 1979) consists of a T_3^+ centre (Fig. 6.12(c)) and a negatively charged NBO centre; in the second (Lucovsky 1979a,b, 1980), a $C_3^+(T_3)$ centre (Fig. 6.12(d)) is formed by the relaxation of a T_3^+ centre, with the consequent formation of a dative bond using the lone-pair electrons on a neighbouring oxygen atom. Although the paramagnetic state has essentially the same configuration in both cases (see Fig. 6.12(c,d)), there is considerable theoretical support for the $C_3^+(T_3)$ centre being the precursor state (Robertson 1988). One other type of diamagnetic defect in a mixed covalent glass such as SiO_2 is a homopolar bond, e.g. Si–Si, arising either from compositional considerations in, say, oxygen-deficient material, or as the oxygen vacancy member of a Frenkel defect pair created in thermal equilibrium in the melt. Using sub-band-gap (6.4 eV) UV radiation from an ArF excimer laser, Tsai *et al.* (1988) have shown that the generation of E′ centres is a two-photon absorption process in this case, and they ascribe the defect creation to the non-radiative, Si—Si bond-breaking recombination of photo-generated excitons.

Vacancy defects can also occur in principle in amorphous *metallic* systems. However, computer simulation studies reported by Chaudhari *et al.* (1983) have shown that, although an atomic vacancy can of course be created in a (DRP) model representing the structure of a metallic glass, it is *not* stable with respect to structural relaxation using a Lennard-Jones potential leading to a lowering of the strain energy; instead, it is found that the void volume essentially becomes distributed almost homogeneously throughout the structure, i.e. as 'free volume'. This instability of vacancies in (metallic) amorphous systems subject to central interatomic interactions is in marked contrast to the results of similar modelling studies involving atomic vacancies in covalent CRN structures subjected to a Keating potential, in which such vacancies were found to be stable with respect to structural relaxation.

6.2.4 Density defects

The structural defects that we will discuss in this section cannot be categorized as point defects, such as dangling bonds, mono- or divacancies, for example. These density defects range from voids (of size say <100 Å), which may be regarded as an aggregate of a non-fixed number of vacancies, to fluctuations in density (or free volume), which may be regarded as vacancies which have become distributed throughout the material and which therefore cannot be regarded any more as well-defined point defects. We have mentioned in passing both of these forms of density defect earlier in this book.

It is rather a moot point whether one ought to regard these density variations as defects in the sense of our previous definition, namely departures from an 'ideal' amorphous structure, or whether they are an inevitable consequence of producing an amorphous solid for the material in question. For example, are voids – which are a universal feature of evaporated a-Si or a-Ge thin films – an inevitable consequence of the fact that these tetrahedrally bonded materials are 'over-constrained' (Phillips 1979 – see sect. 2.5.4), and do they therefore act as strain-relief centres for the excess strain otherwise incorporated into the random network if only perfect tetrahedral coordination is allowed? In addition, we have already seen that in the free-volume model (Cohen and Turnbull 1959 – see sect. 2.3.5) the inclusion of surplus free volume, principally in materials which have a significant degree of non-directional bonding (metals, ionic solids, polymers), is one of the factors that distinguish amorphous materials from their crystalline counterparts and is an unavoidable feature of the amorphous state.

The presence of voids may be detected directly using electron microscopy for relatively large voids, and small-angle scattering (sect. 3.6.2) for smaller sized voids, and indirectly by a variety of techniques including density measurements (e.g. evaporated a-Ge and a-Si can be 3–15% less dense than their crystalline forms, whereas CRN models which are void free only have 1–3% density deficits). The presence of such voids influences to a great extent the physical properties of such materials, and although annealing the samples to elevated temperatures does reduce the void concentration to a certain extent, they cannot be removed entirely before the crystallization temperature T_{cr} is reached. An illustration of the widespread control that voids have on a variety of physical properties is shown in Fig. 6.13, where the unpaired electron spin density, the porosity and the internal stress of evaporated a-Ge films all decrease in a similar fashion on annealing, and these can be associated with a decrease in the void concentration; only the diamagnetic susceptibility retains its enhancement relative to the crystalline value until very close to T_{cr} indicating that it is associated instead with the bulk bonding

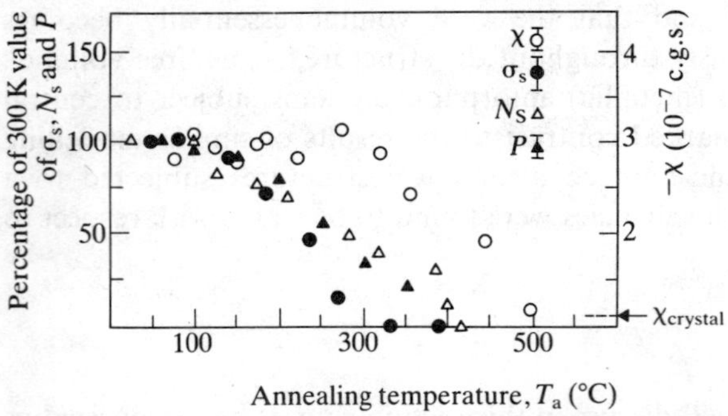

Fig. 6.13 Annealing properties of void-related properties of a-Ge. The properties are porosity (P), free spin density N_s, internal stress σ_s and diamagnetic susceptibility χ (Paesler *et al.* 1974).

states (Paesler *et al.* 1974). It was deduced that for freshly evaporated films, voids with an average radius of $\simeq 30$ Å were present in a concentration of $\simeq 10^{18}$ voids cm^{-3}. The presence of these voids has a profound influence on electronic properties too, because of the dangling bonds that are expected at the void surfaces, even allowing for the fact that many may reconstruct via one or other of the mechanisms mentioned in sections 6.2.2 and 6.2.3. Thus, a.c. conductivity and variable-range hopping d.c. conduction involving these states is observed, as is a large ESR spin signal ($\simeq 10^{19}$ spins cm^{-3}), although isolated dangling bonds in the bulk (sect. 6.2.1) may make a contribution to these behaviours as well. The sensitivity of such evaporated films to changes in ambient gaseous conditions is also understandable, given the high degree of porosity characteristic of void aggregates and thus the ease with which gaseous impurities may interact with the electronically active electron states; this may explain the non-reproducibility of experimental results that bedevilled the field of amorphous semiconductors in the early days.

We turn now to the other form of density defect under consideration, namely those fluctuations in density which represent a distribution of free volume throughout the bulk. We have already dealt in section 2.3.5 with the free-volume concept and how it can account for mass-transport phenomena (e.g. viscosity) in the liquid phase, in particular in being able to account for the Vogel–Tammann–Fulcher relation for the viscosity ([2.8]). These free-volume defects can also control mass-transport phenomena in glasses below T_g, such as diffusion, stress-induced flow (creep) and structural relaxations; Spaepen (1982) and Chaudhari *et al.* (1983) have reviewed these aspects for the case of amorphous metals.

There is much experimental evidence, e.g. from similarities in respective energies, that there is a close similarity, if not identity, between the defects responsible for diffusion and flow, but that the behaviour governing structural relaxation is rather different. For example, the viscosity of metallic glasses increases linearly with time during structural relaxation for $T < T_g$ (Chaudhari *et al.* 1983), indicating that the flow defects responsible for the viscosity are being annihilated at the sites of other types of 'defects' – for example, free volume defects can be annihilated on meeting an appropriate stress concentration in the material. The plausibility of an identity between diffusion and flow defects may be confirmed by consideration of microscopic models for the defects. A diffusion defect must allow a change in nearest neighbours of a given atom, whereas a flow defect, upon the application of a shear stress, must allow a local shear transformation (which is then transferred elastically to the surface of the solid to produce a macroscopic strain). The free volume may be associated with a single atom (see Fig. 6.14(a)), which may hop from one site to another (accomplishing diffusive motion) with a concomitant collapse of the original cage of neighbouring atoms (producing a local shear strain). Alternatively, the free-volume defect may be associated with more than a single atom, and we show schematically in Fig. 6.14(b) the motions of two atoms which can undergo a shear transformation and hence also contribute to diffusion and flow. It is to be noted that these diffusion and flow defects are rather localized, in contrast to relaxation defects which require long-range elastic stress fields to act as sinks in order that they may annihilate free volume to accomplish the irreversible process of structural relaxation (see Fig. 2.1).

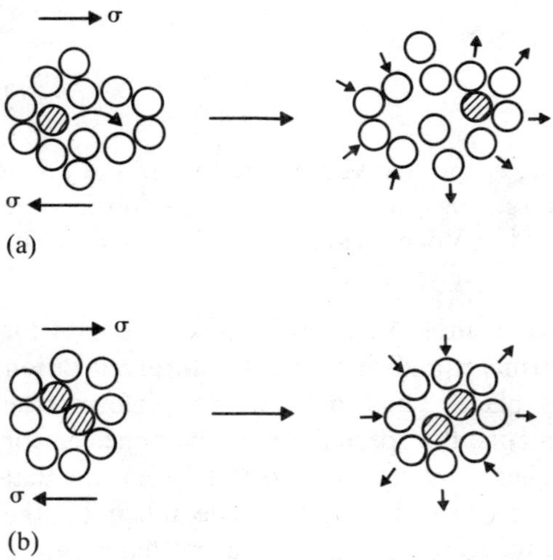

Fig. 6.14 Schematic illustration of the motion of free-volume defects which can give rise to diffusion or flow under the action of a shear stress σ:
(a) single atom motion;
(b) two atom motion.

In order to discuss the influence that volume defects exert on the process of plastic flow we must distinguish between homogeneous and inhomogeneous flow: *homogeneous* flow occurs when each volume element of the sample contributes to the strain, and in amorphous metals this occurs for low stress levels at high temperatures; *inhomogeneous* flow, on the other hand, occurs for very high stress levels at low temperatures, and the strain is localized in a few shear bands. The plastic flow rate of a system subjected to a shear stress σ can be written for the flow defect model as (Spaepen 1981, Chaudhari *et al.* 1983):

$$\dot{\gamma} = n_f \gamma_0 v_0 \nu_f \qquad [6.16]$$

where $\dot{\gamma}$ is the shear strain rate, n_f the concentration of flow defects, v_0 the volume of a defect, γ_0 the local shear strain resulting from the rearrangement of a defect and ν_f the net jump frequency of a flow defect under stress. (Equation [6.16] is in fact completely general, and can equally apply to the case of dislocation motion in crystals.)

Three temperature regimes for homogeneous flow may be distinguished (Spaepen 1981, Chaudhari *et al.* 1983):

(a) $T > T_m$. In the fluid phase, the number of defects is of the order of the number of atoms, and single atom motion is dominant (cf. Fig. 6.14(a)) for which $v_0 \simeq \Omega$ (the atomic volume) and $\gamma_0 \simeq 1$. Furthermore, the activation energy for motion is small because of the minimal structural rearrangements that are necessary, and hence the overall temperature dependence of the viscosity is weak.

(b) $T \simeq T_g$. In this very viscous region ($\eta \simeq 10^{13}$ poise) the dominant parameter in [6.16] is the increase in the defect concentration with increasing temperature. This may be written as:

$$n = \frac{1}{\Omega}\exp\left(-\frac{cv^*}{v_f}\right) \qquad [6.17]$$

where v^* is the minimum hole size, v_f the free volume per atom and c is a geometric factor ($\simeq 1$). This equation, together with the expression for free volume derived in section 2.3.5, yields a Vogel–Tammann–Fulcher form for the viscosity. The flow should still be Newtonian.

(c) $T < T_g$. Behaviour in this region is complicated, since unlike the case for $T > T_g$, the glassy solid is not in thermal equilibrium and structural relaxation to a higher viscosity state takes place. Thus, in order to perform true isoconfigurational creep measurements, the specimen must be annealed for a long time at a temperature higher than that of the test to ensure that structural relaxation does not take place during the time taken by the measurements. Since v_f continues to decrease with decreasing temperature because of the non-configurational part of the thermal expansion, the defect concentration continues to decrease below T_g ([6.17]), and the viscosity should have a non-Arrhenius temperature dependence. Such isoconfigurational experiments are difficult, but have been performed on a number of transition-metal–metalloid glasses (see Chaudhari *et al.* 1983). The results of such experiments indicate that the activation energy of the temperature dependence of the isoconfigurational viscosity is rather large, $\simeq 2$ eV.

At very high stress levels, the strain is confined to the small volumes of shear bands which nucleate at edge cracks and propagate in a direction at 45° to the tensile axis. Fracture occurs along the shear band, and the fracture surfaces after failure display a 'vein' pattern characteristic of the Taylor instability developed when two solid surfaces separated by a fluid layer are pulled apart. These observations can be reconciled within the free-volume model by noting that dilatation at a crack tip will increase the free volume locally and hence dramatically lower the viscosity, rendering it almost liquid-like. This softening mechanism then allows flow on a 45° plane (corresponding to the direction of maximum shear stress), and the deformation can propagate through the material (Polk and Turnbull 1972).

6.2.5 Line defects

Dislocations are a ubiquitous form of line defect in crystalline solids, controlling many of their mechanical properties. It has been suggested that dislocations, perhaps of a generalized form, might also exist in amorphous solids. In particular, the similarity of the morphology of the fracture surfaces of glassy metals subjected to high-stress plastic flow to the slip bands resulting from plastic flow in single crystals suggests that dislocations may be operative in the amorphous phase as well. However, we have seen in the previous section that this morphology can be accounted for by free-volume theory in which high stress softens the material locally on the shear plane by the introduction of extra free volume.

The lack of a long-range periodic lattice in amorphous materials means that it is difficult to conceive of an extended line defect by reference to such a structure; no such problem arises of course with point defects which are defined with respect to the *local* structure. Nevertheless, (dislocation) line defects can be created in amorphous structures following the Volterra method for continuous media, namely by the simple expedient of removing a strip of material with a width equal to an atomic diameter (Chaudhari *et al.* 1983). Two questions then arise: is such a defect stable with respect to structural relaxation, particularly if subject to central forces, and how can such a line defect be detected and characterized in the absence of a periodic lattice if it is indeed stable? Chaudhari *et al.* (1983) have undertaken computer simulation investigations of this problem for defects in amorphous metallic (DRP) structures. They found that, unlike atomic-vacancy point defects (sect. 6.2.3), both screw and edge dislocation defects were stable against structural relaxation. Such line defects were visualized using a 'reference net', that is an ideal 3D lattice (a cubic lattice for an edge dislocation, or a lattice with cylindrical symmetry for a screw dislocation); each atom in the cut but unrelaxed structure is associated with the nearest point on the reference net, and after relaxation of the amorphous structure each point of the net is displaced according to the average displacement of the atom associated with it.

The Burgers vector characteristic of a dislocation in a relaxed amorphous (metallic) structure can also be determined by use of the reference net construction. Chaudhari *et al.* (1983) found that a screw dislocation is unstable with respect to structural relaxation for values of initial displacement b_i smaller than a critical amount, i.e. the final displacement b_f is zero. At the critical value of $b_i = 0.8$ atomic diameter, b_f increases discontinuously, and thereafter increases linearly (Fig. 6.15(a)). This behaviour can be contrasted with that for a screw dislocation in a crystalline matrix, where the critical initial displacement for subsequent stability of the defect in an f.c.c. crystal is only $b_i = 0.3$; moreover, b_f remains constant beyond the threshold (Fig. 6.15(b)).

Rivier (1979) has proposed that an additional type of line defect may exist in amorphous solids. This is termed a 'disclination' and is associated with rotational displacements in contrast with the case of dislocations already mentioned which are associated with translational displacements. A disclination may be viewed as

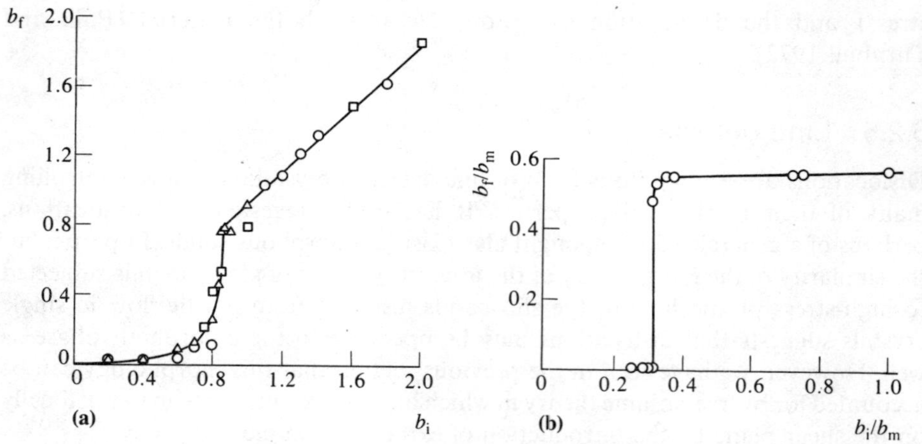

Fig. 6.15 Plot of the final screw displacement (Burgers) vector b_f (after relaxation) versus the initial screw displacement vector b_i (introduced before relaxation) for
(a) amorphous dense random packed structures, and
(b) an f.c.c. crystalline lattice (Chaudhari *et al.* 1983).
(b_m is the atomic diameter).

a procedure for changing the local curvature of a structure. In two dimensions, a disclination *point* may be generated by making a cut in the plane and then either adding or removing matter, thereby creating negative or positive local curvature, respectively. In a 2D hexagonal lattice, therefore, a fivefold ring is the site of a disclination causing positive curvature (Fig. 6.16(a)), and a sevenfold ring is the converse (Fig. 6.16(b)). In three dimensions, the disclination is a *line* and, as with dislocations, there are two types: wedge and screw disclinations (Fig. 6.16(c,d)). In a tetrahedrally bonded amorphous material, e.g. Si, whereas the (diamond cubic) crystalline form consists only of six-membered rings, the amorphous structure contains a large variety of ring sizes, both odd and even (unless precautions are taken to include only even rings, as with the Connell–Temkin (1974) CRN model). By analogy with the situation shown in Fig. 6.16(a,b), the odd-membered rings can be regarded as the cores of disclination lines, which therefore thread the structure, avoiding all six-membered rings and forming closed loops in the body of the structure or terminating at the surface (Rivier 1979). Rivier has further suggested that such disclination defects are found throughout the structure of amorphous solids and could be the origin of many of the universal properties exhibited by glasses, such as two-level systems giving rise to the low-temperature thermal anomalies (sect. 4.4) (see Rivier 1987 for a review).

Although structural features in, for example, tetrahedrally coordinated CRN models of amorphous structures can be identified with the cores of disclination lines, disclination defects arise in another context in connection with non-crystalline structures. Aperiodic structures in (Euclidean) three-dimensional space may be regarded as projections of ordered 'polytopes' in higher dimensions ('curved space') (for a review, see Sadoc and Rivier 1987). As shown by Sadoc and Mosseri (1985), such polytopes can be 'decurved', i.e. projected onto Euclidean space, by iteratively introducing sets of disclination lines. As a consequence of decreasing the local

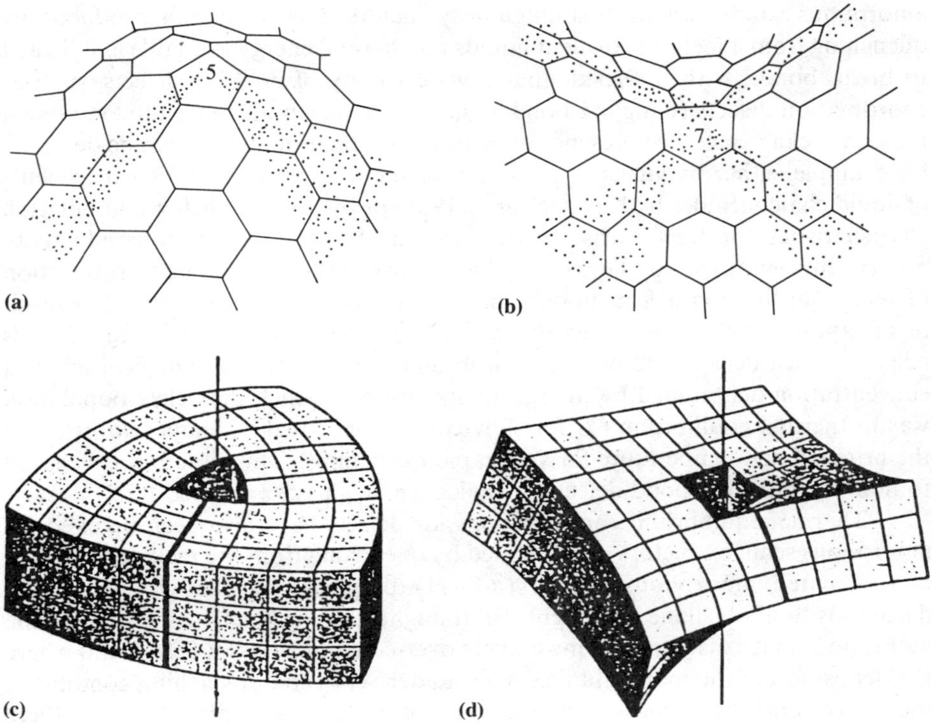

Fig. 6.16 Types of disclination in two dimensions for a honeycomb lattice (a, b), and in three dimensions, showing (c) a wedge disclination and (d) a screw disclination.

curvature, disclinations produce changes in the local structure and topology: if the disclination lines pass through *atoms*, the coordination number increases; if the lines pass instead through *interstices*, the size of the interstices (namely the number of atoms surrounding them) increases, whilst the atomic coordination number remains unchanged.

6.2.6 Thermodynamic aspects of defects

In crystals, intrinsic defects (e.g. Schottky or Frenkel pairs) are in thermal equilibrium with the host structure. The thermal generation of such defects is in general favourable because of the entropy of mixing so gained. Thus, for instance, surfaces, both external and internal (at grain boundaries, dislocation cores, etc.), of ionic crystals are electrically charged owing to the thermal generation of defects (e.g. interstitial ions) at such surfaces. In principle, there is no reason why such thermal generation of defects in the solid state should not occur in amorphous materials as well, although the higher entropy of the amorphous state with respect to the crystal would tend to render the process less favourable. However, in practice it would be difficult to distinguish such thermally generated defects from those arising from the topological disorder.

Nevertheless, one instance in which the thermal generation of defects in amorphous solids can be unambiguously identified is in glasses produced by quenching from a melt. In covalent liquids the thermal energy available is sufficient to break bonds, with the result that a wide variety of bonding defects, such as coordination defects (dangling bonds) and homopolar bonds in multicomponent materials, can occur. The presence of neutral coordination defects can be detected, for example, either by virtue of their paramagnetism (as in the ESR experiments of liquid chalcogenides by Kawazoe *et al.* 1988) or by probing their recombination characteristics for free charge carriers (as in the transient photoconductivity experiments on As_2Se_3 by Thio *et al.* (1984) – see sect. 6.2.2). Although a proportion of such coordination defects may be paramagnetic in the liquid state (as a result of an appreciable reverse reaction in [6.12]) if the correlation energy U_{eff} is negative, such defects will be frozen-in in diamagnetic form on vitrification, at a concentration determined by the last temperature at which the defect population was in thermal equilibrium (T_g for slow cooling or, possibly, temperatures up to the original melt temperature for very rapid quenching). This situation is believed to hold for dangling-bond defects in chalcogenide glasses (see sect. 6.2.2).

Thermal-equilibrium considerations for defects have also been invoked for amorphous semiconductors not obtained by quenching from the melt, for example hydrogenated amorphous silicon (a-Si:H) prepared by the glow-discharge decomposition of silane. It might be thought that in such cases, where the amorphous matrix is rigid and most likely over-constrained (sect. 2.5.4), and where the deposition of the material takes place under very rapid quenching conditions, thermally controlled processes would not play a dominant role. However, there is now a substantial body of experimental evidence in support of thermal-equilibrium processes controlling electronically active defects and the density of states in a-Si:H (see e.g. Street *et al.* 1987a,b).

One theoretical approach which treats the entire electronic structure of amorphous semiconductors in general thermodynamic terms has been proposed by Bar-Yam *et al.* (1986) and Wang *et al.* (1988) (see also Adler 1987), in which the atomic structure and the electronic structure are interdependent: energetics determine the structure, which in turn controls the electronic density of states. The starting point is an ideal disordered network (not necessarily a CRN), with optimally coordinated atoms and equal bond lengths and angles, which sets the zero of energy for electronic excitations; such a hypothetical structure need not (and for large coordination numbers, cannot) be constructed in flat (Euclidean) three-dimensional space. Defects, such as coordination defects, wrong bonds or even deviant bond lengths or angles, can then be (thermally) generated in the ideal network, thereby 'decurving' it onto normal 3D space, and it is assumed that there is a temperature, T^*, below which such structural deviations are frozen. (One possible value for T^* for overall defect creation might be the film deposition temperature, normally about 250°C for a-Si:H (Street 1982); experimental results for various electronic properties of a-Si:H indicate that $T^* \simeq 100$°C, depending on the type of doping, for rearrangement reactions involving existing defects (Street *et al.* 1987a,b).

Thus, the electronic DOS above, say, the valence-band edge is determined by the energy required by the system to create, say, a filled electronic state at energy $E - E_v$ from the valence band. If the (free) energy involved is $F_v(E - E_v)$, the valence-band DOS therefore must have a tail decaying as $\exp[-F_v(E-E_v)/k_B T^*]$, since thermal equilibrium is lost at T^* and a Boltzmann factor determines the thermal excitation probability for $T \geqslant T^*$. Since $E = F(n) - F(n-1)$, where n is the number of electrons and F the total energy of the system, it can be shown that (Bar-Yam *et al.* 1986)

$$F_v(E-E_v) = (1+\delta_v)(E-E_v)$$

where δ_v is the fractional change in $F(n-1)$ due to a structural fluctuation raising the energy of the unoccupied state by $E - E_v$. Thus, the valence-band DOS can be written as

$$g_v(E) = g^* \exp[-(E-E_v)/k_B T_v] \qquad [6.18]$$

where $T_v = T^*/(1+\delta_v)$ and g^* is the value extrapolated to the band edge of the ideal network. Similarly, it can be shown that, for the conduction band, the tail-state DOS is given by

$$g_c(E) = g^* \exp[(E-E_c)/k_B T_c] \qquad [6.19]$$

where $T_c = T^*/\delta_c$. This model therefore predicts *exponential* band tails, arising from relatively minor structural deviation such as fluctuations in bond lengths or angles, in agreement with experimental observations for a-Si:H, for which the characteristic temperatures are found to be $T_c \simeq 300$ K and $T_v \simeq 500$ K (Tiedje 1984).

Similar considerations can be applied to coordination defects as well. If it is assumed that $U_{eff} \simeq 0$ (Bar-Yam and Joannopoulos (1986) have shown that the effective correlation energy for a-Si:H should be rather insensitive to disorder), the thermal-equilibration model of Bar-Yam *et al.* (1986) and Wang *et al.* (1988) predicts that at $T \geqslant T^*$ the concentration of neutral dangling-bond defects will be given by

$$n(T_3^0) = 2n_0 \exp(-F_d/k_B T) \qquad [6.20]$$

where F_d is the formation energy of a T_3 defect, and n_0 is the Si atomic density. The formation energy of the charged defects T_3^+ and T_3^- depends on the Fermi level (Wang *et al.* 1988):

$$F_{d^-} = F_d + \varepsilon^- - E_F \qquad [6.21(a)]$$

$$F_{d^+} = F_d - \varepsilon^+ + E_F \qquad [6.21(b)]$$

where ε^+ and ε^- are the thermal energies required to convert the charged centres from T_3^0. The effective correlation energy is given by $U_{eff} = \varepsilon^- - \varepsilon^+$, so if $U_{eff} \simeq 0$ then $\varepsilon^+ \simeq \varepsilon^- = \varepsilon$. The *total* concentration of defects will be frozen-in for $T \leqslant T^*$, but electronic interconversion between the various charge states can still take place. It is assumed that the material is inhomogeneous, so that the values of ε vary throughout the sample; in certain regions where ε is small, the formation energy of T_3^- centres is small, and in those regions where ε is large, the formation energy of T_3^+ centres is also small; these defects can therefore exist in thermal equilibrium. Thus, a gap-state density of states is predicted, as shown in Fig. 6.17 (Adler 1987, Wang *et al.* 1988), with two peaks in the DOS expected either side of E_F. The

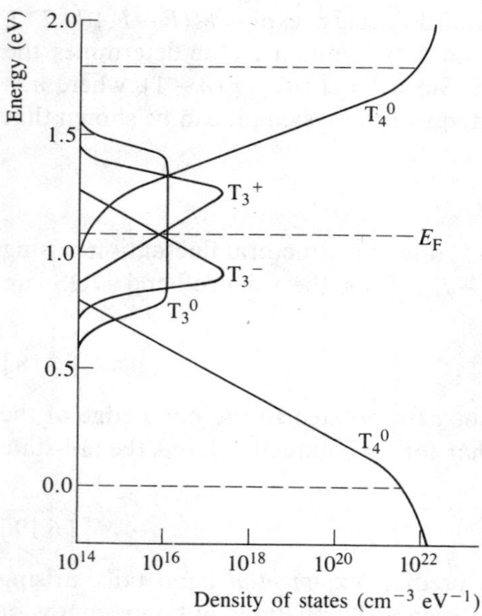

Fig. 6.17 Schematic illustration of the density of electron states for undoped a-Si:H at a temperature greater than the equilibration temperature T^* (Adler 1987).

peak representing (filled) T_3^- centres is predicted to lie *below* that representing (empty) T_3^+ centres, which is analogous to the behaviour of *negative* U_{eff} centres (see Fig. 6.10), even though here $U_{eff} \geqslant 0$; however, in the thermal-equilibrium picture the DOS does not change with electron occupancy or excitation, as it does in the negative-U_{eff} model (see Fig. 6.10).

Bar-Yam *et al.* (1986) used their thermal-equilibrium model to show that the T_3^- levels lie *below* the T_3^+ levels, without having to invoke a negative U_{eff}, thus satisfactorily resolving a previously puzzling paradox about the features of the DOS of a-Si:H (see e.g. Adler 1987, Branz 1989). The optical gap (of intrinsic a-Si:H) appears to be of the order of 1.8 eV, and in n-type material (where only T_3^- centres exist if E_F lies above the defect levels and near E_c) an optical absorption peak is also seen at $E_c - 1.1$ eV, corresponding to the transition $T_3^- \rightarrow T_3^0 + e_{cb}$, whereas in p-doped material a peak is seen near $E_v + 1.3$ eV (i.e. $E_c - 0.5$ eV) and has been ascribed to $T_3^+ \rightarrow T_3^0 + h_{vb}$. Thus, from this experimental evidence it can be inferred that T_3^- centres lie some 0.6 eV below the T_3^+ (or equivalently T_3^0) centres, i.e. $U_{eff} \simeq -0.6$ eV, even though ESR measurements have indicated that the correlation energy of the paramagnetic defects (presumably T_3^0) is *positive*: $U_{eff} \simeq 0.25$ eV (Dersch *et al.* 1981, Stutzmann *et al.* 1987). Previous attempts to explain this paradox have assumed that the electronic levels of charged dangling bonds shift by about 0.5 eV deeper into the gap in doped a-Si:H as a result of pairing of the defects with ionized dopants, e.g. $T_3^- - P_4^+$ and $T_3^+ - B_4^-$ (Kočka 1987). However, the fact that optical transitions obey the Franck–Condon principle (i.e. that the electronic transitions occur at fixed configuration) means that relaxation at the defects is neglected; if the relaxation energies of $\simeq 0.25$ eV (Bar-Yam and Joannopoulos 1986) are considered, the observed experimental optical absorption data can be reconciled with small but positive values of U_{eff} (Branz 1989).

Thermal equilibration ideas have also been applied specifically to coordination defects (dangling bonds) in a-Si:H. Smith and Wagner (1985) and Smith *et al.* (1986) have proposed that the T_3^0 dangling-bond defects observed in intrinsic a-Si:H at room temperature arise from a quenched-in distribution of such centres, created in thermal equilibrium by the breaking of (weak) Si—Si bonds by non-radiative recombination of thermally generated electron–hole pairs at an elevated annealing (or preparation) temperature. (This model of bond scission is identical to that proposed by Stutzmann *et al.* (1985) to explain the photo-generated increase in the ESR spin density in a-Si:H associated with the Staebler–Wronski (1980) effect, except that there the recombining electrons and holes are photo-generated.) In addition, it is assumed that there is a competing (first-order) temperature-activated annealing process in which such thermally generated defects reconstruct as a result of thermal excitation of electrons out of the dangling-bond levels. Thus, the overall kinetics of the dangling-bond concentration N are given by (Smith *et al.* 1986):

$$\frac{dN}{dt} = ckn_i^2 - \nu N f(E)\exp(-E/k_B T) \qquad [6.22]$$

where c is the fraction of electron–hole recombinations producing T_3^0 centres, k is the rate constant for non-radiative inter-band transitions, $n_i(T)$ is the temperature-dependent intrinsic carrier concentration, ν is an annealing attempt frequency (of the order of a phonon frequency) and $f(E)$ is the distribution of annealing activation energies, E. If the quenching rate is sufficiently high, therefore, the annealing term in [6.22] cannot maintain the dangling-bond density in dynamic equilibrium, and an excess spin density is correspondingly frozen in.

Smith *et al.* (1986) have evaluated [6.22], taking $f(E)$ to be a gaussian function centred at $E_0 = 1.1$ eV with a width of 0.1 eV, assuming that the temperature dependence of the intrinsic carrier density is given by

$$n_i^2(T) = (T/T_0)^3 n_i(T_0)\exp(-E_g/2k_B T)$$

where T_0 is some reference temperature (e.g. 300 K), and using the value for the rate of recombination-induced bond breaking $ck = 1.5 \times 10^{-15}\ \text{cm}^3\,\text{s}^{-1}$ (Stutzmann *et al.* 1985); the quenching curves for N are as shown in Fig. 6.18. It can be seen that the calculated dangling-bond densities, even for moderate cooling rates, lie appreciably above the equilibrium spin-density curve, $N_s^{eq}(T)$, which is thermally activated with an activation energy given approximately by $E_a = E_g - E_0 \simeq 0.6$ eV (cf. [6.22]). ESR experiments on rapidly cooled samples of intrinsic a-Si:H indeed show an enhancement of the spin density after quenching (Smith *et al.* 1986).

Thermal equilibration involving defects has also been invoked in a discussion of electrical doping in amorphous semiconductors by electronically active impurities. The doping behaviour in a-Si:H is very different from that found for substitutional doping in *crystalline* semiconductors, e.g. c-Si (see e.g. Street 1985 for a review). For instance, in a-Si:H the doping efficiency is very low (i.e. the fraction of P atoms acting as donors is only about 1%, whereas it is 100% for c-Si).

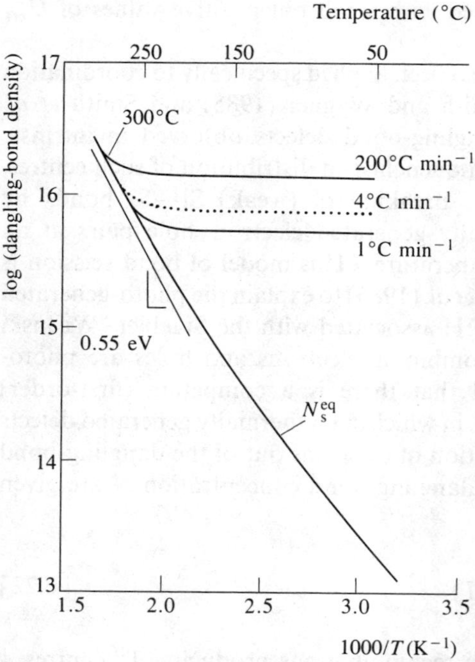

Fig. 6.18 Calculated temperature dependence of the dangling-bond density in a-Si:H as a function of quenching rate according to the model of Smith *et al.* (1986), showing the predicted metastable concentrations, compared with the equilibrium spin density N_s^{eq}.

Furthermore, the doping of a-Si:H causes an *increase* in the dangling-bond density, as monitored by e.g. optical absorption, luminescence and the ESR line at $g = 2.0055$ (corresponding to T_3^0 centres), as well as the expected increase in the density of occupied band-tail states resulting from carriers from ionized donors or acceptors, as monitored by charge sweep-out experiments (Street and Zesch 1984) or by the ESR lines at $g = 2.0044$ and 2.0085, corresponding to conduction band-tail and valence band-tail states (for n- and p-type doping), respectively (Street 1985, Stutzmann *et al.* 1987). Figure 6.19 shows this doping dependence for both dangling-bond and band-tail states. In addition to these ESR-active centres, a small proportion of *unionized* donors (e.g. P_4^0 or As_4^0 centres) can also be distinguished at low temperatures via the hyperfine splitting observed in the ESR spectra that results from unpaired electrons localized on the dopant atoms (Stutzmann *et al.* 1987). Thus, doping in a-Si:H appears to be associated with the simultaneous creation of (dangling-bond) defects.

It was originally believed that doping in amorphous semiconductors would not be favourable at all because all atoms (including impurities) would tend to adopt an atomic coordination in accordance with their normal valence (the so-called $8 - N$ rule (Mott 1967)) in the absence of crystallographic constraints, and consequently dopants would not occupy substitutional sites and be electronically active. This is a dominant effect, as can be seen from the fact that most phosphorus impurity atoms, say, retain non-doping P_3^0 configurations.

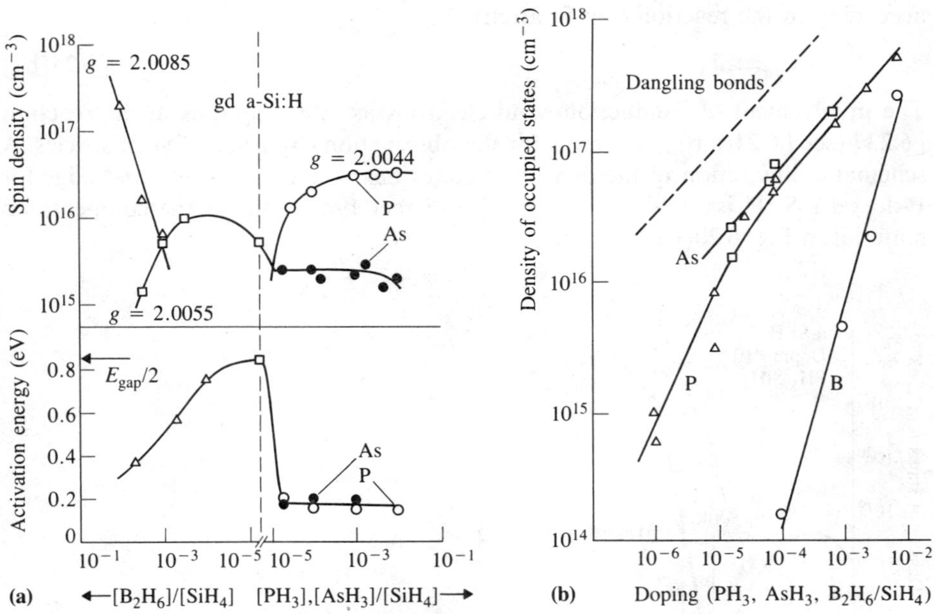

Fig. 6.19 Doping-induced defect creation in a-Si:H.
(a) ESR spin densities and dark-conductivity activation energies as a function of both p-type and n-type dopant (gas-phase) concentrations (Stutzmann *et al.* 1987).
(b) The occupied tail-state densities obtained from charge sweep-out measurements as a function of P, As or B dopant concentrations, compared with the dopant-induced dangling-bond density (Street 1985).

Street (1982) proposed a model for the doping process in a-Si:H which extends the $8-N$ rule to the charged states of the dopants and compensating configurations of the host matrix. Noticing that the ionized state of say the n-type donor, phosphorus (P_4^+), has the same electronic configuration as a normally bonded silicon atom (Si_4^0), he suggested that either the impurity atoms obey the $8-N$ rule in an uncharged state, i.e. the majority of phosphorus atoms adopt a normally bonded, non-doping P_3^0 configuration, or they exist as ionized donors; the unionized donor state is therefore predicted to be energetically unfavourable. The incorporation of phosphorus dopant atoms can therefore be represented by the reaction (Street 1982)

$$Si_4^0 + P_3^0 \rightleftharpoons Si_3^- + P_4^+ \qquad [6.23]$$

i.e. the charged donor state is compensated by the creation of a (negatively charged) dangling-bond defect.

This model was subsequently extended by Street *et al.* (1985) to incorporate an intermediate role played by (conduction-band) tail states. It was assumed that there is an equilibrium between normally bonded P atoms and charged donor configurations with the electrons released into the conduction band:

$$P_3^0 \rightleftharpoons P_4^+ + e_{cb} \qquad [6.24(a)]$$

The conduction-band electron can then subsequently cause a Si—Si bond to break according to the reaction (per Si atom)

$$Si_4^0 + e_{cb} \rightleftharpoons Si_3^- \qquad [6.24(b)]$$

The involvement of conduction-band electrons as intermediaries in the reaction [6.23] (see [6.24(a,b)]) accounts for the observation (by ESR) of such species. A schematic illustration of the density of states near the conduction-band edge for P-doped a-Si:H is shown in Fig. 6.20(a) (Street 1985) and for the compensated material in Fig. 6.20(b) (Stutzmann *et al.* 1987).

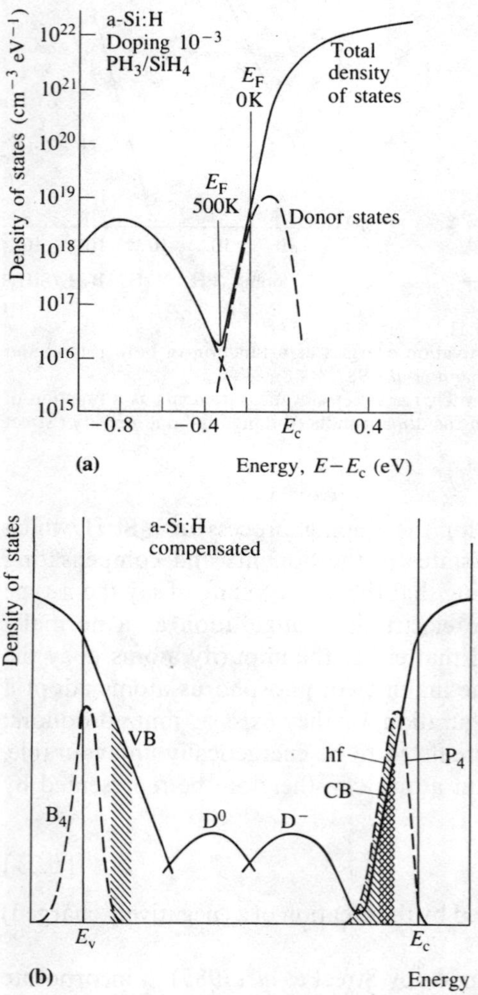

Fig. 6.20 (a) Density of electron states near the conduction-band edge for P-doped n-type a-Si:H, showing the donor-state distribution and the positions of the Fermi level at $T = 0$ and 500 K (comparable to the deposition temperature) (Street 1985).
(b) Schematic density of states for compensated a-Si:H in the gap region between the valence band (VB) and conduction band (CB), showing the active donor (P_4) and acceptor (B_4) bands, together with the proportion of the neutral donor states responsible for the hyperfine-split ESR signal (hf); the hatched regions show the electronic states contributing to the light-induced ESR (Stutzmann *et al.* 1987).

Applying the law of mass action to the equilibrium given in [6.24(a,b)] leads to (Street *et al.* 1985):

$$p_3 = \alpha p_4 n_e \quad [6.25(a)]$$

$$N_0 n_e = \beta d \quad [6.25(b)]$$

where p and d are the concentrations of the corresponding P-related and Si dangling-bond states, respectively, n_e is the density of conduction-band electrons, N_0 is the total Si atomic density, and α and β are rate constants. In addition, charge neutrality dictates that

$$p_4 = n_e + d \quad [6.26]$$

Solving [6.25] and [6.26] leads to (Street *et al.* 1985)

$$n_e = (\beta/N_0)d = p_3^{1/2}[\beta/\alpha(N_0+\beta)]^{1/2} \quad [6.27]$$

Thus, this model predicts that both the dopant-induced dangling-bond density d and the conduction-band electron density n_e should vary as $p_3^{1/2}$ (or approximately as $p^{1/2}$, where p is the total phosphorus concentration. This behaviour is observed experimentally if the P content is taken to be the *gas-phase* dopant concentration (see Fig. 6.19(b)), but the correlation does not seem to hold if the dopant concentration in the solid phase is considered instead (Stutzmann *et al.* 1987).

In the doping model for a-Si:H proposed by Street (1982) and Street *et al.* (1985), it is assumed that thermal equilibration exists between the various dopant-related species and Si dangling-bonds ([6.23] and [6.24]) at the surface of the growing film as it is being deposited at elevated temperatures ($\simeq 250°C$). Although it explains several novel features observed experimentally, such as the dangling bonds created on doping and the square-root dependence on dopant concentrations (Fig. 6.19(b)), the model would predict that this correlation would exist for the actual dopant concentration incorporated into the a-Si:H film rather than the concentration in the gas phase. Thus, the precise mechanism for doping in a-Si:H is still somewhat unclear, although Stutzmann *et al.* (1987) have recently suggested that the dopant-related charge-induced bond breaking, leading to dangling-bond creation, occurs only at *weak* Si–Si bonds, which represent only a small proportion of the total and have bonding and antibonding states in the tails of the valence and conduction bands, respectively.

Our final discussion of thermal equilibration processes also involves a-Si:H. Street *et al.* (1987a) have shown that the electronic properties of n-type or p-type a-Si:H exhibit thermal equilibration characteristics at temperatures below the film deposition temperature. As an example, we show in Fig. 6.21 the behaviour of the d.c. conductivity, from which it can be seen that above a certain temperature ($T_e = 130°C$ for n-type material and $T_e \simeq 90°C$ for p-type material), the material is characterized by a unique value of σ_{dc}, whereas below T_e different behaviour is observed depending on the cooling rate from T_e. Similar behaviour has also been observed in the band-tail charge concentration $n_{bt} = n_d - n_{db}$, where n_d is the ionized dopant (donor or acceptor) concentration and n_{db} is the dangling-bond density; n_{bt} has been monitored by charge sweep-out measurements (Street *et al.* 1987a).

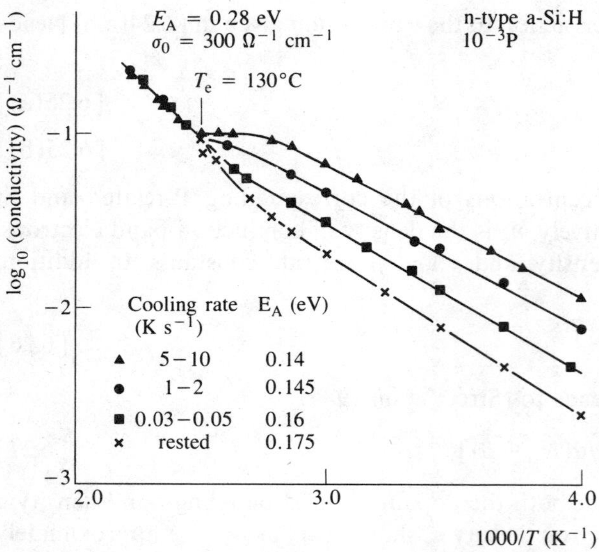

Fig. 6.21 Quenched-in excess d.c. conductivity as a function of temperature for n-type a-Si:H (Street *et al.* 1987).

The behaviour illustrated in Fig. 6.21 is characteristic of that exhibited by a *glass*: (structural) relaxation processes which are in thermal equilibrium above a certain temperature (the glass-transition temperature) in the liquid state, and kinetically controlled processes which control the behaviour at lower temperatures, in the glassy state. Further evidence for 'glass-like' behaviour of the electronic properties of a-Si:H at low temperatures comes from studies of the time-dependent relaxation of the density n_{bt} of occupied shallow band-tail states in n-type a-Si:H samples rapidly cooled to room temperature from an annealing temperature 210°C above T_e, thereby freezing-in a non-equilibrium concentration of tail-state carriers. Kakalios *et al.* (1987) found clear evidence for stretched-exponential (Kohlrausch–Williams–Watts) behaviour (see Fig. 6.22(a)):

$$\Delta n_{bt}(t) = n_0 \exp[-(t/\tau)^{\beta}] \qquad [6.28]$$

Behaviour of this type is generally characteristic of relaxation processes in glassy materials (see sect. 2.3.3). In the light of these experimental results, Street *et al.* (1987a) and Kakalios *et al.* (1987) have proposed the 'hydrogen–glass' model for a-Si:H, wherein the Si framework is regarded as being rigid, but the hydrogen sub-lattice is relatively free to move at elevated temperature $T > T_e$, but becomes a slowly relaxing frozen-in state for $T < T_e$.

The hydrogen diffusion coefficient D_H, measured using a deuterium tracer monitored by SIMS, has an activation energy of 1.4–1.5 eV in intrinsic a-Si:H, and somewhat lower values in doped material $E_a \simeq 1.2$–1.3 eV; interestingly, p-type or n-type doping causes an increase in the value of D_H by up to three orders of magnitude (Street *et al.* 1987b). However, it is found that D_H is observed to be *time-dependent* when samples are annealed at a fixed temperature; this behaviour follows an approximate power-law dependence (Kakalios *et al.* 1987):

$$D_H \simeq D_H^0 (\nu t)^{-\alpha} \qquad [6.29]$$

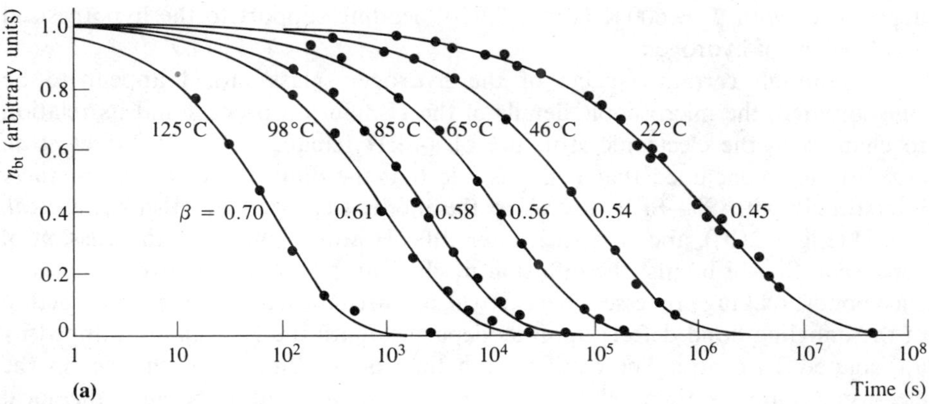

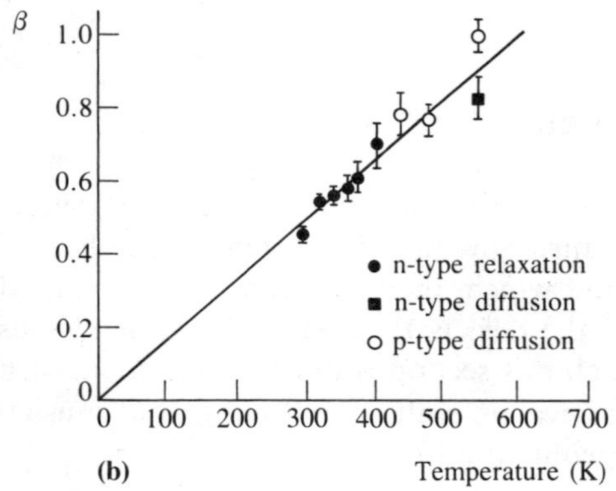

Fig. 6.22 (a) Kohlrausch–Williams–Watts behaviour of the annealing-time dependence of shallow occupied band-tail states in n-type a-Si:H for various temperatures.
(b) Temperature dependence of the Kohlrausch exponent β for n-type a-Si:H obtained from the data shown in (a), (●). The other data points are obtained from the time dependence of the H diffusion coefficient for n-type (■) and p-type (○) material (Kakalios *et al.* 1987).

where v is an attempt frequency and $\alpha \simeq 0.2$–0.5 at 200°C. Stretched-exponential relaxation behaviour (cf. [6.28]) arises when the rate law, for example

$$\frac{d\Delta n_{bt}}{dt} = k\Delta n_{bt} \qquad [6.30]$$

has a time-dependent rate constant $k(t)$ obeying a power law. If it is assumed that H migration is responsible for the relaxation of band-tail carriers, then $k(t) \propto D_H$, with D_H given by [6.29]. Thus, integrating [6.30] with this assumption gives a stretched-exponential expression as in [6.28] with $\beta = 1 - \alpha$. If the dispersion in D_H arises from an exponential distribution of site energies $p \propto \exp(-E/k_B T_0)$, then $\beta = 1 - \alpha = T/T_0$. The data for the band-tail carrier relaxation kinetics and the dispersive hydrogen diffusion both show this predicted linear temperature dependence with $T_0 \simeq 600\,\text{K}$ (Fig. 6.22(b)), lending support to the hypothesized involvement of hydrogen.

Although certain aspects of the hydrogen–glass model* appear to be substantiated, the microscopic details of the H-diffusion process and its relation to changes in the electronic structure of a-Si:H remain uncertain. Street *et al.* (1987b) have concluded that it is possible that the diffusing hydrogen interacts interstitially with Si—Si bonds, as in the model suggested by Hishikawa *et al.* (1987) (Fig. 6.2(b)), and that such interstitial H atoms can cause the scission of very weak Si—Si bonds; the diffusion mechanism of H then involves a series of such bond-breaking processes on its arrival, followed presumably by reconstruction of the dangling-bond defects upon its departure until the H is finally trapped by an isolated T_3^0 centre. The equilibration time of the kinetics of changes in the electronic states is then related to the H diffusive motion because the local rearrangements leading to equilibrium are facilitated by the presence of the hydrogen.

6.3 Defect-controlled properties

So far, we have described some of the more common forms of structural defects that can occur in amorphous solids. We turn now to a discussion of the physical properties determined by such defects and which, therefore, can in turn be used as the bases for experimental probes of the defects themselves. For amorphous semiconductors and insulators, with which this section is mostly concerned, such properties are generally optical or electrical in nature, reflecting the unusual electronic characteristics of many defect configurations.

6.3.1 Gap-state spectroscopies for amorphous semiconductors

There are a number of techniques for investigating not just the presence of (electronically active) defect states, but also their distribution in energy with respect to the gap of an amorphous semiconductor. Since hydrogenated silicon is the most important amorphous material technologically, many of the relevant results have been obtained for it, since the defect population is often the limiting factor in determining its electronic performance. Some of the techniques to be discussed here are generally applicable to all amorphous semiconducting (insulating) materials (i.e. optical absorption), whilst others, notably the junction techniques, require the amorphous material to have such a low background density of states in the gap that the Fermi level can be moved freely relative to the band edges, which restricts the applicability of such techniques essentially to a-Si:H.

A. Optical absorption spectroscopy

Measurement of the optical absorption as a function of photon energy is a versatile technique since the density of electron states can be probed both at the band edges and, in principle, deep within the gap, although sensitivity is a problem for materials with very low absorption coefficients if conventional absorption spectroscopy is used. Interband transitions, giving rise to a power-law behaviour of the optical absorption coefficient $\alpha(\omega)$ for photon energies comparable to the energy gap E_g (or equivalently for values of $\alpha(\omega) \geqslant 10^4\,\mathrm{cm}^{-1}$) have already been discussed in sect. 5.6.2. Instead, we commence this section with a discussion of the absorption at lower energies.

If the interband absorption mechanisms discussed in sect. 5.6.2 were the only ones operating, we would expect a rather sharp 'absorption edge' marking the onset of optical transitions (at an energy corresponding, say, to $\alpha(\omega) \simeq 10^4\,\mathrm{cm}^{-1}$), much as is observed in crystalline semiconductors. Instead, amorphous semiconductors almost always exhibit an absorption edge which is considerably less sharp and which consequently tails well into the gap, as shown schematically in Fig. 5.29, and which invariably obeys an exponential dependence on photon energy:

$$\alpha(\omega) = \alpha_0 \exp[-\Gamma(E_0' - \hbar\omega)] \qquad [6.31]$$

where E_0' is an energy comparable to the threshold energy E_0 involved in interband transitions ([5.124]) and Γ is a temperature-dependent constant, typically having values in the range 10–25 eV^{-1}. Similar behaviour is also observed in some crystalline solids, notably the silver halides, where such exponential absorption edges are termed 'Urbach edges' after their discoverer; some examples, primarily for the chalcogenide glasses, are shown in Fig. 6.23.

The temperature variation of the Urbach edge in amorphous materials is rather complicated, but somewhat resembles that found in crystalline solids. For temperatures between, say, $\simeq 77$ K and T_g, the slope Γ is temperature-independent, although the edge itself shifts in a parallel fashion to lower energies with decreasing temperature, presumably reflecting the underlying temperature shift of the optical gap E_0. Below $\simeq 77$ K, the edge becomes practically temperature-independent, while above T_g the T^{-1} temperature dependence of Γ, as observed in crystalline materials above room temperature, is also found in amorphous materials.

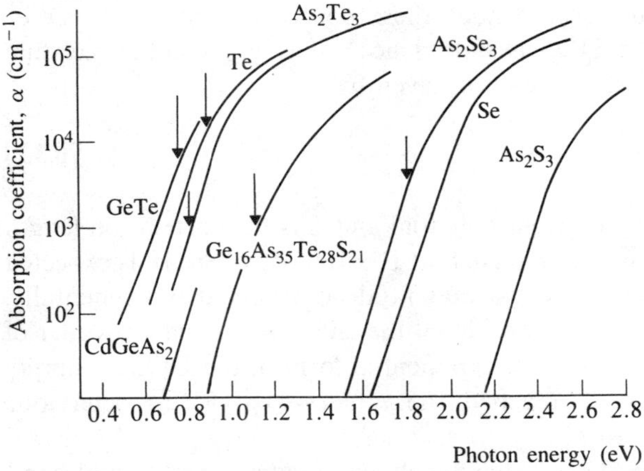

Fig. 6.23 Optical absorption edges for a variety of amorphous semiconductors at room temperature showing Urbach behaviour. The arrows mark the value of $2\Delta E_\sigma$ for those materials for which the electrical conductivity is activated with energy ΔE_σ (after Mott and Davis 1979).

The seemingly ubiquitous observation of exponential absorption edges in disordered materials implies that there might be an underlying common mechanism. There are two possible general causes: either the generality lies in the form of the density of electron states at the band edges, which must be of exponential form to account for the observed absorption behaviour (if matrix-element variations are negligible), or the behaviour arises from a particular universal absorption mechanism that gives the exponential behaviour observed.

The idea of universal exponential density-of-states tails to the valence and conduction bands in disordered materials at first seemed somewhat unlikely, but recent theoretical developments have given the proposal some credence. An early treatment by Halperin and Lax (1966, 1967) assumed that gaussian potential-energy (disorder-related) fluctuations, causing similar fluctuations in the (binding) energies of the electron states, result in a gaussian shape for the tails of the electronic DOS modified by a term which includes the effects of the kinetic energy of localization of the electrons in wells of size a. If it is assumed that this energy $|E| = c/a^2$ (where c is a constant and $|E|$ is the electron energy measured from, say, the bottom of the conduction band or the mobility edge, E_c) and the spatial extent of the fluctuations scales as $|E|^{-d/2}$, the DOS behaves as

$$N(E) \propto \exp(-\gamma |E|^{2-(d/2)}) \qquad [6.32]$$

where d is the space dimensionality. Thus, for $d=3$, [6.32] predicts that $N(E) \propto (-\gamma E^{1/2})$, in disagreement with the Urbach behaviour observed experimentally. Soukoulis *et al.* (1984) modified the Halperin–Lax approach by noting that, according to the scaling localization ideas of Thouless (1977), from [5.41] the localization energy in 3D is given by $\Delta E = Bg/a^3 N(E)$, where B is a constant and g is a dimensionless conductance accounting for the coupling between blocks in the scaling scheme (see sect. 5.3.3); ΔE is much smaller than the quantity c/a^2 since it costs much less energy to compress a highly fluctuating wavefunction than a uniform one. Soukoulis *et al.* (1984) showed that under certain conditions $g \simeq 1/\pi^3$ and $B = 2\pi^2 c\zeta_0 N(E)$, where ζ_0 is the critical value of the correlation length for amplitude fluctuations of extended states at the energy where the wavefunctions just become uniform. Under these conditions, Soukoulis *et al.* (1984) showed that in 3D the DOS is indeed accurately exponential, i.e. that $N(E) \sim \exp(-\Gamma |E|)$, with the slope function given by

$$\Gamma = \frac{16c\zeta_0}{3\pi W^2 L^3} \qquad [6.33]$$

where W^2 is the variance of the random potential and L is the correlation length of the potential fluctuations; [6.33] predicts that $\Gamma \simeq 100/W$, where W is expected to be $\simeq 1$–2 eV, i.e. Γ is of the same order of magnitude as observed experimentally. Abe and Toyozawa (1981) have also considered the effect of gaussian disorder of site energies on the DOS and regard the exponential form of the DOS as simply an intermediate region before is reached the $N(E) \sim \exp(|E|^{1/2})$ behaviour characteristic of the gaussian disorder.

Another general theoretical approach which also predicts exponential band tails was proposed more recently by Bar-Yam *et al.* (1986). As discussed in sect. 6.2.6, their model is based on the hypothesis of thermal equilibration of the structure and defect creation above some temperature T^* (which might be T_g for glasses or, the deposition temperature for amorphous thin films). They ascribe the origin of the exponential band tails in the DOS to the probability of excitation of the system (atomic and electronic structure) in creating either a filled electron state above E_v or an empty state below E_c; the resulting distribution ([6.18], [6.19]) is frozen-in for temperatures $T < T^*$.

A completely different approach to understanding the origin of the Urbach absorption edge was proposed by Dow and Redfield (1970) in terms of the absorption process itself. They suggested that the exponential edge arises from an electric-field-induced broadening of an exciton line, analogous to the Franz–Keldysh effect (photon-assisted interband tunnelling of an electron). Normally, the shape of an exciton absorption line is gaussian, but Dow and Redfield showed in a numerical study that in the presence of a uniform (internal) electric field ($E \simeq 10^6\ \mathrm{V\,cm^{-1}}$), the absorption edge becomes accurately exponential over appreciable ranges of energy. The basic mechanism is one of field ionization, in which the electron can tunnel out of its quasi-bound state in the Coulomb well caused by the hole into the deep, wide potential outside (resulting from the local microfield) which allows new states to exist in the previously forbidden gap. If the outer potential is sufficiently deep and wide, the electron wavefunction will have most of its amplitude there, with an *exponentially* small amplitude near the hole. Since the optical absorption is proportional to the square modulus of this amplitude, Urbach-like behaviour results. There are a variety of possible sources for the internal electric fields: static spatial potential fluctuations, density variations, charged defect centres or longitudinal optic phonon displacements frozen-in at T_g.

Since it is believed that the Urbach edge is disorder-induced (through bonding fluctuations), its behaviour can be used as a monitor of the amount of disorder present in an amorphous material. Cody (1984) has emphasized that although the Urbach-edge slope parameter Γ^{-1} depends on the total amount of disorder in the system, thermal and static disorder contributions cannot be separated in this way. For instance, Fig. 6.24(a) shows the absorption edge in the Urbach region of a-Si:H (dominated by the valence-band DOS which has the broader width), both as a function of measurement temperature, and at a fixed measurement temperature but after a series of high-temperature anneals in which hydrogen effused, thereby causing an increase in the structural disorder of the films. A change in slope of the Urbach edge is clearly evident as the amount of disorder changes. Figure 6.24(b) shows a correlation between the optical gap E_o (obtained with the use of [5.124]) and the slope parameter of the Urbach edge, Γ^{-1}, for the data from Fig. 6.24(a); it is apparent that thermal and static disorder contributions cannot be distinguished in this way.

Although conventional transmission optical absorption spectroscopy can be used to investigate the band-tail DOS, the technique is not sufficiently sensitive to be used as a probe for the defect-related DOS near mid-gap. For this purpose, photothermal deflection spectroscopy (PDS) may be used (see Amer and Jackson (1984) for a review). This technique is based on the detection of the heat absorbed by a sample when illuminated by a chopped pump beam of light (typically a high-intensity arc lamp). A low-intensity probe laser beam, almost grazing the surface of the material, is deflected by the thermally induced modulation of the

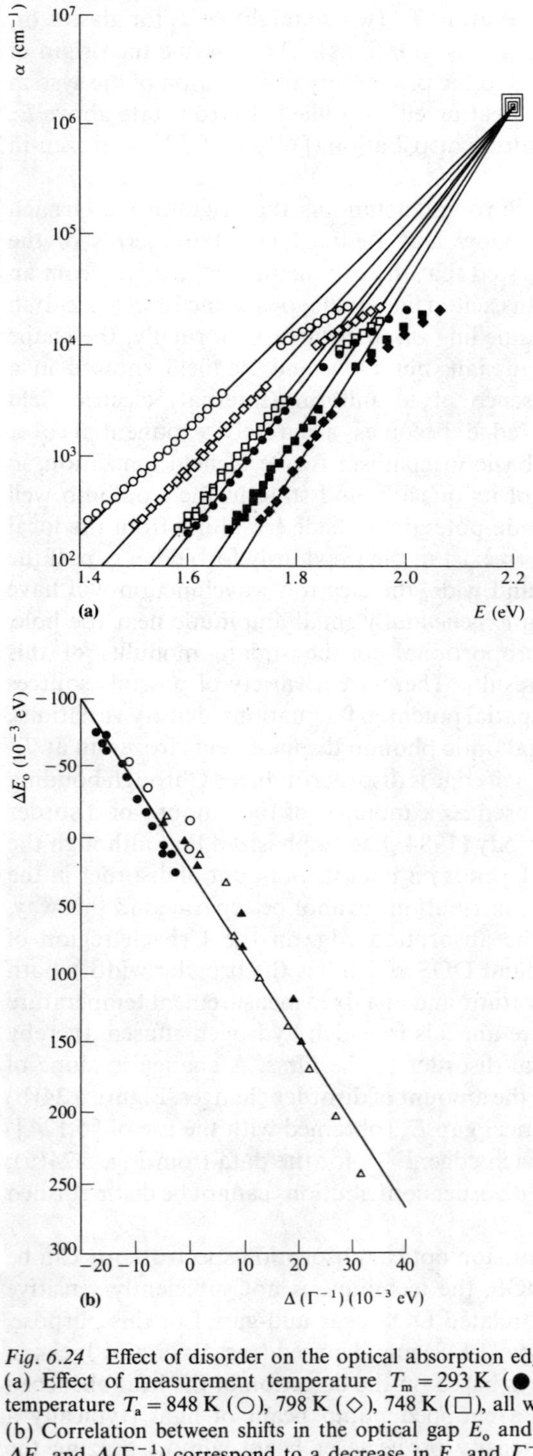

Fig. 6.24 Effect of disorder on the optical absorption edge in a-Si:H (Cody 1984).
(a) Effect of measurement temperature $T_m = 293$ K (●), 151 K (■), 12.7 K (◆); and annealing temperature $T_a = 848$ K (○), 798 K (◇), 748 K (□), all with $T_m = 300$ K.
(b) Correlation between shifts in the optical gap E_o and the Urbach edge parameter Γ^{-1}: positive ΔE_o and $\Delta(\Gamma^{-1})$ correspond to a decrease in E_o and Γ^{-1}. The circles refer to different T_M whereas the triangles are for different T_A with $T_M = 300$ K.

refractive index of a surrounding deflecting medium in thermal contact with the sample, and is measured using a position-sensitive detector. The optical absorption coefficient α is then simply related to the PDS signal S by the relation (Amer and Jackson 1984)

$$1 - \exp(-\alpha l) = \langle S/S_{sat} \rangle \qquad [6.34]$$

where l is the thickness of the sample film, S_{sat} is the saturated PDS signal (for a photon energy where the sample is completely absorbing), and the brackets $\langle \; \rangle$ indicate an average taken over interference fringes. PDS is a highly sensitive technique: for a thin film with a thickness of 1 μm, absorption coefficients as low as 0.1 cm^{-1} can be measured at a photon energy of 0.8 eV.

As an example of the applicability of the PDS technique, we show in Fig. 6.25(a,b) the optical absorption profiles in the region of the gap for n-type and p-type doped a-Si:H (Amer and Jackson 1984); the increase in the mid-gap absorption with increasing doping levels, associated with dopant-induced defect creation, is clearly evident. An estimate for the sign and magnitude of the effective correlation energy of the dangling bonds can also be obtained from such PDS studies. In intrinsic (undoped) a-Si:H the dangling bonds are mainly neutral (T_3^0), whilst in n-type P-doped material the defects are negatively charged (T_3^-) if E_F moves to a position between the defect levels and the conduction band. Consequently, if U_{eff} is positive, so that the T_3^- levels lie above those of T_3^0, the defect-related mid-gap absorption will lie shallower in the gap relative to E_c. This behaviour is observed experimentally in the PDS spectra (Fig. 6.25(c)), where the defect absorption bands can be superimposed by shifting the spectra for the intrinsic material down in energy, i.e. $U_{eff} \simeq 0.35$ eV.

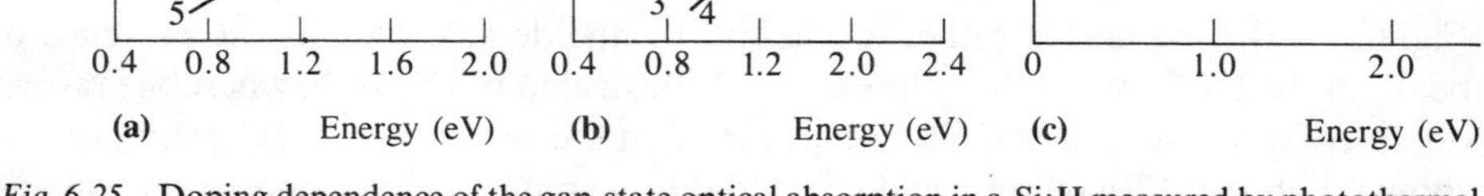

Fig. 6.25 Doping dependence of the gap-state optical absorption in a-Si:H measured by photothermal deflection spectroscopy (Amer and Jackson 1984) for (a) n-type and (b) p-type material with varying dopant concentrations (curve 1 corresponding to the highest concentration, curve 2 the next highest, etc.). An estimate for the effective correlation energy, $U_{eff} = 0.38$ eV, can be obtained by laterally shifting to the right the absorption curves for n-type doped samples with respect to those for intrinsic material so that the defect absorption bands, produced by D^- and D^0 centres, respectively, superpose, as in (c).

B. Junction-related spectroscopies

A number of techniques that have been developed to investigate the DOS in the gap region of amorphous semiconductors exploit the fact that band bending can occur at an interface between the semiconductor and some other material, and the magnitude of these junction-related effects is determined by the DOS. Thus, techniques such as field-effect methods, various capacitance-related methods and deep level transient spectroscopy (DLTS) have all been used in this way, particularly in the study of a-Si:H (for a review, see e.g. Cohen 1984).

Such techniques can be divided into three categories. Field-effect and capacitance–voltage (C–V) measurements essentially probe the space-charge region at the interface under *steady-state* conditions, and information on the DOS is obtained by repeating the measurement at a number of applied voltages. The other techniques involve dynamic measurements, in which the capacitance is measured as a function of frequency or temperature, which therefore set the time or energy scale on which the space charge is probed; the DOS is determined by varying the band bending while keeping the space-charge distribution essentially static. The DLTS techniques involve a substantial change in the properties of the interface region by (optical) perturbation of the gap states to give a non-equilibrium occupancy; during the recovery process the capacitance or current is measured, the time or temperature dependence of which is used to give an estimate for $N(E)$. This category also includes techniques such as thermally stimulated current (TSC) or capacitance (TSCAP) methods.

The principles underlying the field-effect method are illustrated schematically in Fig. 6.26, together with a diagram showing the configuration of a field-effect transistor. The method essentially involves solving Poisson's equation for the space-charge region produced by the application of a gate voltage V_g; the magnitude of the space charge so induced is controlled by the DOS within the gap region. Although the field-effect method was one of the first techniques to be applied to the study of gap-state DOS distributions in a-Si:H, it has a number of disadvantages which severely limit its usefulness. The method assumes that no charge is trapped at the interface (or in the bulk of the material) being studied, and since a region extending typically only 20–100 Å from the interface surface is probed, the field-effect method is an extremely surface-sensitive technique. As a result, inhomogeneities and charge trapping centres at the interface between semiconductor and insulator can affect or even dominate the results obtained. Furthermore, since details of the DOS distribution can be obtained only by an iterative process (see Fig. 6.26), the method is rather insensitive to fine details, such as the presence of peaks in the DOS profile. This limitation is illustrated in Fig. 6.27, where the results obtained by Goodman and Fritzsche (1980) on the sensitivity of the technique are shown. Figure 6.27(a) shows field-effect 'data', i.e. plots of sheet conductance versus gate voltage, calculated from the DOS profile for a-Si:H originally proposed on the basis of such experiments by Madan *et al.* (1976) (and shown as the solid curve in Fig. 6.27(b)). Three different computer fits to these 'data', as shown in Fig. 6.27(a), yield the very different DOS profiles illustrated in Fig. 6.27(b). It can be concluded on the basis of such studies, therefore, that the field-effect method is not capable of detecting fine-scale features, such as peaks corresponding to defect levels, in the DOS deep within the gap of amorphous semiconductors, although the existence of a deep minimum in the DOS near mid-gap does seem to be a genuine feature.

Capacitance measurements of the space-charge region in e.g. Schottky

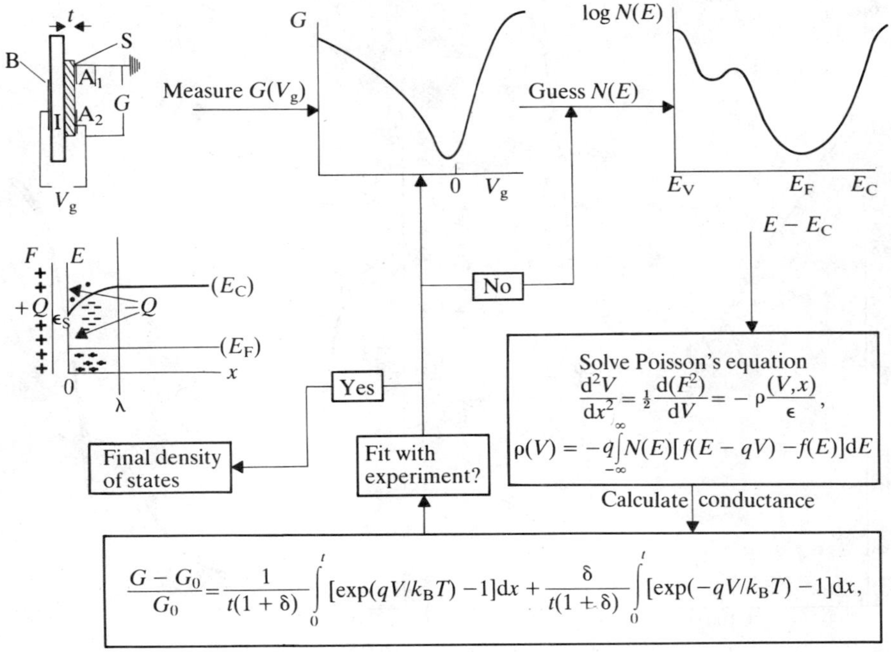

Fig. 6.26 Block diagram showing the principles of the determination of the density of states using field-effect measurements. The field-effect transistor configuration is shown, where S is the sample film (thickness $t \simeq 1\ \mu$m); I, the insulator; A_1, A_2, gap (drain) electrodes (gap $\simeq 100\ \mu$m); B, field electrode; V_g, gate voltage; G, surface conductance. The electron energy profile through the sample film when a positive potential is applied to B is also shown: the space charge $-Q$ in the film is composed of excess electrons in extended and localized states (W. E. Spear and P. G. Le Comber (1972) *J. Non-Cryst. Sol.* **8–10**, 727). In Poisson's equation, ρ is the charge density, F is the field, $f(E)$ is the Fermi–Dirac function. In the expression for the surface conductance, δ is the ratio of hole to electron current, and G_0 is the conductance, both under flat-band conditions (Powell 1981).

barriers can also be used to obtain information on the gap-state DOS. D.C. voltage-dependent capacitance (C–V) measurements are easier to interpret than those in which the complex admittance is measured as a function of temperature and frequency (C–T–ω measurements). At high frequencies, the junction capacitance is controlled by the thermal emission and capture rates of charge trapped in deep-lying gap states, with the result that states (at E_F) lying deeper in the gap than E' cannot respond to the applied a.c. voltage of frequency ω, where

$$E' = k_B T \ln(\nu/\omega) \qquad [6.35]$$

and ν is the exponential prefactor term for majority carrier emission and capture.

The change in the total barrier charge of a Schottky diode with a small change in applied voltage is given by (Cohen 1984)

$$\frac{\delta Q}{\delta V} \equiv C = \varepsilon A \int_0^\infty \delta\rho(x)\,\mathrm{d}x \bigg/ \int_0^\infty x\delta\rho(x)\,\mathrm{d}x = \frac{\varepsilon A}{\langle x \rangle} \qquad [6.36]$$

where A is the area of the diode, and $\langle x \rangle$ is the first moment of the charge distribution $\delta\rho$, which in turn is related to the DOS, $N(E)$ (see Fig. 6.26). The

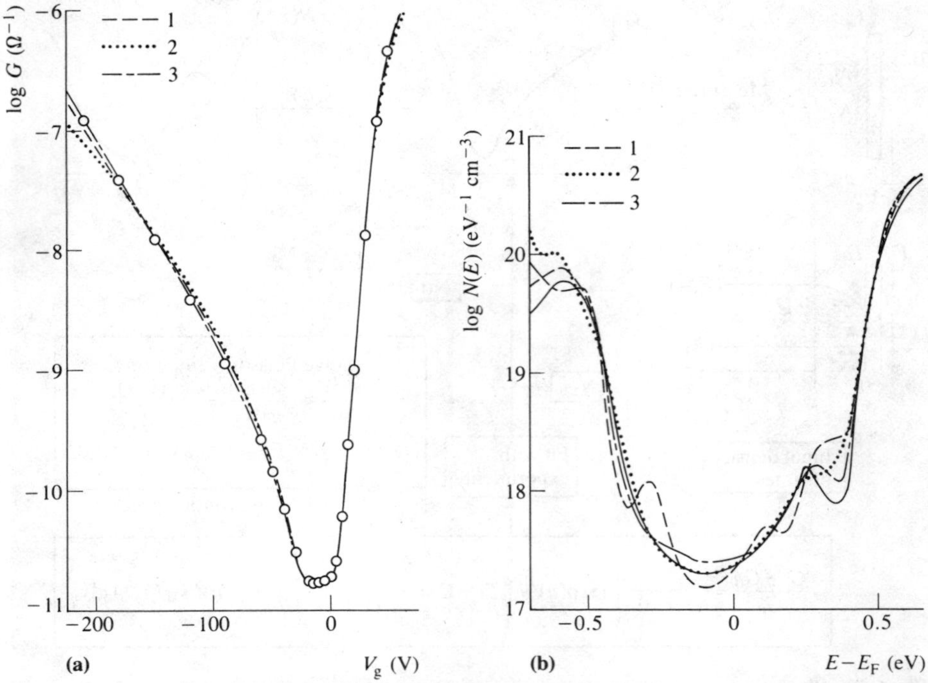

Fig. 6.27 (a) Field-effect 'data' points generated from the DOS profile for a-Si:H proposed by Madan *et al.* (1976) (shown in (b)), together with fits which generate the other DOS profiles shown in (b). (b) DOS profiles which fit the field-effect 'data' shown in (a), derived from the profile given by the solid curve (Goodman and Fritzsche 1980).

DOS is obtained from C–V data by an iterative process similar to that used for field-effect measurements. Figure 6.28(a) shows hypothetical C–V 'data' generated from the DOS profile proposed by Madan *et al.* (1976), together with the fits that correspond to the DOS distributions shown in Fig. 6.28(b), which are characterized by different mobility gaps. It can be seen that all the profiles are essentially the same, with possible evidence for a feature at an energy of 0.4 eV below E_c, and with a deep minimum in the DOS near mid-gap in all cases. Capacitance–voltage measurements are less surface-sensitive than field-effect measurements, are so are more representative of bulk properties.

The DLTS technique is a dynamic variant of the capacitance techniques, but is limited to doped (highly conducting) samples because of constraints imposed by the dielectric relaxation time. The DOS in the upper half of the gap is probed by monitoring the response of the space-charge layer in the material following a voltage (filling) pulse (or a laser pulse for deeper states) in which equilibrium is restored by thermal emission of gap-state electrons to the conduction band (or holes to the valence band for laser excitation). This results in a demarcation energy (for half-occupancy) between filled and empty states of

$$E' = k_B T \ln(\nu t_e / \ln 2) \qquad [6.37]$$

where t_e is the evolution time. A DLTS spectrum consists of a series of

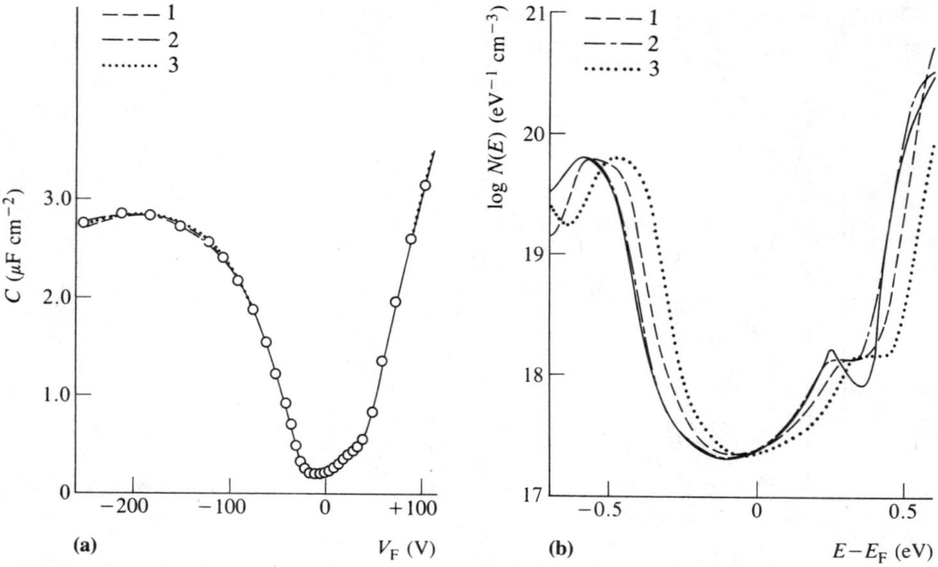

Fig. 6.28 (a) Capacitance–voltage 'data' points generated from the DOS profile for a-Si:H proposed by Madan *et al.* (1976) (shown in (b)), together with fits which generate the other DOS profiles shown in (b).
(b) DOS profiles which fit the capacitance–voltage 'data' shown in (a), derived from the profile given by the solid curve (Goodman and Fritzsche 1980).

measurements of capacitance transients following voltage pulses as a function of temperature to probe the energy dependence of the DOS distribution through [6.37].

The (capacitance) DLTS technique has the advantage that it is rather insensitive to surface states at the semiconductor interface. However, it has the disadvantage that the energy scale of the DOS distribution is determined by [6.37], and the frequency factor ν may not be a constant, as assumed in a simple interpretation of the DLTS spectrum. Indeed, Okushi *et al.* (1982) have found from isothermal capacitance transient spectra (ICTS) that the capture cross-sections of the gap-state traps are energy-dependent; inclusion of this feature in the DOS determination procedure results in a peak instead of the very deep minimum at an energy $\simeq 0.4$ eV below E_c otherwise predicted by DLTS (see Fig. 6.29).

C. Photoluminescence

It must be stressed at the outset that the phenomenon of luminescence is *not* solely associated with the presence of structural defects; we have already seen (sect. 5.6.3) that luminescence can result from intraionic electronic transitions of transition metal (3d) or rare earth (4f) atoms in amorphous matrices. However, photoluminescence is also observed in many amorphous materials which do not contain such luminescent impurities, and therefore the phenomenon must also have a more general origin. Band-to-band photoluminescence could occur; electron–hole pairs

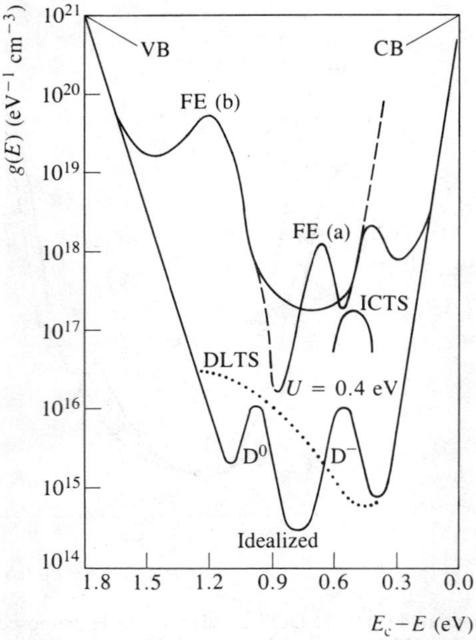

Fig. 6.29 Compilation of densities of states in a-Si:H obtained from a number of experimental techniques (Madan and Shaw 1988).

(excitons) created optically by photons having energies comparable to the band gap would then recombine radiatively directly emitting a photon of energy also comparable to the band gap. Often, however, photoluminescence spectra are observed at energies considerably *lower* than the energy of the exciting photons, and electron states in the gap are often involved. These states can be band-tail states, but are more often associated with structural defects (dangling bonds); moreover the deeper into the gap a level lies, the stronger will be the electron–phonon coupling and this will greatly influence the luminescence process. Extensive reviews on luminescence in amorphous semiconductors have been given by Street (1976) for chalcogenides, Fischer (1979) and Street (1981) for a-Si:H.

Various mechanisms for radiative recombination which give rise to photons emitted at significantly lower energies than those of the exciting radiation are illustrated in Fig. 6.30. One possibility is that an electron excited from the valence band into the conduction band will recombine with defect states deep in the gap (Fig. 6.30(a)); another is that a similarly excited electron will thermalize down in energy through deep-lying localized tail states (e.g. in the Cohen–Fritzsche–Ovshinsky model – sect. 5.3.2) until recombination occurs with a hole trapped at a comparably deep valence band tail state (Fig. 6.30(b)). Both these mechanisms assume negligible electron–lattice coupling. Alternatively, this coupling may be strong, and the energy of the level changes with its electron occupation, with the consequence that if an electron is excited to the conduction band from a defect gap state (at Y in Fig. 6.30(c)), relaxation at the centre takes place after excitation and radiative recombination at the *same* centre (X) will occur at a *different* energy; the difference in excitation and luminescence energies (in this case only) is termed

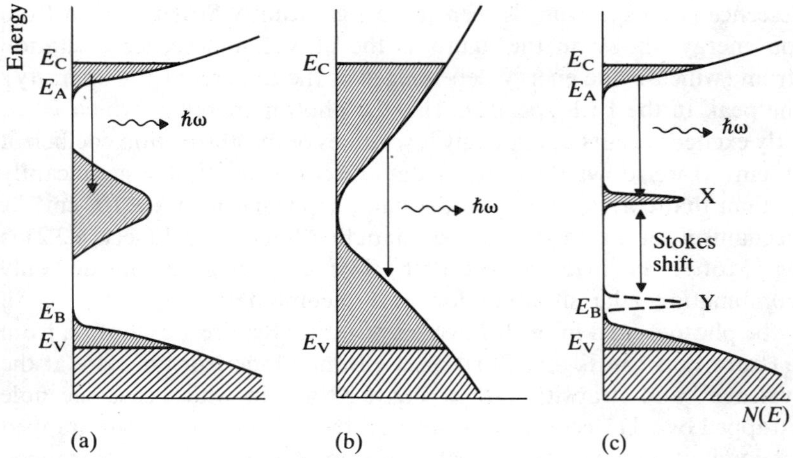

Fig. 6.30 Possible radiative recombination channels which give rise to a luminescent energy much less than the excitation energy: (a) and (b) assume negligible electron–phonon coupling, whereas (c) assumes strong electron–phonon coupling (Street 1976).

the 'Stokes shift'. In particular, note that optical absorption can in principle take place at the *same* energy as the luminescence energy for mechanisms (a) and (b), but *not* for mechanism (c) (this is more clearly demonstrated in Fig. 6.31) and this is the test for a true Stokes shift.

Experimental steady-state (c.w.) photoluminescence spectra for typical chalcogenide glasses (As_2Se_3, As_2S_3 and SiO_2) are shown in Fig. 6.31 where the

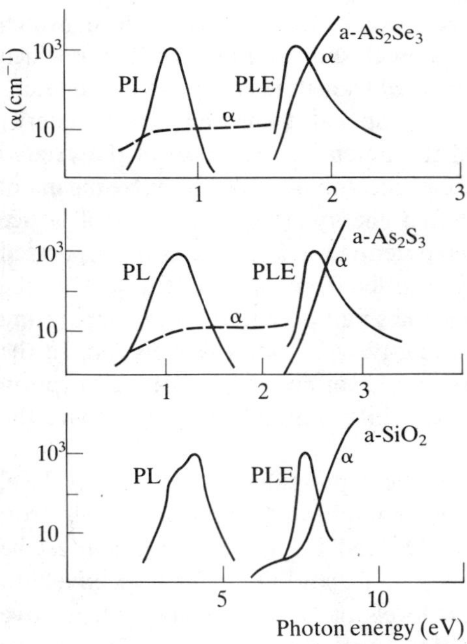

Fig. 6.31 Comparison of steady-state photoluminescence (PL), photoluminescence excitation spectra (PLE), absorption (α) and photo-induced absorption (dashed line) for three chalcogenide glasses, As_2Se_3, As_2S_3 and SiO_2. Note the similarity of PL and PLE spectra in all cases (after Robertson 1982).

photoluminescence (PL) spectrum is seen to be significantly Stokes-shifted from the excitation energy, shown in the figure as the photoluminescence excitation (PLE) spectrum (which is the energy dependence of the integrated PL intensity). Note that the peak in the PLE spectrum (i.e. the photon energy at which PL is most efficiently excited) occurs at relatively low values of the absorption coefficient ($\alpha \simeq 10^2$–10^3 cm^{-1}), well down the Urbach edge and consequently at a significantly lower energy than that corresponding to the band gap (for which $\alpha \simeq 10^5$ cm^{-1}).

The mechanism for PL in the charged dangling bond model (sect. 6.2.2) is the following (Mott *et al.* 1975a, Street 1976). For simplicity we consider only electron recombination (the situation for hole recombination is similar). An electron may be photoexcited in one of two ways; either by direct excitation from a D^- centre (transition 1 in Fig. 6.32(a)) or else from a lone-pair tail state at the top of the valence band (transition 1′ in Fig. 6.32(a)), in which case the hole produced is trapped by a D^- centre. In either case, the result is an electron trapped in a localized conduction band state, capable of recombining with a hole trapped at a D^- centre (transition 2 in Fig. 6.32(a)). The D^- with a trapped hole transforms into a D^0 centre, and therefore suffers a change in configuration. The overall luminescent process can thus be represented as:

$$D^- + h\nu_E \rightarrow D^0 + e \rightarrow D^- + h\nu_{PL} \qquad [6.38]$$

where the excitation energy is greater than that of luminescence, i.e. $h\nu_E > h\nu_{PL}$. This process may be illustrated more graphically by means of a configuration–coordinate diagram (Fig. 6.32(b)), in which the potential curves shown are a sum of electronic and lattice components, and the distortion is clearly evinced by the shift in q of the potential minima. Photo-excitation occurs *vertically* in such a diagram at constant q, in accord with the Franck–Condon principle in which *optical* excitation of electrons takes place in a much shorter time than that in which the structure can relax (this is not so for *thermal* transitions as we shall see later). The excited D^0 centre will relax after the optical transition, losing energy non-radiatively by phonon emission until the potential minimum of the excited state is reached, whereupon the electron may recombine radiatively by means of another vertical transition at a Stokes-shifted energy (Fig. 6.32(c)), or if lattice distortion in the system is strong and the two potentials in Fig. 6.32(b) are separated in q by a large amount, the system may transfer from the excited state to the ground-state configuration by passage over the small potential barrier separating them, and consequently lose energy non-radiatively by phonon emission. In the former case, the luminescence band occurs at a lower energy than the absorption band, and the difference in energy, the Stokes shift, is given by $2W$ (i.e. twice the polaron energy W).

Optical (and thermal) transitions can be represented on an energy level diagram such as in Fig. 6.32(a), but matters are complicated since only *one*-electron transitions can be shown thus, and so D^-/D^0 and D^0/D^+ transitions must be shown separately. In Fig. 6.32(a) the levels A^-, T^- and O^- refer respectively to the D^- level, and the thermal and optical levels of D^0; the equivalent positive quantities refer to the corresponding D^+/D^0 levels. The process [6.38] can be represented on this diagram as an optical absorption transition from A^- to the conduction band edge (transition 1) whereupon the D^0 level relaxes to the optical

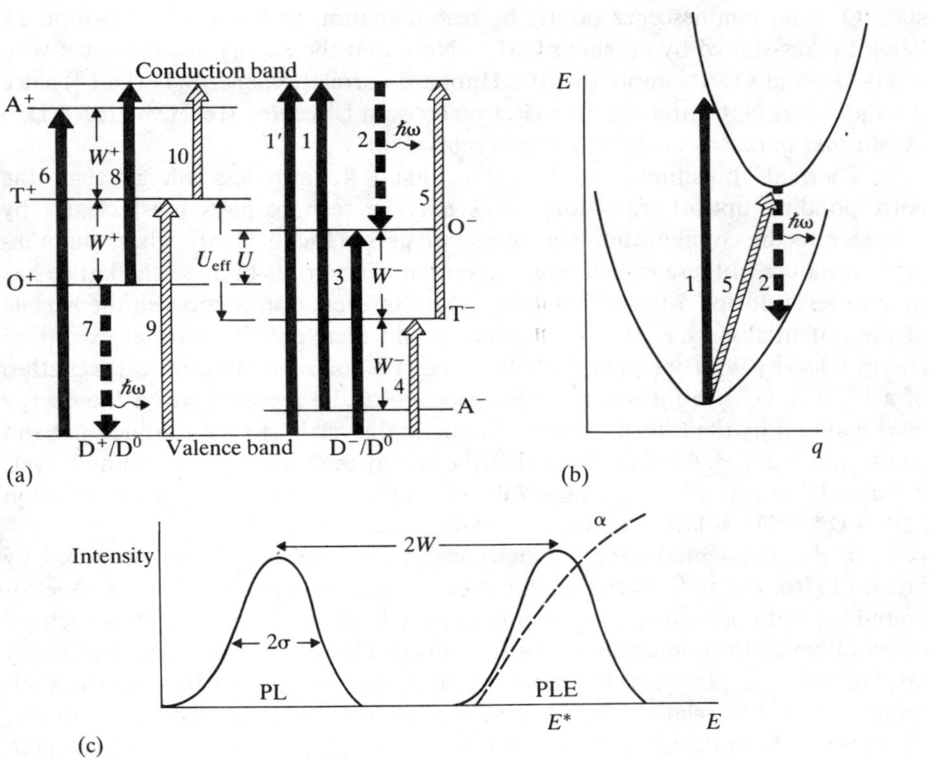

Fig. 6.32 Optical and thermal transitions between states in the gap arising from negative U_{eff} defects, e.g. in chalcogenide glasses.
(a) Levels in the gap and transitions between them. Levels labelled O and T are optical and thermal levels, respectively, and $A^+(A^-)$ is the level of $D^+(D^-)$ brought into the gap from the conduction band (valence band) by its charge. Solid arrows represent optical absorption transitions, dashed arrows represent optical emission (PL) transitions and hatched arrows represent thermal transitions. Define E_1 and E_2 by

$$D^+ + e \xrightarrow{(+E_1)} D^0,\ D^0 + e \xrightarrow{(+E_2)} D^-;\ \text{then } U_{\text{eff}} = E_2 - E_1 < 0.$$

Transition D^-/D^0

Optical:

(1) $D^- \xrightarrow{(E_g - E_2 + W^-)} D^0 + e_{cb}$

(3) $D^0 + e_{vb} \xrightarrow{(E_2 + W^-)} D^-$

Thermal:

(4) $D^0 + e_{vb} \xrightarrow{E_2} D^-$

(5) $D^- \xrightarrow{(E_g - E_2)} D^0 + e_{cb}$

Transition D^+/D^0

Optical:

(6) $D^+ + e_{vb} \xrightarrow{(E_1 + W^+)} D^0$

(8) $D^0 \xrightarrow{(E_g - E_1 + W^+)} D^+ + e_{cb}$

Thermal:

(9) $D^+ + e_{vb} \xrightarrow{E_1} D^0$

(10) $D^0 \xrightarrow{(E_g - E_1)} D^+ + e_{cb}$

(b) Configuration-coordinate diagram for excitation of D^- to D^0. Vertical arrows represent optical transitions; non-vertical arrow represents thermal transition.
(c) Schematic illustration of the photoluminescence (PL) and photoluminescence excitation (PLE) spectra in a chalcogenide glass. The PL is Stokes-shifted by energy $2W$. The absorption edge is also shown for comparison.

state O^- and luminescence occurs by recombination to this level (transition 2), being Stokes-shifted by an energy $2W^-$. Note that the energy difference between levels O^- and O^+ is simply the true Hubbard correlation energy U ([6.1]) since it is the energy for the transfer of an electron between D^0 centres (i.e. $2D^0 \rightarrow D^+ + D^-$) at *constant* q, that is in the absence of relaxation.

Thermal transitions, on the other hand, require less energy than the corresponding optical transitions. This may be seen perhaps most clearly by reference to the configuration-coordinate diagram (Fig. 6.32(b)), where the non-vertical nature of these transitions is apparent. This arises because the lattice has time to relax during a thermal transition and so excitation is between the *minima* of the potential curves, at non-constant q, and therefore the thermal excitation energy is less by W^- (or equivalently W^+) than the corresponding optical excitation of a D^- (or D^+) to form a D^0. This process can be represented on the energy level diagram by the transition from the thermal level T^- to the conduction band (transition 5 in Fig. 6.32(a)). Note that the energy separation of the thermal levels T^- and T^+ is given by U_{eff}, since this now represents the energy of the reaction $2D^0 \rightarrow D^+ + D^-$ if lattice relaxation takes place.

In the defect model for luminescence in chalcogenide glasses proposed by Street (1976), the radiative recombination process is supposed to be a 'free-to-bound' transition involving one carrier loosely bound to a pre-existing structural defect at which the conjugate carrier is trapped. This model explained reasonably satisfactorily the early c.w. luminescence data, but subsequent time-resolved PL results and other related studies have cast doubt on its general applicability; Kastner (1985) has given a more up-to-date review of this problem. In particular, studies of the photoluminescent properties of the chalcogenide *crystals* As_2S_3 and As_2Se_3 have proved to be very useful as the behaviour of glasses and crystals in this case is very similar.

Time-resolved (transient) PL measurements, for example, by Higashi and Kastner (1983) and Robins and Kastner (1984) have revealed a rich behaviour of the PL decay as a function of excitation energy and measurement temperature. It can be seen from Fig. 6.33 that the behaviour of amorphous and crystalline As_2S_3 is very similar, with the PL decay exhibiting a knee in a double-logarithmic representation corresponding to a time of $\simeq 10^{-4}$ s at $T \simeq 5$ K; for temperatures $T \geqslant 5$ K, the decay data for c-As_2S_3 can be fitted by a single exponential function, $\exp(-t/\tau)$, with the decay rate, τ^{-1}, being thermally activated above $\simeq 10$ K (Robins and Kastner 1984). The PL decay behaviour for chalcogenide glasses can be divided into two regimes (see Fig. 6.34), one being dominated by processes characterized by short times, $t \lesssim 10^{-6}$ s, and the other by longer times, $t \gtrsim 10^{-4}$ s; the short-time behaviour is more pronounced in a-As_2S_3 (Higashi and Kastner 1983). The long-time processes account for most of the PL quantum efficiency and therefore they dominate the steady-state (c.w.) behaviour.

The origin of these two decay processes is still not entirely clear, but the following models seem to be plausible. The short-time behaviour seems to be consistent with a mechanism of 'donor–acceptor pair' recombination (Higashi and Kastner 1981, Depinna *et al.* 1983) in which radiative recombination occurs when an electron of a neutralized donor is transferred to a neutralized acceptor (the centres are oppositely charged in the ground state and neutral in the excited

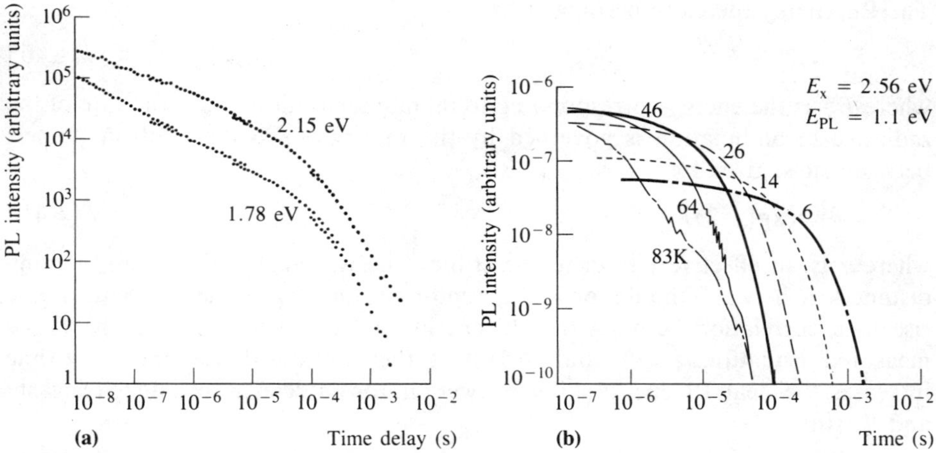

Fig. 6.33 Photoluminescence time decays in chalcogenide materials:
(a) amorphous As_2Se_3 (Higashi and Kastner 1983);
(b) crystalline As_2Se_3 (Robins and Kastner 1984).

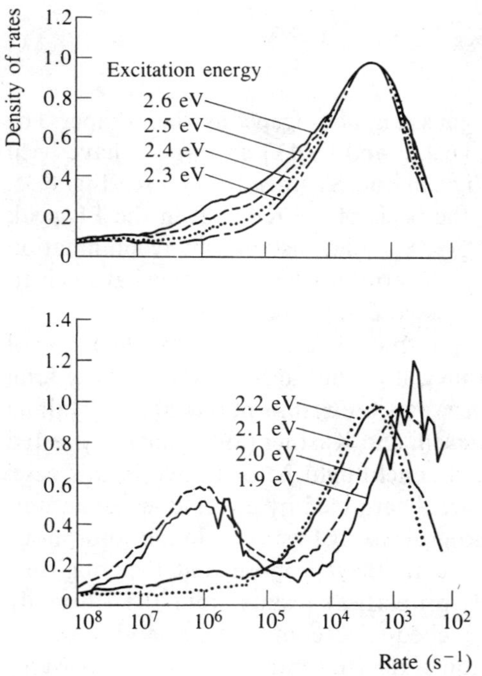

Fig. 6.34 Density of photoluminescence rates for a-As_2S_3 at various excitation energies (Higashi and Kastner 1983).

state, e.g. D^+, D^- pairs). There is an extra contribution E_c to the PL recombination energy, resulting from a Coulomb interaction between ionized donor and acceptor, which is given by

$$E_c = e^2/4\pi\varepsilon\varepsilon_0 R \qquad [6.39]$$

The PL energy therefore becomes

$$E_p = E_p^\infty + E_c \qquad [6.40]$$

where E_p^∞ is the energy corresponding to infinite separation. Since the rate of the radiative recombination is governed by the rate at which the electron tunnels between sites, given by

$$\nu = \nu_0 \exp(-2\alpha R) \qquad [6.41]$$

where α is an effective parameter describing the wavefunction overlap, varying distances R between the donor and acceptor sites acting as the PL centres give rise to a correlation between the PL energy and the time at which the PL is measured. Equations [6.40] and [6.41] together indicate that as the delay time increases, the mean PL energy shifts to lower photon energies according to (Higashi and Kastner 1981)

$$\langle E_p(t)\rangle = E_p^\infty + \frac{2\alpha e^2}{4\pi\varepsilon\varepsilon_0 \ln(\nu_0 t)} \qquad [6.42]$$

and the *width* of the PL peak is also predicted to narrow with increasing time delay:

$$\Delta^2(t) = \sigma^2 + \left[\frac{2\alpha e^2}{4\pi\varepsilon\varepsilon_0 \ln^2(\nu_0 t)}\right]^2 \qquad [6.43]$$

where σ is the width of an assumed gaussian homogeneous line shape. The time-dependent behaviour embodied in [6.42] and [6.43] appears to have been observed experimentally for a-As_2S_3 (Higashi and Kastner 1981) – see Fig. 6.35. Depinna *et al.* (1983) also conclude, on the basis of the red shift in the PL peak energy with excitation energy E_{ex} (for $E_{ex} < E_g$), that distant-pair recombination processes might be involved. The exact PL centre involved is somewhat unclear, but could be isolated VAP defects, as discussed earlier (see Fig. 6.32).

Although a simple donor–acceptor pair model can account for the range of PL decay rates observed at very small times, it cannot account for the very long times ($t \geq 10^{-4}$ s) observed in the PL decay of both amorphous and crystalline chalcogenide materials (see Fig. 6.34). Higashi and Kastner (1983) have suggested that a different radiative recombination channel might be operative in this case, where the slow rates (long decay times) are determined by a forbidden quantum-mechanical selection rule; the only selection rule likely to survive in the amorphous state is that pertaining to spin. In particular, they propose that the long-time behaviour (and hence also the c.w. PL properties) results from the (formally forbidden) transition between a *triplet* excited-state configuration and a singlet (i.e. spin-paired) ground-state configuration of the PL centre (Higashi and Kastner 1983). The radiative PL transition can occur in such a case only if a small amount of singlet character is mixed in with the triplet wavefunction, as a result of spin–orbit coupling. Further evidence in support of the triplet excited state comes from optically detected magnetic resonance (ODMR) measurements (see sect. 6.3.2).

There are two possible candidates for PL centres which could involve triplet excited states (Kastner 1985). One is a 'self-trapped exciton' (STE) state, where the initially photo-created free exciton can subsequently lower its energy by forming

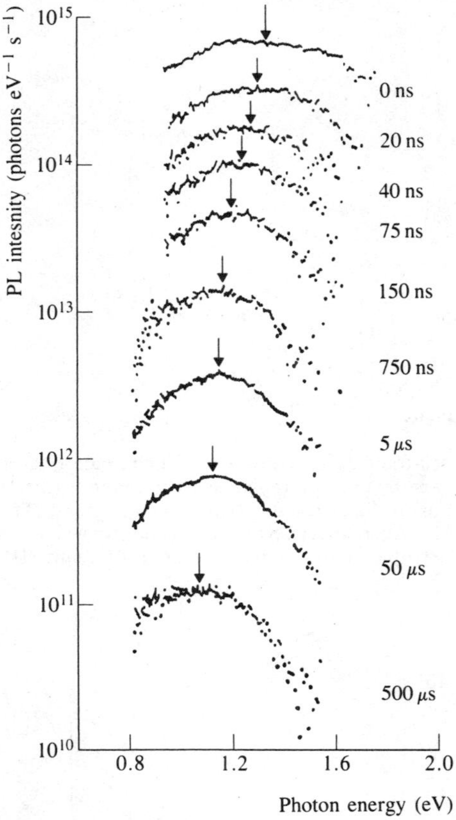

Fig. 6.35 Photoluminescence spectrum of a-As_2S_3 sampled at various times after pulsed excitation (Higashi and Kastner 1981).

a self-trapped (localized) state in which the lattice is locally distorted (Fig. 6.36), as a result of an appreciable exciton–lattice interaction, in an analogous manner to the formation of small polarons by single carriers (sect. 5.5) (Emin 1980, Murayama *et al.* 1980). The significance of this model is that STE states can form at any site in the structure. Street (1977) has further suggested that a bond-switching process, following the formation of a free exciton, can give a *non-radiative* recombination channel in which a metastable STE state occurs in the form of an intimate valence-alternation pair of coordination defects (D^+, D^-) (see Fig. 6.36). Alternatively, the triplet exciton state could be associated with previously existing coordination (D^+, D^-) defects, as suggested by Higashi and Kastner (1981). In *crystalline* As_2S_3, the PL intensity does not saturate with increasing optical excitation density (Higashi and Kastner 1983), lending support to the STE model, but the situation for the amorphous chalcogenides is less clear.

We turn now to a discussion of the PL behaviour exhibited by a-Si:H; several reviews of this subject have appeared (Street 1981, 1984; Wang and Fritzsche 1988). Unlike amorphous chalcogenides (discussed above), where only a single broad PL band is observed, a-Si:H exhibits somewhat richer behaviour. Figure 6.37(a) shows the c.w. PL spectrum of intrinsic a-Si:H as a function of temperature;

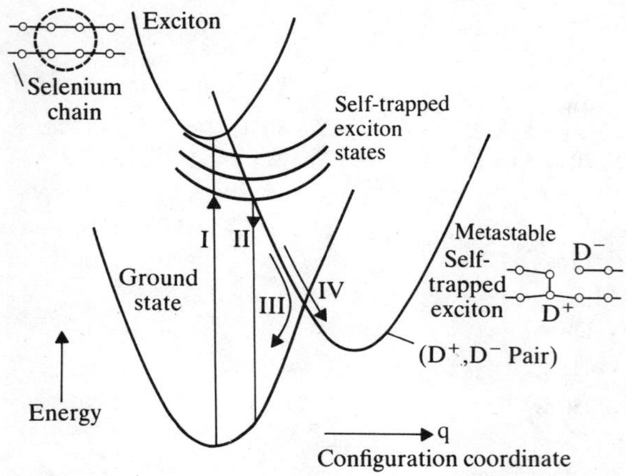

Fig. 6.36 Schematic illustration of the model for self-trapping of excitons in chalcogenide glasses. Optical excitation (I) to an exciton state can be followed by two non-radiative decay channels (Street 1977), either directly back to the ground state (III) or to the metastable self-trapped exciton (STE) state, which can be regarded as a D^+-D^- pair (IVAP). Alternatively, STE states can form where the lattice distortion is less severe, and may then act as the (triplet) radiative recombination PL centre (II).

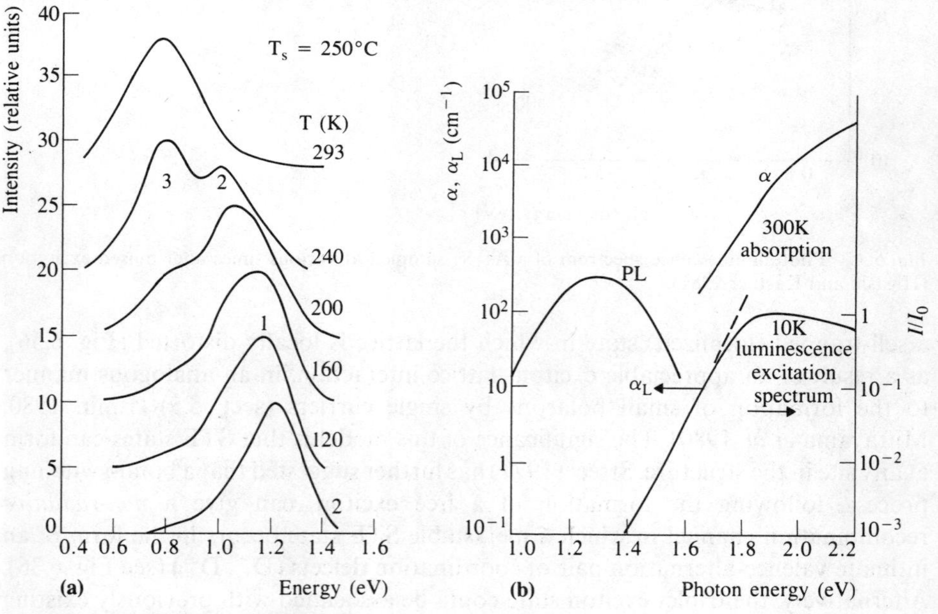

Fig. 6.37 Photoluminescence in a-Si:H.
(a) Temperature dependence of the PL, showing the emergence of the 0.9 eV defect band at high temperatures (Engemann and Fischer 1977).
(b) Photoluminescence and excitation spectra compared with the room-temperature optical absorption spectrum (Street 1984).

it can be seen that the spectrum is characterized by a peak at $\simeq 1.3$ eV at very low temperatures, whilst a lower-energy peak emerges at higher temperatures (Engemann and Fischer 1977); the relationship of the 1.3 eV PL band to the photoluminescence excitation (PLE) spectrum and the optical absorption edge is shown in Fig. 6.37(b) (Street 1984). Although the displacement of the PL band with respect to the PLE spectrum might indicate the existence of an appreciable Stokes shift (see Fig. 6.30(c)), the experimental observation of a wide range of PL decay times (Fig. 6.38(a)), as in the chalcogenide glasses (Fig. 6.33(a)), has been taken to imply that the difference in PL and PLE energies results from thermalization of electrons and holes in band-tail states after optical excitation but before radiative recombination (as in Fig. 6.30(b)); the time dependence of the PL decay as shown in Fig. 6.38(a) is then determined by these thermalization kinetics.

A model for the recombination processes giving rise to the 0.9 and 1.3 eV PL bands is shown in Fig. 6.38(b) (Street 1984). The 1.3 eV peak is postulated to arise from the (not necessarily geminate) recombination between thermalized electrons and holes trapped deep in the conduction-band and valence-band states, respectively. The so-called defect band at 0.9 eV involves (defect) states deep within the gap. It is assumed that the process of electron capture by a neutral dangling bond,

$$T_3^0 + e \rightarrow T_3^-$$

is *non-radiative*; radiative recombination is then presumed to arise from the capture of a valence-band tail-state hole by the resulting negatively charged T_3^- centre:

$$T_3^- + h \rightarrow T_3^0 + h\nu$$

This assumption is supported by the fact that n-type doped a-Si:H, in which all T_3^0 dangling bonds are converted to T_3^- centres (including those dangling bonds created by the doping process – sect. 6.2.6), exhibits only the 0.9 eV PL band.

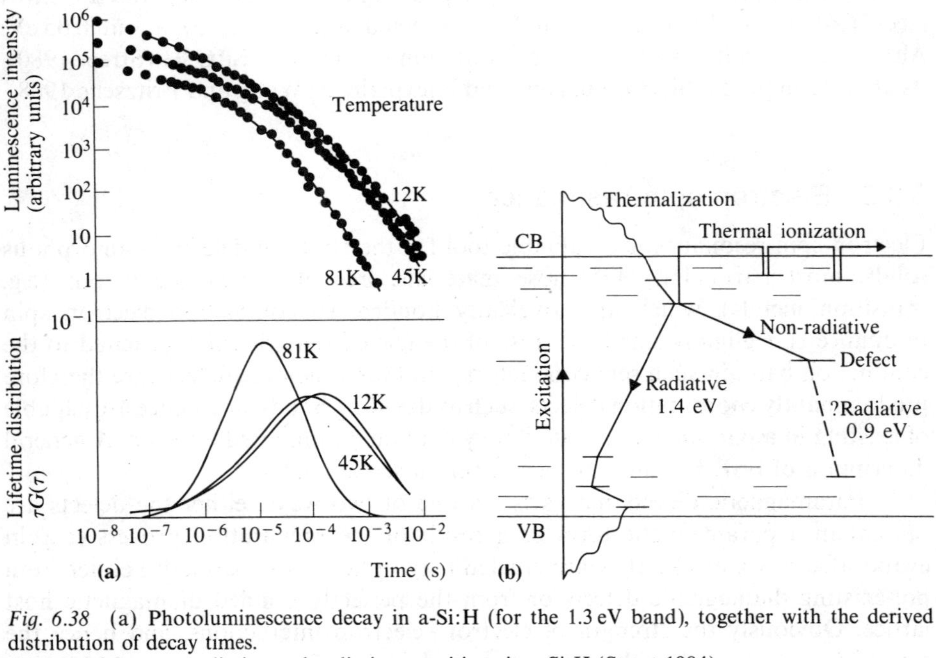

Fig. 6.38 (a) Photoluminescence decay in a-Si:H (for the 1.3 eV band), together with the derived distribution of decay times.
(b) Model for non-radiative and radiative transitions in a-Si:H (Street 1984).

The temperature dependence of the PL spectrum of intrinsic a-Si:H shown in Fig. 6.37(a), in which the 1.3 eV band is thermally quenched, can be understood in terms of the thermal excitation and separation of the trapped tail-state carriers (Fig. 6.38(b)): non-radiative capture of the now mobile carriers by dangling bonds elsewhere in the material is more likely than tail-to-tail radiative recombination.

However, not all aspects of photoluminescence in a-Si:H are yet fully understood. Thus, the relative importance of *geminate* recombination (i.e. involving only electrons and holes created as pairs by photon absorption) is still somewhat unclear. It appears that geminate (radiative) recombination is dominant at low temperatures and for low photo-excitation fluxes. At higher temperatures, an electron and a hole forming a pair can drift apart during the thermalization process at the band edges (although the rather small extended state drift mobility, $\mu_e \simeq 10\,\mathrm{cm^2\,V^{-1}\,s^{-1}}$, would tend not to favour such a separation). If the recombination is indeed geminate, the application of an external electric field would also tend to separate the carriers, leading to a *decrease* in the PL quantum efficiency and an *increase* in the photoconductivity yield. Although this field-induced effect appears to be observed (Mort *et al.* 1980), it is likely that, under normal photo-excitation conditions, non-geminate radiative recombination plays an appreciable role, the electron or hole of a newly created pair recombining with an unrecombined conjugate carrier remaining from a previous absorption event (Dunstan 1984).

A further problem relates to the Coulomb effect manifested in the time-resolved PL spectra. For oppositely charged species (electrons and holes) recombining to form a neutral entity (i.e. the opposite of what happens in chalcogenide glasses – see above), the sign of the Coulomb term ([6.39]) is expected to be *negative*, so that with increasing time delay (thereby probing more widely separated centres (see [6.41])) the PL energy would be expected to *increase*, by about 0.05 eV. Although this shift *is* seen in oxygen-containing material, a-Si:O:H (Street 1980), its absence in pure a-Si:H is puzzling and unexplained (Wang and Fritzsche 1988).

6.3.2 Electron spin resonance

Electron spin resonance is a valuable tool for the study of defects in amorphous solids, most particularly for those materials, free of paramagnetic ions (e.g. transition metals), which are covalently bonded, so conduction electron spin resonance is negligibly small (because of the few carriers thermally excited to the conduction band in such semiconductors), and for which the defects are therefore predominantly coordination defects, such as dangling bonds, and hence are capable of existing in a paramagnetic state if they contain an unpaired electron. A general description of ESR has already been given in section 3.2.4.

Paramagnetic defects may arise in one of two ways: either the defects are present in a paramagnetic form as a result of the preparation process (e.g. in evaporated a-Si), or else they are created (say optically or thermally) either from pre-existing diamagnetic defects or from the perfectly bonded diamagnetic host lattice. Obviously the strength of electron–electron interactions, and hence the sign of U_{eff}, is crucial to the existence or otherwise of paramagnetic centres.

Electron spin resonance signals in pure chalcogenide glasses are generally observed only after optical excitation (see Bishop *et al.* 1977), the exception being Ge–S alloys; in the ground state, these materials are invariably diamagnetic (Agarwal 1973). It was this observation that prompted Anderson (1975) to propose his model for spin pairing. All chalcogenide materials studied so far, e.g. Se, and As and Ge chalcogenide alloys, exhibit a resonance near $g = 2$, which is reasonably broad ($\simeq 500$ gauss wide) and which has been ascribed to an unpaired spin predominantly localized on a single chalcogen atom, viz. a C_1^0 centre. In addition, As-containing glasses, as well as a-As itself, exhibit an additional, much broader resonance ($\simeq 1500$ gauss wide), which has been ascribed to an electron trapped in an As orbital containing a significant amount of s-character (perhaps a P_4^0 centre); the width of the resonance arises from the unresolved structure of the hyperfine interaction with the As ($I = 3/2$) nucleus. In all cases, identification of the nature of the centre is made possible by simulation of the ESR spectrum by means of powder-pattern modelling (see sect. 3.2.4).

Identification of these ESR centres with their precursor charged dangling bond states is made possible by the close correlation between ESR and photoluminescence. Experimentally, it is found that if the PL intensity is monitored under continuous optical excitation using near-band-gap light (for which $\alpha \simeq 10^3\ cm^{-1}$), the intensity *decreases* with continuing excitation: i.e. the luminescence 'fatigues'. Concomitant with this PL fatigue is a growth of the ESR signal. The explanation for this behaviour follows on from the discussion of the photoluminescence mechanism given in the last section. If it is assumed that the luminescent centre is a D^-, the hole of an optically generated electron–hole pair will be trapped, resulting in an electron loosely bound to a D^0 centre. In the model for the temperature dependence of PL due to Street (1976), it is the hopping away of the electron before recombination takes place that is responsible for the decrease in PL intensity with increasing temperature. It has been proposed (see e.g. Bishop *et al.* 1977) that this is also the mechanism by which the PL intensity fatigues under continuous excitation, and the end result of this process is a metastable D^0 centre which, being paramagnetic, will give rise to an ESR signal. Just as the PL signal has disappeared by $\simeq 150$ K, so too does the ESR signal vanish on warming the sample to approximately the same temperature, presumably by the diffusion of excess carriers to the D^0 centres followed by non-radiative recombination. At the same time as the ESR signal grows at the expense of the PL intensity, an optical absorption band appears at energies near mid-gap. The presence of this photo-induced absorption can be understood by reference to Fig. 6.32(a) with the supposition that metastable D^0 states are created by the fatiguing process. It can be seen that the optical levels for D^0 lie near the middle of the gap, and so the two transitions involving this centre, namely $D^0 + e_{vb} \rightarrow D^-$ and $D^0 \rightarrow D^+ + e_{cb}$ (transitions 3 and 8 respectively), involve energies much less than that of the band gap. Note also that optical transitions of this sort involving the mid-gap absorption band *regenerate* luminescent centres (i.e. D^- or D^+), and so illumination of a fatigued sample with *half-band-gap* light should restore the luminescent intensity and concomitantly reduce ('bleach') the ESR signal, as indeed is observed (Bishop *et al.* 1977).

There is a certain degree of controversy surrounding the density of photo-induced spins observed in such experiments. There is some evidence that the spin density saturates (at $\simeq 10^{18}$ cm^{-3}), indicating that pre-existing centres are being excited (viz. D^+ or $D^- \rightarrow D^0$), but other experiments show no sign of saturation even for very long illumination times, and the continuous increase in spin signal seems clear evidence for defect creation (perhaps by the mechanism illustrated in Fig. 6.36), and this behaviour has been indirectly linked with the photostructural changes observed in chalcogenide glasses. This matter therefore remains somewhat uncertain as indeed does the link between ESR and PL fatigue, since there is some evidence that the fatigue process involves some non-radiative channel other than that which produces D^0 centres. An additional interesting complication is that while very similar PL behaviour is observed in *crystalline* chalcogenide materials, they do *not* exhibit an ESR spectrum, photo-induced or otherwise.

Electron spin resonance has played a significant role in determining the nature of radiation-induced centres in other materials too, notably silica. Again two types of radiation-induced process can be visualized: irradiation by neutrons or charged particles such as protons or alpha particles results in a high degree of damage and defects are created through atomic displacement by knock-on processes; on the other hand, irradiation with electromagnetic radiation or electrons can create (paramagnetic) defects by electronic excitation, e.g. by bond rupture, or can transform pre-existing diamagnetic defects into a paramagnetic state by the trapping of the photo-electrons or holes. We have discussed previously (sects 6.2.2 and 6.2.3) an example of a defect in silica, namely the E′ centre (see Fig. 6.12). This has axial symmetry, and hence the g-tensor has just two values $g_{\parallel}$ *and* $g_{\perp}$; $\Delta g = g - g_s$, the shift from the free-spin value g_s ([3.70]), is small and negative with $|\Delta g_{\perp}| > |\Delta g_{\parallel}|$.

Another example, very much related to the chalcogen ESR centres described above, are the so-called 'oxygen hole centres' (OHCs) observed in γ-irradiated silica-based glasses (Stapelbroek *et al.* 1979); these have also been discussed briefly in section 6.2.2. The g-tensor in this case is completely anisotropic, with $g_1 \simeq g_s$, $g_2 \simeq 2.01$, but with a broad distribution of g_3-values (with $\langle g_3 \rangle \simeq 2.07$) found from detailed simulation of the line shape. The distribution in g_3-values is presumed to arise from gaussian distributions in the energy level splittings. The nature of such centres can be determined by consideration of the g-values, in conjunction with the observation of the hyperfine splittings characterized by the six-line spectrum observed in ^{17}O (I = 5/2) enriched v-SiO$_2$, which arises because of the interaction between electron and nuclear spins.

In general, the wavefunction of an unpaired spin can be written as:

$$|\psi\rangle = \alpha|ns\rangle + \beta|np\rangle + \delta_i|k_i\rangle \qquad [6.44]$$

where $|ns\rangle$ and $|np\rangle$ are atomic orbitals of the central atom, $|k_i\rangle$ are other orbitals of the central atom and ligand orbitals, and the coefficients α and β are given by:

$$\alpha^2 = \frac{A_{\mathrm{iso}}}{A_{\mathrm{ns}}} = \frac{(A_{\parallel} + 2A_{\perp})}{3A_{\mathrm{ns}}} \qquad [6.45]$$

$$\beta^2 = \frac{A_{\mathrm{aniso}}}{A_{\mathrm{np}}} = \frac{(A_{\parallel} - A_{\perp})}{3A_{\mathrm{np}}} \qquad [6.46]$$

Here A_{ns} and A_{np} are atomic hyperfine coupling constants, and $A_{\parallel}$ and $A_{\perp}$ are the hyperfine coupling constants obtained from the experimental spectrum for the centre in question (in the case of axial symmetry of the hyperfine tensor). Furthermore, the wavefunction coefficients are related via a normalization condition $\alpha^2+\beta^2+\gamma^2=1$, where $\gamma^2=\Sigma_i\delta_i^2$. In addition, for the case of axial symmetry of the centre, such as exists for a dangling bond sp^3 orbital, there is a relation between the angle θ subtended at the central atom by the dangling bond orbital and a normally bonded orbital and the hyperfine coefficients:

$$\tan\theta = -[2(1+\beta^2/\alpha^2)]^{1/2} \qquad [6.47]$$

Hence detailed analysis of the ESR spectrum yields values for the wavefunction coefficients α and β and therefore the nature and environment of the orbital in which the unpaired spin resides.

In this manner, the OHCs are found to be predominantly associated with oxygen p-orbitals, and therefore the centres arise when a hole is trapped either in a non-bridging (C_1) orbital, or else in a lone-pair orbital of a normally bonded, i.e. bridging, atom (C_2^0). However, there is strong evidence for the presence of *two* distinct oxygen hole centres in any sample of vitreous silica, a 'wet' OHC associated predominantly with high —OH-containing material, and a 'dry' OHC observed mainly in low —OH-containing silica (Stapelbroek *et al.* 1979). The detailed nature of the differences of such defects are not known, however; it is an obvious assumption that the wet OHC is associated with a proton (in an —OH group), but the absence of an observable proton hyperfine interaction dictates that the unpaired spin and the proton must be separated by a large distance ($\geqslant 3$ Å).

A related defect is the so-called aluminium centre which is believed to be an unpaired spin on a bridging oxygen bonded to a Si atom and a substitutional Al atom; the hyperfine interaction with ^{27}Al ($I=5/2$) atoms gives rise to a six-line spectrum. This centre is of interest since it is also found in crystalline α-quartz, and the optical absorption associated with it is responsible for the colour of smoky quartz.

An interesting correlation exists between the ESR observed in evaporated a-Si or Ge and another defect-related property, namely the d.c. hopping conductivity (see sect. 6.3.3). It is found that the ESR line width (defined as the peak-to-peak separation ΔH_{pp}), measured in terms of the magnetic sweep field, i.e. in gauss) can be deconvoluted into two components, a temperature-independent term, $\Delta H_{pp}(0)$ (caused mainly by g-value broadening through the spin–orbit coupling, but also in part by unresolved hyperfine structure) and a strongly temperature-dependent term, $\Delta H_{pp}(T)$. This latter term is approximately proportional to the hopping conductivity, $\sigma_h(T)$, which results from the hopping of electrons between localized defect states (see sect. 6.3.3). The line broadening then results from a decrease in lifetime caused by an enhancement of spin–lattice relaxation due to the electron motion.

Finally we consider a technique which links ESR and PL, and which has been extensively used in the study of crystalline semiconductors, but only very recently has been applied to the study of amorphous materials. This is 'optically detected magnetic resonance', ODMR – for an extensive review of the technique see Cavenett (1981). In essence, the method consists of monitoring the photoluminescence intensity at the same time as microwave radiation excites resonance between the Zeeman levels of the centre, split by an external magnetic field: instead of the absorption of the microwaves being used as a monitor of resonance as in a conventional ESR experiment, the PL intensity is measured instead. The resonance condition is found as usual by sweeping the magnetic field, and the g-value for the resonant centre is determined from the position of the maximum or minimum of the PL intensity. The advantage of ODMR is that, since microwave absorption is due to electronic transitions to levels otherwise little populated in thermal equilibrium, an *increase* in the PL intensity at resonance implies that transitions from such a level are radiative, whereas the observation of a *decrease* in PL intensity means that a non-radiative channel is being explored. In this way, therefore, identification of a PL emission band with a given centre becomes possible.

For a-Si:H, a number of ODMR studies have been carried out (see Morigaki 1984 for a review). Several ODMR lines are observed, but authors differ in the detailed interpretation of them, partly because of difficulties encountered in deconvoluting the spectra. Nevertheless, a quenching (non-radiative) line is observed with a g-value of $\simeq 2.005$, which is ascribed to the T_3^0 dangling-bond centre, and an enhancing (radiative) line at $g \simeq 2.01$, which is thought to involve band-tail carriers.

Finally, note that ODMR is just one example of spin-dependent recombination. The ability of microwaves to invert (flip) the spin rate of a free electron or hole can dramatically influence the recombination rate, manifested also in photoconductivity experiments (see Solomon 1979).

6.3.3 D.C. hopping conductivity

We have previously considered the mechanism for the d.c. electrical conductivity of 'ideal' amorphous semiconductors, i.e. in the absence of structural defects when the gap is therefore free of states except perhaps for tail states (sect. 5.4.1). In real materials, which may contain a high concentration of defects and consequently have a high density of states in the gap, electron transport can take place via such defect states, and the magnitude of this defect-controlled conductivity may greatly exceed that due to conventional band conduction. Moreover, the temperature dependence of this form of conduction can have a distinctive form, easily distinguishable from the simple activated behaviour characteristic of band conduction. It is therefore of interest to discuss such behaviour in some detail.

We have seen earlier in this chapter that structural (bonding) defects give rise to electron states lying within the gap of amorphous semiconductors and these will therefore be localized. However, one of the definitions of localization is that the configurationally averaged conductivity associated with the states at energy E is zero, $\langle \sigma_E \rangle = 0$; i.e. unlike extended states, localized states by themselves cannot carry current. Thus, conduction involving localized states (of whatever origin) can only take place by means of transitions of electrons from full states to neighbouring empty states, usually with phonon assistance. It is this transfer of electrons between sites localized at different positions in space that is commonly termed 'hopping', although we prefer to reserve the usage for transitions which involve energy activation.

The mechanism of hopping conduction was first proposed to account for the electrical conduction that takes place in crystalline semiconductors (e.g. Si) which have been doped (say n-type) and compensated. 'Impurity conduction' occurs due to the hopping of electrons between filled localized donor states (which are randomly situated), and states which are empty because of the compensation by acceptors: the acceptors also create a coulomb potential which perturbs the donor energy levels, meaning that phonons are necessary for the carriers to overcome the energy differences introduced. It is an obvious extension of this model to apply it to the case of conduction in amorphous semiconductors, in which *all* the states lying in the mobility gap are localized if the disorder in potentials is sufficiently great, rather than for the case of impurity conduction where only the states lying in the impurity (e.g. donor) band are localized in the Anderson sense.

The simplest approach is perhaps that given by Mott (1969) – see also Mott and Davis (1979). Consider two localized sites, one filled and at or slightly below the Fermi energy E_F, and the other empty and above E_F; their energy and spatial separations are Δ and R, respectively. The hopping transition rate γ will be determined by three factors: (i) the probability of the existence of a phonon of energy Δ, given by the Boltzmann factor $\exp(-\Delta/k_B T)$ – the presence of the phonon is necessary to conserve energy in the hopping process; (ii) the probability of electron transfer between the sites – if it is assumed that each state is exponentially localized, with the same localization length α^{-1}, $\psi \sim \exp(-\alpha r)$, and if an electron transfer is by means of tunnelling, the probability is given by the overlap, i.e. $p \propto \exp(-2\alpha R)$; (iii) a rate term which can be regarded as an attempt frequency ν_0, and which will depend on the strength of the electron–phonon coupling and the phonon density of states, but will be only weakly dependent on R or Δ. The hopping transition rate is then given by:

$$\gamma = \nu_0 \exp(-2\alpha R - \Delta/k_B T) \qquad [6.48]$$

(Note that for hops downward in energy the probability is just $\nu_0 \exp(-2\alpha R)$ by the principle of detailed balance.) The conductivity may be calculated by making use of the Einstein relation between mobility and diffusion coefficient $\mu = eD/k_B T$, where $D = \gamma R^2/6$ if the conduction process can be thought of as arising from a random, diffusive motion, and if the density of states at the Fermi level $N(E_F)$ is constant, so that $k_B T N(E_F)$ is the concentration of electrons contributing:

$$\sigma = \frac{e^2}{6} R^2 \nu_0 \exp(-2\alpha R - \Delta/k_B T) \qquad [6.49]$$

At low temperatures, where the number and energy of phonons are both small, hopping to near neighbours (thereby minimizing the $\exp(-2\alpha R)$ term) is unfavourable because of the large energy separations encountered on average. Instead, it is more favourable for the electron to tunnel to more distant sites, at the expense of the $\exp(-2\alpha R)$ term, but at a saving in the $\exp(-\Delta/k_B T)$ term since there is a higher probability that more distant sites will have smaller energy separations. Mott (1969) therefore considered that the factor $(2\alpha R + \Delta/k_B T)$ should be optimized, subject to the condition that there is at least one state at a given spatial and energy separation, i.e.

$$1 = \frac{4\pi}{3} N(E_F)\Delta R^3 \qquad [6.50]$$

If [6.50] is substituted into the factor $(2\alpha R + \Delta/k_B T)$, maximization of the exponent yields for the optimum hopping distance:

$$\bar{R} = \left[\frac{9}{8\pi\alpha N(E_F)k_B T}\right]^{1/4} \qquad [6.51]$$

Substitution of $\bar{R}$ into [6.50] then gives a value for the optimized energy separation $\bar{\Delta}$, and if these are both inserted into [6.49], this yields the final expression for the conductivity:

$$\sigma = \sigma_0 \exp[-A/T^{-1/4}] \qquad [6.52]$$

where

$$\sigma_0 = \frac{e^2}{6}\nu_0 N(E_F)\left[\frac{9}{8\pi\alpha N(E_F)k_B T}\right]^{1/2} = \frac{e^2\nu_0}{2(8\pi)^{1/2}}\left[\frac{N(E_F)}{\alpha k_B T}\right]^{1/2} \qquad [6.53]$$

and

$$A = 2.1[\alpha^3/k_B N(E_F)]^{1/4} \qquad [6.54]$$

Thus, we see that 'variable-range hopping' in a band of constant density of states is characterized by an $\exp(-AT^{-1/4})$ temperature dependence of the d.c. conductivity.

This behaviour has been observed in several amorphous semiconductors at low temperatures, where the activated nature of band conduction (sect. 5.4.1) ensures that it is smaller than the contribution from variable-range hopping: Fig. 6.39 shows schematically the change in conduction mechanism on lowering the temperature. Variable-range hopping has been observed principally in materials which contain positive U defects (see sect. 6.2.1), e.g. a-Si or a-Ge. Note that using [6.52] it is possible in principle to obtain independent estimates for $N(E_F)$ and α from experimental values of both the pre-exponential factor σ_0, i.e. the $T = 0$

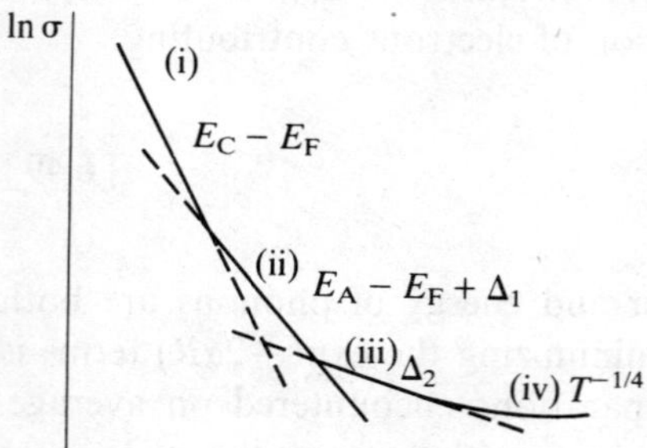

Fig. 6.39 Schematic illustration of the temperature dependence of conductivity expected for an amorphous semiconductor. The activation energies found in the various regimes are indicated (after Mott and Davis 1979).

intercept ([6.53]) and the exponent A([6.54]). However, almost universally it is found that the pre-exponential term is anomalously high, yielding unphysically large values for $N(E_F)$ or ν_0, depending on which is given. This behaviour is not fully understood, but may have to do with single-phonon theories that have been developed to obtain values for ν_0. Nevertheless, the exponent A appears to be better defined, and much use has been made of it to extract values for $N(E_F)$. However, $N(E_F)$ cannot be obtained independently of α, but given assumed values for α^{-1} (usually $\simeq 10$ Å), estimates in the range 10^{18}–$10^{19}\ \text{cm}^{-3}\ \text{eV}^{-1}$ are commonly obtained. It is of interest to note that if $N(E_F) = 10^{19}\ \text{cm}^{-3}\ \text{eV}^{-1}$ and $\alpha^{-1} = 10$ Å then [6.51] gives a hopping distance of $\bar{R} = 80$ Å at $T = 100$ K.

Many different theoretical approaches dealing with the hopping problem have been developed since Mott's early treatment. Apsley and Hughes (1974, 1975) considered the hops to take place in a four-dimensional space (consisting of one energy and three spatial coordinates), and using a different averaging technique to optimize the 'range', obtained an expression of the same form as [6.52] for the case of a constant density of states at low temperature and electric fields, but with some differences in the values of the pre-exponential and exponent factors. They also obtained an expression for the electric field dependence of the hopping conductivity (i.e. non-ohmic behaviour) which is observed at high fields for systems in the hopping regime. The effect of the applied field F is to modify the energy of a hop between two sites i and j, separated by a distance R_{ij}, by an amount $eFR_{ij}\cos\theta$, where θ is the angle between the directions $\boldsymbol{F}$ and $\boldsymbol{R}_{ij}$. At high fields, the quantity eFR_{ij} becomes greater than the average hopping energy, and an electron then only moves by means of downward hops with the emission of a phonon with each hop. In this case, *no* temperature dependence of the hopping conductivity is expected, and it takes the limiting form:

$$\sigma = \sigma_0^{\text{HF}} \exp(-B/F^{1/4}) \qquad [6.55]$$

(since the critical field is $F \simeq 3/(4\pi e N(E_F) R^4)$ and the conductivity is proportional to $\exp(-2\alpha R)$). In fact Apsley and Hughes (1975) obtained for the pre-exponential and exponent factors respectively:

$$\sigma_0^{\text{HF}} = \frac{Ne^2\nu_0}{\alpha^2}\left[\frac{\alpha^4}{4\pi e N(E_F)}\right]^{1/4}\frac{1}{F^{1/4}} \qquad [6.56]$$

and

$$B = \left[\frac{64\alpha^4}{\pi e N(E_F)}\right]^{1/4} \qquad [6.57]$$

Thus by making measurements at low fields (in the $T^{-1/4}$ regime) and at high fields (in the $F^{-1/4}$ regime), independent estimates for α and $N(E_F)$ may be obtained from the exponents alone ([6.54] and [6.57]).

A more rigorous treatment of the hopping process is given by percolation theory (see sect. 5.3.6). The electron states having energies in the gap are randomly distributed in an amorphous solid, and the transition rate γ_{ij} between a pair of such sites has the form given in [6.48]. This situation can be represented as a series of conductances G_{ij} connecting any given pair of sites whose value is given by:

$$G_{ij} \propto \frac{e^2}{k_B T} \gamma_{ij}(R_{ij}, E_i, E_j)$$

However, because of the random distribution of site separations and energies, there is an enormous spread in transition rates or equivalent conductances in such a system. The overall conductivity of the amorphous solid modelled as a resistor network can be written as $\sigma = \tilde{G}/\tilde{R}$, where $\tilde{G}$ is some characteristic value of the conductances in the network and $\tilde{R}$ is some characteristic length-scale of the system. The essential feature of the percolation approach is to assert that $\tilde{G}$ is given by G_c, the critical percolation conductance, defined as the largest conductance such that there is a connected path through the network for all $G_{ij} > G_c$. Thus it is neither the abundant small conductances, nor the relatively scarce large conductances which determine the conductivity, but the critical conductance which establishes a percolation path: resistors for which $G_{ij} \ll G_c$ are effectively shorted out and those for which $G_{ij} \gg G_c$ form isolated regions which cannot therefore contribute to current throughout the sample.

Thus, solution of the hopping problem hinges on the evaluation of the critical percolation conductance for the network in question. If we rewrite [6.48] for a pair of sites i and j, introducing the energies of each site explicitly, the condition $G_{ij} > G_c$ may be written as (Ambegaokar *et al.* 1971):

$$[2\alpha R_{ij} + (|E_i| + |E_j| + |E_i - E_j|)/2k_B T] < \ln\left(\frac{\nu_0}{\gamma_c}\right) \qquad [6.58]$$

where $\gamma_c = k_B T G_c / e^2$. The inequality 6.58 may be re-expressed in dimensionless terms as:

$$\frac{R_{ij}}{R_{max}} + \frac{(|E_i| + |E_j| + |E_i - E_j|)}{2E_{max}} < 1 \qquad [6.59]$$

where

$$R_{max} = \frac{1}{2\alpha} \ln\left(\frac{\nu_0}{\gamma_c}\right) \qquad [6.60(a)]$$

$$E_{max} = k_B T \ln\left(\frac{\nu_0}{\gamma_c}\right) \qquad [6.60(b)]$$

Thus the inequalities $R_{ij} < R_{max}$ and $|E_i| \leqslant E_{max}$ must independently be satisfied for [6.58] (or [6.59]) to hold. The total number of sites per unit volume such that $|E_i| \leqslant E_{max}$ is:

$$n = 2N(E_F) E_{max} \qquad [6.61]$$

assuming $N(E)$ to be constant in the region of E_F. The n sites are randomly distributed in space and the term (E_i/E_{max}) is randomly distributed between -1 and 1. The critical path is established if [6.59] is satisfied, but part at least of this network must span the sample. This requirement can be expressed in the form:

$$nR_{max}^3 = \eta_c \qquad [6.62]$$

where η_c is a constant of order unity. Ambegaokar *et al.* (1971) estimate $\eta_c \simeq 4$ from a consideration of the percolation limit for 4-D (three spatial and one energy) hyperspheres. (Other authors, e.g. Pike and Seeger (1974), use Monte Carlo techniques to obtain similar answers.) Combining [6.60], [6.61] and [6.62] gives:

$$\ln\left(\frac{\nu_0}{\gamma_c}\right) = \left(\frac{4\alpha^3\eta_c}{k_B T N(E_F)}\right)^{1/4} \qquad [6.63]$$

which gives an expression for the hopping conductivity of the Mott form [6.52], where,

$$A = \left(\frac{4\alpha^3\eta_c}{k_B N(E_F)}\right)^{1/4} \qquad [6.64]$$

Thus far we have considered exclusively hopping in three dimensions where, if the density of states is constant, the conductivity has the Mott $T^{-1/4}$ form. Consider now, however, the case when the length of one dimension of the system is reduced below that of the hopping length, $\bar{R}$ ([6.51]). In this case, hopping motion is effectively confined to *two* dimensions and it can be shown (see problem 6.1) that a different temperature dependence results. In fact, the equation for variable-range hopping in this regime is:

$$\sigma = \sigma_0 \exp[-C/T^{1/3}] \qquad [6.65]$$

where

$$C = \left(\frac{3\alpha^2}{d k_B N(E_F)}\right)^{1/3} \qquad [6.66]$$

where d is the film thickness and the constant varies slightly between different treatments. Note carefully that it has been assumed that 2-D behaviour results because the hopping is forced to be two dimensional because the film thickness is less than the hopping distance, and in this case the density of states in [6.66] is the *bulk* value ($\text{cm}^{-3}\,\text{eV}^{-1}$). If, however, a truly two-dimensional random system is considered (as for example at the 'inversion layer' in MOSFETs – see Mott *et al.* 1975*b*), the 2-D density of states ($\text{cm}^{-2}\,\text{eV}^{-1}$) must be used, and there is therefore no thickness term d in [6.66]. Shown in Fig. 6.40 are experimental data for thin films of a-Ge of different thicknesses, deposited by evaporation in ultra-high vacuum (Knotek *et al.* 1974). Clear evidence is seen for a change in temperature dependence from $T^{-1/4}$ to $T^{-1/3}$ for films thinner than $\simeq 500$ Å, lending support to the notion of 2-D variable-range hopping. Further confirmation of this comes from the film thickness dependence of the slope C of $\ln\sigma$ vs. $T^{-1/3}$ plots (Fig. 6.40(c)): the slope varies as $d^{-1/3}$ as expected from [6.66], but no variation of the slope A is found in the 3-D regime. It should be noted from [6.54] and [6.66] that α and $N(E_F)$ appear together in a different functional form in the exponents for 3-D and 2-D hopping, and so varying the thickness of the films and measuring the temperature dependence of σ in the two regimes offers another method of obtaining the quantities α and $N(E_F)$ independently, in addition to that using the field dependence mentioned earlier.

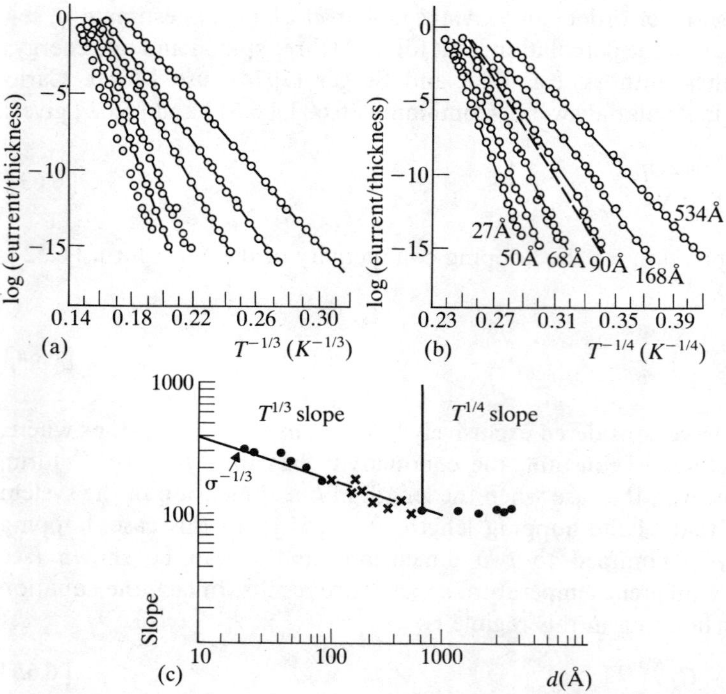

Fig. 6.40 Illustration of variable-range hopping in a-Ge and the transition from 3-D to 2-D behaviour. Data are plotted versus (a) $T^{-1/3}$ and (b) $T^{-1/4}$ (Knotek *et al.* 1974); the dashed lines are straight. The thickness dependence of the slope of such plots is shown in (c), clearly showing the transition from 2-D to 3-D hopping (Knotek 1974).

6.3.4 Frequency-dependent conductivity

We have seen in the previous section that upon the application of a d.c. electric field, conduction can take place among localized gap states in amorphous semiconductors by a mechanism of phonon-assisted hopping, which is simply activated with a constant activation energy if hopping is between (nearest) neighbours at constant range ([6.49]), but with effectively a temperature-dependent activation energy (i.e. $T^{-1/4}$ behaviour) if the hopping is between neighbours at variable ranges ([6.52]). This d.c. behaviour could be thought of as the $\omega = 0$ limit of a *frequency-dependent* (a.c.) conductivity, $\sigma(\omega)$, observed upon the application of an alternating (sinusoidal) field of frequency ω if the hopping mechanism is the *same* in both cases. We will see that for some situations this is a valid approach, but in others, notably for those materials which do *not* exhibit d.c. hopping conductivity but instead only band conduction, it is not valid and the a.c. conductivity must therefore result from a completely different mechanism.

Frequency-dependent conductivity is also interesting in that it has been observed in every amorphous semiconductor or insulator measured, and also because in the frequency range $\simeq 10$–10^6 Hz in almost every case the qualitative behaviour is the same, i.e.:

$$\sigma(\omega) = C\omega^s \tag{6.67}$$

where the exponent s is near unity ($\leqslant 1$) and may be weakly temperature dependent, and C is independent of frequency and also only weakly dependent on temperature. Thus a.c. conduction is, like the thermal anomalies observed at low temperatures (sect. 4.5) unique to the amorphous phase, and furthermore it seems to be a general property shared by most if not all amorphous semiconductors and insulators. We will therefore at the outset establish those general principles which mainly determine this behaviour, and leave until later a fuller discussion of the various mechanisms that have been proposed to account in detail for the phenomenon. Extensive reviews of this topic have been given by Long (1982) and Elliott (1987a).

The concept of a frequency-dependent conductivity should not appear strange since it has previously appeared in another guise, albeit for a different frequency range. Thus, a.c. conductivity is related to the optical absorption coefficient, $\alpha(\omega)$ (cf. [5.23]), since both involve energy loss within the solid, by:

$$\sigma(\omega) = \frac{n_0 \alpha(\omega)}{377} \qquad [6.68]$$

where n_0 is the refractive index, and the units of $\sigma(\omega)$ and $\alpha(\omega)$ are $\Omega^{-1}\,\text{cm}^{-1}$ and cm^{-1}, respectively. As an example of this relationship, we show in Fig. 6.41 data for a-As_2Se_3 plotted in this fashion for a frequency range spanning 13 decades. Although a plot of this sort is interesting, it does not imply of course that the same mechanism responsible for the loss (a.c. conductivity or optical absorption) is operative for all frequencies. In this section dealing with defect-related properties, we consider only the linear region in the frequency range $\simeq 10$–10^6 Hz, since this is determined by electronic hopping between defects.

We will demonstrate first the origin of a linear frequency-dependent conductivity. Consider the application of a harmonically varying electric field $E(t) = E_0 \sin \omega t$ to a sample, producing a polarization $P(t)$. These are related via

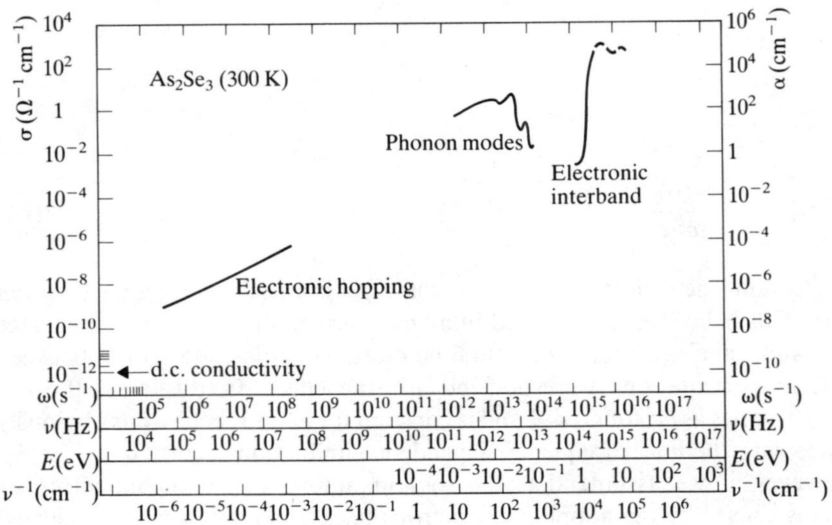

Fig. 6.41 Comparison of absorption (or equivalently conductivity) in different frequency regimes for a-As_2Se_3. The equivalence of various frequency and energy scales is also shown (Mott and Davis 1979).

Fourier transforms to the complex quantities $E(\omega)$ and $P(\omega)$ in the frequency domain, where:

$$P(\omega) = \varepsilon_0 \chi(\omega) E(\omega) \tag{6.69}$$

where ε_0 is the free-space permittivity. The dielectric susceptibility is in general a complex function of frequency:

$$\chi(\omega) = \chi'(\omega) + i\chi''(\omega) \tag{6.70}$$

The imaginary part, $\chi''(\omega)$, is termed the 'dielectric loss', since the current due to this is in phase with the applied field. The susceptibility is further related to the dielectric constant ([5.17]) by:

$$\varepsilon_1(\omega) = 1 + \chi'(\omega) \tag{6.71(a)}$$

$$\varepsilon_2(\omega) = \chi''(\omega) \tag{6.71(b)}$$

Finally the a.c. conductivity is related to the dielectric loss by:

$$\sigma(\omega) = \varepsilon_0 \omega \chi''(\omega) \tag{6.72}$$

We now assume that the dielectric response of the amorphous solid to the applied alternating field is Debye-like, i.e. the polarization, which is caused, say, by the transition of a charge between two preferred sites, decays with first-order kinetics on removal of the exciting field, with a characteristic relaxation time, τ, viz. $dP/dt = -P/\tau$. This has the solution $P(t) = P_0 \exp(-t/\tau)$, and Fourier transforming $f(t) = [-dP/dt] = (P_0/\tau)\exp(-t/\tau)$ yields finally the frequency dependence of the susceptibility:

$$\chi(\omega) = \frac{\chi(0)}{(1 + i\omega\tau)}$$

or

$$\chi'(\omega) = \frac{\chi(0)}{(1 + \omega^2\tau^2)} \tag{6.73(a)}$$

$$\chi''(\omega) = \frac{\omega\tau\chi(0)}{(1 + \omega^2\tau^2)} \tag{6.73(b)}$$

If τ is a constant, the dielectric loss $\chi''(\omega)$ will exhibit a peak at a frequency given by $\omega\tau = 1$. This behaviour is observed in many glassy hydrocarbon polymers (see e.g. Wong and Angell 1976), where rotational motion of polar side-groups between two preferred orientations is responsible for the effect. In this case, the a.c. conductivity varies as ω^2 for $\omega\tau \ll 1$ and is constant for $\omega\tau \gg 1$, behaviour obviously at variance with the linear frequency dependence under consideration.

However, if we assume that the loss in amorphous semiconductors or insulators is caused by the hopping of electrons back and forth between localized defect states, which will be randomly situated in space and energy, a broad distribution of relaxation times $n(\tau)$ must exist. In this case, the a.c. conductivity may be written as:

$$\sigma(\omega) = \int_0^\infty a(\omega)n(\tau)\frac{\omega^2\tau}{(1+\omega^2\tau^2)}\,d\tau \qquad [6.74]$$

where $a(\omega)$ is the polarization for a pair of sites, and we have assumed that the total conductivity may be obtained by summing over the individual contributions to the dielectric response made independently by *pairs* of sites, each exhibiting a Debye response.

If the distribution of relaxation times is such that $n(\tau) \propto 1/\tau$ and $a(\omega)$ is frequency independent, a linear a.c. conductivity results, since

$$\sigma(\omega) \propto \int \frac{\omega\, d(\omega\tau)}{(1+\omega^2\tau^2)} \propto \omega$$

For the condition $n(\tau) \propto 1/\tau$ to hold, the relaxation time must be an exponential function of some random variable ζ:

$$\tau = \tau_0 \exp(\zeta) \qquad [6.75]$$

where τ_0 may be regarded as being equal to an inverse phonon frequency, and ζ has a flat distribution, $n(\zeta) = \text{constant}$ (i.e. $n(\tau) = n(\zeta)(d\zeta/d\tau) \propto \tau^{-1}$). In practice, both $n(\zeta)$ and $a(\omega)$ may be weakly dependent on frequency and this results in the frequency exponent s in [6.67] being slightly sublinear.

The essential physics relating to the precise mechanism of charge transport giving rise to a near-linear a.c. conductivity is contained in [6.75], in particular in the quantity ζ. Two different mechanisms that could give rise to the form of [6.75] are the following:

(a) Classical activation of a carrier *over* a potential barrier W separating two sites, in which case:

$$\zeta = W/k_B T \qquad [6.76]$$

(b) Quantum-mechanical tunnelling (QMT) of a carrier *through* the potential barrier between the sites separated by a distance R, in which case:

$$\zeta = 2\alpha R \qquad [6.77]$$

The former mechanism in its simplest form, when W is independent of the spatial separation R, can be operative in both the simple Debye-loss case ([6.73(b)], when W is fixed and the frequency ω_{max} at which the maximum loss occurs (given by $\tau\omega_{max} = 1$) is activated, $\omega_{max} \propto \exp(-W/k_B T)$, and also in the case when there is a distribution of barrier heights in an amorphous material.

Pollak and Pike (1972) suggested that the latter case holds generally in all glassy systems exhibiting low-temperature thermal anomalies, and that the loss mechanism is due to the motion of *atoms* (carrying at least a partial charge) between two sites forming a two-level system (see Fig. 4.19). In this case, the polarizability can be shown to depend on R and Δ via (Pollak and Geballe 1961):

$$a = \frac{e^2R^2}{12k_B T\cosh^2(\Delta/2k_B T)} \qquad [6.78]$$

The total conductivity can be evaluated using [6.74], and assuming the activation energies W to be uniformly distributed as $p(R, \Delta, W) = \delta(R - R_0)/(R^2 W_0 \Delta_0)$ gives:

$$\sigma(\omega) = \frac{\pi\omega e^2 R_0^2 N k_B T \tanh(\Delta_0/2k_B T)}{6\Delta_0 W_0} \qquad [6.79]$$

where N is the concentration of sites (cm^{-3}). Note that this atomic hopping model predicts that the frequency dependence of $\sigma(\omega)$ is exactly linear and temperature independent, at variance with the experimental observation that in glasses s is generally sub-linear and increases with decreasing temperature. Note also that both this model and the simple Debye-loss are examples of *fixed-range* hopping mechanisms, in which carrier motion takes place between sites separated by a constant distance, but only those sites which have activation energies such that $\omega\tau = 1$, i.e. $W_\omega = k_B T \ln(1/\omega\tau_0)$, contribute significantly to the total conductivity. In the random case, this may be regarded as 'variable-energy' hopping, since measurement at a frequency ω selects only those sites having energies W_ω.

If one considers now the other extreme model for carrier motion, namely quantum-mechanical tunnelling (QMT) ([6.77]), it is generally assumed that the defect centres are randomly distributed, and the probability of finding a centre at distance R is:

$$P(R)\,dR = 4\pi N R^2\,dR \qquad [6.80]$$

Now, however, the hopping *is* variable range in nature, since the requirement that $\omega\tau = 1$ imposes the condition that only those pairs of centres separated by the hopping distance R_ω, contribute significantly to the conductivity at a frequency ω, where:

$$R_\omega = \frac{1}{2\alpha} \ln(1/\omega\tau_0) \qquad [6.81]$$

Equations [6.78] and [6.80] taken with [6.81] lead to the expression for $\sigma(\omega)$ ([6.74]) containing an R_ω^4 term. If this is evaluated, the following equation for the a.c. conductivity resulting from quantum-mechanical tunnelling is obtained:

$$\sigma(\omega) = CN^2(E_F) k_B T e^2 \alpha^{-5} \omega \ln^4(1/\omega\tau_0) \qquad [6.82]$$

where the constant C is given by $\pi/3$ (Austin and Mott 1969) or $\pi^4/96$ (Pollak 1971), and $N(E_F)$ is the density of states at the Fermi level ($cm^{-3}\,eV^{-1}$). The frequency exponent of [6.82] can be written as:

$$s = 1 - 4\ln^{-1}(1/\omega\tau_0) \qquad [6.83]$$

whereupon it can be seen that in this model s should be temperature independent but frequency dependent, having the value $s = 0.8$ for $\omega = 10^4\,s^{-1}$ and $\tau_0 = 2 \times 10^{-13}$ s. These predictions are again contrary to experimental findings, even for those systems (a-Si, a-Ge) for which the tunnelling model is expected to apply.

These considerations have led the author (Elliott 1977, 1987a) to propose an alternative model for a.c. conduction, at least in chalcogenide glasses where the defects suffer a negative Hubbard U (sect. 6.2.2). This model is a combined

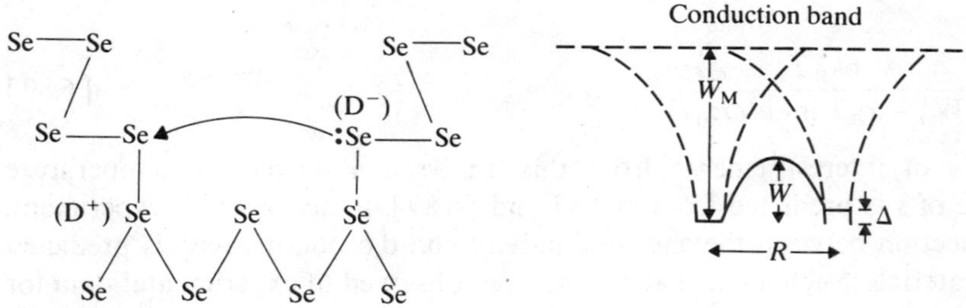

Fig. 6.42 (a) Schematic illustration of the hopping process envisaged to take place in the bipolaron correlated barrier hopping model for a.c. conduction in chalcogenide glasses (Elliott 1977). Two electrons transfer from a D^- to a D^+ centre, which thereby exchange places (the new bond formed is shown by the dotted line). (b) Diagram of the potentials experienced by carriers distance R apart. W_M is the energy required to remove two electrons from a D^- centre to form a D^+; W is the height of the barrier over which the carriers must hop. The disorder energy is shown as Δ.

variable-energy and variable-range hopping model, since the height of the potential barrier is correlated with the intersite separation, i.e. 'correlated barrier hopping', CBH. The detailed mechanism is shown in Fig. 6.42(a); *two* electrons are assumed to hop together from a D^- centre to a neighbouring D^+ centre over the potential barrier between them which, because the centres are charged, is reduced by a Coulomb term:

$$W = W_M - \frac{8e^2}{4\pi\varepsilon\varepsilon_0 R} \qquad [6.84]$$

and this reduction is shown schematically in Fig. 6.42(b). The value of the maximum barrier height (at infinite separation) can be shown by a consideration of defect energies to be approximately equal to the optical band gap E_o

$$W_M \simeq E_o$$

Using [6.75], [6.76], [6.80] and [6.84], together with the equation for the polarizability ([6.78]) corrected for two-electron transport, [6.74] can be evaluated to give (Elliott 1987a)

$$\sigma(\omega) = \frac{\pi^3}{3\Delta_0} N^2 \varepsilon\varepsilon_0 k_B T \omega R_\omega^6 \tanh(\Delta_0/2k_B T) \qquad [6.85]$$

where N is the concentration of charged defects, and the hopping distance is given by:

$$R_\omega = \left(\frac{2e^2}{\pi\varepsilon\varepsilon_0}\right)\frac{1}{(W_M - W_\omega)} \qquad [6.86]$$

The frequency exponent is defined by:

$$s = \partial \ln \sigma / \partial \ln \omega \qquad [6.87]$$

and hence

$$s = 1 - \beta \qquad [6.88]$$

where

$$\beta = \frac{6k_B T}{[W_M - k_B T \ln(1/\omega\tau_0)]} \qquad [6.89]$$

Two points of interest emerge from this analysis: a decreasing temperature dependence of s is predicted from [6.88] and [6.89], in accord with experiment, and a connection between the a.c. conductivity and d.c. conductivity is predicted for these materials. Such a correlation has been observed in experimental data for chalcogenide glasses (see Mott and Davis 1979).

Although the pair approximation is adequate at sufficiently high frequencies and/or low temperatures, it must be recognized that it fails completely to reproduce the behaviour in the d.c. limit, i.e. as $\omega \to 0$. This is because *if* the same mechanism is operative in both a.c. and d.c. regimes, as the frequency is lowered, carrier motion between an increasing number of sites takes place, until the d.c. limit or a percolation path is established. Thus in this case, any theory must be capable of accounting for the conductivity behaviour in all frequency regimes. Summerfield and Butcher (1982) developed an 'extended pair approximation' (EPA) model which takes into account, in an averaged way, that the relaxation rate of a given pair of sites is affected by all the other sites in the network. Using this approach, Summerfield and Butcher (1985) have been able to calculate the total conductivity response, ranging from d.c. to the high-frequency (pair-approximation) limit for a number of different microscopic transport mechanisms, including QMT and CBH. Movaghar *et al.* (1980a,b,c) have also evaluated $\sigma(\omega)$ over the entire frequency scale using a different approach, namely using the continuous-time random walk model, first proposed by Scher and Montroll (1975).

In contrast to the above approaches for treating the a.c. conductivity, which are based on the premise that the conduction occurs by the motion of charge carriers between defect states lying in the gap, and that the overall response arises from a summation of the contributions of independent relaxation events acting in parallel, two other causes of the a.c. conductivity observed in amorphous semiconductors have been suggested. One alternative approach asserts that the frequency dependence of the conductivity arises instead from conduction in the bands (or in the band tails) in materials showing potential fluctuations; this model has already been discussed in sect. 5.4.4. The other approach assumes that the a.c. response results from relaxation processes acting in *series*, i.e. a given relaxation event constrains subsequent events (Palmer *et al.* 1984). This approach has been employed in a discussion of a.c. *ionic* conductivity in glasses (see sect. 4.5.3), but it is not clear whether it could also be applied to electronic conductivity in amorphous semiconductors.

6.3.5 Drift mobility

In this chapter and the last we have discussed electrical transport, in general under steady-state conditions, omitting from the discussion much mention of transient phenomena. However one form of transient electrical transport has been very valuable in giving information about states in the gap of amorphous semiconductors and the mechanism of charge transport, namely drift mobility (or transient photoconductivity) measurements. In these experiments, carriers are injected at one position in a sample and in the presence of an applied field E their transit time t_t to another point at a distance L is determined (see Spear 1969, Marshall 1983, 1985, and Tiedje 1984 for reviews). The drift mobility is then defined as:

$$\mu_d = L/Et_t \qquad [6.90]$$

If the circuit time constant is much less than the experimental time-scale (i.e. $RC \ll t_t$) then the field in the sample is approximately uniform during the transit and [6.90] is valid. Highly resistive samples ($\rho > 10^7\,\Omega\,\text{cm}$) need to be used so that the dielectric relaxation time is much greater than t_t and therefore local charge neutralization does not occur to a significant extent during the charge transit. Samples can be arranged with the electrodes in a sandwich configuration, one of which may be semi-transparent if optical excitation is to be employed. Carrier injection is then effected either by the application of a voltage pulse (of duration less than t_t) of the appropriate polarity to an electrode, in which case carriers of a single type (electron or hole) are created, or else by the application of a short pulse of strongly absorbed light or of an electron beam; in either case electron–hole pairs are created close to one electrode and, according to the polarity of the applied field, carriers of only one sign can transit, the others being dissipated at the electrode.

In the ideal case, a perfect current pulse whose leading profile is given by a step-function, would drift unaltered across the sample and the time at which the carrier sheet reached the back electrode would define the transit time, t_t. In practice, dispersion of the charge sheet takes place to a greater or lesser extent. In the simplest case, the charge sheet spreads out as it drifts across the sample but retains its identity as a well-defined pulse. The broadening of the pulse arises because the drift velocity of the carriers obeys gaussian statistics; the dispersion of the gaussian pulse as measured by the FWHM, σ, increases with time as $\sigma \propto t^{1/2}$, and the mean displacement of the pulse of course is proportional to the elapsed time, $l \propto t$. This process leads to a rounding of the previously sharp current transit profile, but nevertheless t_t may easily be determined: this process is illustrated in Fig. 6.43(a).

In the case shown in Fig. 6.43(a), where the transit time is clearly defined, it is pertinent to ask what is the drift mobility as deduced from [6.90]: is it equal to the conductivity mobility μ_e ([5.78], [5.79]) characteristic of conduction in extended states? In the ideal case where the carriers transport in a band in the absence of traps, μ_d and μ_e are identical, but in a real amorphous semiconductor traps are invariably present, either shallow-lying as localized tail states below the mobility edge or deeper lying in the form of 'defect' states in the gap. In the simplest case, when the carrier motion occurs at an effective mobility edge (say E_c) and is 'trap limited' by occasional trapping by a single well-defined set of traps at energy E_t below E_c, the drift mobility is, by definition:

$$\mu_d = \mu_e \frac{n_{\text{free}}}{n_{\text{free}} + n_{\text{trap}}} \qquad [6.91]$$

which can be written approximately as:

$$\mu_d = \frac{N_c}{N_t} \mu_e \exp(-E_t/k_B T) \qquad [6.92]$$

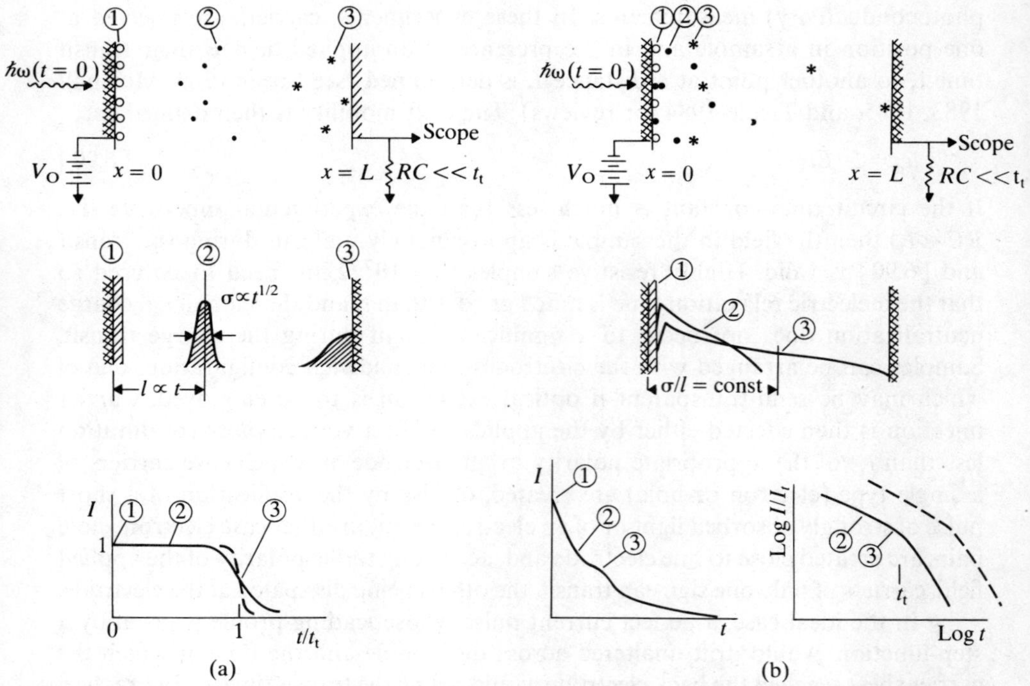

Fig. 6.43 Schematic illustration of carrier propagation under (a) gaussian and (b) non-gaussian conditions. In all cases, (1) refers to the situation at $t = 0$, (2) to $t < t_t$, (3) to $t \simeq t_t$. The current pulse in the external circuit induced by the charge displacement is plotted linearly versus time, and also double-logarithmically for the non-gaussian case (where the dashed curve represents transients at a lower applied bias field, i.e. a longer transit time) (after Pfister and Scher 1978).

where N_c $(= k_B T N(E_c))$ and N_t are the concentration of states at the (conduction band) mobility edge and of the trapping levels, respectively: a similar equation holds for the case of hole transport. Thus, for this case the trap level E_t may be obtained from the temperature dependence of μ_d, and the density of trap states can be deduced from a combination of drift mobility and d.c. conductivity studies: since the pre-exponential factor of the d.c. conductivity for conduction in extended states is $e\mu_e N_c$ (cf. [5.76]), [6.92] gives, together with [5.83],

$$N_t = \frac{\sigma}{e\mu_d}(-\gamma/k_B)\exp[(\Delta E_\sigma - E_t)/k_B T] \qquad [6.93]$$

An interesting example of drift mobility is that seen in a-SiO_2 films, where electron and hole transport exhibit startlingly different behaviour. *Electrons* have a remarkably high drift mobility at room temperature, $\mu_d^e \simeq 20\,\text{cm}^2\,\text{V}^{-1}\,\text{s}^{-1}$ (Hughes 1973), which is characteristic of conduction mobilities in extended states; this observation means that the width of the localized tail states at the conduction band edge must be exceedingly small. Mott has ascribed this to the s-like character of the states (derived from both Si and O) at the bottom of the conduction band; the considerable degree of short-range order combined with the s-like nature of the electron states means that the disorder potential will be small thereby restricting the range of the localized tail states, which is the same argument invoked to explain similar behaviour in liquid rare gases (Mott and Davis 1979). The situation for the valence band edge will be completely different however; the directional character of the predominant p-states means that even small amounts of topological disorder can give rise to a considerable range in energy of localized tail states. Indeed, the *hole* drift mobility in a-SiO_2 is much smaller than for electrons at room temperature ($\mu_d^h \simeq 10^{-5}\ cm^2\ V^{-1}\ s^{-1}$), indicative of considerable interaction between the hole and these (or other) states. Furthermore, the transport mechanism for holes appears to be different in two time regimes: for times $\lesssim 10^{-6}$ s, the activation energy of μ_d is $\simeq 0.16$ eV, but this increases to $\simeq 0.37$ eV at longer times (Hughes 1977). The behaviour at very short times, the so-called 'prompt mobility', has been ascribed to the formation and subsequent hopping of a hole-like small polaron in the lone-pair p-states at the top of the valence band, and the polaronic nature of the carrier is supported by the observation of a marked decrease in activation of energy of μ_d at low temperatures ($\simeq 200$ K); precisely this behaviour is expected for the polaron model, it will be recalled (sect. 5.5), where the multiphonon hopping processes, dominant at high temperature, become frozen-out with decreasing temperature leading ultimately to polaron band conduction. The drift mobility in a-SiO_2 at times longer than $\simeq 10^{-6}$ s cannot be described in this way, and is in fact a manifestation of the extremely dispersive transits often observed in amorphous semiconductors and which will be discussed next.

So far, we have discussed drift mobility under the conditions where some dispersion of the carrier pulse may be evident, but the pulse nevertheless remains well defined and characterized by gaussian statistics. In many cases, however, the dispersion of the transit pulse is so severe that the pulse shape is no longer recognizable and defining the transit time becomes very difficult (see Fig. 6.43(b)); this situation has been termed 'anomalous transit dispersion' (Pfister and Scher 1978). This is characterized by the pulse spread, σ, and its mean displacement, l, having the *same* time dependence, i.e. $\sigma/l = \text{constant}$ (cf. $\sigma/l \propto t^{-1/2}$ for gaussian transport). In order for anomalous transit dispersion to occur, the microscopic processes which control the transport must have a wide distribution of characteristic times which extends into the experimental time domain determined by t_t (which measures the transit time of a certain fraction of the fastest injected carriers), and only then will the carrier packet grow asymmetrically as shown schematically in Fig. 6.43(b). In this anomalous regime, the drift mobility is field dependent and also is an apparent function of the film thickness.

This intriguing behaviour was first accounted for by a theory developed by Scher and Montroll (1975); the reader is referred to Pfister and Scher (1978) for an extensive review. The model simulates the actual distribution of transition rates which characterize the carrier transit (due to either multiple trapping events in a distribution of trap levels or to tunnelling between randomly separated traps) by means of a single probability distribution $\psi(t)$ for a carrier moving randomly on a *regular* lattice, where $\psi(t)$ represents the probability of a carrier hopping to its next site in time t, having arrived at the original site at $t = 0$; such a process is referred to as a 'continuous time random walk' (CTRW). Scher and Montroll (1975) proposed that the distribution function $\psi(t)$ should behave as

$$\psi(t) \propto t^{-(1+\alpha_d)} \qquad [6.94]$$

where the disorder parameter α_d (not to be confused with the inverse localization length) takes values between 0 and 1, and is related in a complicated way to microscopic parameters such as trap deaths, radii of localized states, etc., decreasing as the amount of disorder increases. Although the power-law dependence of $\psi(t)$ embodied in [6.94] does reproduce many of the features of anomalous dispersive transport, it is not a unique distribution (and cannot be so since all the 'randomness' of the hopping system is folded into this one function); note, however, the difference between the distribution function characteristic of CTRW ([6.94]) and that which characterizes *gaussian* transport, i.e.

$$\psi(t) = \tau^{-1} \exp(-t/\tau) \qquad [6.95]$$

which is associated with a *single* event time τ.

The general result of the CTRW model can be summarized in the following equations for the time dependence of the current:

$$I(t) \propto t^{-(1-\alpha_d)} \quad t < t_t \qquad [6.96(a)]$$

$$I(t) \propto t^{-(1+\alpha_d)} \quad t > t_t \qquad [6.96(b)]$$

Figure 6.44(a) shows the temperature dependence of the hole drift velocity in a-Se for different values of electric field; note particularly that as the temperature is reduced below $\simeq$200 K the approximately rectangular current profile typical of

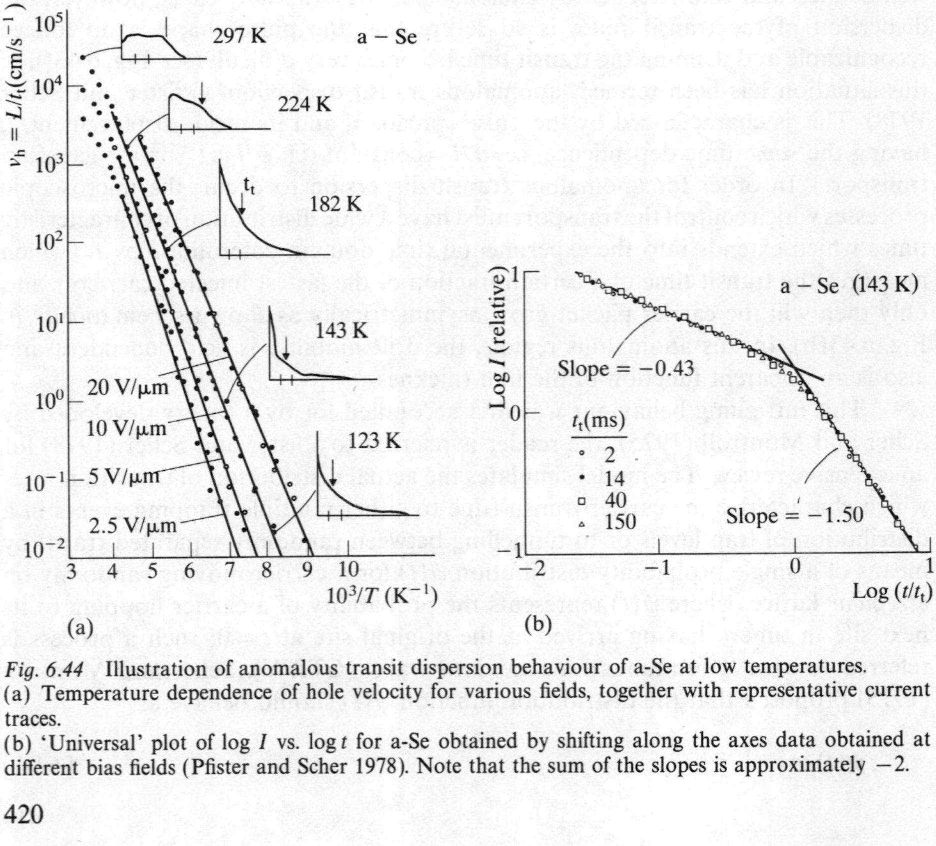

Fig. 6.44 Illustration of anomalous transit dispersion behaviour of a-Se at low temperatures. (a) Temperature dependence of hole velocity for various fields, together with representative current traces.
(b) 'Universal' plot of log I vs. log t for a-Se obtained by shifting along the axes data obtained at different bias fields (Pfister and Scher 1978). Note that the sum of the slopes is approximately −2.

gaussian transport broadens out, until at the lowest temperatures the transit time cannot be distinguished on a simple plot of I vs. t. However, if the current transient is plotted double-logarithmically (Fig. 6.44(b)), a clear kink is observed defining t_t and the sum of the two slopes is -1.93, i.e. close to the theoretical prediction of -2. Figure 6.44(b) clearly illustrates another aspect of non-gaussian dispersive transit, namely that since α_d is approximately field independent, changing the experimental conditions (e.g. E) at constant temperature merely results in a parallel shift of $I(t)$ along the axes; thus normalizing the curves relative to t_t should produce 'universal' curves, as is observed for a-Se, a-As_2Se_3 and several organic systems. This universality is in fact observed in a-Se only in the temperature range $140 < T < 170$ K; above these temperatures the transition to gaussian transport takes place which is not accounted for by the CTRW model, and below these temperatures the sum rule for the slopes above and below t_t fails, perhaps because a simple algebraic expression for $\psi(t)$ no longer holds.

Finally, we consider the question of the thickness and field dependence of the transit time. To a first approximation $t_t \propto (L/E)^{1/\alpha_d}$, where L is the thickness and E the field. If this is substituted into [6.90], it is seen that the drift mobility is a superlinear function of both field and thickness. Figure 6.45 shows a plot of $\log t_t$ vs. $\log(E/L)$ for a-As_2Se_3 which has a slope of $1/\alpha_d = -1.85$, and which is

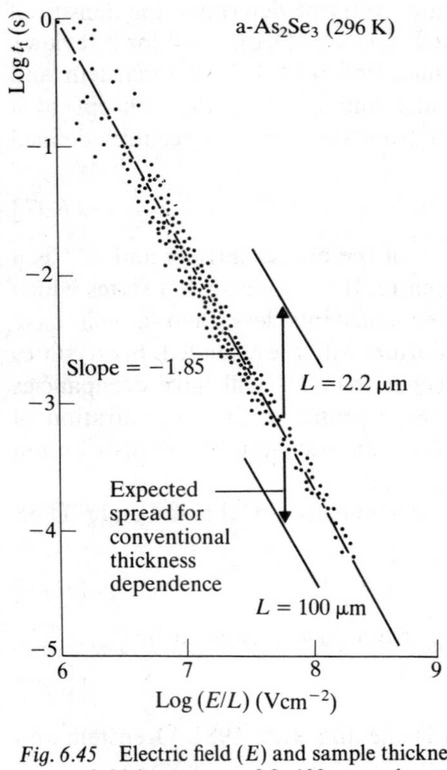

Fig. 6.45 Electric field (E) and sample thickness (L) dependence of the transit time for a-As_2Se_3. The range of thicknesses was 2.2–100 μm and range of fields 2–10 V μm^{-1}. Also shown is the spread in transit times expected if conventional gaussian statistics were controlling the transit (Pfister and Scher 1978).

independent of sample thickness and applied field; the thickness dependence expected for conventional gaussian transport is much greater than is actually observed.

Although the CTRW theory of Scher and Montroll (1975) is based on a hopping mechanism for the carriers, *any* transport mechanism which gives rise to a sufficiently broad dwell-time distribution function $\psi(t)$ will also lead to anomalous dispersion in transient carrier transport. Thus Schmidlin (1977) and Noolandi (1977) have shown that results identical to those given by the CTRW approach are obtained from a 'multiple-trapping' (MT) model in which carriers are trapped by a set of trapping levels (which may be discrete or form a continuous distribution in energy) and which are subsequently released into the bands where transport takes place. The (quasi-)universality of behaviour observed in anomalous-transit data (see Figs. 6.44(b) and 6.45) can be reproduced by the MT models if a sufficient number of trapping levels with different characteristics are considered. The observation that the MT model can also give rise to anomalous transit behaviour thereby removes a central difficulty associated with the CTRW model, namely that isolated sites which have particularly long release-time constants (and which are therefore supposed to control the transient photocurrent decay at very long times) have correspondingly low *entry* probabilities (Pollak 1977), and therefore should not exert a controlling influence on the transport behaviour.

The results of the time-of-flight transient photocurrent experiments described above have been used by a number of authors to try to determine the density of states at the band edges, particularly in a-Si:H (see e.g. Tiedje 1984 for a review). One such approach, proposed by Tiedje and Rose (1981) and Orenstein and Kastner (1981), is based on the MT model and employs the concept of a *thermalization energy* E_{th}, separating 'shallow' from 'deep' trapping centres, defined as

$$E_{th} = k_B T \ln(\nu t) \qquad [6.97]$$

where t is the elapsed time after the creation of the excess carriers and ν^{-1} is a characteristic release-time constant for the centre. It is assumed that states which are shallower than E_{th}, having release-time constants less than t, will have occupancies which are in quasi-thermal equilibrium with the extended (band) states (as in [6.91]), whereas states which are deeper than E_{th} will have occupancies governed only by capture statistics. The energy profile of the concentration of trapped carriers will then have a peak near E_{th} if the (tail-state) DOS distribution decreases sufficiently rapidly in energy.

For an assumed *exponential* energy dependence of the tail-state DOS distribution,

$$N(E) = N_0 \exp(-E/k_B T^*) \qquad [6.98]$$

the distribution of trapped charge at $E = E_{th}$, from [6.97], is given by

$$N_t(E_{th}) = N_0(\nu_0 t)^{-\alpha_d} \qquad [6.99]$$

where the power-law exponent is given by (Tiedje and Rose 1981, Orenstein and Kastner 1981)

$$\alpha_d = T/T^* \qquad [6.100]$$

The waiting-time distribution function, $\psi(t)$, can be obtained by assuming that all the trapped charge is released after a dwell time defined by [6.97], i.e. $t = \nu_0^{-1} \exp(E_{th}/k_B T)$, in which case

$$\psi(t)\,dt \simeq [N_t(E_{th})/N_t^0]\,dE$$

where N_t^0 is the total trapped charge, and since $dE = (dE/dt)\,dt$, where $dE/dt = k_B T/t$ from [6.97], then

$$\psi(t) \simeq (N_0/N_t^0) k_B T t^{-(1+\alpha_d)} \qquad [6.101]$$

Thus, the assumption of an exponential DOS distribution in the MT model also leads to a power-law behaviour for $\psi(t)$, as in the CTRW model ([6.94]). In a discussion of MT and CTRW models for dispersive transit, Pollak (1977) obtained another expression for the dispersion parameter α_d:

$$\alpha_d = k_B T \frac{d}{dE}[\ln N_t(E)]\bigg|_{E = k_B T \ln(\nu_0 t)} \qquad [6.102]$$

which, for an exponential DOS ([6.98]), also yields $\alpha_d = T/T^*$.

For a-Si:H, electron drift-mobility studies indicate that the transport is non-dispersive at high temperatures ($T \geqslant 150$ K), with a value $\mu_d \simeq 1$ cm^2 V^{-1} s^{-1} at room temperature and an activation energy for μ_d given by $\Delta E \simeq 0.13$ eV (Spear 1983). However, at lower temperatures the electron transport becomes dispersive, whereas hole transport appears to be dispersive for all temperatures. This has been interpreted on the basis of the MT model in terms of exponential band tails, the width of the valence-band DOS distribution being larger ($T_v^* \simeq 490$ K) than that for the conduction-band edge ($T_c^* \simeq 310$ K) (Tiedje 1984). In this interpretation, the width of the valence-band tail, $k_B T_v^* \simeq 42$ meV, is very similar to the inverse slope $\Gamma^{-1} \simeq 48$ meV obtained from the Urbach edge of the optical absorption spectrum [6.31], implying that the Urbach edge in a-Si:H could be determined by the valence-band tail-state profile (Tiedje 1984).

However, the validity of the Tiedje–Rose–Orenstein–Kastner approach has been called into question (see e.g. Marshall 1988). In particular, the role of the thermalization energy E_{th} ([6.97]) acting as a demarcation energy is questionable, particularly since it is based solely on trap-release criteria, even though the trapping probability is intimately involved in the thermal equilibration process. Recently, Michiel *et al.* (1986) have investigated time of flight characteristics in both the dispersive and non-dispersive regimes by calculating the drift mobility for an arbitrary DOS distribution, approximated in terms of a set of discrete levels. In this case, the transit time can be written as (Schmidlin 1977)

$$t_{tr} \simeq t_0 + \sum_i M_i \nu^{-1} \exp(E_i/k_B T) \qquad [6.103]$$

where t_0 is the trap-free transit time ($t_0 = L/\mu_e E$), ν the trap-release rate and M_i the number of trapping events for a carrier in trap i at energy E_i during the time t_0. Michiel *et al.* (1986) show that in this case the drift mobility can be written as

$$\mu_d = \mu_e \left[1 + \sum_i (N_i/N_o) \exp(E_i/k_B T)\right]^{-1} \qquad [6.104]$$

where N_i is the density of the ith trap and N_0 is the electron density in the band. Using this approach, these authors have shown that the form of the experimental time-of-flight data for a-Si:H does not necessarily imply exactly exponential band tails.

Problems

6.1 Show by using an analysis similar to that employed in section 6.3.3 that the temperature dependence of the d.c. hopping conductivity in the case when the thickness of the film is less than the hopping length obeys the law $\ln \sigma \propto T^{-1/3}$ ([6.65]). Hence estimate the thickness of a film for which a transition from 2-D to 3-D behaviour is expected at 300 K (take $N(E_F) \simeq 10^{19}\ \text{cm}^{-3}\,\text{eV}^{-1}$ and $\alpha^{-1} \simeq 10$ Å).

6.2 Discuss how thermopower and conductivity measurements may be used to decide whether the conduction channel in an amorphous semiconductors is through the extended states at a mobility edge or through the localized states at the Fermi level.

6.3 Derive the expressions which relate the position of the Fermi level to the electron occupancy of a defect, suffering both positive and negative U_{eff} ([6.14] and [6.15]).

Bibliography

R. Balian, M. Kleman and J.-P. Poirier (eds) (1981), *Physics of Defects* (North-Holland).

M. H. Brodsky (ed.) (1979), *Amorphous Semiconductors* (*Topics in Applied Physics*, vol. 36) (Springer-Verlag).

N. E. Cusack (1988), *Physics of Structurally Disordered Matter* (Adam Hilger).

N. F. Mott and E. A. Davis (1979), *Electronic Processes in Non-Crystalline Materials* (OUP).

J. Robertson (1982), *Phys. Chem. Glasses* **23**, 1.

7 Applications

7.1 Introduction

Amorphous solids have been used in a variety of applications for thousands of years. Until very recently, however, silicate glasses were the only type of material commonly used; the relatively recent discovery that very many other types of material can be produced in amorphous form, either as bulk glasses or as thin films, has led to a very rapid expansion in the use of such materials in, for example, electronic, magnetic and optical applications.

Non-crystalline materials have many advantages over their crystalline counterparts where applications are concerned. In general, amorphous solids are relatively easy to prepare: large-area, homogeneous amorphous thin films can be prepared, for example of a-Si:H for solar-cell or thin-film transistor applications; bulk glasses can often be readily formed from the melt by relatively slow quenching procedures (or by sol–gel methods); more importantly, near the glass-transition temperature the material remains workable (i.e. the viscosity is relatively low) over a range of temperatures so that it can easily be fashioned into various shapes or drawn into fibres. Furthermore, amorphous materials, particularly bulk glasses, are often structurally homogeneous and isotropic on a macroscopic length scale. As a result, their physical properties are also isotropic and homogeneous, unlike crystalline materials for which the intrinsic behaviour of single crystals may be anisotropic, and the presence of grain boundaries in polycrystalline samples can dominate the overall behaviour. Thus, large-scale optical transparency is readily achievable in, for instance, silicate glasses used for optical components (including windows) and fibre-optic cables. The absence of structural defects, such as grain boundaries or dislocations, also has a dramatic effect on the mechanical behaviour and use of amorphous materials in mechanical engineering applications. For example, because amorphous materials are often structurally homogeneous and do not have crystal-like line defects, they can have mechanical strengths approaching ideal values, and consequently fibres of silica glass or ribbons of metallic glasses can be used as fibre-reinforcement elements in composite materials. Amorphous materials have a further, distinct advantage over their crystalline counterparts in that homogeneous amorphous phases can often be formed in mixed-component systems over rather wide ranges of composition, not necessarily restricted to stoichiometric values. For this reason, physical properties of such amorphous materials can often be varied continuously simply by varying the composition. Thus, the intensity of colouring in ornamental (silicate) glasses can be controlled by varying the concentration of colouring agents (e.g. transition-metal ions) in the glassy matrix; likewise, the intrinsic optical properties, namely the optical band gap, of amorphous semiconductors can be varied systematically by changing the composition of an alloy system, e.g. a-$Si_{1-x}C_x$:H.

In this chapter some of the more important technological applications of amorphous materials are surveyed; the more functional uses of glasses, e.g. as utensils, or their use in straightforward mechanical applications, e.g. in fibre-reinforced composites, are not considered. For ease of presentation, the various applications are categorized according to the physical property which is exploited.

7.2 Electronic applications

7.2.1 Introduction

A number of electronic applications of amorphous materials, almost all of them semiconductors, have been proposed or put into practice, especially since the discovery by Spear and Le Comber (1976) that a-Si:H could be doped either n-type or p-type by incorporating PH_3 or B_2H_6 in the silane glas flow of a glow-discharge decomposition reactor (see sect. 6.2.6). In many cases, amorphous materials are used simply as passive elements in electronic devices made from conventional crystalline semiconductors (e.g. Si): for example, a-SiO_2 or a-Si_3N_4

is used as the insulator layer in various thin-film transistor configurations. Of interest here, however, are the uses of amorphous materials as the *active* elements in electronic devices. Some reviews of recent work in this area, particularly on electronic device applications of a-Si:H, have been given by Hamakawa (1987) and in the volume edited by Pankove (1984); a-Si:H and chalcogenide glass-based applications are covered by Madan and Shaw (1988).

7.2.2 Electrophotographic applications

One of the most widespread current applications of amorphous semiconductors is in the field of electrophotography, or *xerography* (the name being derived from the Greek, and meaning 'dry writing'), in document copying machines. The material first used in this context was a-Se in thin-film form (perhaps alloyed with a few per cent of As to improve the mechanical properties of the films, and to reduce the tendency for crystallization to occur), although more recently a-Si:H films have been utilized instead (Shimizu 1985).

The first demonstration of xerography was by Carlson and Kornei on 22 October 1938 in Astoria, NY. These first experiments used sulphur instead of selenium for the electrophotographic medium, but the principle is the same. Carlson described the discovery thus (Pfister 1979):

> I went to the lab that day and Otto [Kornei] had a freshly-prepared sulphur coating on a zinc plate. We tried to see what we could do toward making a visible image. Otto took a glass slide and printed on it in Indian ink the notation '10-22-38 Astoria'.
>
> We pulled down the shade to make the room as dark as possible, then he rubbed the sulphur surface vigorously with a handkerchief to apply an electrostatic charge, laid the slide on the surface and placed the combination under a bright incandescent lamp for a few seconds. The slide was then removed and lycopodium powder was sprinkled on the sulphur surface. By gently blowing on the surface, all the loose powder was removed and there was left on the surface a near-perfect duplicate in powder of the notation which had been printed in the glass slide.
>
> Both of us repeated the experiment to convince ourselves that it was true, then we made some permanent copies by transferring the powder image to wax paper and heating the sheets to melt the wax. Then we went out to lunch and to celebrate.

The principles underlying the modern xerographic process using a-Si:H (or an organic photo-receptor) are precisely the same as those described by Carlson (see Fig. 7.1). The surface of the photo-conducting film is first charged positively to a surface potential $V_0 \simeq 700$ V by moving the film under a corona charging device (a). The charged photo-receptor is now exposed to light reflected from the document to be copied; light is reflected preferentially from the white areas of the page and these photons on striking the film are strongly absorbed near the surface and create electron–hole pairs (b). The photo-electrons neutralize the positive charges locally at the surface, while the holes drift across the film under the action of the field caused by the surface potential and neutralize the negative charge induced on the Al substrate (c). Thus a negative image is formed in the surface charge distribution. The latent image is developed by means of negatively, triboelectrically charged toner particles (each being a carbon particle $\simeq 10\ \mu$m in diameter surrounded by a low-melting plastic carrier bead $\simeq 100\ \mu$m in diameter); these are attracted to the remaining positively charged areas on the film (d). Finally, the developed image

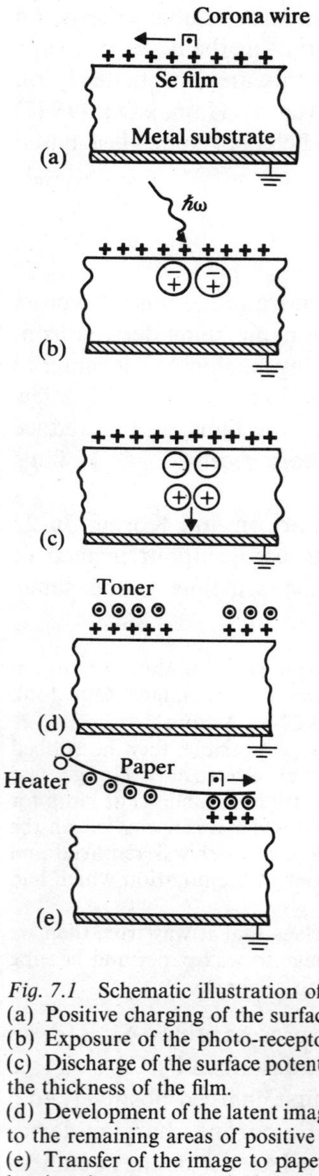

Fig. 7.1 Schematic illustration of the Xerox process, using a-Se as the photo-receptor.
(a) Positive charging of the surface by a corona discharge.
(b) Exposure of the photo-receptor by photons reflected from the document to be copied.
(c) Discharge of the surface potential locally by the photo-generated electrons, the holes drifting across the thickness of the film.
(d) Development of the latent image by means of negatively charged toner particles which are attracted to the remaining areas of positive surface charge.
(e) Transfer of the image to paper by means of a second corona discharge and subsequent fixing by heating the paper to melt the toner particles into place (after Mott and Davis 1979).

is transferred by attracting the toner particles to a sheet of paper, which has been corona charged to the opposite polarity of that of the toner, whereupon the paper is heated which melts the toner particles and fixes them in position (e).

Certain conditions must be met for a material to be a suitable photo-receptor. It must be able to be produced in defect-free large areas, and the material must be mechanically robust and chemically inert in order to withstand, respectively, both abrasion and chemical degradation in the presence of intense light and electric fields; an additional advantage is for the film to be flexible so belt operation may be used. As far as electrical properties are concerned, the material must be highly resistive ($\rho \gtrsim 10^{12}\ \Omega$ cm) in order that the surface potential V_0 does not decrease appreciably before or after light exposure, and that image contrast is not lost by lateral charge flow on the surface. Furthermore, the carrier generation or 'quantum efficiency', η_q (which is the number of *free* carriers generated per incident photon) must be high at the output wavelengths of suitable lamps. The carrier generation rate can be written in terms of the quantum efficiency as (Mott and Davis 1979):

$$G = \eta_q \left[\frac{I_0(1-R)\{1-\exp(-\alpha d)\}}{d} \right] \qquad [7.1]$$

where the quantity in square brackets is the number of photons absorbed per second in a sample of thickness d in the direction of the incident radiation, I_0 is the incident light flux density, R the reflectivity and α the absorption coefficient.

Although the d.c. resistivity of a-Se is high enough to satisfy the charge-leakage criterion, that of intrinsic a-Si:H is insufficient (see Table 5.4). In this case, a passivation layer of, for example, a-$Si_{1-x}N_x$:H or a-$Si_{1-x}C_x$:H must be deposited on the top surface of the a-Si:H photoreceptor layer so as to immobilize the charge injected into it from the photoconducting a-Si:H layer below. With a-Si:H, this disadvantage is outweighed by the advantages it has over previous a-Se photoreceptors, namely in having extremely high surface hardness, non-toxicity and high optical sensitivity (even for the IR region).

The device described above utilizes a copying drum, on the outside surface of which is deposited the amorphous photoreceptor; in this case, the width and circumference of the drum must correspond to the size of document normally being copied (e.g. A4); different-sized documents are accommodated by using (de)magnifying optics. One solution to this problem is to use a linear image sensor which scans the document being copied lengthways (Kaneko 1984); such a device based on a-Si:H photodiodes is shown schematically in Fig. 7.2. The scanner

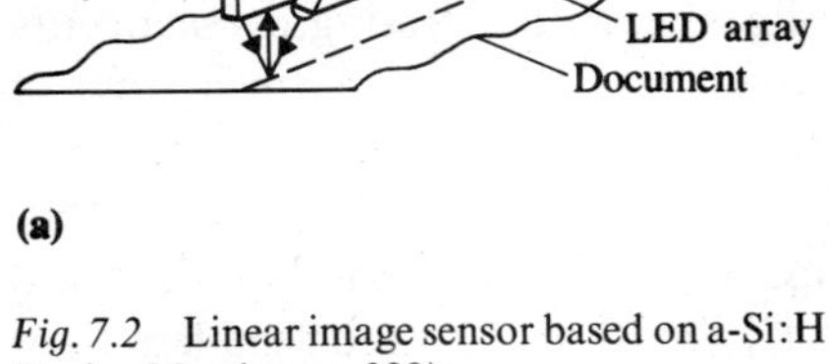

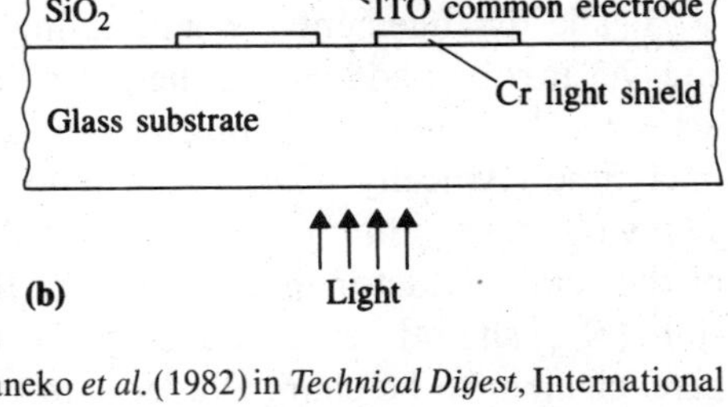

Fig. 7.2 Linear image sensor based on a-Si:H (S. Kaneko *et al.* (1982) in *Technical Digest*, International Device Meeting, p. 328):
(a) overall view of sensor;
(b) photodiode element using a-Si:H as the photosensor with Si_3N_4 and p-type a-Si:H blocking contacts.

consists of a long photosensor array (Fig. 7.2(a)) made up from individual photodiodes having a-Si:H as the active component (Fig. 7.2(b)). The semi-transparent indium–tin–oxide (ITO) bottom contact is positively biased, and of the carriers photogenerated near the interface between the silicon nitride and a-Si:H layers upon absorption of photons reflected from the document being copied, the holes drift up to the aluminium top electrode and the currrent is monitored. The bottom Si_3N_4 and top p-type a-Si:H blocking layers prevent the (dark) injection of holes and electrons, respectively, into the a-Si:H photosensor.

7.2.3 Thin-film transistors

There is currently a considerable demand for large-area (say 30 cm × 30 cm) flat displays, for example for use as small colour television or lap-top computer screens. The active optical components in such devices are generally twisted-nematic liquid crystals, operating in transmission mode with rear illumination. However, since liquid crystals are not themselves highly non-linear optical elements, individual elements in the display can be addressed only by incorporating a non-linear switching element, such as a thin-film transistor (TFT), at each picture element (pixel). Thus, in order to fabricate large-area flat displays, an array of TFT devices must be constructed, and for this the thin-film deposition technology of a-Si:H is very suitable. Reviews of activity in this area have been given by Le Comber *et al.* (1981) and Madan and Shaw (1988).

A schematic illustration of an a-Si:H-based TFT element and an array of such TFT drivers corresponding to that of the picture elements of a display are shown in Fig. 7.3. The staggered configuration TFT shown in Fig. 7.3 is simply that of a field-effect transistor, as discussed in sect. 6.3.1. Normally, a positive gate voltage V_g is applied, causing the bands of the a-Si:H to bend down at the interface with the silicon nitride insulator, and negative charge to accumulate at the interface, thereby leading to an enhancement of the sheet conductance (n-channel mode). For the opposite polarity of V_g, positive charge is induced at the interface, forming a depletion layer; for sufficiently large values of V_g an 'inversion layer' is produced with hole accumulation (p-channel mode), which also leads to an enhancement of the conductance. Figure 7.4 shows the dependence of the source–drain current on V_g for a TFT device operating in both n- and p-channel modes; the field-effect mobilities for electrons and holes obtained from such devices are $\simeq 10^{-1}$ and $\simeq 10^{-3}$ cm^2 V^{-1} s^{-1}, respectively.

The arrangement for addressing individual pixels in the array shown in Fig. 7.3 involves addressing the gates line by line, individual pixel elements being activated by applying voltages to the source lines; each addressing event occurs in a time (typically 60 μs), which must be much less than the refresh time of 17 ms (corresponding to a frame rate of 60 Hz). Thus, the material requirements of the semiconductor in the TFT are that the resistance of the TFT in the 'on' state (R_{on}) should be small enough for the capacitor formed by the liquid-crystal pixel element to be charged adequately; since $C_{LC} \simeq 10$ pF, R_{on} must be less than $\simeq 6$ MΩ. Moreover, the resistance of the TFT in the 'off' state should be sufficiently high for the stored charge in C_{LC} not to decay appreciably during the refresh time, and also to give an adequate grey scale. Amorphous hydrogenated silicon fulfils

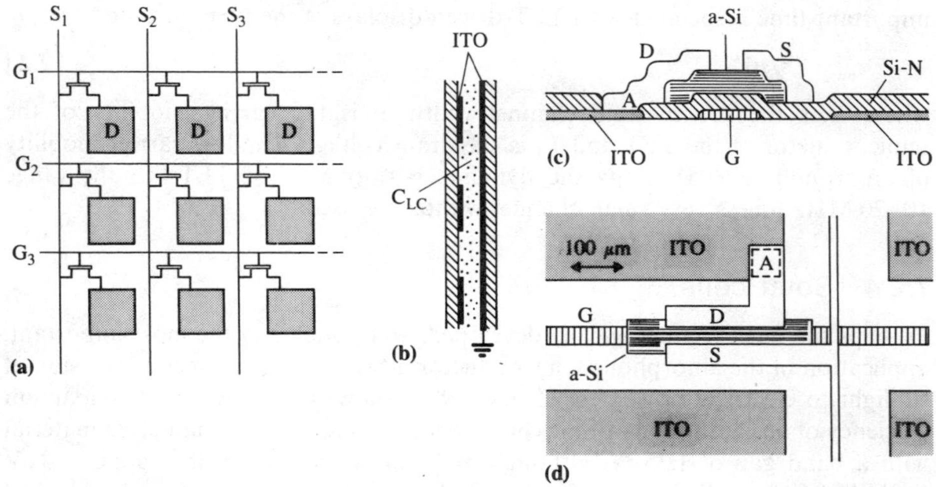

Fig. 7.3 Schematic illustration of an addressable liquid-crystal display panel with a-Si:H TFT elements (Le Comber *et al.* 1981).
(a) Plan view of the TFT array, showing the drain contact (D) and gate (G) and source (S) lines.
(b) Side view, showing the liquid crystal sandwiched between two semi-transparent glass substrates coated with indium–tin oxide (ITO); C_{LC} is the capacitance of the liquid crystal.
(c) Section through a TFT element, showing the staggered arrangement of source, drain and gap electrodes.
(d) Illustration of the configuration of a TFT driver in the addressable matrix array.

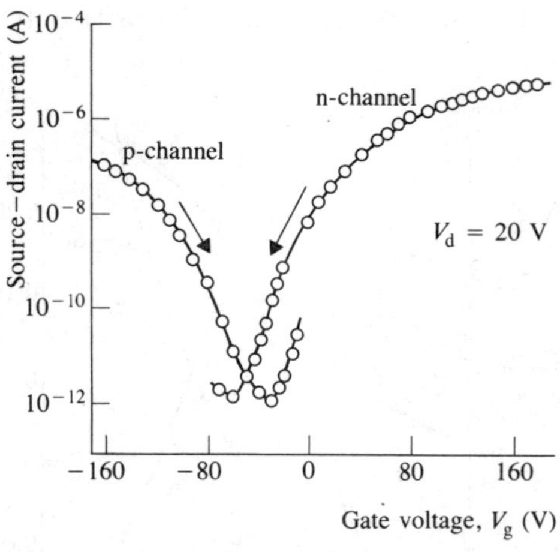

Fig. 7.4 The dependence of the source–drain current on gate voltage for a TFT device operating in both n- and p-channel modes. The arrows indicate the direction of the change in voltage (H. Matsumura and Y. Nara (1980), *J. Appl. Phys.* **51**, 6443).

these requirements adequately (see Fig. 7.4), although there are some disadvantages associated with its use. Although the *dark* conductivity of a-Si:H is low enough to preclude charge-leakage effects, the material has such a high photoconductivity (see sect. 7.2.4) that the TFT element needs to be masked from any illumination such as that used in the TV operation of flat liquid-crystal displays. One other important time associated with TFT-driven displays is the turn-on time,

$$t_{on} = L^2/\mu V_d \qquad [7.2]$$

where L is the source-drain channel width, μ is the carrier mobility of the semiconductor of the TFT and V_d is the drain voltage. The low carrier mobility of amorphous a-Si:H limits the dynamic performance of TFTs to the range 10–20 MHz unless very small channel widths are used.

7.2.4 Solar cells

Perhaps the most technologically developed, and potentially the most important, application of the amorphous semiconductor a-Si:H is in the direct conversion of sunlight to electrical power. As Henry (1980) showed, the theoretical maximum efficiency of an ideal photovoltaic cell reaches 31% for a semiconducting material with a band gap of 1.35 eV, although materials with gaps in the range 1–2 eV (including c-Si, a-Si:H, c-GaAs and c-$CuInSe_2$) are predicted to have ideal photovoltaic efficiencies higher than 25%; these calculations are based on AM1.5 ('air mass 1.5') illumination conditions. The solar illumination intensity, P_{in}, is 100 mW cm^{-2} at sea-level on a clear sunny day with the sun at the zenith, and this is termed AM1 illumination; the intensity outside the atmosphere ($P_{in} = 135$ mW cm^{-2}) is termed AM0. The AM1 spectrum peaks at a wavelength of 0.5 μm (Fig. 7.5(a)), which matches the absorption edge of a-Si:H better than that of c-Si (Fig. 7.5(b)): since the edge of c-Si is due to indirect optical transitions, it lies considerably below that of a-Si:H, which means that a film of a-Si:H only

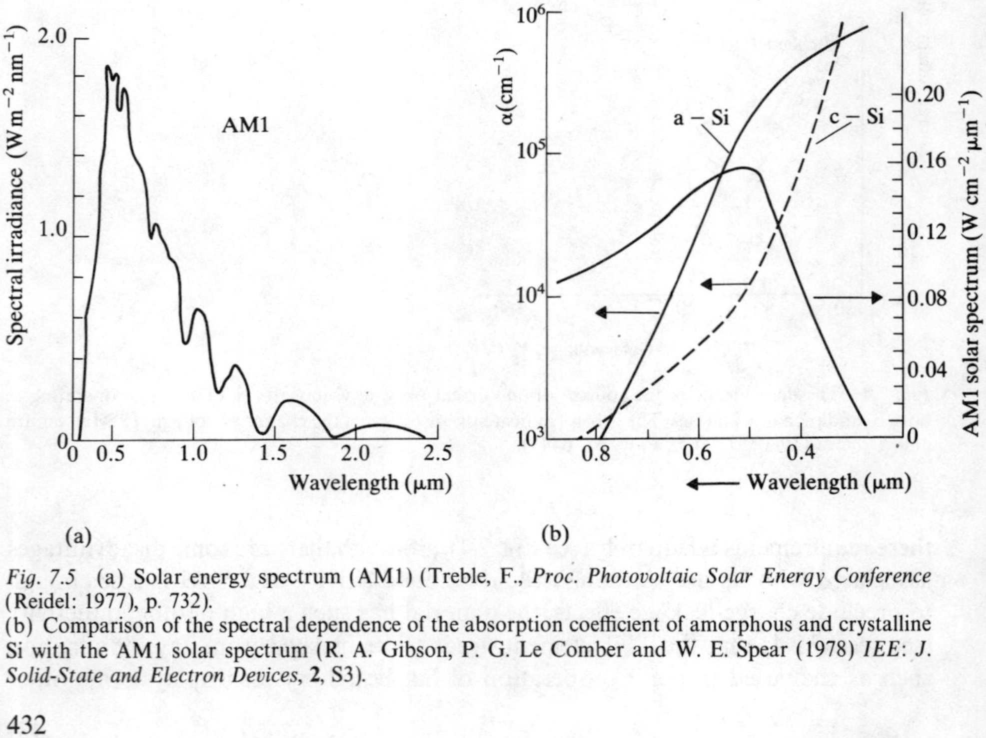

Fig. 7.5 (a) Solar energy spectrum (AM1) (Treble, F., *Proc. Photovoltaic Solar Energy Conference* (Reidel: 1977), p. 732).
(b) Comparison of the spectral dependence of the absorption coefficient of amorphous and crystalline Si with the AM1 solar spectrum (R. A. Gibson, P. G. Le Comber and W. E. Spear (1978) *IEE: J. Solid-State and Electron Devices*, **2**, S3).

a few micrometres thick will absorb most of the useful sunlight, whereas a thickness of several hundred micrometres of c-Si is required to do the same.

Although AM1 photovoltaic efficiencies of c-Si solar cells of 20% have been achieved, i.e. approaching the ideal limit, this material is unsuitable for large-scale applications for both economic and practical reasons: the maximum size of a cell array is determined by the maximum diameter of slice available, currently about 15 cm. Amorphous hydrogenated silicon films, however, can readily be deposited onto large-area substrates (as for TFT applications – sect. 7.2.3), and can even be deposited onto a *continuous* flexible substrate (see e.g. Madan and Shaw 1988). Thus, although the photovoltaic efficiency of a-Si:H ($\lesssim 10\%$) is considerably lower than that of its crystalline counterpart, this deficiency is ameliorated to a considerable extent by the fact that large-area films of a-Si:H can be produced relatively cheaply.

The energy conversion efficiency, η, of a solar cell can be deduced from a knowledge of certain of its operating characteristics (see Fig. 7.6), and is defined by the relation

$$\eta = J_{sc} V_{oc} FF / P_{in} \qquad [7.3]$$

where J_{sc} is the short-circuit current density, V_{oc} is the open-circuit voltage and FF is the 'fill factor', defined by the relation

$$FF = J_m V_m / J_{sc} V_{oc} \qquad [7.4]$$

where J_m and V_m are the values of current density and voltage at the point of maximum power (see Fig. 7.6). In addition to the intrinsic conversion loss experienced even by an ideal cell as a result of mismatch between the optical absorption characteristics of the semiconductor and the solar spectrum, actual solar cells suffer a number of additional loss processes such as reflection losses,

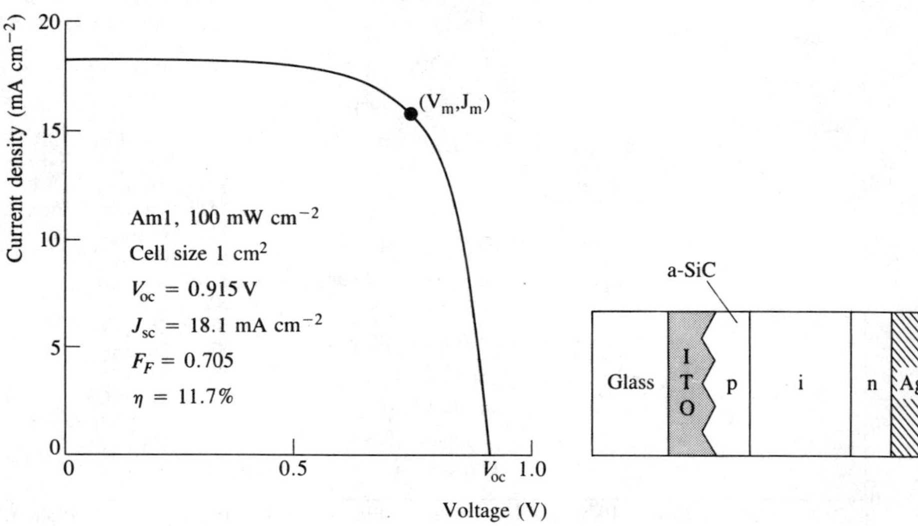

Fig. 7.6 Current–voltage output characteristics of an a-SiC/a-Si:H p^+–i–n^+ heterojunction solar cell having an energy conversion efficiency $\eta = 11.5\%$ (Hamakawa 1987).

incomplete extraction or non-radiative recombination of photogenerated electrons and holes, and series resistance effects. This limits the maximum efficiency of simple a-Si:H solar cells to $\eta \simeq 10\%$, with values of fill factor $FF \lesssim 0.7$, unless more complicated cell geometries (e.g. tandem cells) are used. Figure 7.7 shows how the conversion efficiency of a-Si:H-based solar cells has increased with time.

Solar cells operate by separating photo-induced electron–hole pairs by means of the internal electric fields present in the vicinity of rectifying junctions formed in the semiconductor device, and extracting these carriers so that they can do work

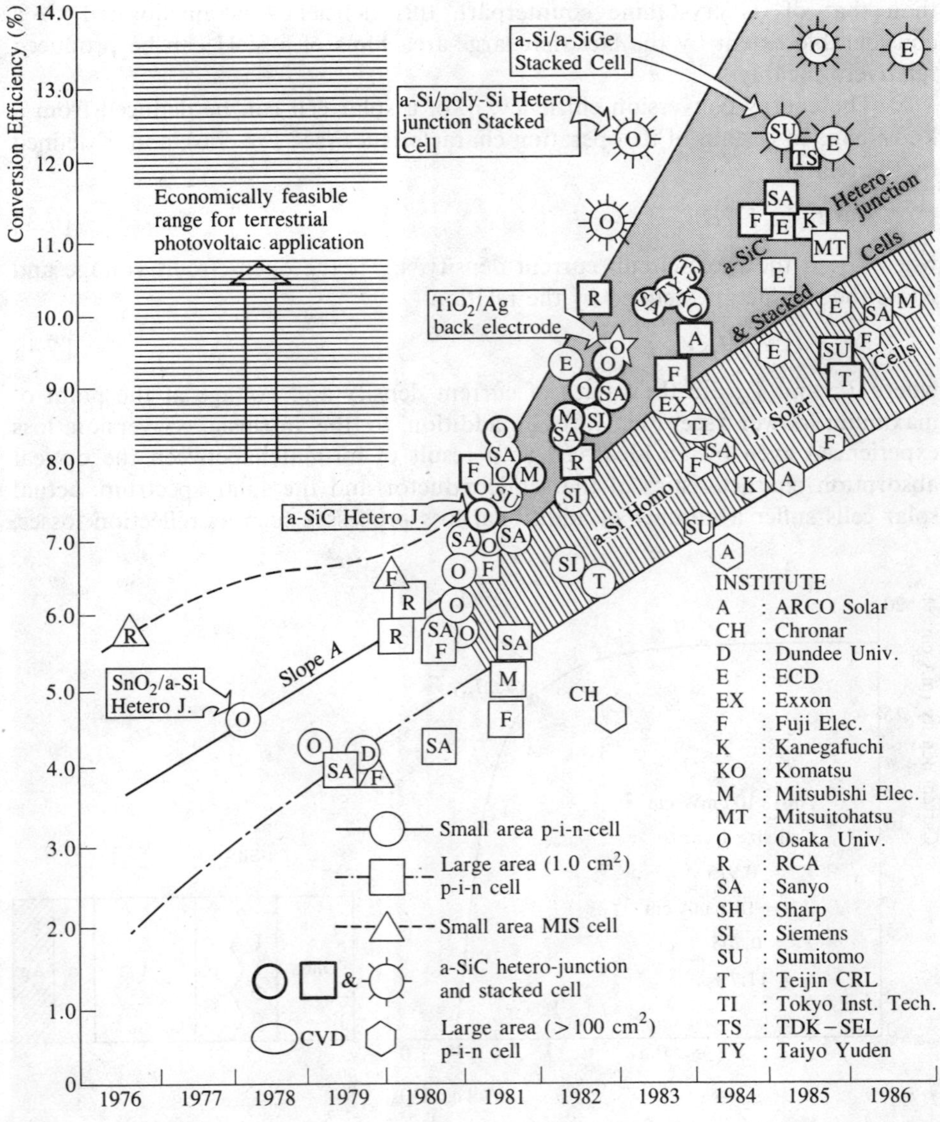

Fig. 7.7 Improvement in the conversion efficiency of a-Si:H-based solar cells with time (Hamakawa 1987).

in an external circuit. Such rectifying junctions occur at metal–semiconductor junctions (Schottky barriers), formed by depositing either a high-work-function metal onto an n-type semiconductor, or a low-work-function metal onto a p-type semiconductor, or at p–n junctions in the semiconductor itself. Neither of these configurations leads to high-efficiency solar cells made from a-Si:H, principally because of problems with interface states at the metal–semiconductor junction in the former case, and doping-induced defect creation in the latter. For these reasons, the best performance has been achieved using a p^+–i–n^+ configuration (in which the p^+–a-Si:H layer is deposited first onto a conducting substrate). Here, the field-induced charge separation takes place mainly in the intrinsic i-layer, thereby obviating any effect of increased trapping of the photogenerated carriers (leading to a decrease in the product $\mu\tau$ of the carrier mobility and lifetime, principally of the minority carriers), due to the dopant-induced increase in the dangling-bond density (see sect. 6.2.6).

7.3 Electrochemical applications

7.3.1 Introduction

The ability of certain ions, notably Li^+ and Ag^+, to diffuse readily in oxide or chalcogenide glasses in the presence of a concentration gradient or an electric field (see sect. 4.5) opens up a range of possible electrochemical applications, for example in the fields of energy storage (batteries), displays or chemical sensors. Amorphous materials can offer a number of advantages in such applications over their crystalline or liquid counterparts: for example, their ionic diffusion properties are isotropic, and are not affected by grain-boundary or preferred pathway effects as in crystalline electrolytes, and amorphous electrolytes can often be prepared in thin-film form for microelectronic or display applications.

7.3.2 Solid-state batteries

There is much current interest in the all-solid-state battery, employing a solid electrolyte, for example for high-temperature applications, for use in integrated microelectronic devices and to avoid the leakage and packaging problems inherent in batteries based on conventional liquid electrolytes. Several requirements need to be met if ionically conducting solids are to be used as electrolytes in batteries. First, there are constraints on the conductivity: the ionic conductivity must be high and comparable to that of liquid electrolytes ($\sigma_i \simeq 10^{-2}\,\Omega^{-1}\,\mathrm{cm}^{-1}$ at room temperature), i.e. the material must be 'superionic', so that the internal resistance of the cell is not significant. Conversely, the electrolyte material should have a negligible *electronic* conductivity, so that the battery is not short-circuited internally, shortening its shelf-life. Secondly, there are chemical constraints: the electrolyte material should be stable, particularly with respect to the electrode reactions occurring at the interfaces with the anode and the cathode. Thirdly, there are mechanical considerations: ideally the electrolyte should be sufficiently flexible for intimate mechanical contact, and hence electrical contact, to be maintained throughout the discharge process of a primary cell when material is removed from the anode area, and throughout the charging and discharging cycles of a secondary cell where a volume change may occur if intercalation-type cathode materials are used. Ionically conducting glasses can fulfil most, if not all, of these requirements, and consequently such solid electrolytes are starting to be used in certain battery applications. They have an additional advantage over conventional liquid electrolytes in that generally only one type of ion is mobile; the other, being part of the structural framework of the glass, is immobile. Gabano (1985) and Souquet (1986, 1988) have given general reviews of this field.

Lithium-based batteries are promising energy-storage-systems where a high specific-energy capability (E, measured in Wh kg^{-1}) is required. One such type of cell which has been developed consists of a Li (or LiAl) metal anode, a crystalline TiS_2 intercalation-type cathode and, as the electrolyte, a LiI-doped $Li_2S-P_2S_5$ glass, which has a room-temperature Li^+-ion conductivity of $2 \times 10^{-3}\,\Omega^{-1}\,cm^{-1}$ (Gabano 1985). This battery has a high specific-energy characteristic of $E \simeq 150$ Wh kg^{-1} (cf. $E \simeq 40$ Wh kg^{-1} for a lead–acid accumulator cell) and an output voltage of $\simeq 2$ V is obtained, with a rather high current density which ranges from 0.1 to 1 mA cm^{-2} for ambient temperature and 100°C, respectively (Gabano 1985, Akridge and Vourlis 1986).

Another battery application which employs a glassy electrolyte is the sodium–sulphur battery. It is operated at a high enough temperature ($\simeq 300$°C), at which both the sodium and sulphur are in the liquid state, for Na_2S to dissolve well in the sulphur. The electrolyte is a sodium borosilicate glass, in the form of hollow fibres, which therefore serves to separate the molten electrode materials. This cell also has a high specific energy characteristic ($E \simeq 120$ Wh kg^{-1}), and a voltage of $\simeq 2$ V, and can provide current densities up to 4 mA cm^{-2} (Souquet 1986).

Although considerable effort has been expended in developing electrolytes in glassy form suitable for battery applications, less work has been done to date on studying amorphous cathode materials which might allow the construction of monolithic batteries. One type of cathode material which offers the prospect of constructing rechargeable (secondary) all-solid-state batteries consists of insertion compounds, such as crystalline layered materials (e.g. TiS_2), in which the cell ions (e.g. Li^+) can be intercalated; the metal intercalation is accompanied by charge transfer to the host metal d-band (e.g. $Ti^{4+} + e^- \rightarrow Ti^{3+}$). These types of cathode material have the disadvantage that appreciable volume changes can occur during the (de-)intercalation process, leading to contact problems at the cathode–electrolyte interface and the loss of reversibility in the charging/discharging cycle. These problems may be obviated by using amorphous cathode materials within which ions such as Li^+ may be reversibly introduced, and which are 'mixed' conductors, i.e. they exhibit an appreciable intrinsic electronic conductivity (thereby allowing a cathode made from such a material to be in electrical contact with the external circuit) and also an extrinsic ionic conductivity when containing inserted ions. Examples of such possible amorphous cathode materials are glasses containing transition metals such as $V_2O_5-P_2O_5$ mixtures, and amorphous molybdenum sulphides (MoS_2, MoS_3). In these cases, the electronic conductivity arises from *small-polaron* conduction (sect. 5.5) amongst transition-metal ions with different charge states (e.g. V^{4+}, V^{5+}). These materials are attractive since the volume changes occurring upon ion insertion can be smaller than those characteristic of crystalline insertion compounds.

A remaining goal is to find a glass-forming material (preferably sulphide-based) which can be used as the basis of a superionic glassy solid electrolyte, and also to form an electronically conducting cathode material. In such a monolithic cell, the mobile ionic species would not sense the interface between electrolyte and cathode, and consequently over-potentials associated with such interfaces would be greatly reduced (Souquet 1988).

7.3.3 Electrochemical sensors

The electrochemical processes taking place at the interfaces between cathode or anode and an electrolyte can be used to measure the concentration (or activity) of the electrochemically active ionic species in the electrolyte by simply monitoring the e.m.f. of the cell so formed. Thus, if glassy materials are to be used as ion-selective electrochemical sensors, they must satisfy the requirements for their use in electrochemical energy-storage (battery) applications (sect. 7.3.2).

The most widespread use of an amorphous material as an electrochemical sensor is in the so-called 'glass electrode' commonly used to monitor proton activity, i.e. pH, of aqueous solutions (see Eisenman 1967 and Baucke 1985 for reviews). In this sensor a cell is constructed from a glass membrane in the form of a sealed tube containing an internal buffer solution and reference electrode (the glass electrode), together with another reference electrode, both in contact with the solution being measured. The glass used for the active membrane of the glass electrode can be an alkali silicate if hydrogen ion concentration (pH) is to be measured, or one of various alkali aluminosilicate or borosilicate glasses if alkali-ion selectivity, or pM (M = alkali ion), is required; whatever the glass, the type of (alkali) ion used as network modifier in the glass electrode must be different from that which it is desired to monitor in solution.

The mechanism underlying the operation of the glass electrode can be understood as follows (Baucke 1985); we consider a simple alkali silicate glass, for simplicity. The surface of the glass electrode will contain $\rangle$Si—O—M′ groups (where M′ is the modifier in the glass) and, in contact with an aqueous solution containing M^+ ions ($M \neq M'$), these groups will dissociate completely since there are no M′ ions in solution. The resulting surface siloxy groups ($\rangle Si—O^-$) can then establish an equilibrium with protons in solution, forming silanol groups:

$$SiOH + H_2O \rightleftharpoons SiO^- + H_3O^+ \qquad [7.5]$$

for which the e.m.f. of the cell reaction can be written as (Baucke 1985)

$$E = -\frac{RT}{F} \ln K_d + \frac{RT}{F} \ln\left(\frac{a_{H_3O^+} a_{SiO^-}}{a_{SiOH}}\right) \qquad [7.6]$$

where R is the gas constant, F is the Faraday's constant, K_d is the dissociation equilibrium constant for [7.5], and a_X is the activity of species X. Thus, [7.6] indicates that the e.m.f. of the glass electrode should be linearly proportional to pH, and this behaviour is observed experimentally over a pH range of 10–11 units (Baucke 1985). The metal-ion (pM) response of the glass electrode can be regarded as arising from an association equilibrium between siloxy groups at the surface and metal ions in solution:

$$>\text{Si—O}^- + \text{M}^+_{\text{solution}} \rightleftharpoons >\text{Si—O}^- \ldots \text{M}^+ \qquad [7.7]$$

In this case, an expression similar to [7.6] can be derived, replacing K_d by the association constant k_a for the equilibrium [7.7], and where $a_{H_3O^+}$ is replaced by a_{M^+}; thus the e.m.f. should be proportional to pM.

Although ion-selective glass membranes are normally fabricated from oxide glasses, Owen (1980) and Tohge and Tanaka (1986) have demonstrated that chalcogenide glasses (e.g. As_2Se_3) doped with the appropriate metal can be used as the basis for electrochemical sensors for Cu^{2+} or Pb^{2+} ions and Hg^{2+} or Cd^{2+} ions, respectively. These sensors are attractive because the glasses are predominantly electronically, rather than ionically, conducting, and so problems with reversibility at the interface between the membrane and the back contact are therefore minimized.

7.3.4 Electrochromic optical devices

The phenomenon of electrochromism – that is, an electric-field-induced change in the optical transmission – is well known in oxides of polyvalent metals (e.g. W, Mo, V, In and Ti), where dark blue 'bronzes' are formed on reaction of the oxide with either hydrogen or an alkali metal. Amorphous polyvalent metal oxides, e.g. a-WO_3, also exhibit electrochromic effects, and these materials have been proposed as the basis of a number of optical devices, including electrochromic mirrors (or 'smart' windows) and flat displays. In general, hydrogen is used as the electrochromic agent since the reaction involving it is both reversible and the most rapid.

A schematic illustration of the operation of an electrochromic mirror, functioning in two modes (diffusion-controlled and electric-field-controlled) is shown in Fig. 7.8. In the diffusion-controlled mode (Fig. 7.8(a)), the a-WO_3 layer is in contact with an inert reflector electrode which can either be a source of, or a sink for, hydrogen. Application of an electric field between the two electrodes sandwiching a reservoir of H (e.g. in the form of H_2O in a storage layer) causes hydrogen to be generated, which then diffuses as neutral H^0 via a grain-boundary mechanism through the metallic layer forming the reflective electrode, and ultimately enters the a-WO_3 layer. In a field-controlled system (Fig. 7.8(b)), H^+ ions generated by the application of a positive potential to the back electrode pass through the reflector layer and an electrolyte before entering the WO_3 film; electrons injected at the same time from the other interface from the negatively charged semitransparent ITO electrode then neutralize the H^+ ions, allowing the hydrogen bronze H_xWO_3 ($x \leqslant 0.3$) to be formed. In both cases, since H_xWO_3 is very strongly coloured, the effective reflectance of the mirror is greatly reduced when the bronze is formed electrochemically.

The coloration of the bronzes is believed to be caused by photo-induced

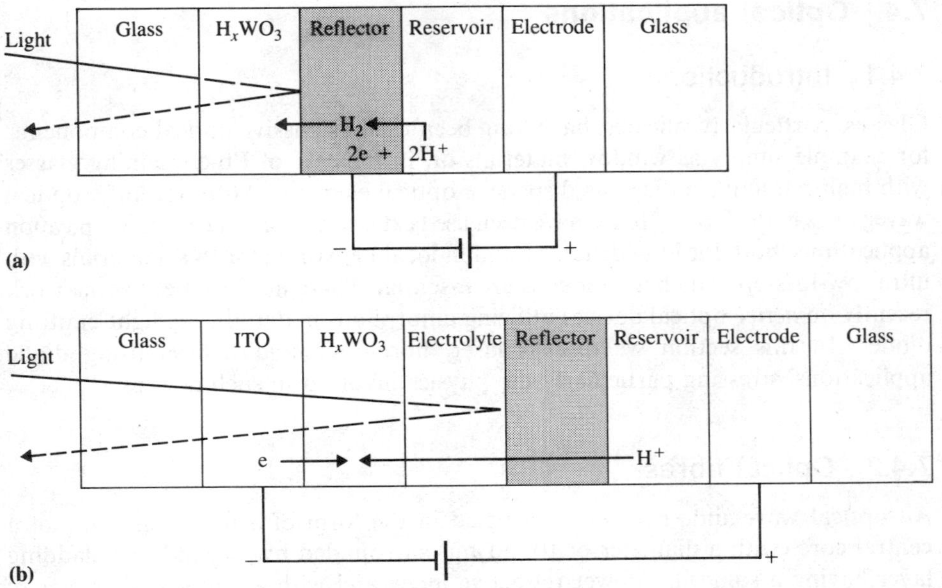

Fig. 7.8 Schematic illustration of electrochromic mirrors utilizing a-WO_3 as the active element (Baucke 1985):
(a) diffusion-controlled mirror;
(b) electric field-controlled mirror.

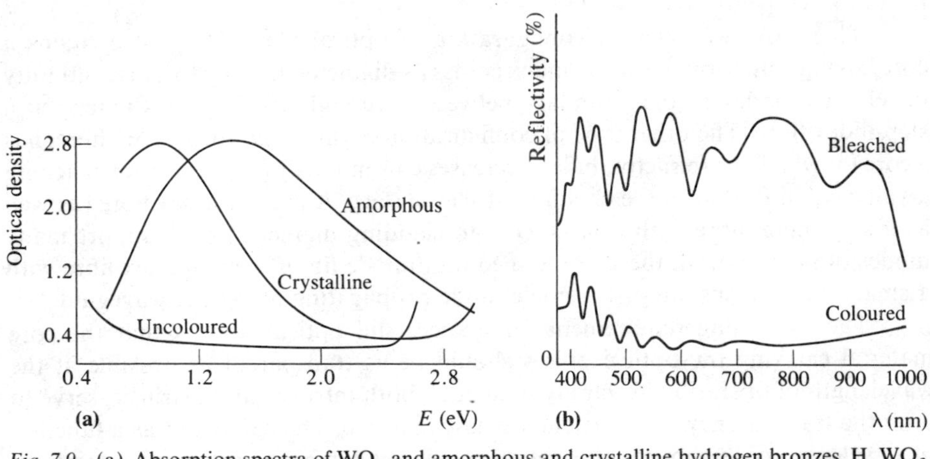

Fig. 7.9 (a) Absorption spectra of WO_3 and amorphous and crystalline hydrogen bronzes, H_xWO_3. (b) Change in reflectance of an electric-field-controlled electrochromic mirror. The oscillations are caused by thin-film interference effects (Baucke 1985).

interionic electron transfer (Baucke 1985); strong absorption bands are seen in the optical region (see Fig. 7.9(a)). The change in reflectivity for a field-controlled electrochromic mirror is indicated in Fig. 7.9(b). Such continuously adjustable reflectance mirrors would be of use in automobile applications, where a change in reflectance of $\simeq 20\%$ could be accomplished in about 5 s with a driving voltage of ± 1.5 V.

7.4 Optical applications

7.4.1 Introduction

Glasses, particularly silicates, have long been used as passive optical components, for example simply as window materials or, in the case of Pb-containing glasses with high refractive indices, as dispersive optical elements. More recently, optical waveguides in the form of fibres have been used extensively for optical communication applications, both for long distances and in local networks; for this, materials with ultra-low-loss optical characteristics are essential. There has also been some work recently on *active* optical devices utilizing amorphous materials, e.g. light-emitting diodes. In this section we discuss laser, fibre-optic and light-emitting diode applications, stressing particularly the physics involved in such devices.

7.4.2 Optical fibres

An optical waveguide can be constructed in the form of a fibre consisting of a central core (with a diameter of 10–50 μm) surrounded by a cylindrical cladding layer having a somewhat lower refractive index and with an outside diameter of 100–150 μm. A light pulse introduced into the core at one end of such an optical fibre is guided along the fibre, being confined to the core region through a series of total internal reflections at the interface between core and cladding. Such optical waveguides can readily be fabricated by pulling fibres from an appropriate preform held at a temperature above T_g.

There are two types of configuration of optical fibre. The first contains a core having a uniform refractive index across its diameter, leading to a discontinuity in refractive index at the boundary between core and cladding; this is termed a step-index fibre. The other type of configuration, termed a graded-index fibre, has a core for which the refractive index decreases continuously from its centre, reaching a value equal to that of the cladding at its interface. A graded-index fibre (as well as a step-index fibre with a large core-to-cladding diameter) can support many modes of propagation; these are termed multimode fibres; a step-index fibre with a small core can sustain just a single mode propagating along the waveguide.

The overriding requirement for a successful optical fibre is that the core material carrying the optical pulses should be as transparent as possible at the wavelength of operation. A variety of factors, both intrinsic and extrinsic, serve to limit the transparency of a particular glassy material. The behaviour as a function of wavelength of the optical absorption coefficient can be regarded as arising from three distinct terms:

$$\alpha(\lambda) = A/\lambda^4 + B_1 \exp(B_2/\lambda) + C_1 \exp(-C_2/\lambda) \qquad [7.8]$$

where A, B_1, B_2, C_1 and C_2 are constants. The first term describes Rayleigh scattering from microscopic density and compositional fluctuations in the glass, frozen-in at T_g; the second describes the Urbach absorption edge, lying in the UV region for silicate or fluoride glasses; the third describes the net multiphonon absorption response in the far-IR region, which approximates to an exponential behaviour. Taken together, the overall absorption response for a glass as described by [7.8] is a V-shaped curve delineated by the two exponential terms, with the Rayleigh scattering term, which is a less rapidly varying function of wavelength, determining the minimum level of absorption in the gap between the UV and IR edges if extrinsic absorption bands, due to impurities, are absent. Figure 7.10 shows experimental optical absorption profiles for pure v-SiO_2; the overall V-shaped profile is apparent. The absorption is shown on a scale of optical loss, measured in dB km^{-1}, the conventional unit used in describing attenuation losses in optic fibres.

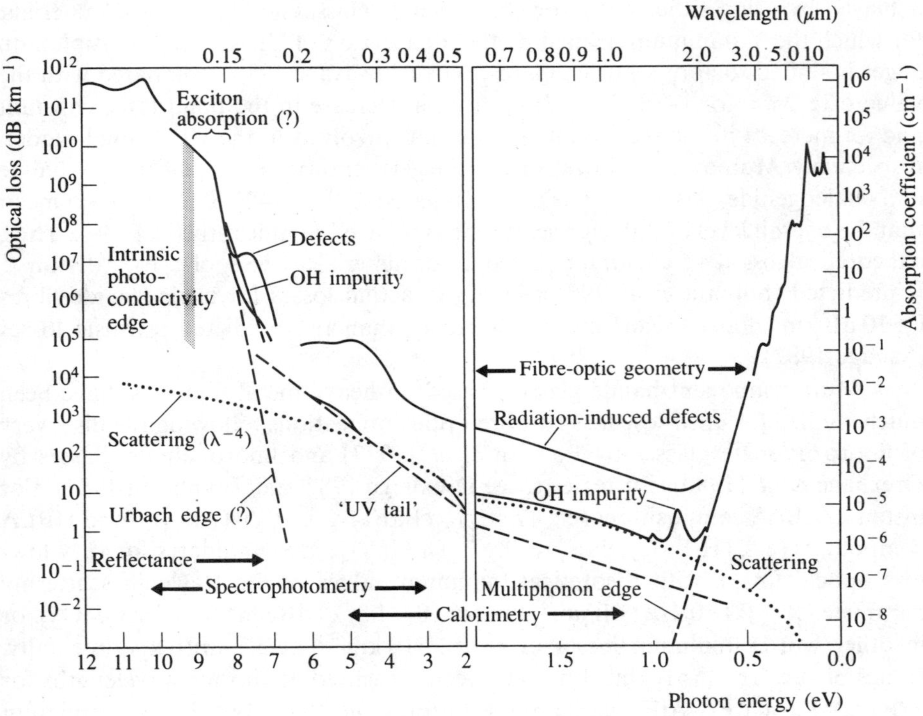

Fig. 7.10 Optical absorption spectra for pure v-SiO_2, based on results from a number of experiments, showing the V-shaped profile delineated by the multiphonon and Urbach edges at low and high photon energies, respectively (D. L. Griscom (1985), in *Glass: Current Issues*, (A. F. Wright and J. Dupuy, eds), NATO Series E: Applied Sciences, No. 92 (Martinus Nijhoff), p. 362).

For pure v-SiO_2 used as a core material, the minimum loss, limited by Rayleigh scattering in the absence of defect or impurity absorption, is predicted to be $\simeq 0.1$ dB km^{-1} at a wavelength of 1.6 μm (Poignant 1981), and losses close to this intrinsic limit ($\simeq 0.2$ dB km^{-1}) have been attained in actual fibres of v-SiO_2 (Miya *et al.* 1979).

Two factors can be optimized to reduce still further the residual attenuation losses due to Rayleigh scattering in glasses used in fibre-optic applications. From the particular wavelength dependence characteristic of Rayleigh scattering ([7.8]), scattering losses can be reduced significantly by operating at longer wavelengths, say 2–3 μm, if a particular glass is chosen which has a multiphonon absorption edge also lying at a longer wavelength. Furthermore, since the constant A in [7.8] describing the magnitude of the Rayleigh scattering can be written as (Shibata *et al.* 1981)

$$A \simeq 8\pi^3(n^2-1)^2\kappa_T k_B T_g \qquad [7.9]$$

where n is the refractive index and κ_T the isothermal compressibility, the level of scattering loss at a particular wavelength can be reduced by using a glass with a lower glass-transition temperature.

Both these ways of reducing the level of Rayleigh scattering loss can be achieved by using either halide or chalcogenide glasses as fibre-optical materials, for which the V-minimum defined by the intersection of Urbach and multiphonon edges is shifted to about 4 or 6 μm, respectively (Savage 1987), compared with the value of $\simeq 2\ \mu$m for v-SiO_2, resulting from a decrease in the bond force constant and an increase in the reduced mass of atoms involved in the vibrational modes, respectively. Moreover, the low values of the glass-transition temperature for halide and chalcogenide glasses ($T_g \simeq 600$ K, compared with $\simeq 1400$ K for v-SiO_2) mean that the overall level of Rayleigh scattering is reduced significantly (cf. [7.9]). Thus, a theoretical loss at 4.5 μm for the chalcogenide material a-$GeSe_3$ of $\simeq 10^{-2}$ dB km^{-1} is predicted (Shibata *et al.* 1981); however, actual losses for chalcogenide fibres ($\geqslant 10$ dB km^{-1} at $\simeq 3\ \mu$m) are much higher than the predicted intrinsic losses (Savage 1987).

Multicomponent halide glasses, based on heavy metal fluorides, have been much studied for ultra-low-loss IR fibre-optic applications following the discovery of fluorozirconate glasses by Poulain *et al.* (1977) and fluorohafnate glasses by Drexhage *et al.* (1980) (for reviews, see Drexhage 1987 and Takahashi 1987). For instance, ZBLA compositions (e.g. $(ZrF_4)_{57}(BaF_2)_{36}(LaF_3)_3(AlF_3)_4$) and HBLA compounds (e.g. $(HfF_4)_{57}(BaF_2)_{36}(LaF_3)_3(AlF_3)_4$) are candidates for very-low-loss optical fibres, with theoretical minimum values of the Rayleigh scattering parameter $A = 0.1–0.4/\lambda^4$ dB μm^4/km (cf. $A = 1.0/\lambda^4$ dB μm^4/km for v-SiO_2), or in other words minimum losses of $\simeq 10^{-3}$ dB km^{-1} at 3.5 μm; experimentally, values of $A \simeq 0.6/\lambda^4$ dB μm^4/km have been measured at shorter wavelengths for ZrF_4–BaF_2–GdF_3–AlF_3 glass fibres (Hattori *et al.* 1987), but the true minimum limit has yet to be realized. Fluoride glasses are prone to crystallization, particularly during the drawing process (the crystallites from which give rise to a characteristic λ^{-2} scattering loss behaviour), and in order to attain hypothetical (Rayleigh-scattering-limited) losses, concentrations of impurities in the glass such as hydroxyl groups, transition metals and other rare-earth elements must be reduced below the parts per billion level (Poignant 1987).

7.4.3 Laser materials

Ions, containing incomplete electronic shells (e.g. 3d or 4f ions), can undergo intraionic radiative recombination (luminescence) transitions when dissolved in either glassy or crystalline hosts (see sect. 5.6.3). In certain circumstances, when the electronic energy levels of the excited states of such ions are suitably arranged, a ‘population inversion’ can take place (i.e. with more species in an optically pumped excited state than in the ground state), from which a stimulated radiative decay may take place – i.e. 'lasing' action occurs.

In order to achieve a population inversion by optical pumping, a two-level system (i.e. the ground state and a single excited state) is insufficient since, in this case, the maximum ratio of the population of the excited state to that of the ground state is attained under saturation conditions, i.e. the ratio is 1:1. What is required is at least one additional, intermediate excited state (2) to which electrons can relax (non-radiatively) after initial optical excitation to a higher-lying excited state (3), and from which they can subsequently decay radiatively to the ground state (1); a population inversion between levels 2 and 1 is built up if the transition rate for $2 \rightarrow 1$ is much slower (e.g. by being spin-forbidden) that that for $3 \rightarrow 2$. Examples of lasers having such a three-level system are the Er^{3+}–glass system and the ruby laser (Cr^{3+} in crystalline Al_2O_3). The most intensively studied rare-earth ion/glass lasing system is that involving Nd^{3+}, in which *four* levels participate – an additional level (1′), lying close in energy to the ground state, is the one to which stimulated radiative (lasing) decay occurs. When incorporated in glassy matrices, Nd^{3+} ions can lase at a variety of wavelengths (0.92–1.37 μm) including the line at 1.06 μm resulting from the $^4F_{3/2} \rightarrow {}^4I_{11/2}$ transition. The interest in the Nd^{3+}–glass system stems from its possible use as a high-power laser source for inertial-confinement fusion applications. Weber (1983, 1987) has given reviews of glass-laser systems.

Although the active component in glass lasers is invariably a rare-earth ion (usually Nd^{3+}), the use of a *glassy* matrix for such lasing ions brings a number of advantages in such applications, including the obvious one of ease of fabrication: the laser material can readily be cast, cut or pulled into the desired shape (e.g. rod, disc or fibre). Glasses are also advantageous as hosts in view of their optical transparency over a wide range of wavelengths: transparency at short wavelengths is important so as to optimize the efficiency of optical pumping of the lasing ions; transparency at longer wavelengths is necessary in order to maximize the output intensity of the laser radiation from the glass. Thus, although oxide glasses (e.g. silicates) have a wide transmittance range of about 0.2–4 μm (see Fig. 7.10), the use of halide glasses such as HBLA or ZBLA can extend the long wavelength cut-off to nearly 10 μm (see sect. 7.4.2) for IR laser applications. Furthermore, the fact that certain fluoride glass compositions based on BeF_2 have an appreciable optical transition below 150 μm offers the prospect of fluoroberyllate glasses being used as hosts in UV or far UV laser applications (Weber 1983).

Glasses also have an important advantage for use as laser hosts from a compositional point of view: generally, impurities (e.g. lasing ions) can be incorporated homogeneously over rather large ranges of impurity contents, and the glassy host composition itself can be systematically varied (e.g. by varying the relative composition of mixed-anion glasses) so as to change the lasing characteristics of the matrix. One such parameter is the cross-section for stimulated emission, given by the relation (Weber 1983, 1987)

$$\sigma = \frac{8\pi^3 e^2 (n^2+2)^2 \lambda S(J, J')}{3hc(2J+1) n \, \Delta\lambda} \qquad [7.10]$$

where n is the refractive index, λ and $\Delta\lambda$ are respectively the wavelength and the width of the line resulting from transitions between states J and J′, and S is the line strength of the transition. Of the parameters controlling σ in [7.10], $S(\mathrm{J}, \mathrm{J}')$, n and $\Delta\lambda$ are strongly dependent on the type of glassy host or, in other words, on the nature of the ligand environment of the lasing ions. The transition probability term, $S(\mathrm{J}, \mathrm{J}')$, is essentially determined by the strength of the odd-order terms in the expansion of the local electric field at the rare-earth ion site, in turn determined by the type of anion and cation nearest neighbours, but the overall behaviour is rather complicated. Thus, for alkali disilicate glassy hosts, S (and hence σ) decreases (by a factor of 2) with increasing size of the alkali ion ($\mathrm{Li} \rightarrow \mathrm{K}$), whereas the converse is observed for alkali (and alkaline earth) phosphate and fluoroberyllate glasses (Weber 1983). The linewidth $\Delta\lambda$ of the laser emission is mainly determined by the nature of the ligands surrounding the rare-earth lasing ions; in general, the smaller the anionic electric field strength, the smaller the Stark splitting and the narrower the linewidth. Thus, the smallest values of $\Delta\lambda$ are obtained by using monovalent halide instead of divalent oxide anionic species in the glassy matrix. In addition, $\Delta\lambda$ generally increases with increasing charge and decreasing size of the network modifier for a given network former. Finally, the narrowest linewidths occur when the lasing rare-earth ions have very symmetric and chemically homogeneous coordination shells (Weber 1983, 1987).

Two laser applications which depend on the use of glassy hosts for their operation are optical-fibre lasers and high-power lasers. Small-scale fibre-optic laser oscillators and amplifiers are of interest for integrated optics and telecommunications applications. In these, silica optical fibres doped with $\simeq 1000$ ppm Nd^{3+} ions retain their high optical transparency, yet at the same time exhibit low-threshold ($\simeq 1$ mW) lasing characteristics when pumped with a GaAlAs diode laser (Weber 1987). The output power performance of high-power lasers suitable for use in inertial-confinement fusion applications is limited by 'self-focusing' effects within the glassy laser matrix, the very high light intensities in which are sufficiently large for non-linear optical behaviour to arise, for example in the refractive index

$$n = n_0 + n_2 I \qquad [7.11]$$

where n_2 is the non-linear refractive-index coefficient and I is the light intensity. This effect is minimized by using as a laser host a glass having a low value of n_2, i.e. a low-index, low-dispersion material such as a BeF_2-based glass (Weber 1983).

7.4.4 Light-emitting diodes

As discussed in sect. 7.2, a-Si:H can be doped n- or p-type and so, in principle, diodes based on this material could be made to exhibit electroluminescence (EL) resulting from the radiative recombination of electrically injected carriers (Kukimoto 1984). In such devices fabricated by Kruangam *et al.* (1985, 1987), p–i–n diode structures made from the wide-band-gap material amorphous silicon carbide (a-SiC:H) emit EL in the visible region at room temperature; the optical gap of the luminescent active i-layer (in the range 2.48–2.64 eV) determines the wavelength of EL emission (in the corresponding range 1.65–1.85 eV) (Kruangam *et al.* 1985). The efficiency of EL emission from such light-emitting diodes can be improved by increasing the optical gap E_o of the injecting n- and p-layers of a-SiC:H (Kruangam *et al.* 1987); a tenfold increase in light output has been achieved by increasing E_o

by $\simeq 0.5$ eV from 2 eV by increasing the hydrogen dilution of SiH_4 and CH_4 used in the plasma CVD preparation process. However, increasing the E_0 of the p- and n-layers causes an increase in the energy of EL emission; furthermore, increasing the H dilution of the reactant gases results in an increasing proportion of microcrystalline material being produced within the amorphous phase of a-SiC:H, as happens also with pure a-Si:H.

7.5 Magnetic applications

7.5.1 Introduction

Many applications have been proposed to exploit the unique magnetic properties exhibited by metallic glasses. In fact, it was originally thought that the existence of an ordered ferromagnetic state required the presence of long-range translational order, as in crystals. However, the exchange interaction between two spins which gives rise to ferromagnetism is short-range. Furthermore, since the *local* structural environment in an amorphous metal is very similar to that of the corresponding crystal (see sect. 3.3), the discovery that ferromagnetism *can* exist in the amorphous state should not, perhaps, have been too surprising. The first amorphous metallic ferromagnet found was an unstable Co–Au alloy produced by vapour quenching; but the study of magnetism in amorphous materials really began with the discovery of ferromagnetism in a *glassy* material, $Fe_{83}P_{10}C_7$ (Duwez and Lin 1967), produced by quenching from the melt, a method which afforded the prospect of producing stable materials in a usable form, albeit as thin foils. It has since been found that a given transition-metal–metalloid (TMM) alloy, the most common studied type of amorphous magnetic material, is generally ferromagnetic if the corresponding *crystalline* alloy (minus the metalloid component) is itself ferromagnetic.

Glassy TMM alloys exhibit a unique combination of magnetic (and mechanical) properties. They are generally 'soft' ferromagnets, i.e. they require relatively small coercive fields, $\boldsymbol{H}_c$, in the opposite direction to that of the magnetization, $\boldsymbol{M}$, in order to destroy all the magnetization; in other words, the B–H (or M–H) hysteresis loop is very narrow. Values of H_c in the range 0.05–0.1 Oe are common for glassy TMM ferromagnets, comparable to those observed in soft crystalline ferromagnets (see Table 7.1). These low values reflect the appreciable structural and chemical homogeneity of such glasses (meaning that grain boundaries or precipitates, which can act as pinning centres for domain walls, are essentially absent). Another measure of the softness of the magnetic properties is that the magnetic anisotropy of such glasses is very low, and in some cases zero. Metallic glasses have extremely high, almost ideal, values of mechanical strength, hardness and toughness, which is of considerable advantage in applications involving wear (e.g. recording heads).

Glassy metallic alloys have another important advantage over their crystalline counterparts, namely that important magnetic properties, such as saturation magnetic moment (or magnetization), Curie temperature and magnetostriction, can be varied smoothly simply by changing the composition, e.g. the proportion

Table 7.1 Some properties of amorphous and crystalline ferromagnets (from Luborsky 1977)

Alloy	*Coercive field* H_c (mOe)	*Saturation induction (298 K)* $4\pi M_s$ (kG)	*Curie temp.* θ_c (°C)	*Magnetostriction* λ_s ($\times 10^6$)
Amorphous				
$Fe_{80}B_{20}$ (M*2605)	40	16.0	374	30
$Fe_{80}P_{16}B_1C_3$ (M*2615)	50	14.9	292	30
$Fe_{40}Ni_{40}P_{14}B_6$ (M*2826)	19	8.2	247	11
$Fe_{29}Ni_{49}P_{14}B_4Si_2$ (M*2826B)	11	4.2	382	5
$Fe_3Co_{72}P_{16}B_6Al_3$	15	6.3	260	~0
Crystalline				
$Ni_{80}Fe_{16}Mo_4$ (Mumetal)	25	7.8	460	~0
$Ni_{80}Fe_{20}$ (Hymu '800')	5	8.2	400	~0
$Fe_{96.8}Si_{3.2}$ (oriented transformer steel)	300–500	20.3	730	4

M* = Metglas®

of the early transition metal in an early–late transition-alloy or, in TMM alloys, by changing the relative proportion of either two transition metals or of two metalloid elements in ternary or quaternary alloys, whilst keeping the overall glass-forming composition $T_{80}M_{20}$ more or less unchanged (O'Handley 1983); Fig. 7.11 shows these variations for a variety of TMM alloys. The pronounced changes of magnetic moment with composition have been ascribed to charge transfer (or possibly chemical bonding interactions) between metalloid and metal.

It is intuitively understandable that in crystalline ferromagnets there should be certain crystallographic directions in which magnetization is 'easy' (i.e. very low applied field strengths are required to achieve magnetic saturation), and others for which it is 'hard'; thus in single crystal b.c.c. Fe, $\langle 100 \rangle$ is the easy direction and $\langle 111 \rangle$ the hard direction, and for f.c.c. Ni, $\langle 111 \rangle$ and $\langle 100 \rangle$ are the easy and hard directions, respectively. This intrinsic magnetic anisotropy will not be preserved in amorphous ferromagnets, but an extrinsic magnetic anisotropy can nevertheless be induced, which may be of two kinds: that due to residual structural inhomogeneities or strains introduced during the quenching process, or that due to annealing in the presence of an applied magnetic field. These magnetic anisotropies are *uniaxial* in the sense that the easy direction is only along the inducing direction, and not along all structurally equivalent directions as is possible for anisotropy in crystalline materials. The degree of uniaxial magnetic anisotropy is commonly expressed in terms of the excess energy, E_u, which is related to the angle ϕ between the domain magnetization direction and for example the external magnetic field direction by:

$$E_u = -K_u \cos^2 \phi \qquad [7.12]$$

where K_u is called the uniaxial anisotropy constant (Chen 1977). K_u has a value of typically $\simeq 10^3$ J m^{-3} for an as-quenched glassy alloy, whereas for magnetic-field-induced anisotropy it is typically $\simeq 10^2$ J m^{-3} (the same order of magnitude

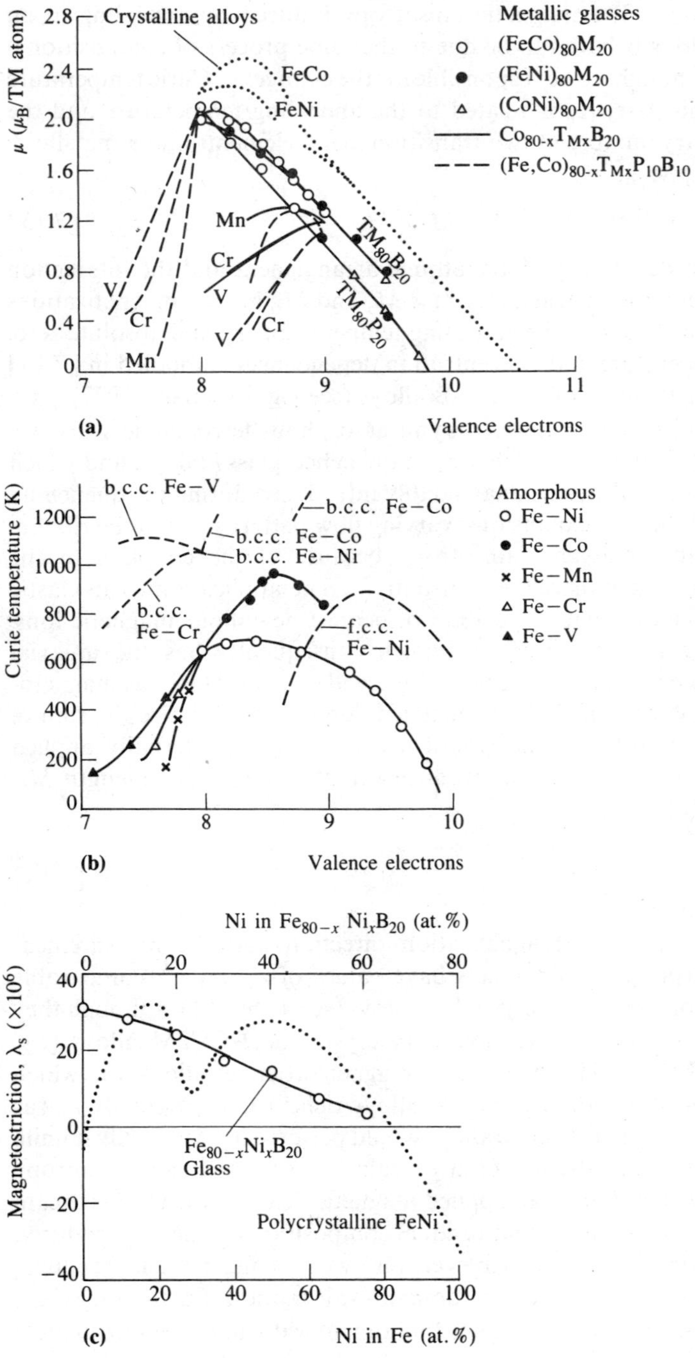

Fig. 7.11 Compositional variations of magnetic properties in glassy transition-metal–metalloid alloys (O'Handley 1983):
(a) mean saturation magnetic moment;
(b) Curie temperature;
(c) magnetostriction.

as for crystalline alloys). The magnetic anisotropy induced by annealing in an external magnetic field is believed to be due to the same process of compositional short-range ordering thought to be responsible for the changes of Curie temperature with annealing. In this case, K_u is related to the annealing temperature and the composition of a binary mixture of two transition metal elements and a metalloid, e.g. $(T(1)_x T(2)_{1-x})_{80} M_{20}$, by:

$$K_u = ANx^2(1-x)^2(M_T/M_0)^2(M_a/M_0)^2/T_a \qquad [7.13]$$

where A is a constant depending on the atomic arrangement and the interaction range, N is the atom number density, and M_T, M_a and M_0 are the magnetizations at the measuring temperature, T, the annealing temperature, T_a, and absolute zero, respectively. The temperature and concentration dependences embodied in [7.13] have been verified for the case of Fe–Ni–B alloys (see e.g. Luborsky 1977).

The final form of magnetic anisotropy in amorphous ferromagnets that we wish to discuss is that found in unannealed, as-quenched glassy alloys, and which arises because of the internal strains unavoidably introduced during the quenching process by differential thermal expansions, varying flow patterns within the ribbon, etc. These strains cause magnetic anisotropy because of the coupling to the magnetization via the magnetostrictive constant, λ_s. The application of an elastic stress will slightly alter the relative position of nearest-neighbour magnetic ions, resulting in a change in the anisotropy which consequently has the uniaxial symmetry of the induced strain. Conversely, there will be a spontaneous magnetostrictive strain within each uniformly magnetized domain to minimize the sum of elastic and magnetic anisotropy energies. Thus, rotating an externally applied magnetic field by 90°, for instance, will produce a fractional change in length $\Delta l/l$ in a magnetostrictive alloy, given by

$$\frac{\Delta l}{l} = \frac{3}{2}\lambda_s(\cos^2\theta - \tfrac{1}{3}) \qquad [7.14]$$

where θ is the angle between magnetization direction and the measurement direction. Many amorphous ferromagnets have values of $\lambda_s \simeq 10^{-5}$, comparable to those observed in corresponding crystalline alloys (see Table 7.1), although they fall well below the enormous values found in crystalline RE–TM alloys (e.g. $\lambda_s = 1.7 \times 10^{-3}$ in c-$TbFe_2$). However, 'zero magnetostrictive' alloys, for which $\lambda_s = 0$, have been produced, notably Fe–Co alloys rich in Co. Originally it was thought that such zero magnetostrictive alloys would perforce have virtually infinite permeabilities because of the absence of any strain-induced magnetic anisotropy which impedes remagnetization by an applied magnetic field. However, this dream has been shattered by the discovery that residual compositional ordering produced either during quenching or annealing (in even very weak magnetic fields) induces sufficient magnetic anisotropy to impede domain-wall motion. Furthermore, the surface of a ribbon itself acts as a pinning site for domain walls, and so unfortunately the ideal anisotropy-free glassy ferromagnet cannot be realized in practice.

Reviews of the magnetic behaviour of amorphous metals have been given by Luborsky (1977, 1980, 1983), Durand (1983) and O'Handley (1983), and of their technological applications by Raskin and Smith (1983), Luborsky (1985) and Gambino (1985).

7.5.2 Transformer cores

The greatest impact of the use of glassy metallic alloys has been in transformer-core applications, for operation both at low frequencies (50–60 Hz) in distribution transformers, and at high frequencies in a variety of special applications. In the USA and Japan, where power distribution transformers are widely used to effect the final voltage step-down of the electricity supply to local users, enormous savings in cost are anticipated through the reduction of transformer-core losses if metallic glass cores are used instead of the currently used grain-oriented crystalline Fe–Si alloys. In use, the primary coils of these transformers are energized continuously, and it has been estimated that, in the USA alone, the total annual core losses amount to $\simeq 5 \times 10^{10}$ kW h (Ramanan *et al.* 1987) which could be reduced by up to 75% if all cores were replaced with glassy material, resulting in an annual saving of about $250 million (Luborsky 1985).

The power loss in a magnetic material is given by the area enclosed by the hysteresis loop, or of the quantity $\int H\, dB/dt$, where the width of the loop is determined by the coercive field H_c. Losses in low-frequency (50–60 Hz) transformer operation for mains (line) applications can therefore be reduced by minimizing H_c, and the coercivity is determined by a number of factors which lead to the pinning of domain walls, such as intrinsic fluctuations in the glass, short-range compositional fluctuations (atomic clustering), surface irregularities and bulk (free-volume) defects causing strain-related pinning in magnetostrictive materials. Fluctuation-related hysteresis losses may be reduced by annealing the metallic glass ribbons, particularly in the presence of a longitudinal magnetic field which establishes domain walls more or less parallel to the direction of the operating magnetic flux (Ramanan *et al.* 1987). However, free-volume defects cannot be annealed out in this way, and the only way to reduce the effect of stress-induced anisotropies is to use an alloy with zero magnetostriction, for example by alloying Ni (see Fig. 7.11(c)) in Fe–B-based glasses, as in commercial glassy transformer-core materials.

Such metallic-glass-based transformer cores exhibit a 75% reduction in losses compared with conventional grain-oriented crystalline $Fe_{97}Si_3$ laminated core materials (Luborsky 1985) and are being used to replace the traditional cores. It has been estimated (Raskin and Smith 1983) that core losses in chokes (ballasts) used in fluorescent lamps are comparable to those in distribution transformers, and so in principle the use of glassy Fe–B cores could effect a similar saving; however, it may be more advantageous to switch to higher-frequency operation than to redesign cores to enable them to be built from metallic glass materials.

At higher frequencies, losses in transformer cores are mainly through eddy currents, and here the use of metallic glasses is again advantageous since these materials have appreciably higher electrical resistivities than their crystalline permalloy counterparts, thereby significantly reducing eddy losses. Furthermore, the saturation induction of glasses such as Fe(Cr, Mo)–(B, Si) alloys ($B_s \simeq 0.6$–1.4 T) are substantially higher than in ferrites ($B_s \simeq 0.4$ T), and Co-based glasses in particular exhibit extremely high permeabilities.

7.5.3 Device applications

The unique magnetic characteristics exhibited by metallic glasses make them attractive materials for use in a variety of small-scale device applications.

Thus, Co-based, zero-magnetostriction alloys are used as record/playback heads in audio and computer magnetic-tape systems, where the requirements of high saturation induction, high permeability and high surface hardness are admirably met by the use of amorphous metals.

If, instead of annealing-in a longitudinal domain pattern as in transformer-core applications (sect. 7.5.2), the magnetic field is applied transversely to the ribbon length during annealing, a transverse domain structure is readily established in a glassy metal material owing to its low inherent anisotropy. This configuration means that the rotation of domains, and hence the magnetostrain, is maximized upon subsequent longitudinal excitation; the very large magnetomechanical coupling coefficients characteristic of such systems can give rise to extremely high efficiencies (up to 90%) for converting magnetic energy into mechanical energy, and vice versa, which are much larger than those associated with, say, conventional piezoelectric ceramic materials. Such metallic-glass components can therefore be used as the basis of very sensitive transducers.

References

(Numbers in parentheses refer to section numbers within the text.)

Abe, S. and Toyozawa, Y. (1981) *J. Phys. Soc. Jap.* **50**, 2185. (5.6.2, 6.3.1)

Abrahams, E., Anderson, P. W., Licciardello, D. C. and Ramakrishnan, T. V. (1979) *Phys. Rev. Lett.* **42**, 673. (5.3.4, 5.3.5, 5.4.1)

Adam, G. and Gibbs, J. H. (1965) *J. Chem. Phys.* **43**, 139. (2.3.3)

Adler, D. (1987) *Physics of Amorphous Semiconductor Devices* (*SPIE Proc.* vol. 763), p. 2. (6.2.6)

Adler, D. and Yoffa, E. J. (1976) *Phys. Rev. Lett.* **36**, 1197. (6.2.2)

Agarwal, S. C. (1973) *Phys. Rev.* **B7**, 685. (6.3.2)

Akridge, J. R. and Vourlis, H. (1986) *Solid St. Ionics* **18–19**, 1082. (7.3.2)

Alben, R., Weaire, D., Smith, J. E. and Brodsky, M. H. (1975) *Phys. Rev.* **B11**, 2271. (3.2.6, 4.3.2)

Alexander, S. and Orbach, R. (1982) *J. Phys.* (*Paris*) *Lett.* **43**, L625. (4.4.6)

Alexander, S., Laermans, C., Orbach, R. and Rosenberg, H. M. (1983) *Phys. Rev.* **B28**, 4615. (4.4.6)

Allen, J. W., Wright, A. C. and Connell, G. A. N. (1980) *J. Non-Cryst. Sol.* **42**, 509. (2.5.2)

Altshuler, B. L. and Aronov, A. G. (1985) in *Electron–Electron Interactions in Disordered Systems* A. L. Efros and M. Pollak (eds) (North-Holland), p. 1. (5.4.1)

Ambegaokar, V., Halperin, B. I. and Langer, J. S. (1971) *Phys. Rev.* **B4**, 2612. (6.3.3)

Amer, N. M. and Jackson, W. B. (1984) in *Semiconductors and Semimetals*, J. Pankove (ed.), vol. 21B (Academic), p. 83. (6.3.1)

Amini, M., Mitra, S. K. and Hockney, R. W. (1981) *J. Phys.* **C14**, 3689. (3.5.6)

Anderson, O. L. and Stuart, D. A. (1954) *J. Am. Ceram. Soc.* **37**, 573. (4.5.2)

Anderson, P. W. (1958) *Phys. Rev.* **109**, 1492. (5.3.2, 5.3.4)

Anderson, P. W. (1975) *Phys. Rev. Lett.* **34**, 953. (6.2.2, 6.2.3, 6.3.2)

Anderson, P. W. (1979) in *Ill-condensed Matter*, R. Balian, R. Maynard and G. Toulouse (eds) (North-Holland). (2.3.3, 2.3.5, 6.2.2)

Anderson, P. W., Halperin, B. I. and Varma, C. M. (1972) *Phil. Mag.* **25**, 1. (4.4.5)

Anderson, P. W., Abrahams, E. and Ramakrishnan, T. V. (1979) *Phys. Rev. Lett.* **43**, 718. (5.4.1)

Anderson, P. W., Thouless, D. J., Abrahams, E. and Fisher, D. S. (1980) *Phys. Rev.* **B22**, 3519. (5.3.3)

Andrew, E. R. (1981) *Phil. Trans. Roy. Soc.* (*Lond.*) **A299**, 505. (3.2.4)

Angell, C. A. (1986) *Sol. St. Ionics* **18–19**, 72. (4.5.1)

Angell, C. A. (1988) *J. Non-Cryst. Sol.* **102**, 205. (2.2, 2.3.3)

Angell, C. A., Cheeseman, P. A., Clarke, J. H. R. and Woodcock, L. V. (1977) in *The Structure of Non-Crystalline Materials*, P. H. Gaskell (ed.) (Taylor and Francis), p. 191. (3.5.6)

Applebaum, J. A. and Hamann, D. R. (1976) *Rev. Mod. Phys.* **48**, 479. (6.2.2)

Apsley, N. and Hughes, H. P. (1974) *Phil. Mag.* **30**, 963; (1975) ibid. **31**, 1327. (6.3.3)
Arai, M., Price, D. L., Susman, S., Volin, K. J. and Walter, U. (1988) *Phys. Rev.* **B37**, 4240. (4.3.4)
Atzmon, M., Unruh, K. and Johnson, W. L. (1985) *J. Appl. Phys.* **58**, 3865. (1.3.10)
Aur, S., Kofalt, D., Waseda, Y., Egami, T., Wang, R., Chen, H. S. and Teo, B. K. (1983) *Solid St. Comm.* **48**, 111. (3.2.2)
Austin, I. G. and Mott, N. F. (1969) *Adv. Phys.* **18**, 41. (5.5, 6.3.4)

Bale, H. D. and Schmidt, P. W. (1984) *Phys. Rev. Lett.* **53**, 596. (3.6.2)
Bamford, C. R. (1977) *Colour Generation and Control in Glass* (Elsevier). (5.6.3)
Baraff, G. A., Kane, E. O. and Schlüter, M. (1980a) *Phys. Rev.* **B21**, 3563; (1980b) *Phys. Rev.* **B21**, 5662. (6.2.3)
Bar-Yam, Y. and Joannopoulos, J. D. (1986) *Phys. Rev. Lett.* **56**, 2203. (6.2.1, 6.2.6)
Bar-Yam, Y., Adler, D. and Joannopoulos, J. D. (1986) *Phys. Rev. Lett.* **57**, 467. (6.2.2, 6.2.6, 6.3.1)
Baucke, F. G. K. (1985) in *Glass*: *Current Issues*, A. F. Wright and J. Dupuy (eds) (NATO Series E: *Applied Sciences*, no. 92) (Martinus Nijhoff), p. 481. (7.3.3, 7.3.4)
Baum, J., Gleason, K. K., Pines, A., Garroway, A. N. and Reimer, J. A. (1986) *Phys. Rev. Lett.* **56**, 1377. (3.2.4)
Baum, J., Munowitz, M., Garroway, A. N. and Pines, A. (1977) *J. Chem. Phys.* **83**, 2015. (3.2.4)
Beeman, D. and Alben, R. (1977) *Adv. Phys.* **26**, 339. (4.2.4, 4.3.2, 4.3.3)
Beeman, D. and Bobbs, B. L. (1975) *Phys. Rev.* **B12**, 1399. (3.5.3, 3.5.5)
Beeman, D., Tsu, R. and Thorpe, M. F. (1985) *Phys. Rev.* **B32**, 874. (4.3.3)
Beier, W. and Frishchat, G. H. (1985) *J. Non-Cryst. Sol.* **73**, 113. (4.5.2)
Bell, R. J. (1972) *Rep. Progr. Phys.* **35**, 1315. (4.2.3, 4.3.2)
Bell, R. J. and Dean, P. (1966) *Nature* (*Lond.*) **212**, 1354. (3.5.3)
Bell, R. J. and Dean, P. (1972) *Phil. Mag.* **25**, 1381. (4.3.2)
Bengtzelius, U. (1986) *Phys. Rev.* **A34**, 5059. (2.3.4)
Bengtzelius, U., Götze, W. and Sjölander, A. (1984) *J. Phys.* **C17**, 5915. (2.3.4, 2.3.5)
Bennett, C. H. (1972) *J. Appl. Phys.* **43**, 2727. (3.5.2)
Bergmann, G. (1983) *Phys. Rev.* **B28**, 2914. (5.3.4, 5.4.1)
Bernal, J. D. (1959) *Nature* (*Lond.*) **183**, 141; (1964) *Proc. Roy. Soc.* (*Lond.*) **A280**, 299. (3.5.2)
Bhatia, A. B. and Thornton, D. E. (1970) *Phys. Rev.* **B2**, 3004. (3.2.2)
Bishop, S. G. and Shevchik, N. J. (1974) *Sol. St. Comm.* **15**, 629. (3.7.1)
Bishop, S. G., Strom, U. and Taylor, P. C. (1977) *Phys. Rev.* **B15**, 2278. (6.3.2)
Boolchand, P. (1985) in *Physical Properties of Amorphous Materials*, D. Adler, B. B. Schwartz and M. C. Steele (eds) (Plenum), p. 221. (3.2.5)
Bordas, J. and West, J. B. (1976) *Phil. Mag.* **34**, 501. (5.2.2)
Bordewijk, P. (1975) *Chem. Phys. Lett.* **32**, 592. (4.5.3)
Born, M. and Huang, K. H. (1954) *Dynamical Theory of Crystal Lattices* (Oxford University Press). (4.1)
Bosse, J. and Krieger, U. (1986) *J. Phys.* **C19**, L609. (2.3.4)
Boucher, B., Chieux, P., Convert, P. and Tournarie, M. (1983) *J. Phys.* **F13**, 1339. (3.6.2)
Boudreaux, D. S. and Gregor, J. M. (1977) *J. Appl. Phys.* **48**, 152. (3.5.2)
Boulon, G., Moine, B. and Bourcet, J. C. (1980) *Phys. Rev.* **B22**, 1163. (5.6.3)
Boutron, P. and Alben, R. (1975) *J. Chem. Phys.* **62**, 4848. (3.5.3)
Boyce, J. B. and Stutzmann, M. (1985) *Phys. Rev. Lett.* **54**, 565. (3.2.4, 6.2.1)

Bragg, W. L. and West, J. (1930) *Phil. Mag.* **10**, 823. (3.2.2)
Branz, H. M. (1989) *Phys. Rev.* **B39**, 5107. (6.2.6)
Bray, P. J. (1978) in *Borate Glasses: Structure, Properties, Applications.* L. D. Pye, V. D. Frechette and N. J. Kreidl (eds) (Plenum). (3.2.4)
Bray, P. and O'Keefe, J. G. (1963) *Phys. Chem. Glasses* **4**, 37. (3.2.4)
Brecher, C. and Riseberg, L. A. (1976) *Phys. Rev.* **B13**, 81. (5.6.3)
Brodsky, M. H. (1975) in *Light Scattering in Solids*, M. Cardona (ed.) (*Topics in Applied Physics*, vol. 8) (Springer-Verlag). (4.3.3)
Brodsky, M. H. (1978) *Thin Sol. Films* **50**, 57. (1.3.3)
Buchenau, U. (1985) *Z. Phys.* **B58**, 181. (4.4.4)
Buchenau, U. (1987) *Phys. Scripta* **T19**, 350. (4.4.4, 4.4.5)
Buchenau, U., Nücker, N. and Dianoux, A. J. (1984) *Phys. Rev. Lett.* **53**, 2316. (4.4.4)
Buchenau, U., Prager, M., Nücker, N., Dianoux, A. J., Ahmad, N. and Phillips, W. A. (1986) *Phys. Rev.* **B34**, 5665. (4.4.4, 4.4.5, 4.4.6)
Buchenau, U., Zhou, H. M., Nücker, N., Gilroy, K. S. and Phillips, W. A. (1988a) *Phys. Rev. Lett.* **60**, 1318. (4.4.4, 4.4.5)
Buchenau, U., Prager, M., Kamitakahara, W. A., Shanks, H. R. and Nücker, N. (1988b) *Europhys. Lett.* **6**, 695. (4.4.4)
Bullett, D. W. and Kelly, M. J. (1979) *J. Non-Cryst. Sol.* **32**, 225. (5.2.1)
Busse, L. E. (1984) *Phys. Rev.* **B29**, 3639. (3.4.4)
Busse, L. E. and Nagel, S. R. (1981) *Phys. Rev. Lett.* **47**, 1848. (3.4.4)

Cahn, J. W. (1965) *J. Chem. Phys.* **42**, 93 (3.7.2)
Cahn, J. W. and Charles, R. H. (1965) *Phys. Chem. Glasses* **6**, 181. (3.7.2)
Cahn, R. W. (1980) *Contemp. Phys.* **21**, 43. (1.3.5)
Car, R. and Parinello, M. (1985) *Phys. Rev. Lett.* **55**, 2471 (3.5.6)
Cardona, M. and Ley, L. (eds) (1978) *Photoemission in Solids I and II* (*Topics in Applied Physics*, vols. 26, 27) (Springer-Verlag). (5.2.2)
Cargill, G. S. (1975) in *Solid State Physics*, vol. 30, F. Seitz, D. Turnbull and H. Ehrenreich (eds) (Academic), p. 227. (3.2.2, 3.5.2)
Cargill, G. S. (1981) in *Diffraction Studies on Non-Crystalline Substances*, I. Hargittai and W. J. Orville-Thomas (eds) (Elsevier) p. 783. (3.5.2)
Cargill, G. S. and Spaepen, F. (1981) *J. Non-Cryst. Sol.* **43**, 91. (3.3)
Carlos, W. E. and Taylor, P. C. (1982) *Phys. Rev.* **B25**, 1435. (3.2.4)
Carpenter, J. M. (1985) *J. Non-Cryst. Sol.* **76**, 1. (3.2.2)
Carpenter, J. M. and Pelizzari, C. A. (1975a) *Phys. Rev.* **B12**, 2397. (4.3.4, 4.4.4)
Carpenter, J. M. and Pelizzari, C. A. (1975b) *Phys. Rev.* **B12**, 2391. (4.4.4)
Cavenett, B. C. (1981) *Adv. Phys.* **30**, 475. (6.3.2)
Cervinka, L. (1988) *J. Non-Cryst. Sol.* **106,** 291. (3.4.4)
Chandra, S. (1981) *Superionic Solids: Principles and Applications* (North-Holland). (4.5.1)
Charles, R. J. (1961) *J. Appl. Phys.* **32**, 1115. (4.5.2)
Chaudhari, P., Spaepen, F. and Steinhardt, P. J. (1983) in *Glassy Metals II* H. Beck and H. J. Güntherodt (eds) (*Topics in Applied Physics*, vol. 53) (Springer-Verlag), p. 127. (6.2.3, 6.2.4, 6.2.5)
Chen, L. C. and Spaepen, F. (1988) *Nature* (*Lond.*) **336**, 366. (1.2)
Chmelka, B. F., Mueller, K. T., Pines, A., Stebbins, J., Wu, Y. and Zwanziger, J. W. (1989) *Nature* (*Lond.*) **339**, 42. (3.2.4)
Clement, V., Ravaine, D., Deportes, C. and Billat, R. (1988) *Sol. St. Ionics* **28–30**, 1572. (5.4.3)
Cochrane, R. W. and Strom-Olsen, J. O. (1984) *Phys. Rev.* **B29**, 1088. (5.4.1)

Cody, G. D. (1984) in *Semiconductors and Semimetals*, J. Pankove (ed.), vol. 21B (Academic), p. 11. (6.3.1)
Cohen, J. D. (1984) in *Semiconductors and Semimetals*, J. Pankove (ed.), vol. 21C (Academic), p. 9. (6.3.1)
Cohen, M. H. and Grest, G. S. (1979) *Phys. Rev.* **B20**, 1077. (2.3.5)
Cohen, M. H. and Turnbull, D. (1959) *J. Chem. Phys.* **31**, 1164. (2.3.5, 6.2.4)
Cohen, M. H., Fritzsche, H. and Ovshinsky, S. R. (1969) *Phys. Rev. Lett.* **22**, 1065. (5.3.2, 5.4.1, 6.2.1, 6.2.2, 6.3.1)
Cole, K. S. and Cole, R. H. (1941) *J. Chem. Phys.* **9**, 341. (4.5.3)
Compaan, K. and Haven, Y. (1956) *Trans. Faraday Soc.* **52**, 786; (1958), ibid. **54**, 1498. (4.5.2)
Connell, G. A. N. (1975) *Sol. St. Comm.* **16**, 109. (3.5.3)
Connell, G. A. N. (1979) in *Amorphous Semiconductors*, M. H. Brodsky (ed.) (*Topics in Applied Physics*, vol. 36) (Springer-Verlag). (5.2.2)
Connell, G. A. N. and Temkin, R. J. (1974) *Phys. Rev.* **B9**, 5323. (3.5.5, 5.2.1, 6.2.5)
Cooper, A. R. (1978) *Phys. Chem. Glasses* **19**, 60. (2.5.1)
Copley, J. R. D. and Rowe, J. M. (1974) *Phys. Rev. Lett.* **32**, 49. (4.3.4)
Corb, B. W., Wei, W. D. and Averbach, B. L. (1982) *J. Non-Cryst. Sol.* **53**, 29. (3.4.1)
Cote, P. J. and Meisel, L. V. (1981) in *Glassy Metals I* (*Topics in Applied Physics*, vol. 46) (Springer-Verlag), p. 141. (5.4.1)
Cotton, F. A. and Wilkinson, G. (1980) *Advanced Inorganic Chemistry* (Wiley). (5.6.3)
Courtens, E., Pelous, J., Phalippou, J., Vacher, R. and Woignier, T. (1987) *Phys. Rev. Lett.* **58**, 128. (4.4.6)
Courtens, E., Vacher, R., Pelous, J. and Woignier, T. (1988) *Europhys. Lett.* **6**, 245. (4.4.6)
Davidson, D. W. (1961) *Can. J. Chem.* **39**, 571. (4.5.3)
Davidson, D. W. and Cole, R. H. (1951) *J. Chem. Phys.* **19**, 1484. (4.5.3)
Davis, E. A. (1979) in *Amorphous Semiconductors*, M. H. Brodsky (ed.) (*Topics in Applied Physics*, vol. 36) (Springer-Verlag). (6.1)
Davis, E. A., Wright, H., Doran, N. J. and Nex, C. M. M. (1979) *J. Non-Cryst. Sol.* **32**, 257. (4.3.4)
Day, D. E. (1976) *J. Non-Cryst. Sol.* **21**, 343. (4.5.4)
Dean, P. (1972) *Rev. Mod. Phys.* **44**, 127. (4.2.3)
Dean, P and Martin, J. L. (1960) *Proc. Roy. Soc.* (*Lond.*) **A259**, 409. (4.2.3)
de Jong, B. H. W. S., Keefer, K. D., Brown, G. E. and Taylor, C. M. (1981) *Geochim. Cosmochim. Acta.* **45**, 1291. (3.2.4)
de Leeuw, S. W., He, H. and Thorpe, M. F. (1985) *Sol. St. Comm.* **56**, 343. (4.2.1)
De Neufville, J. P. and Rockstad, H. K. (1974) in *Proc. 5th Int. Conf. on Amorphous and Liquid Semiconductors*, J. Stuke and W. Brenig (eds) (Taylor and Francis), p. 419. (2.4)
Depinna, S. P., Cavenett, B. C. and Lamb, W. E. (1983) *Phil. Mag.* **B47**, 99. (6.3.1)
Dersch, H., Stuke, J. and Beichler, J. (1981) *Phys. Stat. Sol.* **105b**, 265. (6.2.6)
Dersch, U. and Thomas, P. (1987) in *Disordered Semiconductors*, M. A. Kastner, G. A. Thomas and S. R. Ovshinsky (eds) (Plenum), p. 697. (5.4.1)
Dersch, U., Grünewald, M., Overhof, H. and Thomas, P. (1987) *J. Phys.* **C20**, 121. (5.6.2, 6.2.1)
DiCarli, P. S. and Jamieson, J. C. (1959) *J. Chem. Phys.* **31**, 1675. (1.3.10)
Döhler, G. H. (1979) *Phys. Rev.* **B19**, 2083. (5.4.1, 5.4.2, 5.4.3)
Döhler, G. H., Dandoloff, R. and Bilz, H. (1980) *J. Non-Cryst. Sol.* **42**, 87. (2.5.4)
Doolittle, A. K. (1951) *J. Appl. Phys.* **22**, 1471. (2.3.5)

Dove, D. B. (1973) *Phys. Thin Films* **7**, 1. (3.2.2)
Dow, J. D. and Redfield, D. (1970) *Phys. Rev.* **B1**, 3358. (6.3.1)
Downing, H. L., Peterson, N. L. and Jain, H. (1982) *J. Non-Cryst. Sol.* **50**, 203. (4.5.4)
Drexhage, M. G. (1987) in *Halide Glasses for Infrared Fibreoptics*, R. M. Almeida (ed.) (NATO Series E: *Applied Sciences*, no. 123) (Martinus Nijhoff), p. 219. (7.4.2)
Drexhage, M. G., Moynihan, C. T. and Saleh Boulos, M. (1980) *Mat. Res. Bull.* **15**, 213. (7.4.2)
Dubois, J. M., Gaskell, P. H. and LeCaer, G. (1985) *Proc. Roy. Soc. (Lond.)* **A402**, 323. (3.3, 3.4.3, 3.5.3)
Dunstan, D. J. (1984) *Sol. St. Comm.* **49**, 395. (6.3.1)
Dupree, R. and Pettifer, R. F. (1984) *Nature (Lond.)* **308**, 523. (3.2.4)
Dupree, R., Holland, D., McMillan, P. W. and Pettifer, R. F. (1984) *J. Non-Cryst. Sol.* **68**, 399. (3.2.4)
Dupree, R., Holland, D. and Mortuza, M. G. (1987) *Nature (Lond.)* **328**, 416. (3.2.4)
Durand, J. (1983) in *Glassy Metals II*, H. Beck and H. Guntherodt (eds) (*Topics in Applied Physics*, vol. 53) (Springer-Verlag), p. 343. (7.5.1)
Durham, P. J., Pendry, J. B. and Hodges, C. H. (1981) *Sol. St. Comm.* **38**, 159. (3.2.3)
Duwez, P. and Lin, S. C. H. (1967) *J. Appl. Phys.* **38**, 4096. (7.5.1)
Dyre, J. C. (1986) *J. Non-Cryst. Sol.* **88**, 271. (4.5.3)

Efros, A. L. and Pollak, M. (eds) (1985) *Electron–Electron Interactions in Disordered Systems* (North-Holland). (5.3.4)
Egami, T. and Waseda, Y. (1984) *J. Non-Cryst. Sol.* **64**, 113. (2.5.1)
Eisenman, G. (ed.) (1967) *Glass Electrodes for Hydrogen and Other Cations: Principle and Practice* (Dekker). (7.3.3)
Elliott, S. R. (1977) *Phil. Mag.* **36**, 1291. (6.3.4)
Elliott, S. R. (1978) *Phil. Mag.* **B37**, 435. (3.5.3)
Elliott, S. R. (1987a) *Adv. Phys.* **36**, 135. (6.3.4)
Elliott, S. R. (1987b) *J. Non-Cryst. Sol.* **97–98**, 159. (3.4)
Elliott, S. R. (1988) *Sol. St. Ionics* **27**, 131. (4.5.2, 4.5.3)
Elliott, S. R. (1989) *Adv. Phys.* **38**, 1. (3.2.4)
Elliott, S. R. and Owens, A. P. (1989) *Phil. Mag.* B. (to be published). (4.5.3)
Emin, D. (1973) in *Electronic and Structural Properties of Amorphous Semiconductors*, P. G. Le Comber and J. Mort (eds) (Academic). (5.5)
Emin, D. (1976) in *Physics of Structurally Disordered Solids*, S. S. Mitra (ed.) (Plenum). (5.4.1, 5.5)
Emin, D. (1977) *Phil. Mag.* **35**, 1189. (5.4.3)
Emin, D. (1980) *J. Non-Cryst. Sol.* **35–36**, 969. (6.3.1)
Engemann, D. and Fischer, R. R. (1977) *Phys. Stat. Sol.* **79b**, 195. (6.3.1)
Etherington, G., Wright, A. C., Wenzel, J. T., Dore, J. C., Clarke, J. H. and Sinclair, R. N. (1982) *J. Non-Cryst. Sol.* **48**, 265. (3.5.3)
Evans, R., Greenwood, D. A. and Lloyd, P. (1971) *Phys. Lett.* **A35**, 57. (5.4.1)

Fasol, G., Cardona, M., Honle, W. and von Schnering, H. G. (1984) *Sol. St. Comm.* **52**, 307. (3.4.4)
Fenz, P., Müller, H., Overhof, H. and Thomas, P. (1985) *J. Phys.* **C18**, 3191. (5.4.1, 5.4.2, 5.4.3)
Feuston, B. P. and Garofalini, S. H. (1988) *J. Chem. Phys.* **89**, 5818. (3.5.6)
Fiegl, F. J., Fowler, W. B. and Yip, K. L. (1974) *Sol. St. Comm.* **14**, 225. (6.2.3)
Finney, J. L. (1970) *Proc. Roy. Soc. (Lond.)* **A319**, 479. (3.5.2)

Fischer, R. (1979) in *Amorphous Semiconductors*, M. H. Brodsky (ed.) (*Topics in Applied Physics*, vol. 36) (Springer-Verlag). (6.3.1)
Flory, P. J. (1953) *Principles of Polymer Chemistry* (Cornell University Press). (3.5.3)
Fox, T. G. and Flory, P. J. (1950) *J. Appl. Phys.* **21**, 581; (1951) *J. Phys. Chem.* **55**, 221; (1954) *J. Polym. Sci.* **14**, 315. (2.3.5)
Frank, F. C. (1952) *Proc. Roy. Soc.* (*Lond.*) **A215**, 43. (3.3)
Frederickson, G. H. (1988) *A. Rev. Phys. Chem.* **39**, 149. (2.3.4)
Frick, B., Richter, D., Petry, W. and Buchenau, U. (1988) *Z. Phys.* **B70**, 73. (2.3.4)
Friedman, L. (1971) *J. Non-Cryst. Sol.* **6**, 329. (5.4.3)
Friedman, L. (1980) *Phil. Mag.* **B41**, 347. (5.4.3)
Friedman, L. (1987) in *Non-Crystalline Semiconductors*, M. Pollak (ed.) vol. I (CRC Press), p. 111. (5.4.3)
Friedman, L. and Holstein, T. (1963) *Ann. Phys.* (*New York*), **21**, 494. (5.5
Frischat, G. H. (1975) *Ionic Diffusion in Oxide Glasses* (*Diffusion Monograph Series*) (Trans Tech). (4.5.2)
Fritzsche, H. (1973) in *Electronic and Structural Properties of Amorphous Semiconductors*, P. G. Le Comber and J. Mort (eds) (Academic). (5.3.6)
Frye, R. C. and Adler, D. (1981) *Phys. Rev.* **B24**, 5812. (6.2.2)
Funke, K. (1986) *Ber. Bunsenges, Phys. Chem.* **90**, 661. (4.5.3)
Fuoss, P. H. (1980) PhD. Thesis, Stanford Synchrotron Radiation Laboratory Report 80/06. (3.2.2, 3.4.4)
Fuoss, P. H., Eisenberger, P., Warburton, W. K. and Bienenstock, A. (1981) *Phys. Rev. Lett.* **46**, 1537. (3.2.2)

Gabano, J. P. (1985) in *Glass*: *Current Issues*, A. F. Wright and J. Dupuy (eds) (NATO Series E: *Applied Sciences*, no. 92) (Martinus Nijhoff), p. 457. (7.3.2)
Galeener, F. L. (1982) *Sol. St. Comm.* **44**, 1037. (3.4.4, 4.2.1)
Galeener, F. L. (1985) *Phil. Mag.* **B51,** L1. (3.5.3)
Galeener, F. L. and Geissberger, A. E. (1983) *Phys. Rev.* **B27**, 6199. (4.2.1)
Galeener, F. L. and Lucovsky, G. (1976) *Phys. Rev. Lett.* **37**, 1474. (4.3.2)
Galeener, F. L. and Thorpe, M. F. (1983) *Phys. Rev.* **B28**, 5802. (4.2.1)
Galeener, F. L., Leadbetter, A. J. and Stringfellow, M. W. (1983) *Phys. Rev.* **B27**, 1052. (4.3.2, 4.3.4)
Galeener, F. L., Barrio, R. A., Marinez, E. and Elliott, R. J. (1984) *Phys. Rev. Lett.* **53**, 2429. (4.2.1)
Gambino, R. J. (1985) in *Glass*: *Current Issues*, A. F. Wright and J. Dupuy (eds) (NATO Series E: *Applied Sciences*, no. 92) (Martinus Nijhoff), p. 153. (7.5.1)
Gaskell, P. H. (1975) *Phil. Mag.* **32**, 211. (3.3.3)
Gaskell, P. H. (1979a) *J. Phys.* **C12**, 4337. (3.1)
Gaskell, P. H. (1979b) *J. Non-Cryst. Sol.* **32**, 207; (1981) *Nature* (*Lond.*) **289**, 474. (3.3, 3.5.3)
Gaskell, P. H. (1987) in *Metallic and Semiconducting Glasses*, A. K. Bhatnagar (ed.) (Trans Tech), p. 71. (3.4.3)
Gaskell, P. H. and Tarrant, I. D. (1980) *Phil. Mag.* **B42**, 265. (3.5.3)
Gaskell, P. H., Gibson, J. M. and Howie, A. (1977) in *The Structure of Non-Crystalline Materials,* P. H. Gaskell (ed.) (Taylor and Francis), p. 181. (3.5.4)
Gee, G. (1970) *Contemp. Phys.* **11**, 313. (2.3.2, 2.3.5, 2.4)
Geszti, T. (1983) *J. Phys.* **C16**, 5805. (2.3.4)

Gibbs, J. H. (1960) in *Modern Aspects of the Vitreous State*, J. D. MacKenzie (ed.) (Butterworth), Ch. 7. (2.3.2)
Gibbs, J. H. and Di Marzio, E. A. (1958) *J. Chem. Phys.* **28**, 373. (2.3.2)
Gilman, J. J. (1980) *Science* **208**, 856. (2.5.2)
Giordano, N. (1980) *Phys. Rev.* **B22**, 5635. (5.3.3)
Gladden, L. F. and Elliott, S. R. (1987) *Phys. Rev. Lett.* **59**, 908. (3.4.3, 3.4.4)
Gladden, L. F., Carpenter, T. A. and Elliott, S. R. (1986) *Phil. Mag.* **B53**, L81. (3.2.4, 3.5.3)
Gladden, L. F., Elliott, S. R., Sinclair, R. N. and Wright, A. C. (1988) *J. Non-Cryst. Sol.* **106**, 120. (3.4.2, 4.3.4)
Gleason, K. K., Petrich, M. A. and Reimer, J. A. (1987) *Phys. Rev.* **B36**, 3259. (3.2.4)
Goldstein, M. (1969) *J. Chem. Phys.* **51**, 3728. (2.3.3)
Goldstein, M. (1973) *J. Phys. Chem.* **77**, 667. (2.3.1)
Gonser, U. and Preston, R. (1983) in *Glassy Metals II*, H. Beck and H. J. Güntherodt (eds) (*Topics in Applied Physics*, vol. 53) (Springer-Verlag), p. 93. (3.2.5)
Goodman, C. H. L. (1975) *Nature (Lond.)* **257**, 370. (2.5.3)
Goodman, N. B. and Fritzsche, H. (1980) *Phil. Mag.* **B42**, 149. (6.3.1)
Götze, W. (1986) *Phys. Scripta* **34**, 66. (2.3.4)
Götze, W. and Sjögren, L. (1987) *Z. Phys.* **B65**, 415. (2.3.4)
Graebner, J. E. and Allen, L. C. (1984) *Phys. Rev.* **B29**, 5626. (4.4.4)
Graebner, J. E., Allen, L. C. and Golding, B. (1985) *Phys. Rev.* **B31**, 904. (6.2.1)
Greaves, G. N. (1973) *J. Non-Cryst. Sol.* **11**, 427. (5.5)
Greaves, G. N. (1976) in *Structure and Excitations of Amorphous Solids*, G. Lucovsky and F. L. Galeener (eds) (*AIP Conf. Proc.* no. 31) p. 136, (3.5.3)
Greaves, G. N. (1978) *Phil. Mag.* **B37**, 447. (6.2.2, 6.2.3)
Greaves, G. N. (1979) *J. Non-Cryst. Sol.* **32**, 295. (6.2.2, 6.2.3)
Greaves, G. N. and Davis, E. A. (1974) *Phil. Mag.* **29**, 1201. (3.5.3, 4.2.4)
Greaves, G. N. (1985) *J. Non-Cryst. Sol.* **71**, 203. (3.4.3)
Greegor, R. B., Lytle, F. W., Sandstrom, D. R., Wong, J. and Schultz, P. (1983) *J. Non-Cryst. Sol.* **55**, 27. (3.2.3)
Grigorovici, R. and Manaila, R. (1969) *J. Non-Cryst. Solids* **1**, 371. (3.5.4)
Griscom, D. L. (1978) in *The Physics of Silicon Dioxide and its Interfaces*, S. T. Pantelides (ed.) (Pergamon). (6.2.3)
Griscom, D. L. (1985) *J. Non-Cryst. Sol.* **73**, 51 (6.2.2, 6.2.3)
Grünewald, M., Müller, H., Thomas, P. and Wuertz, D. (1981) *Sol. St. Comm.* **38**, 1011. (5.5)
Guinier, A. and Fournet, G. (1955) *Small-Angle Scattering of X-rays* (Wiley). (3.6.2)
Güntherodt, H. J. (1977) in *Festkörperprobleme* **17** (Vieweg), p. 25. (5.4.1)

Hafner, J. (1981) in *Glassy Metals I*, H. J. Güntherodt and H. Beck (eds) (*Topics in Applied Physics Series*, vol. 46) (Springer-Verlag). (2.5.5)
Hafner, J. (1983a) *Phys. Rev.* **B27**, 678; (1983b) *J. Phys.* **C16**, 5773. (4.2.2, 4.3.4)
Hafner, J., Tegze, M. and Jaswal, S. S. (1988) *J. Non-Cryst. Sol.* **106**, 270. (2.5.5, 3.5.6)
Hales, S. (1727) *Vegetable Staticks* (Imrys and Woodward: London). (3.5.2)
Halperin, B. I. and Lax, M. (1966) *Phys. Rev.* **148**, 722; (1967) ibid. **153**, 802. (6.3.1)
Hamakawa, Y. (1987) in *Non-Crystalline Semiconductors*, M. Pollak (ed.), vol. I (CRC Press), p. 229. (7.2.1)
Haneman, D. (1961) *Phys. Rev.* **121**, 1093. (6.2.2)
Hannon, A. C., Sinclair, R. N., Blackman, J. A., Wright, A. C. and Galeener, F. L. (1988) *J. Non-Cryst. Sol.* **106**, 116. (3.4.2, 4.3.3, 4.3.4)

Hansen, J. P. and McDonald, I. R. (1976, 1986) *Theory of Simple Liquids* 1st and 2nd edns. (Academic). (2.3.4)

Harde, H., Bunde, H. and Dieterich, W. (1986) *J. Chem. Phys.* **85**, 4123. (4.5.4)

Harrison, W. A. (1980) *Electronic Structure and the Properties of Solids. The Physics of the Chemical Bond* (Freeman). (6.2.2)

Hattori, H., Kanamori, T., Sakaguchi, S. and Ohishi, Y. (1987) *Appl. Opt.* **26**, 650. (7.4.2)

Heikes, R. R. and Ure, R. W. (1961) *Thermoelectricity* (Interscience). (5.5)

Heine, V. (1980) in *Solid State Physics*, vol. 35, H. Ehrenreich, F. Seitz and D. Turnbull (eds) (Academic), p. 1. (4.2.5)

Hemley, R. J., Jephcoat, A. P., Mao, H. K., Ming, L. C. and Manghnani, M. H. (1988) *Nature (Lond.)* **334**, 52. (1.3.10)

Henderson, B. (1972) *Defects in Crystalline Solids* (Edward Arnold). (6.1, 6.2.3)

Hendrickson, J. R. and Bray, P. J. (1972) *Phys. Chem. Glasses* **13**, 43; (1972) ibid. **13**, 107. (4.5.4)

Henry, C. H. (1980) *J. Appl. Phys.* **51**, 4494. (7.2.4)

Hess, H. F., De Conde, K., Rosenbaum, T. F. and Thomas, G. A. (1982) *Phys. Rev.* **B25**, 5578. (5.3.5, 5.3.6)

Hickey, B. J. and Morgan, G. J. (1986) *J. Phys.* **C19**, 6195. (5.2.1)

Higashi, G. S. and Kastner, M. A. (1981) *Phys. Rev.* **B24**, 2295. (6.3.1)

Higashi, G. S. and Kastner, M. A. (1983) *Phil. Mag.* **B47**, 83. (6.3.1)

Hishikawa, Y., Watanabe, K., Tsuda, S., Nakano, S., Ohnishi, M. and Kuwano, Y. (1987) *J. Non-Cryst. Sol.* **97–98**, 399. (6.2.1, 6.2.6)

Höhr, A., Neumann, H. B., Schmidt, P. W., Pfeiffer, P. and Avnir, D. (1988) *Phys. Rev.* **B38**, 1462. (3.7.1)

Holstein, T. (1959) *Ann. Phys. (New York)* **8**, 325. (5.5)

Holstein, T. (1973) *Phil. Mag.* **27**, 225. (5.4.3)

Howson, M. A. (1984) *J. Phys.* **F14**, L25. (5.4.1)

Howson, M. A. and Greig, D. (1984) *Phys. Rev.* **B30**, 4805. (5.4.1)

Hughes, R. C. (1973) *Phys. Rev. Lett.* **30**, 1333; (1977) *Phys. Rev.* **B15**, 2012. (6.3.5)

Hyde, B. G., Andersson, S., Bakker, M., Plug, C. M. and O'Keefe, M. (1979) *Progr. Sol. St. Chem.* **12**, 273. (3.4.3)

Ichikawa, T. (1975) *Phys. Stat. Sol.* **29a**, 293. (3.5.2)

Imai, H., Arai, K., Saito, T., Ichimura, S., Nonaka, H., Vigouroux, J. P., Imagawa, H., Hosono, H. and Abe, Y. (1988) in *The Physics and Technology of Amorphous* SiO_2, R. A. B. Devine (ed.) (Plenum), p. 153. (6.2.3)

Imbusch, G. F. (1987) *Phys. Scripta* **T19**, 354. (5.6.3)

Ingram, M. D. (1980) *J. Am. Ceram. Soc.* **63**, 248. (4.5.4)

Ingram, M. D. (1987) *Phys. Chem. Glasses* **28**, 215. (4.5.1, 4.5.2, 4.5.4)

Isard, J. O. (1969) *J. Non-Cryst. Sol.* **1**, 235. (4.5.4)

Ishii, K. (1973) *Suppl. Progr. Theor. Phys.* **53**, 77. (5.3.3)

Jäckle, J. (1986) *Rep. Progr. Phys.* **49**, 171. (2.3, 2.3.3)

Jäckle, J. (1989) *J. Phys. Cond. Mat.* **1**, 267. (2.3.1)

Jackson, W. B. (1988) *Phys. Rev.* **B38**, 3595. (5.4.1)

Jackson, W. B., Oh, S. J., Tsai, C. C. and Allen, J. W. (1984) *Phys. Rev. Lett.* **53**, 1481. (5.2.2)

Jacobs, R. L. (1986) *J. Phys.* **C19**, L119. (2.3.4)

Jain, H., Peterson, N. L. and Downing, H. L. (1983) *J. Non-Cryst. Sol.* **55**, 283. (4.5.2, 4.5.4)

Joannopoulos, J. D. (1979) *J. Non-Cryst. Sol.* **32**, 241. (4.4.2, 5.2.1)

Joannopoulos, J. D. and Cohen, M. L. (1976) in *Solid State Physics*, vol. 31, H. Ehrenreich, F. Seitz and D. Turnbull (eds) (Academic), p. 71. (5.2.1, 5.3.2)

Johari, G. P. and Goldstein, M. (1970) *J. Chem. Phys.* **53**, 2372; (1971) ibid. **55**, 4245. (2.3.3)

Johnson, W. L. (1986) *Progr. Mat. Sci.* **30**, 81. (1.3.9, 1.3.10, 1.3.11, 2.5.4)

Jones, D. P., Thomas, N. and Phillips, W. A. (1978) *Phil. Mag.* **B38**, 271. (4.4.2)

Jones, G. O. (1956) *Glass* (Methuen). (2.2)

Jonscher, A. K. (1983) *Dielectric Relaxation in Solids* (Chelsea Dielectrics Press). (4.5.3)

Kakalios, J., Street, R. A. and Jackson, W. B. (1987) *Phys. Rev. Lett.* **59**, 1037. (6.2.6)

Kamitakahara, W. A., Shanks, H. R., McClelland, J. F., Buchenau, U., Gompf, U. and Pintschovius, L. (1984) *Phys. Rev. Lett.* **52**, 644. (4.1, 4.3.4)

Kamitakahara, W. A., Soukoulis, C. M., Shanks, H. R., Buchenau, U. and Grest, G. S. (1987) *Phys. Rev.* **B36**, 6539. (4.1, 4.3.4)

Kaneko, S. (1984) in *Semiconductors and Semimetals*, J. I. Pankove (ed.), vol. 21D (Academic), p. 139. (7.2.2)

Kastner, M. (1972) *Phys. Rev. Lett.* **28**, 355. (6.2.2)

Kastner, M. A. (1985) *J. Non-Cryst. Sol.* **77–78**, 1173. (6.3.1)

Kastner, M., Adler, D. and Fritzsche, H. (1976) *Phys. Rev. Lett.* **37**, 1504. (6.2.2)

Kauzmann, W. (1948) *Chem. Rev.* **43**, 219. (2.3.2, 2.4)

Kawazoe, H., Yanagita, H., Watanabe, Y. and Yamane, M. (1988) *Phys. Rev.* **B38**, 5661. (6.2.6)

Keating, P. N. (1966) *Phys. Rev.* **145**, 637. (4.2.2)

Kelly, M. J. (1980) in *Solid State Physics*, vol. 35, H. Ehrenreich, F. Seitz, and D. Turnbull (eds) (Academic), p. 295. (5.2.1)

Kikuchi, M. (1988) *J. Appl. Phys.* **64**, 4997. (5.4.1)

Klein, R. S. (1985) *Phys. Rev.* **B31**, 2014. (5.4.3)

Klement, W., Willens, R. H. and Duwez, P. (1960) *Nature (Lond.)*, **187**, 169. (1.3.5)

Knotek, M. L. (1974) in *Tetrahedrally Bonded Amorphous Semiconductors*, M. H. Brodsky, S. Kirkpatrick and D. Weaire (eds) (*AIP Conf. Proc.* no. 20), p. 297. (6.3.3)

Knotek, M. L., Pollak, M. and Donovan, T. M. (1974) in *Proc. 5th Int. Conf. on Amorphous and Liquid Semiconductors*, J. Stuke and W. Brenig (eds) (Taylor and Francis), p. 225. (6.3.3)

Kobliska, R. J. and Solin, S. A. (1973) *Phys. Rev.* **B8**, 756. (4.3.3)

Koch, C. C., Cavin, O. B., McKamey, C. G. and Scarbrough, J. O. (1983) *Appl. Phys. Lett.* **43**, 1017. (1.3.11)

Kočka, J. (1987) *J. Non-Cryst. Sol.* **90**, 91. (6.2.6)

Kohlrausch, R. (1854) *Poggendorff's Ann. Phys.* **1**, 56, 179. (2.3.3)

Kone, A. and Souquet, J. L. (1986) *Sol. St. Ionics* **18–19**, 454. (4.5.2)

Kone, A., Reggiani, J. C. and Souquet, J. L. (1983) *Sol. St. Ionics* **9–10**, 709. (4.5.4)

Kramer, B. and Weaire, D. (1979) in *Amorphous Semiconductors*, M. H. Brodsky (ed.) (*Topics in Applied Physics*, vol. 36) (Springer-Verlag). (5.2.1, 5.3.2, 5.3.4)

Krieger, U. and Bosse, J. (1987) *Phys. Rev. Lett.* **59**, 1601. (2.3.4)

Kruangam, D., Endo, T., Guang-pu, W., Nonomura, S., Okamoto, H. and Hamakawa, Y. (1985) *J. Non-Cryst. Sol.* **77–78**, 1429. (7.4.4)

Kruangam, D., Toyama, T., Hattori, Y., Deguchi, M., Okamoto, H. and Hamakawa, Y. (1987) *J. Non-Cryst. Sol.* **97–98**, 293. (7.4.4)

Kukimoto, H. (1984) in *Semiconductors and Semimetals*, J. I. Pankove (ed.), vol. 21D (Academic), p. 239. (7.4.4)

Lamparter, P., Sperl, W., Nold, E., Rainer-Harbach, G. and Steeb, S. (1982) in *Proc. 4th Int. Conf. on Rapidly Quenched Metals*, T. Masumoto and J. Suzuki (eds) (Jap. Inst. of Metals), p. 343. (3.3.2)
Lannin, J. S. (1977) *Phys. Rev.* **B15**, 3863. (4.3.3)
Lapka, R., Oelhafen, P., Gubler, U. M. and Güntherodt, H. J. (1985) *Phys. Rev.* **B31**, 7734. (5.6.1)
Leadbetter, A. J. (1973) in *Chemical Applications of Thermal Neutron Scattering*, B. T. M. Willis (ed.) (Oxford University Press), p. 146. (4.1, 4.3.4)
Leadbetter, A. J. and Wright, A. C. (1972) *J. Non-Cryst. Solids*, **7**, 23. (3.2.2)
Leadbetter, A. J., Smith, P. M. and Seyfert, P. (1976) *Phil. Mag.* **33**, 441. (4.2.4, 4.3.4)
Leamy, H. J., Gilmer, G. H. and Dirks, A. G. (1980) in *Current Topics in Materials Science*, vol. 6, E. Kaldis (ed.) (North-Holland), p. 309. (1.3.1, 3.7.1)
LeCaer, G. and Dubois, J. M. (1987) *Key. Eng. Mat.* **13–15**, 555. (3.2.5)
Le Comber, P. G., Jones, D. I. and Spear, W. E. (1977) *Phil. Mag.* **35**, 1173. (5.4.3)
Le Comber, P. G., Snell, A. J., MacKenzie, K. D. and Spear, W. E. (1981) *J. Phys.* (*Paris*) **42**, C4–423. (7.2.3)
Lee, P. A. and Pendry, J. B. (1975) *Phys. Rev.* **B11**, 2795. (3.2.3)
Letcher, J. H. and Schmidt, P. W. (1966) *J. Appl. Phys.* **37**, 649. (3.6.2)
Leutheusser, E. (1982) *J. Phys.* **C15**, 2801; (1982) ibid. **C15**, 2827; (1984) *Phys. Rev.* **A29**, 2765. (2.3.4, 2.3.5)
Levine, D. and Steinhardt, P. J. (1984) *Phys. Rev. Lett.* **53**, 2477. (1.2)
Licciardello, D. C., Stein, D. L. and Haldane, F. D. M. (1981) *Phil. Mag.* **B43**, 189. (6.2.2)
Lin, C., Busse, L. E., Nagel, S. R. and Faber, J. (1984) *Phys. Rev.* **B29**, 5060. (3.4.4)
Lohneysen, H. V., Schink, H. J. and Beyer, W. (1984) *Phys. Rev. Lett.* **52**, 549. (6.2.1)
Long, A. R. (1982) *Adv. Phys.* **31**, 553. (6.3.4)
Long, A. R. (1989) *Phil. Mag.* **B59**, 377. (5.4.4)
Lorch, E. (1969) *J. Phys.* **C2**, 229. (3.2.2)
Lottici, P. P. (1987) *Phys. Rev.* **B35**, 1236. (4.3.5)
Luborsky, F. E. (1977) in *Amorphous Magnetism II*, R. A. Levy and R. Hasegawa (eds) (Plenum). (7.5.1)
Luborsky, F. E. (1980) in *Ferromagnetic Materials*, E. P. Wohlfarth (ed.) (North-Holland). (7.5.1)
Luborsky, F. E. (1983) in *Amorphous Metallic Alloys*, F. E. Luborsky (ed.) (Butterworth), p. 360. (7.5.1)
Luborsky, F. E. (1985) in *Glass: Current Issues*, A. F. Wright and J. Dupuy (eds) (NATO Series E: *Applied Sciences*, no. 92) (Martinus Nijhoff), p. 139. (7.5.1, 7.5.2)
Lucovsky, G. (1972) *Phys. Rev.* **B6**, 1480. (4.3.3)
Lucovsky, G. (1979a) *Phil. Mag.* **B39**, 513; (1979b) ibid. **B39**, 531. (6.2.2, 6.2.3)
Lucovsky, G. (1980) *Phil. Mag.* **B41**, 457. (6.2.2, 6.2.3)
Lucovsky, G. (1985) *J. Non-Cryst. Sol.* **76**, 173. (3.2.6)
Lucovsky, G. and Hayes, T. M. (1979) in *Amorphous Semiconductors*, M. H. Brodsky (ed.) (*Topics in Applied Physics*, vol. 36) (Springer-Verlag), p. 215. (3.2.6)
Lucovsky, G., de Neufville, J. P. and Galeener, F. L. (1974) *Phys. Rev.* **B9**, 1591. (3.2.6)
Lucovsky, G., Galeener, F. L., Geils, R. H. and Keezer, R. C. (1977) in *Structure of Non-Crystalline Materials*, P. H. Gaskell (ed.) (Taylor and Francis), p. 127. (3.3)

Lucovsky, G., Nemanich, R. J. and Knights, J. C. (1979) *Phys. Rev.* **B19**, 2064. (3.2.6)

McCrum, N. G., Read, B. E. and Williams, G. (1967) *Anelastic and Dielectric Effects in Polymeric Solids* (Wiley). (2.3.3)
MacDonald, J. R. (1987) *Impedance Spectroscopy* (Wiley). (4.5.3)
McElfresh, D. K. and Howitt, D. G. (1986) *J. Am. Ceram. Soc.* **69**, C237. (4.5.2)
Mackay, A. L. (1982) *Physica* **A114**, 609. (3.3)
MacKinnon, A. and Kramer, B. (1983) *Z. Phys.* **B53**, 1. (5.3.4, 5.3.6)
Madan, A. and Shaw, M. P. (1988) *The Physics and Applications of Amorphous Semiconductors* (Academic). (7.2.1, 7.2.3, 7.2.4)
Madan, A., Le Comber, P. G. and Spear, W. E. (1976) *J. Non-Cryst. Sol.* **20**, 239. (6.3.1)
Madelung, O. (1978) *Introduction to Solid-State Theory* (Springer-Verlag). (4.1, 5.1)
Maley, N., Beeman, D. and Lannin, J. S. (1988) *Phys. Rev.* **B38**, 10611. (4.3.2, 4.3.3, 4.3.4)
Mandelbrot, B. B. (1982) *The Fractal Geometry of Nature* (Freeman). (3.6.2, 3.7.1)
Mansfield, P. (1981) *Phil. Trans. Roy. Soc. (Lond.)* **A299**, 479. (3.2.4)
Maradudin, A. A., Montroll, E. W., Weiss, G. H. and Ipatova, I. P. (1971) in *Solid State Physics, Advances in Research and Applications*, Suppl. 3, H. Ehrenreich, F. Seitz and D. Turnbull (eds) (Academic). (4.1)
Marshall, J. M. (1983) *Rep. Progr. Phys.* **46**, 1235. (6.3.5)
Marshall, J. M. (1988) in *Amorphous Silicon and Related Materials*, H. Fritzsche (ed.) (World Scientific), p. 799. (6.3.5)
Martin, S. W. (1988) *J. Am. Ceram. Soc.* **71**, 438. (4.5.2)
Martin, S. W. and Angell, C. A. (1986) *J. Non-Cryst. Sol.* **83**, 185. (4.5.2, 4.5.3)
Meek, P. E. (1976) *Phil. Mag.* **33**, 897. (4.3.4)
Meyer, W. and Neldel, H. (1937) *Z. Tech. Phys.* **12**, 588. (5.4.1)
Mezei, F., Knaak, W. and Farago, B. (1987) *Phys. Rev. Lett.* **58**, 571; (1987), *Phys. Scripta* **T19**, 363. (2.3.4)
Michiel, H., Adriaenssens, G. J. and Davis, E. A. (1986) *Phys. Rev.* **B34**, 2486. (6.3.5)
Mishima, O., Calvert, L. D. and Whalley, E. (1984) *Nature (Lond.)* **310**, 393. (1.3.10)
Miya, T., Terunuma, Y., Hosaka, T. and Miyashita, T. (1979) *Electronics Lett.* **15**, 106. (7.4.2)
Mobilio, S. and Filipponi, A. (1987) *J. Non-Cryst. Sol.* **97–98**, 365. (3.2.3)
Mooij, J. H. (1973) *Phys. Stat. Sol.* **17a**, 521. (5.4.1)
Morgan, G. J. and Hickey, B. L. (1985) *J. Phys.* **F15**, 2473. (5.3.4)
Morigaki, K. (1984) in *Semiconductors and Semimetals*, J. Pankove (ed.), vol. 21C (Academic), p. 155. (6.3.2)
Mort, J., Troup, A., Morgan, M., Grammatica, S., Knights, J. and Lujan, R. (1980) *Appl. Phys. Lett.* **38**, 277. (6.3.1)
Moss, S. C. and Price, D. L. (1985) in *Physics of Disordered Materials*, D. Adler, H. Fritzsche and S. R. Ovshinsky (eds) (Plenum), p. 77. (3.4.4)
Mott, N. F. (1967) *Adv. Phys.* **16**, 49. (6.2.6)
Mott, N. F. (1969) *Phil. Mag.* **19**, 835. (6.3.3)
Mott, N. F. (1980) *J. Phys.* **C13**, 5433. (6.1)
Mott, N. F. (1981) *Phil. Mag.* **B44**, 265. (5.3.5)
Mott, N. F. (1987a) *Conduction in Non-Crystalline Materials* (Clarendon Press). (5.3.5)
Mott, N. F. (1987b), *J. Non-Cryst. Sol.* **97–98**, 531. (5.4.1)
Mott, N. F. and Davis, E. A. (1979) *Electronic Processes in Non-Crystalline*

Materials 2nd edn. (Oxford University Press). (2.5.5, 5.2.2, 5.3.3, 5.3.4, 5.3.5, 5.4, 5.4.1, 5.4.2, 5.5, 5.6.2, 6.2.1, 6.3.3, 6.3.4, 6.3.5)

Mott, N. F. and Kaveh, M. (1981) *J. Phys.* **C14**, L659. (5.3.4)

Mott, N. F. and Twose, W. D. (1960) *Adv. Phys.* **10**, 107. (5.3.3)

Mott, N. F., Davis, E. A. and Street, R. A. (1975a) *Phil. Mag.* **32**, 961. (6.2.2, 6.3.1)

Mott, N. F., Pepper, M., Pollitt, S., Wallis, R. H. and Adkins, C. J. (1975b) *Proc. Roy. Soc. (Lond.)* **A345**, 169. (6.3.3)

Moustakas, T. D. (1979) *J. Electronic Mat.* **8**, 391. (6.2.1)

Movaghar, B., Pohlmann, B. and Sauer, G. W. (1980a) *Phys. Stat. Sol.* **b97**, 533. (6.3.4)

Movaghar, B., Pohlmann, B. and Schirmacher, W. (1980b) *Phil. Mag.* **B41**, 49; (1980c) *Sol. St. Comm.* **34**, 451 (6.3.4)

Moynihan, C. T. and Lesikar, A. V. (1981) *J. Am. Ceram. Soc.* **64**, 40. (4.5.4)

Mozzi, R. L. and Warren, B. E. (1969) *J. Appl. Cryst.* **2**, 164. (3.2.4, 3.5.3)

Mukherjee, S. P. (1980) *J. Non-Cryst. Sol.* **42**, 477. (1.3.6)

Müller, D., Berger, G., Grunze, I., Ladwig, G., Hallas, E. and Haubenreisser, U. (1983) *Phys. Chem. Glasses* **24**, 37. (3.2.4)

Müller, H. and Thomas, P. (1984) *J. Phys.* **C17**, 5337. (5.4.1)

Murayama, K., Suzuki, H. and Ninomiya, T. (1980) *J. Non-Cryst. Sol.* **35–36**, 915. (6.3.1)

Myers, M. B. and Felty, E. J. (1976) *Mat. Res. Bull.* **2**, 715. (2.4)

Nagel, S. R. (1977) *Phys. Rev.* **B16**, 1694. (5.4.1)

Nagel, S. R. (1982) *Adv. Chem. Phys.* **51**, I. Prigogine and S. A. Rice (eds) (Wiley), p. 227. (5.4.1)

Nagel, S. R. and Tauc, J. (1975) *Phys. Rev. Lett.* **35**, 380. (2.5.5)

Nagels, P. (1979) in *Amorphous Semiconductors*, M. H. Brodsky (ed.) (*Topics in Applied Physics*, vol. 36) (Springer-Verlag). (5.4, 5.4.3)

Neilson, G. W. and Enderby, J. E. (1983) *Proc. Roy. Soc.* **A390**, 353. (3.2.2)

Nelson, D. R. (1983) *Phys. Rev.* **B28**, 5515. (3.3)

Nelson, D. R. and Spaepen, F. (1989) in *Solid-State Physics*, vol. 42, H. Ehrenreich and D. Turnbull (eds) (Academic), p. 1. (1.2)

Nemanich, R. J., Solin, S. A. and Lucovsky, G. (1977) *Sol. St. Comm.* **21**, 273. (3.4.2)

Newell, R. G., Feuston, B. P. and Garofalini, S. H. (1989) *J. Mat. Res.* **4**, 434. (3.5.6)

Ngai, K. L. (1985) in *Non-Debye Relaxation in Condensed Matter*, T. V. Ramakrishnan (ed.) (World Scientific), p. 23. (2.3.3)

Ngai, K. L., Reinecke, T. L. and Economou, E. N. (1978) *Phys. Rev.* **B17**, 790. (6.2.2)

Ngai, K. L., Rendell, R. W. and Jain, H. (1984) *Phys. Rev.* **B30**, 2133. (4.5.3)

Ninomiya, T. (1977) in *The Structure of Non-Crystalline Materials*, P. H. Gaskell (ed.) (Taylor and Francis). (6.1)

Noolandi, J. (1977) *Phys. Rev.* **B16**, 4466; (1977) ibid. **B16**, 4474. (6.3.5)

Oelhafen, P. (1983) in *Glassy Metals II*, H. Beck and H. J. Güntherodt (eds) (*Topics in Applied Physics*, vol. 53) (Springer-Verlag), p. 283. (2.5.5)

O'Handley, R. C. (1983) in *Amorphous Metallic Alloys*, F. E. Luborsky (ed.) (Butterworth), p. 257. (7.5.1)

Ohtani, E., Taulelle, F. and Angell, C. A. (1985) *Nature (Lond.)* **314**, 78. (3.2.4)

Okushi, H., Tokumaru, Y., Yamasaki, S., Oheda, H. and Tanaka, K. (1982) *Phys. Rev.* **B25**, 4314. (6.3.1)

Orbach, R. (1986) *Science* **231**, 814. (4.4.4, 4.4.6)

Orenstein, H. and Kastner, M. A. (1981) *Phys. Rev. Lett.* **46**, 1421. (6.3.5)

Overhof, H. and Beyer, W. (1981) *Phil. Mag.* **B43**, 433. (5.4.1)
Overhof, H. and Beyer, W. (1983) *Phil. Mag.* **B47**, 377. (5.4.1, 5.4.2)
Overhof, H. and Beyer, W. (1984) *Phil. Mag.* **B49**, L9. (5.4.2)
Ovshinsky, S. R. and Madan, A. (1978) *Nature (Lond.)* **276**, 482. (1.3.3)
Owen, A. E. (1973) in *Electronic and Structural Properties of Amorphous Semiconductors*, P. G. Le Comber and J. Mort (eds) (Academic). (2.2)
Owen, A. E. (1980) *J. Non-Cryst. Sol.* **35–36**, 999. (7.3.3)

Pabst, A. (1952) *Am. Mineralogist* **37**, 137. (1.3.9)
Paesler, M. A., Agarwal, S. C., Hudgens, S. J. and Fritzsche, H. (1974) in *Tetrahedrally Bonded Amorphous Semiconductors*, M. H. Brodsky, S. Kirkpatrick and D. Weaire (eds) (*AIP Conf. Proc.* no. 20), p. 37. (6.2.4)
Palmer, R. G., Stein, D. L., Abrahams, E. and Anderson, P. W. (1984) *Phys. Rev. Lett.* **53**, 958. (4.5.3, 6.3.4)
Pandey, K. C. (1981) *Phys. Rev. Lett.* **47**, 1913. (6.2.2)
Pankove, J. I. (ed.) (1984) *Semiconductors and Semimetals*, vol. 21D (Academic). (7.2.1)
Pantelides, S. T. (1986) *Phys. Rev. Lett.* **57**, 2979. (6.2.1)
Pauling, L. (1985) *Nature (Lond.)* **317**, 512; (1987) *Phys. Rev. Lett.* **58**, 365. (1.2)
Penrose, R. (1974) *Bull. Inst. Math. Applic.* **10**, 266. (1.2)
Pfister, G. and Scher, H. (1978) *Adv. Phys.* **27**, 747. (6.3.5)
Phillips, J. C. (1979) *J. Non-Cryst. Sol.* **34**, 153. (2.5.4, 3.5.3, 6.2.4)
Phillips, J. C. (1981) *J. Non-Cryst. Sol.* **43**, 37. (3.4.2)
Phillips, W. A. (1972) *J. Low Temp. Phys.* **7**, 351. (4.4.5)
Phillips, W. A. (1978) *J. Non-Cryst. Sol.* **31**, 267. (4.4.2, 4.4.5)
Phillips, W. A. (ed.) (1981) *Amorphous Solids: Low Temperature Properties* (Springer-Verlag). (4.4.1, 4.4.2, 4.4.3, 4.4.5)
Pierre, A., Uhlmann, D. R. and Molea, F. N. (1972) *J. Appl. Cryst.* **5**, 216. (3.6.2)
Pike, G. E. and Seeger, C. H. (1974) *Phys. Rev.* **B10**, 1421. (6.3.3)
Pistoulet, B., Roche, F. M. and Abdalla, S. (1984) *Phys. Rev.* **B30**, 5987. (5.4.4)
Poignant, H. (1981) *Electronics Lett.* **17**, 973. (7.4.2)
Poignant, H. (1987) in *Halide Glasses for Infrared Fiberoptics*, R. M. Almeida (ed.) (NATO Series E: *Applied Sciences*, no. 123) (Martinus Nijhoff), p. 35. (7.4.2)
Polk, D. E. (1970) *Scripta Metall.* **4**, 117. (2.5.1, 3.5.2)
Polk, D. E. (1972) *Acta Metall.* **20**, 485. (2.5.2, 3.5.2)
Polk, D. E. and Boudreaux, D. S. (1973) *Phys. Rev. Lett.* **31**, 92. (3.5.3, 5.2.1)
Polk, D. E. and Turnbull, D. (1972) *Acta Metall.* **20**, 493. (6.2.4)
Pollak, M. (1971) *Phil. Mag.* **23**, 519. (6.3.4)
Pollak, M. (1977) *Phil. Mag.* **36**, 1157. (6.3.5)
Pollak, M. (ed.) (1987) *Noncrystalline Semiconductors* (CRC Press). (5.4)
Pollak, M. and Geballe, T. H. (1961) *Phys. Rev.* **122**, 1742. (6.3.4)
Pollak, M. and Pike, G. E. (1972) *Phys. Rev. Lett.* **28**, 1449. (6.3.4)
Poulain, M., Chanthanasinh, M. and Lucas, J. (1977) *Mat. Res. Bull.* **12**, 151. (7.4.2)
Powell, M. J. (1981) *Phil. Mag.* **B43**, 93. (6.3.1)
Pramanik, M. H. A., Butcher, P. N. and Cox, I. D. (1983) *Phil. Mag.* **B47**, 437. (4.5.2)
Price, D. L. and Carpenter, J. M. (1987) *J. Non-Cryst. Sol.* **92**, 153. (4.3.2, 4.3.4)
Price, D. L., Misawa, M., Susman, S., Morrison, T. I., Sheng, G. K. and Grimsditch, M. (1984) *J. Non-Cryst. Sol.* **66**, 443. (3.4.4)
Price, D. L., Susman, S. and Wright, A. C. (1987) *J. Non-Cryst. Sol.* **97–98**, 167. (3.4.4)

Prigogine, I. and Defay, R. (1954) *Chemical Thermodynamics* (Longman Greens). (2.3.1)

Ramanan, V. R. V., Smith, C. H. and Fish, G. E. (1987)*Key. Eng. Mát.* **13–15**, 849. (7.5.2)
Rammal, R. and Toulouse, G. (1983) *J. Phys. (Paris) Lett.* **44**, L13. (5.3.4)
Rao, K. V. (1983) in *Amorphous Metallic Alloys*, F. E. Luborsky (ed.) (Butterworth), p. 401. (5.4)
Ravaine, D. and Souquet, J. L. (1977) *Phys. Chem. Glasses* **18**, 27. (4.5.2)
Ravaine, D., Perera, G. and Poulain, M. (1983) *Sol. St. Ionics* **9–10**, 631. (4.5.1)
Raskin, D. and Smith, C. H. (1983) in *Amorphous Metallic Alloys*, F. E. Luborsky (ed.) (Butterworth), p. 381. (7.5.1)
Rawson, H. (1967) *Inorganic Glass-Forming Systems* (Academic). (2.5.2, 3.7.2)
Reilly, O. J. and Spear, W. E. (1978) *Phil. Mag.* **38**, 295. (5.4.3)
Reimer, J. A., Vaughan, R. W. and Knights, J. C. (1981) *Sol. St. Comm.* **37**, 161. (3.2.4)
Renninger, A. L., Rechtin, M. D. and Averbach, B. L. (1974) *J. Non-Cryst. Sol.* **16**, 1. (3.5.5)
Rivier, N. (1979) *Phil. Mag.* **A40**, 859. (6.2.5)
Rivier, N. (1987) *Adv. Phys.* **36**, 95. (6.2.5)
Robertson, J. (1976) *Phil. Mag.* **34**, 13. (3.4.1)
Robertson, J. (1982) *Phys. Chem. Glasses* **23**, 1. (6.1, 6.2.2, 6.2.3, 6.3.1)
Robertson, J. (1988) in *The Physics and Technology of Amorphous SiO_2*, R. A. B. Devine (ed.) (Plenum), p. 113. (6.2.3)
Robins, L. H. and Kastner, M. A. (1984) *Phil. Mag.* **B50**, 29. (6.3.1)
Ruppersberg, H. (1988) *J. Non-Cryst. Sol.* **106**, 89. (3.3)

Sadoc, J. F. and Dixmier, J. (1976) *Mat. Sci. Eng.* **23**, 187. (3.2.2)
Sadoc, J. F. and Mosseri, R. (1985) *J. Phys. (Paris)* **46**, 1809. (6.2.5)
Sadoc, J. F. and Rivier, N. (1987) *Phil. Mag.* **B55**, 537. (6.2.5)
Sadoc, J. F. and Wagner, C. N. J. (1983) in *Glassy Metals II*, H. Beck and H. J. Güntherodt (eds) (*Topics in Applied Physics*, vol. 53) (Springer-Verlag), p. 51. (3.2.2, 3.3)
Salamon, M. B. (ed.) (1979) *Physics of Superionic Conductors* (Springer-Verlag). (4.5.1)
Samoson, A., Lippmaa, E. and Pines, A. (1988) *Molec. Phys.* **65**, 1013. (3.2.4)
Savage, J. A. (1987) in *Halide Glasses for Infrared Fiberoptics*, R. M. Almeida (ed.) (NATO Series E: *Applied Sciences*, no. 123) (Martinus Nijhoff), p. 367. (7.4.2)
Sayers, D. E., Stern, E. A. and Lytle, F. W. (1971) *Phys. Rev. Lett.* **27**, 1204; (1975) *Phys. Rev. Lett.* **35**, 584. (3.2.3)
Schaefer, D. W., Martin, J. E., Wiltzius, P. and Cannell, D. S. (1984) *Phys. Rev. Lett.* **52**, 2371. (3.7.1)
Scher, H. and Montroll, E. W. (1975) *Phys. Rev.* **B12**, 2455. (6.3.4, 6.3.5)
Scherer, G. W. (1984) *J. Am. Ceram. Soc.* **67**, 504. (2.3.3)
Schlegel, A., Wachter, P., Ackermann, K. P., Liard, M. and Güntherodt, H. J. (1979) *Sol. St. Comm.* **31**, 373. (5.6.1)
Schmidlin, F. (1977) *Phys. Rev.* **B6**, 2362. (6.3.5)
Schmidt, P. W. (1982) *J. Appl. Cryst.* **15**, 567. (3.6.2)
Schwarz, R. B. and Johnson, W. L. (1983) *Phys. Rev. Lett.* **51**, 415. (1.3.11)
Schwarz, R. B., Petrich, R. R. and Saw, C. K. (1985) *J. Non-Cryst. Sol.* **76**, 281. (1.3.11)

Selzer, P. M., Huber, D. L., Hamilton, D. S., Yen, W. M. and Weber, M. J. (1976) *Phys. Rev. Lett.* **36**, 813. (5.6.3)
Sen, P. N. and Thorpe, M. F. (1977) *Phys. Rev.* **B15**, 4030. (3.2.6, 4.2.1)
Sevillano, E., Meuth, H. and Rehr, J. J. (1979) *Phys. Rev.* **B20**, 4908. (4.3.5)
Shapiro, B. and Abrahams, E. (1981) *Phys. Rev.* **B24**, 4025. (5.4.1)
Shectman, D., Blech, I., Gratias, D. and Cahn, J. W. (1984) *Phys. Rev. Lett.* **53**, 1951. (1.2)
Shevchik, N. J. (1977a) *Phil. Mag.* **35**, 805; (1977b) ibid. **35**, 1289. (3.2.2)
Shibata, S., Horiguchi, M., Jinguji, K., Mitachi, S., Kanamori, T. and Manabe, T. (1981) *Electronics Lett.* **17**, 775. (7.4.2)
Shimakawa, K., Long, A. R. and Imagawa, O. (1987) *Phil. Mag. Lett.* **56**, 79. (5.4.4)
Shimizu, I. (1985) *J. Non-Cryst. Sol.* **77–78**, 1363. (7.2.2)
Shuker, R. and Gammon, R. W. (1970) *Phys. Rev. Lett.* **25**, 222. (4.3.3)
Silver, A. H. and Bray, P. J. (1958) *J. Chem. Phys.* **29**, 984. (3.2.4)
Simmons, J. H., Elterman, P. B., Simmons, C. J. and Mohr, R. K. (1979) *J. Am. Ceram. Soc.* **62**, 158. (4.5.3)
Sinclair, R. N. (1985) *J. Non-Cryst. Sol.* **76**, 61. (4.3.4)
Smith, J. V. and Blackwell, C. S. (1983) *Nature* (*Lond.*) **303**, 223. (3.2.4)
Smith, N. V. (1988) *Rep. Progr. Phys.* **51**, 1227. (5.2.2)
Smith, Z. E. and Wagner, S. (1985) *Phys. Rev.* **B32**, 5510. (6.2.6)
Smith, Z. E., Aljishi, S., Slobodin, D., Chu, V., Wagner, S., Lenahan, P. M., Arya, R. R. and Bennett, M. S. (1986) *Phys. Rev. Lett.* **57**, 2450. (6.2.6)
Soklakov, A. I. and Nechaeva, V. V. (1967) *Soviet Phys. Sol. St.* **9**, 715. (3.4.4)
Solomon, I. (1979) in *Amorphous Semiconductors*, M. H. Brodsky (ed.) (*Topics in Applied Physics*, vol. 36) (Springer-Verlag). (6.3.2)
Soukoulis, C. M., Cohen, M. H. and Economou, E. N. (1984) *Phys. Rev. Lett.* **53**, 616. (6.3.1)
Soules, T. F. (1980) *J. Chem. Phys.* **73**, 4032. (3.5.6)
Souquet, J. L. (1981) *A. Rev. Mat. Sci.* **11**, 211. (4.5.2)
Souquet, J. L. (1986) in *Materials for Solid-State Batteries*, B. V. R. Chowdari and S. Radhakrishna (eds) (World Scientific), p. 241. (7.3.2)
Souquet, J. L. (1988) *Sol. St. Ionics* **28–30**, 693. (7.3.2)
Spaepen, F. (1981) in *Physics of Defects*, R. Balian, M. Kléman and J.-P. Poirier (eds) (North-Holland). (6.2.4)
Spear, W. E. (1969) *J. Non-Cryst. Sol.* **1**, 197. (6.3.5)
Spear, W. E. (1974) in *Proc. 5th Int. Conf. on Amorphous and Liquids Semiconductors*, J. Stuke and W. Brenig (eds) (Taylor and Francis), p. 1. (6.2.2)
Spear, W. E. (1977) *Adv. Phys.* **26**, 811. (1.3.3)
Spear, W. E. (1983) *J. Non-Cryst. Sol.* **59–60**, 1. (6.3.5)
Spear, W. E. and Le Comber, P. G. (1975) *Sol. St. Comm.* **17**, 1193. (1.3.3)
Spear, W. E. and Le Comber, P. G. (1976) *Phil. Mag.* **33**, 935. (5.4.1, 6.2.1, 7.2.1)
Spicer, W. E. (1974) in *Proc. 5th Int. Conf. on Amorphous and Liquid Semiconductors*, J. Stuke and W. Brenig (eds) (Taylor and Francis), p. 499. (5.2.2)
Spicer, W. E. and Donovan, T. M. (1970) *J. Non-Cryst. Sol.* **2**, 66. (5.2.2)
Springett, B. E. (1973) *Phys. Rev. Lett.* **31**, 1463. (5.4.4)
Staebler, D. L. and Wronski, C. R. (1980) *J. Appl. Phys.* **51**, 3262. (5.4.1, 6.2.6)
Stapelbroek, M. and Griscom, D. L. (1978) in *The Physics of Silicon Dioxide and its Interfaces*, S. T. Pantelides (ed.) (Pergamon). (6.2.2)
Stapelbroek, M., Griscom, D. L., Friebele, E. J. and Siegel, G. H. (1979) *J. Non-Cryst. Sol.* **32**, 313. (6.3.2)

Stathis, J. H. (1988) in *The Physics and Technology of Amorphous* SiO_2, R. A. B. Devine (ed.) (Plenum), p. 141. (6.2.3)
Stauffer, D. (1985) *Introduction to Percolation Theory* (Taylor and Francis). (5.3.6)
Stebbins, J. F. (1987) *Nature* (*Lond.*) **330**, 465. (3.2.4)
Steinhardt, P. J. (1987) *Science* **238**, 1242. (1.2)
Steinhardt, P. J., Nelson, D. R. and Ronchetti, M. (1983) *Phys. Rev.* **B28**, 784. (3.3)
Stephens, P. W. and Goldman, A. I. (1986) *Phys. Rev. Lett.* **56**, 1168; (1986) ibid. **57**, 2331. (1.2)
Stern, E. A. (1978) *Contemp. Phys.* **19**, 289. (3.2.3)
Stoneham, A. M. (1975) *Theory of Defects in Solids* (Oxford University Press). (6.1)
Street, R. A. (1976) *Adv. Phys.* **25**, 397. (6.3.1, 6.3.2)
Street, R. A. (1977) *Sol. St. Comm.* **24**, 363. (6.3.1)
Street, R. A. (1980) *Sol. St. Comm.* **34**, 157. (6.3.1)
Street, R. A. (1981) *Adv. Phys.* **30**, 593. (6.3.1)
Street, R. A. (1982) *Phys. Rev. Lett.* **49**, 1187. (6.2.6)
Street, R. A. (1984) in *Semiconductors and Semimetals*, J. Pankove (ed.), vol. 21B (Academic), p. 197. (6.3.1)
Street, R. A. (1985) *J. Non-Cryst. Sol.* **77–78**, 1. (6.2.6)
Street, R. A. and Mott, N. F. (1975) *Phys. Rev. Lett.* **35**, 1293. (6.2.2)
Street, R. A. and Zesch, J. C. (1984) *Phil. Mag.* **B50**, L19. (6.2.6)
Street, R. A., Biegelsen, D. K., Jackson, W. B., Johnson, N. M. and Stutzmann, M. (1985) *Phil. Mag.* **B52**, 235. (6.2.6)
Street, R. A., Kakalios, J., Tsai, C. C. and Hayes, T. M. (1987a) *Phys. Rev.* **B35**, 1316. (6.2.6)
Street, R. A., Tsai, C. C., Kakalios, J. and Jackson, W. B. (1987b) *Phil. Mag.* **B56**, 305. (6.2.6)
Stuke, J. (1987) *J. Non-Cryst. Sol.* **97–98**, 1. (5.4.1)
Stutzmann, M., Jackson, W. B. and Tsai, C. C. (1985) *Phys. Rev.* **B32**, 23. (6.2.6)
Stutzmann, M., Biegelsen, D. K. and Street, R. A. (1987) *Phys. Rev.* **B35**, 5666. (6.2.6)
Suck, J. B. and Rudin, H. (1983) in *Glassy Metals II*, H. Beck and H. J. Güntherodt (eds) (*Topics in Applied Physics*, vol. 53) (Springer-Verlag), p. 217. (4.3.4)
Suck, J. B., Rudin, H., Güntherodt, H. J. and Beck, H. (1983) *Phys. Rev. Lett.* **50**, 49. (4.3.4)
Sugai, S. (1987) *Phys. Rev.* **B35**, 1345. (3.4.2)
Summerfield, S. and Butcher, P. N. (1982) *J. Phys.* **C15**, 7003; (1985) *J. Non-Cryst. Sol.* **77–78**, 135. (6.3.4)
Sun, K. H. (1947) *J. Am. Ceram. Soc.* **30**, 277. (2.5.2)
Susman, S., Price, D. L., Volin, K. J., Dejus, R. J. and Montague, D. S. (1988) *J. Non-Cryst. Sol.* **106**, 26. (3.4.4)
Suzuki, A., Sato, H. and Kikuchi, R. (1984) *Phys. Rev.* **B29**, 3550. (4.5.4)
Suzuki, S. and Abe, Y. (1981) *J. Non-Cryst. Sol.* **43**, 141. (2.3.5)

Takahashi, S. (1987) *J. Non-Cryst. Sol.* **95–96**, 95. (7.4.2)
Tanaka, K. (1985) *Sol. St. Comm.* **54**, 867 (2.4)
Tanaka, K. (1987) *J. Non-Cryst. Sol.* **90**, 363. (3.4.4)
Tarnow, E., Payne, M. C. and Joannopoulos, J. D. (1988) *Phys. Rev. Lett.* **61**, 1772. (6.2.2)
Tarnow, E., Joannopoulos, J. D. and Payne, M. C. (1989) *Phys. Rev.* **B39**, 6017. (6.2.2)
Taylor, P. C., Baugher, J. F. and Kriz, H. M. (1975) *Chem. Rev.* **75**, 203. (3.2.4)
Temkin, R. J. (1978) *J. Non-Cryst. Sol.* **28**, 23. (3.5.3)

Temkin, R. J., Paul, W. and Connell, G. A. N. (1973) *Adv. Phys.* **22**, 581. (3.2.2, 3.5.4)
Thathachari, Y. T. and Tiller, W. A. (1982) *J. Appl. Phys.* **53**, 8615. (3.5.3)
Thio, T., Monroe, D. and Kastner, M. A. (1984) *Phys. Rev. Lett.* **52**, 667. (6.2.2, 6.2.6)
Thorpe, M. F. (1976) in *Physics of Structurally Disordered Solids*, S. S. Mitra (ed.) (Plenum). (4.1)
Thorpe, M. F. (1983) *J. Non-Cryst. Sol.* **57**, 355. (2.5.4)
Thorpe, M. F. and de Leeuw, S. W. (1986) *Phys. Rev.* **B33**, 8490. (4.3.2)
Thorpe, M. F. and Galeener, F. L. (1980) *Phys. Rev.* **B22**, 3078. (4.2.1)
Thouless, D. J. (1977) *Phys. Rev. Lett.* **39**, 1167. (5.3.3, 6.3.1)
Thouless, D. J. (1979) in *Ill-Condensed Matter*, R. Balian, R. Maynard and G. Toulouse (eds) (North-Holland). (5.3.2, 5.3.5)
Thouless, D. J. (1980) *Sol. St. Comm.* **34**, 683. (5.3.3)
Tiedje, T. (1984) in *Semiconductors and Semimetals*, J. Pankove (ed.), vol. 21C (Academic), p. 207. (6.2.6, 6.3.5)
Tiedje, T. and Rose, A. (1981) *Sol. St. Comm.* **37**, 48. (6.3.5)
Tohge, N. and Tanaka, M. (1986) *J. Non-Cryst. Sol.* **80**, 550. (7.3.3)
Tsai, T. E., Griscom, D. L. and Friebele, E. J. (1988) *Phys. Rev. Lett.* **61**, 444. (6.2.3)
Turnbull, D. (1969) *Contemp. Phys.* **10**, 473. (1.3.5)
Turnbull, D. and Cohen, M. H. (1961) *J. Chem. Phys.* **34**, 120; (1970) ibid. **52**, 3038. (2.3.5)

Van den Berg, R. and Von Lohneysen, H. (1985) *Phys. Rev. Lett.* **55**, 2463. (4.4.4)
Vanderbilt, D. and Joannopoulos, J. D. (1980a) *Sol. St. Comm.* **35**, 535; (1980b) *Phys. Rev.* **B22**, 2927. (6.2.2)
Vanderbilt, D. and Joannopoulos, J. D. (1981) *Phys. Rev.* **B23**, 2596. (6.2.2)
Vanderbilt, D. and Joannopoulos, J. D. (1983a) *Phys. Rev.* **B27**, 6311; (1983b) *J. Non-Cryst. Sol.* **59–60**, 937. (6.2.2)
Vepřek, S. and Beyeler, H. U. (1981) *Phil. Mag.* **B44**, 557. (3.4.4)
Verrall, D. J. and Elliott, S. R. (1988) *Phys. Rev. Lett.* **61**, 974. (3.4.2, 4.3.4)
Vollhardt, D. and Wölfle, P. (1982) *Phys. Rev. Lett.* **48**, 699. (5.3.4)
Vomvas, A. and Roilos, M. (1984) *Phil. Mag.* **B49**, 143. (5.5)

Wagner, C. N. J. (1978) *J. Non-Cryst. Sol.* **31**, 1. (3.1, 3.2.2)
Wagner, C. N. J. (1980) *J. Non-Cryst. Sol.* **42**, 3. (3.2.2)
Wagner, C. N. J. (1985) *J. Non-Cryst. Sol.* **76**, 29. (3.2.2)
Wagner, C. N. J. and Lee, D. (1980) *J. Phys.* (*Paris*) **41**, C8-242. (3.2.2)
Walter, U., Price, D. L., Susman, S. and Volin, K. J. (1988) *Phys. Rev.* **B37**, 4232. (4.2.1, 4.3.4)
Wang, R. and Merz, M. D. (1976) *Nature* (*Lond.*) **260**, 35. (2.5.3)
Wang, W. and Fritzsche, H. (1988) in *Amorphous Silicon and Related Materials*, H. Fritzsche (ed.) (World Scientific), p. 779. (6.3.1)
Wang, X., Bar-Yam, Y., Adler, D. and Joannopoulos, J. D. (1988) *Phys. Rev.* **B38**, 1601. (6.2.6)
Wang, Y. R. and Overhauser, A. W. (1988) *Phys. Rev.* **B38**, 9601. (4.3.4)
Ward, L. (1988) *The Optical Constants of Bulk Materials and Films* (Adam Hilger). (5.6.1)
Warren, B. E. (1969) *X-ray Diffraction* (Addison-Wesley). (3.2.2, 3.3)
Warren, B. E. and Mavel, G. (1965) *Rev. Sci. Instr.* **36**, 196. (3.2.2)
Waseda, Y. (1980) *Structure of Non-Crystalline Materials: Liquids and Amorphous Solids* (McGraw-Hill). (3.1, 5.4.1)
Watkins, G. D. and Troxell, J. R. (1980) *Phys. Rev. Lett.* **44**, 593. (6.2.3)

Weaire, D. and Alben, R. (1972) *Phys. Rev. Lett.* **29**, 1505. (4.2.1, 4.3.1)
Weaire, D. and Thorpe, M. F. (1971) *Phys. Rev.* **B4**, 2508. (5.1, 5.2.1)
Weaire, D., Thorpe, M. F. and Heine, V. (1972) *J. Non-Cryst. Sol.* **8–10**, 128. (5.1)
Weber, M. (1983) *Wiss. Zeitschr. Friedrich-Schiller Univ.* **32**, 239; (1987) ibid. **37**, 689. (7.4.3)
Weeding, T. L., de Jong, B. H. W. S., Veeman, W. S. and Aitken, B. G. (1985) *Nature* (*Lond.*) **318**, 352. (3.2.4)
Weinstein, F. C. and Davis, E. A. (1973/4) *J. Non-Cryst. Sol.* **13**, 153. (3.5.4)
Weyl, W. A. (1959) *Coloured Glasses* (Society of Glass Technology). (5.6.3)
Widom, M. (1988) in *Introduction to Quasicrystals*, M. V. Jaric (ed.), vol. I (Academic), p. 59. (3.3)
Williams, G. and Watts, D. C. (1970) *Trans. Faraday Soc.* **66**, 80. (2.3.3)
Williams, M. L., Landel, R. F. and Ferry, D. J. (1955) *J. Am. Chem. Soc.* **77**, 3701. (2.3.5)
Williams, S. J. and Elliott, S. R. (1982) *Proc. Roy. Soc.* (*Lond.*) **A380**, 427. (3.4.2, 3.5.3)
Windisch, C. F. and Risen, W. M. (1982a) *J. Non-Cryst. Sol.* **48**, 307. (3.4.2)
Windisch, C. F. and Risen, W. M. (1982b) *J. Non-Cryst. Sol.* **48**, 325. (3.4.3)
Windsor, C. G. (1981) *Pulsed Neutron Scattering* (Taylor and Francis). (3.2.2)
Wong, J. and Angell, C. A. (1976) *Glass: Structure by Spectroscopy* (Dekker). (2.3.2, 3.2.4, 3.2.5, 5.6.3, 6.2.2, 6.3.4)
Woodcock, L. V. and Angell, C. A. (1981) *Phys. Rev. Lett.* **47**, 1129. (2.3.5)
Wooten, F. and Weaire, D. L. (1984) *J. Non-Cryst. Sol.* **64**, 325. (3.5.5, 4.3.2)
Wooten, F., Winer, K. and Weaire, D. L. (1985) *Phys. Rev. Lett.* **54**, 1392. (3.5.5, 4.3.4, 5.2.1)
Wright, A. C. (1974) *Adv. Struct. Res. Diffr. Meth.* **5**, 1. (3.1, 3.2.2)
Wright, A. C. and Leadbetter, A. J. (1976) *Phys. Chem. Glasses* **17**, 122. (3.1, 3.2.2)
Wright, A. C., Connell, G. A. N. and Allen, J. W. (1980) *J. Non-Cryst. Sol.* **42**, 69. (3.5.3)
Wright, A. C., Etherington, G., Erwin Desa, J. A. and Sinclair, R. N. (1982) *J. Phys.* (*Paris*) **43**, C9–31. (3.2.2)
Wright, A. C., Sinclair, R. N. and Leadbetter, A. J. (1985) *J. Non-Cryst. Sol.* **71**, 295. (3.4.4)
Wright, A. C., Price, D. L., Clare, A. G., Etherington, G. and Sinclair, R. N. (1987) *Diffusion Defect Data* **53–54**, 255. (3.5.3)

Yang, C. Y., Paesler, M. A. and Sayers, D. E. (1987) *Phys. Rev.* **B36**, 980. (4.3.5)
Yeh, X. L., Samwer, K. and Johnson, W. L. (1983) *Appl. Phys. Lett.* **42**, 242. (1.3.11)
Yonezawa, F. and Cohen, M. H. (1981) in *Fundamental Physics of Amorphous Semiconductors*, F. Yonezawa (ed.) (Springer-Verlag). (5.1)
Yoshino, S. and Okazaki, M. (1977) *J. Phys. Soc. Jap.* **43**, 415. (5.3.4)

Zachariasen, W. H. (1932) *J. Am. Chem. Soc.* **54**, 3841. (2.5.1, 3.5.3)
Zallen, R. (1983) *The Physics of Amorphous Solids* (Wiley). (3.5.3, 5.3.6)
Zarzycki, J. and Naudin, F. (1969) *J. Non-Cryst. Sol.* **1**, 215. (3.7.2)
Zeller, R. C. and Pohl, R. O. (1971) *Phys. Rev.* **B4**, 2029. (4.4.1, 4.4.2, 4.4.3, 4.4.5, 4.5.3)
Ziman, J. M. (1961) *Phil. Mag.* **6**, 1013. (5.4.1)
Ziman, J. M. (1972) *Principles of the Theory of Solids* (Cambridge University Press). (5.4.1)

Index